BOHMIAN MECHANICS AND QUANTUM THEORY: AN APPRAISAL

BOSTON STUDIES IN THE PHILOSOPHY OF SCIENCE

VOLUME 184

BOHMIAN MECHANICS AND QUANTUM THEORY: AN APPRAISAL

Edited by

JAMES T. CUSHING
Department of Physics, University of Notre Dame

ARTHUR FINE
Department of Philosophy, Northwestern University

SHELDON GOLDSTEIN
Department of Mathematics, Rutgers University

KLUWER ACADEMIC PUBLISHERS
DORDRECHT / BOSTON / LONDON

Library of Congress Cataloging-in-Publication Data

Bohmian mechanics and quantum theory : an appraisal / edited by James
T. Cushing, Arthur Fine, Sheldon Goldstein.
p. cm. -- (Boston studies in the philosophy of science ; v.
184)
Includes bibliographical references and index.
ISBN 0-7923-4028-0 (alk. paper)
1. Quantum theory--Mathematics. I. Cushing, James T., 1937- .
II. Fine, Arthur. III. Goldstein, Sheldon, 1947- . IV. Series.
Q174.B67 vol. 184
[QC174.12]
001'.01 s--dc20
[530.1'2] 96-11973

ISBN 0-7923-4028-0

Published by Kluwer Academic Publishers,
P.O. Box 17, 3300 AA Dordrecht, The Netherlands.

Kluwer Academic Publishers incorporates
the publishing programmes of
D. Reidel, Martinus Nijhoff, Dr W. Junk and MTP Press.

Sold and distributed in the U.S.A. and Canada
by Kluwer Academic Publishers,
101 Philip Drive, Norwell, MA 02061, U.S.A.

In all other countries, sold and distributed
by Kluwer Academic Publishers Group,
P.O. Box 322, 3300 AH Dordrecht, The Netherlands.

Printed on acid-free paper

Printed in the Netherlands

TABLE OF CONTENTS

III. HISTORICAL, CONCEPTUAL AND PHILOSOPHICAL PERSPECTIVES RELATED TO BOHMIAN MECHANICS

IV. COMPARISONS WITH SOME OTHER PROGRAMS

PREFACE

We are often told that quantum phenomena demand radical revisions of our scientific world view and that no physical theory describing well defined objects, such as particles described by their positions, evolving in a well defined way, let alone deterministically, can account for such phenomena. The great majority of physicists continue to subscribe to this view, despite the fact that just such a deterministic theory, accounting for all of the phenomena of nonrelativistic quantum mechanics, was proposed by David Bohm more than four decades ago and has arguably been around almost since the inception of quantum mechanics itself.

Our purpose in asking colleagues to write the essays for this volume has not been to produce a *Festschrift* in honor of David Bohm (worthy an undertaking as that would have been) or to gather together a collection of papers simply stating uncritically Bohm's views on quantum mechanics. The central theme around which the essays in this volume are arranged is David Bohm's version of quantum mechanics. It has by now become fairly standard practice to refer to his theory as Bohmian mechanics and to the larger conceptual framework within which this is located as the causal quantum theory program. While it is true that one can have reservations about the appropriateness of these specific labels, both do elicit distinctive images characteristic of the key concepts of these approaches and such terminology does serve effectively to contrast this class of theories with more standard formulations of quantum theory.

The papers in this volume have been grouped around four main topics: the essentials of Bohmian mechanics; detailed applications and extensions of these basics; historical and conceptual analyses of the program; comparisons and contrasts with other formulations of quantum mechanics. We hope that this book will contribute to a serious and ongoing critical analysis, extension and application of the central ideas of Bohmian mechanics to test them for consistency, empirical adequacy and fruitfulness in generating new avenues of research. By no means are all of the authors represented here enthusiastic supporters of Bohm's approach to quantum mechanics. Nor do the editors necessarily agree with every major position taken in these essays. Points of dissent among authors, even on what are in some cases matters of historical or technical fact, will be evident to the reader. For example, the third essay in this volume suggests that the standard history of early quantum mechanics significantly underrates how complete Louis de Broglie's 1927 pilot-wave theory actually was and the extent to which it anticipated Bohm's 1952 version of quantum mechanics. This type of give-and-take is, of course, not unexpected if a research program

J. T. Cushing et al. (eds.), Bohmian Mechanics and Quantum Theory: An Appraisal, vii–viii.

is to be fairly and productively scrutinized and if there is to be ongoing debate.

We, and many of the other authors of these essays, benefited greatly from having the opportunity to discuss our ideas at a conference, "Quantum Theory Without Observers," conceived of and organized by Professor Detlef Dürr of the Mathematics Institute at the University of Munich and held in Bielefeld, Germany, on July 24–28, 1995. Although this volume is in no sense a proceedings of that conference (which was devoted to more than just Bohmian mechanics), it is nevertheless true that the majority of our authors were participants at this conference, where they were able to subject the work represented in their essays to the critical appraisal of, and to engage in stimulating exchanges with, the many experts in the foundations of quantum mechanics who were gathered for this splendid occasion. For this we are all indebted to Professor Dürr.

JAMES T. CUSHING
Department of Physics
University of Notre Dame

ARTHUR FINE
Department of Philosophy
Northwestern University

SHELDON GOLDSTEIN
Department of Mathematics
Rutgers University

December, 1995

JAMES T. CUSHING

THE CAUSAL QUANTUM THEORY PROGRAM

In this introductory essay[1] about the causal quantum theory program,[2] I summarize some orthodox views on quantum mechanics, sketch David Bohm's 1952 version of quantum theory and today's conventional wisdom about it, examine the origins of this program and the rapid rise of the Copenhagen interpretation to dominance, ask how things might have gone very differently and why they did not, illustrate (mainly by discussing Wolfgang Pauli's own reaction) how in the 1950s Bohm's (1952a,b) theory was rejected out of hand, and, finally, outline a landscape of the renewed interest in foundational problems of quantum mechanics. I begin by recalling another infamous case, from the early part of the twentieth century, of the scientific community's dismissing a challenging theoretical possibility without even seriously considering the evidence for it.

1. AN EARLIER SNUB BY THE SCIENTIFIC COMMUNITY

Arthur Koestler's *The Case of the Midwife Toad* (1971) begins by recounting the contents of a note found beside the body of Dr. Paul Kammerer on an Austrian mountain path in the fall of 1926 and then goes on to comment:

> Thus ended the greatest scientific scandal of the first half of our century. Its hero and victim was one of the most brilliant and unorthodox biologists of his time. He was forty-five years old when the joint pressures of an inhuman Establishment and his own all-too-human temperament drove him to suicide. . . . Yet an obituary article in *Nature*, which is probably the world's most respected scientific journal, called his last book "one of the finest contributions to the theory of evolution which has appeared since Darwin." (Koestler 1971, 13–14)

Kammerer felt that pure Darwinism could not fully account for the all too rapid rate of progressive changes in species and, so, advocated a form of the Lamarckian theory of the inheritance of acquired characteristics. Kammerer was an exceptionally gifted experimenter in breeding several successive generations of lizards and toads in captivity. The amphibian creature (actually, the nuptial pads on the male's fore-limbs) that was at the center of a controversy covering fourteen years was the midwife toad, *Alytes obstetricans*. At the Institute for Experimental Biology in Vienna, Kammerer had made extensive observations, over several years, on lizards and on toads that showed definitively, so Kammerer claimed, that certain acquired (i.e., learned) characteristics become inherited. Other people were unable even to breed in captivity so many successive generations of these creatures, let alone to confirm or refute his published findings.

What is interesting for comparison with the case of the (non) reception of David Bohm's theory that I discuss below is the intensity of the emotion

J. T. Cushing et al. (eds.), Bohmian Mechanics and Quantum Theory: An Appraisal, 1–19.

(in the allegedly objective scientific community) that surrounded the debate over Kammerer's work. The main opponent in England was one of the leading Darwinians of the day, William Bateson – who, incidentally, had begun his own professional life as a Lamarckian. The central reason that the scientific community would not even seriously consider Kammerer's published results – in spite of the fact of his recognized skill as an experimenter – was that those results were in no way compatible with or understandable in terms of the accepted theory of the day (Darwinism). In this episode, vituperation and *ad hominem* attacks replaced logic and objectivity. So, even though the Kammerer case centered on experimental data, while Bohm's concerned rival theories, there are significant parallels in the reactions of the scientific community and in the personal prices that these scientists had to pay for their unorthodox positions. Neither of these challenging theories or conjectures was ever refuted by logic or by experiment, yet they were almost universally derided.

2. THE RECEIVED WISDOM

Today it is generally assumed that the success of quantum mechanics demands that we accept a world view in which physical processes at the most fundamental level are seen as being irreducibly and ineliminably indeterministic. That is, one of the great watersheds in twentieth-century scientific thought is the "Copenhagen" insight that empirical evidence and logic are seen as necessarily implying an indeterministic picture of nature. This is in marked contrast to any classical representation of a clockwork universe. Even though it is difficult to specify precisely what constitutes the traditional, or Copenhagen, interpretation[3] of quantum mechanics (since so many different versions of quantum theory go under this name), there is a fairly easily identifiable core of central commitments that can be listed. The threads of these beliefs are found in the writings of several of the founders of quantum mechanics. In order to avoid a charge of setting up a straw man only then to knock it down or of putting words into the mouths of the founders of the traditional formulation of quantum mechanics, let me actually cite a few of their pronouncements.

In Niels Bohr's early analyses of quantum theory, around the time of his seminal 1927 Como lecture, we read:

> In contrast to ordinary mechanics, the new quantum mechanics does not deal with a space-time description of the motion of atomic particles. . . . The difficulties [in constructing pictures of interactions between atoms and radiation] seem to require just that renunciation of mechanical models in space and time which is so characteristic a feature in the new quantum mechanics . . . . (Bohr 1934, 48, 50)

* * *

> [The quantum] postulate implies a renunciation as regards the causal space-time co-ordination of atomic processes. (Bohr 1934, 53)

* * *

Indeed, only by a conscious resignation of our usual demands for visualization and causality was it possible to make Planck's discovery fruitful in explaining the properties of the elements on the basis of our knowledge of the building stones of atoms. (Bohr 1934, 108)

Werner Heisenberg, in his retrospective recollections of these early days, tells us that:

It should be emphasized, however, that the probability function does not in itself represent a course of events in the course of time. It represents a tendency for events and our knowledge of events. The probability function can be connected with reality only if one essential condition is fulfilled: if a new measurement is made to determine a certain property of the system. (Heisenberg 1958, 46)

* * *

[T]here is no orbit in the ordinary sense. (Heisenberg 1958, 48)

* * *

[T]he idea of an objective real world whose smallest parts exist objectively in the same sense as stones or trees exist, independently of whether or not we observe them . . . is impossible. (Heisenberg 1958, 129)

And Max Born weighs in with his own views on the role of causality in physics:

[I]n quantum theory it is the *principle of causality*, or more accurately that of *determinism*, which must be dropped and replaced by something else. . . . We now have a *new form* of the law of causality. . . . It is as follows: if in a certain process the initial conditions are determined as accurately as the uncertainty relations permit, then the probabilities of all possible subsequent states are governed by exact laws. (Born 1951, 155, 163–164. Emphases in original.)

* * *

[N]o concealed parameters can be introduced with the help of which the indeterministic description could be transformed into a deterministic one. Hence if a future theory should be deterministic, it cannot be a modification of the present one but must be essentially different. How this could be possible without sacrificing a whole treasure of well established results I leave to the determinists to worry about. (Born 1949, 109)

Support for similar beliefs can also be found in John von Neumann's work, while Albert Einstein and Erwin Schrödinger remained incredulous at such certitude on these matters. These positions that I have attributed to Bohr, Heisenberg and Born are not merely the products of a bygone era, but are echoed in modern textbooks on quantum mechanics. As just one example of many that could be cited:

It is clear that [the results of the double-slit experiment] can in no way be reconciled with the idea that electrons move in paths. . . . In quantum mechanics there is no such concept as the path of a particle. (Landau and Lifshitz 1958, 2)

We can characterize these central conceptual commitments of standard versions of quantum mechanics as:

1. No description of quantum phenomena is possible in terms of particle trajectories in a space-time background.
2. Probability enters into quantum theory in an ineliminable and essentially nonclassical way (i.e., probabilities cannot not be merely epistemic).

3. It is in principle impossible to give a deterministic account of quantum phenomena (at the level of the events or observables themselves).

Of course, John Bell very effectively reminded us that it was Bohm's 1952 theory that gave lie to all of these beliefs by explicitly doing the impossible (Bell 1987a, 160).

In order to generate some conceptual space for the argument of the next section, let me parse a scientific theory into two distinct components: its formalism and its interpretation. These are logically separable, even if they are often entangled in practice. Here a formalism means a set of equations and a set of calculational rules for making predictions that can be compared with experiment.[4] Both standard quantum mechanics and Bohm's version use the same set of rules for predicting the values of observables. The (physical) interpretation refers to what the theory tells us about the underlying structure of these phenomena (i.e., a corresponding story about the furniture of the world – an ontology). Hence, one formalism with two different interpretations counts as two different theories.

3. BOHM'S THEORY AND TYPICAL CRITICISMS OF IT

Since the formalism of quantum mechanics is not identical with, and need not include, the Copenhagen interpretation, there exists the possibility of an alternative, observationally equivalent theory based on that formalism. In a sense, David Bohm's 1952 work can be seen as an exercise in logic – proving that Copenhagen dogma was not the only logical possibility compatible with the facts. Bohm accepted the formalism of quantum mechanics and showed that more microstructure is consistent with it than had previously been appreciated. The usual calculational rules of standard quantum mechanics still give average values for observed quantities, but now there are actual particles and trajectories.

Mathematical details aside, what Bohm did was to take the Schrödinger equation

$$(1) \qquad i\hbar \frac{\partial \psi}{\partial t} = H\psi$$

which has the form of a wave equation and hence naturally invites a wave (or, perhaps, a wave-particle) interpretation, and derived from it the equation (similar to Newton's second law of motion)[5]

$$(2) \qquad \frac{\mathrm{d}\boldsymbol{p}}{\mathrm{d}t} = \boldsymbol{F} = -\nabla(V + U)$$

where

$$(3) \qquad \boldsymbol{p} = \nabla S$$

Here U is the so-called quantum potential

$$(4) \qquad U = \frac{\hbar}{2m}\frac{\nabla^2 R}{R}$$

where R and S represent the polar decomposition of the wave function as $\psi = Re^{iS/\hbar}$. Equation (2) naturally invites a particle interpretation in terms of trajectories, with $\boldsymbol{p}$ now identified as the mechanical momentum $m\boldsymbol{v}$ of the particle. The Bohm theory is observationally identical to the standard one provided:

i. The wave function $\psi = Re^{iS/\hbar}$ is a solution to the Schrödinger equation.
ii. The velocity of the particle is given as $\boldsymbol{v} = (1/m)\nabla S$ (the guidance condition).
iii. The precise location of a particle is not predicted or controlled, but has a statistical (ensemble) distribution according to the probability density $P = |\psi|^2$.

This system of Bohmian mechanics (BM) is completely deterministic in the sense that precise specification of the initial position (or "hidden variable") of a particle uniquely determines its future behavior. Because a quantum potential is present in addition to the usual classical potential, these trajectories are very sensitive to the initial positions of the particles. There is a certain benign nonlocality that produces the usual observed quantum effects. The Heisenberg uncertainty relations state practical limitations on the accuracy with which we can know simultaneously the values of certain pairs of variables. One of the most beautiful aspects of Bohm's paper is his treatment of the measurement problem (which now becomes a non-problem). There is no collapse of the wave function and, hence, no measurement problem. The basic idea is that a particle always has a definite position between measurements. There is no superposition of properties and "measurement" (or observation) is an attempt to discover this actual position.

One would also expect no particular conceptual problem in obtaining the classical equations of motion from those of Bohm's theory since the basic exact dynamical equation [Eq. (2)] already has the general form of Newton's second law. This intuition is essentially correct since, loosely speaking, when the quantum potential is negligible,[6] then the dynamical law becomes just the classical equation of motion. Here we have a logically consistent and empirically adequate deterministic theory of quantum phenomena.

Most physicists do not really know much about Bohm's theory and those who are even vaguely aware of its existence usually "know" that it is wrong, although they are not exactly certain just why. When pressed, they often point out that it must be incorrect, since it conflicts with the Copenhagen interpretation of quantum mechanics. Of course, that would be telling only if one had an independent argument that the Copenhagen version were necessarily correct. Among the folklore-wisdom charges against Bohm's program are that it is *ad hoc*, having simply been concocted to yield the same observational consequences as standard quantum mechanics (i.e., saying the same thing in another way – whatever that is supposed to mean); that such a deterministic picture of quantum phenomena would be a return to the darkness of the clockwork universe of Laplace; that it cannot accommodate the "spin" of a particle; that there is no action-reaction symmetry between the influence of the field (ψ) and the particles guided by this field (i.e., ψ acts on the particles, but not vice versa); that there is no suitable relativistic generalization; that there is no extension to (quantum) field theory. Upon closer examination, each of these can be seen to be either specious or inconclusive, as shown, for instance, by several of the papers in this volume. Even worse, there are physicists who believe that Bell's theorem and certain experiments designed to test the Bell inequalities have categorically shown that *all* hidden-variables versions of quantum mechanics are ruled out. Lest the reader be incredulous at such a claim on my part, let me cite the following depressing evidence. The sole evaluation of a research proposal for theoretical work to be done, within the framework of Bohm's version of quantum theory, on some foundational questions in quantum mechanics, consisted essentially *in toto* of this:

> The subject under consideration, the rival Copenhagen and causal (Bohm) interpretations of the quantum theory, has been discussed for many years and in the opinion of several members of the Physics Division of the NSF, the situation has been settled. The causal interpretation is inconsistent with experiments which test Bell's inequalities. Consequently . . . funding . . . a research program in this area would be unwise. (Cushing 1995b, 507)

This provides an interesting insight into one sociological aspect of the actual functioning of the scientific enterprise. The first sentence can be taken as a statement of fact (i.e., that such an opinion does exist among at least some of the physicists at the National Science Foundation). But, it is the second sentence that is the most telling and relevant for my purposes here. It is fact that John Bell was most impressed by Bohm's theory, that it was a primary motivating factor in Bell's own work on his now-famous theorem and that Bell's theorem most specifically does not apply to Bohm's theory (since Bell's theorem covers a certain class of local hidden-variables theories, while Bohm's theory is nonlocal). This example illustrates that, once certain beliefs become ensconced as "objective knowledge" in the collective consciousness of the scientific community, they are exceptionally resistent to being exorcised from that body of "facts".

The simple truth is that Bohm's theory offers us an empirically adequate, conceptually different view of physical phenomena in which there is an objective reality whose existence does not depend upon an observer. Whether this is better or worse than the traditional view of quantum phenomena is a matter that must be decided on the basis of criteria other than empirical adequacy and logical consistency alone. That is, this alternative approach to quantum theory is a perfectly viable one even today. The purpose of many of the essays in this volume is to examine the implications of this theory. Once one appreciates that Bohm's theory cannot simply be dismissed out of hand, some seven decades after the formulation and general acceptance of the Copenhagen version of quantum mechanics, one can then ask the relevant historical and sociological question of how the standard (or Copenhagen) theory established its hegemony. I turn now to a few observations on this question.

4. WHENCE THE COPENHAGEN HEGEMONY?

As early as the mid 1920s, before any coherent quantum mechanics had even been formulated, there already existed an alternative causal quantum theory program in which fundamental physical processes were seen as being irreducibly and ineliminably absolutely deterministic. One can readily trace the origins and eventual fate of this program. My story begins with Louis de Broglie's theory of phase waves, includes various hydrodynamical and velocity-field (essentially "hidden variables") interpretations, such as those due to Erwin Madelung and to Einstein, progresses through de Broglie's theory of the double solution and his (provisional) pilot-wave model, the encounters with the Copenhagen group at the 1927 Solvay Congress, the impact of von Neumann's "impossibility" proof, Bohm's 1952 paper, and subsequent work on stochastic mechanics as a basis for the Schrödinger equation. Today we often see as either specious or at least far from logically compelling many of the arguments that were taken as decisive in an earlier era (such as Heisenberg's, Bohr's and Born's opinions cited earlier in this essay – opinions that soon became transmuted into inviolable dogma – about the impossibility of a picturable and causal quantum theory in a space-time background, Pauli's "refutation" of de Broglie's 1927 pilot-wave model, von Neumann's "proof" of the impossibility of hidden variables in quantum mechanics, Einstein's belief that even the type of nonlocality implied by quantum mechanics could make impossible doing science as we have known it and the apparent "conflict" between quantum nonlocality and relativity).

A causal program would have been a far less radical departure from the then-accepted framework of classical physics than was the so-called Copenhagen version of quantum mechanics that rapidly gained ascendancy by the late 1920s and has been all-but universally accepted ever since. How could this happen? Historical contingency and associated social factors play essential and ineliminable roles in the construction and selection of a successful scientific theory from among its observationally equivalent and unrefuted competitors. In the following section, I indicate that the mere order in which events take place can be a key factor in determining which of two empirically adequate and fruitful, but observationally equivalent, scientific theories is accepted by the scientific community. First, though, I examine the actual course of history. In addition to empirical evidence, several philosophical, sociological and psychological factors were partially, yet importantly, determinative of the Copenhagen version of quantum mechanics being accepted, by the late 1920s, to the virtual exclusion of even a serious consideration of alternative programs. Let me illustrate this by listing some of the relevant factors that affected the emergence of quantum theory and a view of the world based on it.

By the late nineteenth century there were significant (although by no means dominant) philosophical precedents for the concept of indeterminism in nature, as opposed to the straightforward determinism often associated with classical physics. For example, the causality principle for physical processes and the Kantian doctrine that acceptance of causality was a precondition for human understanding were challenged. Contingency and inherent chance in nature's actions were suggested. Such thinking opposed belief in a completely rational universe and proposed instead an element of irrationality in a nature having contingent laws. In a given situation, there would be equally possible alternatives to what does, in fact, occur. These concepts became important elements in the philosophy of Harald Høffding, to whom I shall return shortly.

For individuals who are key to generating and promulgating new theoretical structures, personal, or psychological, factors can be important for the origin of central concepts.[7] As an illustration of these types of influences, consider that of Høffding's philosophy on Bohr's formulation and interpretation of quantum mechanics. In 1966, an influential history of quantum physics sketched the philosophical currents present when quantum theory was formulated in the first quarter of this century (e.g., those I referred to in the preceding paragraph) and intimated a significant influence of Høffding's teaching on Bohr's own work in the foundations of quantum mechanics (Jammer 1989, 173–186). Subsequent research has substantiated the impact of Høffding's philosophy on Bohr (Faye 1991; 1994). Not only did Bohr attend Høffding's lectures as a student, but he also read his works and corresponded with him later in life. One of Høffding's tenets was that in life decisive events proceed through sudden "jerks" or discontinuities. Bohr explicitly acknowledged the influence of Høffding's philosophy on his own formulation of complementarity.

In contrast to this picture, a co-worker, perhaps better a disciple, of Bohr's, in a 1963 encomium published shortly after Bohr's death, in emotional tones categorically denied any external influence of earlier philosophy on Bohr's views of atomic phenomena (Petersen 1963). Rather remarkably, the author claims that, as a young man, Bohr had a strong interest in philosophy and was going to write a book on the subject, but essentially had no time to do so because he devoted his energies to constructing a theory of atomic phenomena. When, years later, Bohr had finally formulated his principle of complementarity and teased out many of its profounder (philosophical and epistemological) consequences, it so happened that these views precisely coincided with, and hence supported quite objectively, Bohr's own earlier philosophical views. This earlier philosophy was not seen as plausibly a (partial, at least) causal factor in the doctrine of complementarity, but instead as a happy coincidence of harmony between philosophy and physics.

Speculative thought, even in science, takes place amidst the ebb and flow of larger intellectual currents, and concepts like the ones mentioned above were available to Bohr and to the other creators of quantum theory. The claim of a close connection between Bohr's approach to atomic physics and these early philosophical influences does not, of course, imply that these alone completely determined Bohr's later position (or that of anyone else) on the interpretation of quantum mechanics. Rather, they were one major factor. It is the intersection of general predilections like those discussed here with the puzzles presented by physical phenomena that results in a definite theory or research program.

Even if one grants the importance of such background factors as a partial explanation of why certain key figures (such as Bohr, Heisenberg, Pauli, Born) found a particular formulation of quantum theory appealing, that in itself would not account for the ascendancy of that theory in the scientific community at large. For that, we must consider broader social factors. It has been over twenty years now since the dramatic and controversial "Forman thesis" was advanced that acausality was embraced by German quantum physicists in the Weimar era as a reaction to the hostile intellectual and cultural environment that existed there prior to and during the formulation of modern quantum mechanics (Forman 1971). The goal was to establish a causal connection between this social/intellectual milieu and the content of science, in this case quantum mechanics. The general structure of this argument is the following. Causality for physicists in the early twentieth century "meant complete lawfulness of Nature, determinism [(i.e., event-by-event causality)]" (Forman 1979, 11). Such lawfulness was seen by scientists as absolutely essential for science to be a coherent enterprise. A scientific approach was also taken to be necessarily a rational one. When, in the aftermath of the German defeat in World War I, science was held responsible (not only by its failure, but even more because of its spirit) for the sorry state of society, there was a reaction against rationalism and a return to a romantic, "irrational" ideal.

Later work on this same theme argued that the new quantum mechanics, which was overwhelmingly a German creation, was taken as supporting certain views (consonant with the prevailing German cultural milieu) that it did not, in fact, support (Forman 1984). These three characteristics are acausality, intuitiveness (or picturability) and individuality. They were valued characteristics of the milieu and quantum theory either exhibited or, in some cases, was made to appear to exhibit, these same features by its creators. The new quantum mechanics was readily enough taken as rejecting causality, although one could (as Einstein did) see it merely as a statistical theory. Only by a virtual inversion of the meaning of the term *anschaulich* as it had apparently applied to his own matrix mechanics was

Heisenberg able, in his influential 1927 indeterminacy paper, to make the formalism (really to define it as) "intuitive" (Forman 1984, 338). Heisenberg discussed the intuitive or picturable content (however little or restricted) of quantum mechanics. It would, however, seem stretching matters to claim that quantum mechanics is overall, as a theory or representation of the world, either intuitive or picturable in any ordinary sense of those terms. Similarly, Bohr spoke of the individuality of atomic processes and even of "the 'indestructible individuality' of material particles" (Forman 1984, 342). These physicists made the most of the acausality in quantum mechanics – which was easy enough – and then contorted any natural reading of that theory's lessons in claiming it to be intuitive (or picturable) and to support a type of individuality.

Let me distinguish between a strong Forman thesis, that would claim a major causal role of the cultural milieu in determining the very form and content of a scientific theory, as opposed to a weak Forman thesis, that would see the cultural milieu as sometimes playing an important role in the acceptance and propagation of an already formulated scientific theory. This weak Forman thesis can remain agnostic on the nature of the factors (i.e., internal and/or external) that are responsible for creating and shaping the theory itself. This is the position I believe is supported by the emergence of the Copenhagen version of quantum theory.

So far I have concentrated on forces tending to produce consensus on what would become the orthodox view. On the other hand, what was to become the Copenhagen interpretation of quantum mechanics certainly had its worthy opponents (e.g., Einstein, de Broglie, Schrödinger). However, the main thing that the latter had in common was their dislike of Copenhagen, without offering a fully formulated alternative to it (except de Broglie who suffered badly at Pauli's hands at the 1927 Solvay Congress). By contrast, the Copenhagen group, while they did have significant differences among themselves on interpretative issues, presented a united front on the theory as a whole. Even more important, they had the talent, organization and drive to fashion a working theory that proved to be phenomenally successful calculationally. Most working scientists really care mainly about that and are willing to leave the "philosophical" issues aside – or even to accept without deep reflection a certain amount of metaphysical baggage.[8]

In addition, there were personal factors that lent urgency to Copenhagen's victory. Not only did Heisenberg have personal ambitions for advancement, but several chairs in theoretical physics were opening up in Germany. There was a conscious realization by members of the Copenhagen school that control of the future direction of theoretical physics was at stake. This conflict can be characterized as one over superiority and professional dominance (i.e., the interpretation and those physicists that would determine the future direction of research) (Beller 1983a). The victors, in Bohr's own characterization, would be able to realize their "'wishes' for the future of physics" (quoted in Cassidy 1992, 225).

If the final decision between two theories were a wholly rational one, independent of these prior metaphysical commitments and of sociological factors (and this is by no means obvious in the case of quantum mechanics), then the origins of and reasons for those commitments, while perhaps interesting in themselves, would actually be irrelevant for the ultimate theory choice and for its "logical" justification. This is just the usual distinction between the logic of discovery and the logic of justification. However, while empirical evidence is essential for rejecting incorrect theories and for suggesting new avenues of speculation and of research, it is by itself insufficient to account fully for the choice in favor of one of these competing interpretations of quantum mechanics over the other. It is precisely because of this fact that we have had to consider broader factors as well.

By the mid-1930s, the option of a causal quantum theory program had generally been rejected as logically impossible and necessarily empirically inadequate when, in fact, it need not have been taken to be either. What the general conceptual arguments of Bohr and his school really constitute are consistency arguments for, rather than proofs of, a possible representation of microphenomena. We might say that the Copenhagen interpretation defined itself as true and strengthened its hold on physics, rewriting history so that Einstein, de Broglie and Schrödinger largely fade from view, thus leaving Copenhagen as the only intelligible version of quantum mechanics (Heilbron 1988, 219).

5. A LITTLE "COUNTERFACTUAL" HISTORY

I am concerned here with the acceptance and rejection of observationally equivalent, alternative and, indeed, incompatible descriptions or theories of our actual world. Specifically, if certain equally plausible conditions, rather than the actually-occurring and highly contingent historical ones, had prevailed and the interpretation of quantum mechanics had initially taken a very different route from the Copenhagen one around 1925–1927, would our world view of fundamental microprocesses necessarily have been brought back, by the "internal" logic of science, to our currently accepted picture of an inherently and irreducibly indeterministic nature? Could our present understanding of the behavior of the fundamental laws of nature in terms of an inherently indeterministic physics have been replaced by the apparently diametrically opposed view of absolute determinism? I have argued elsewhere (Cushing 1994) that the answer to the first question is "no" and to the second an emphatic "yes". An essential and intriguing factor is the ineliminable role of the determinative contingencies that charted our course to the current hegemony. I now sketch how we might have been brought to a very different, more nearly classical, position had there been a reordering in the temporal sequence of a few key historical events. This is not to deny that there were already serious conceptual problems for classical physics, even in its own domain of applicability, or to imply that the scientific community did not have "good reasons" (although, I claim, not uniquely compelling ones) for making the choices it did during the quantum revolution. An alternative choice would not have left us without foundational problems to resolve, but the set would be different.

A highly "reconstructed" but entirely plausible bit of history could run as follows (all around 1925–1927). Heisenberg's matrix mechanics and Schrödinger's wave mechanics are formulated and shown to be mathematically equivalent. Study of a classical particle subject to Brownian motion (about which Einstein surely knew something) leads to a "classical" understanding of the already discovered "Schrödinger" equation, which is then given a realistic interpretation. A stochastic mechanics underpins this interpretation with a visualizable model of microphenomena. This would have made evident the possibility of a largely classical foundation for that key equation. A realistic ontology would still remain a live option. Since stochastic mechanics is quite difficult to handle mathematically, people would likely have tackled the less formidable task of exploring the implications of the (equivalent) linear Schrödinger equation. The Dirac transformation theory and an operator formalism would still have been available as a *convenience* for further development of the formalism to provide algorithms for calculation.

Einstein was deeply committed to a realistic world view in which microentities have a continuous, objective, observer-independent existence. Two other core commitments for him were event-by-event causality (essentially determinism) and locality. These were three *desiderata* that he felt any candidate for a fundamental physical theory should satisfy. Now the entanglement or nonlocality of this stochastic mechanics formalism would soon have become apparent. Einstein would have rejected any model with this property.[9] However, before 1930 Heisenberg (and not he alone) was aware of the essence of the no-signaling theorem (Heisenberg 1930, 39). With sufficient conceptual motivation, the formal tools available in 1927 would have allowed proof of a Bell-type theorem. If that had happened at that time, Einstein and the rest of the quantum physicists would have perceived in sharp relief the choice between determinism and locality in any theory. A causal formulation of quantum theory might then have appeared less unpalatable than the Copenhagen version actually chosen by the scientific community.

Einstein could next have made the transition from stochastic mechanics to Bohmian mechanics since stochastic mechanics turns out to be both indeterministic and nonlocal. The option of keeping (in-principle) determinism with an objective reality and accepting an empirically benign nonlocality, as in Bohm's quantum-potential theory, might have been pursued. However, it does not seem worth demanding that my counterfactual scenario had to go along some highly specific path. What is important is that there were precedents for such moves and that the necessary pieces for them were already in place. If, early on, such a causal quantum theory program had been pursued, rather than the Copenhagen one, it would have had the internal resources to cope with the generalizations essential for a broad-based empirical adequacy. We could today have arrived at a very different world view of micro-phenomena. If someone were then to present the (merely) empirically equally adequate Copenhagen version, with all of its own counterintuitive and mind-boggling aspects, who would have listened?

We have seen that neither internal factors (such as logical consistency and empirical adequacy) alone nor external ones (such as metaphysical or psychological commitments and sociological pressures) alone have been sufficient to account for the wide acceptance of the Copenhagen world view in place of a causal one. Nature provides (often tight) constraints, but there still remains latitude in theory choice – in the present case, an interpretation or a world view. The actual route taken to closure was a complex one with many overlapping factors playing important roles. Science, even in its products or laws, remains historical or contingent in an essential manner. Developments might have gone a very different way at certain critical junctures. Why they did not may be as important as the reasons for the "right" choices that science has made. We are not particularly uncomfortable with a lack of inevitability in other areas of history, but for science such a possibility may strike many as bordering on the sacrilegious, or at least as teetering on the brink of an abyss of rationality skepticism. However, a lack of absolute certainty or inevitability need not result in total arbitrariness in our theories.

6. THE RECEPTION OF BOHM'S THEORY

I turn now to the reception that Bohm's 1952 theory was accorded by the scientific community. In that year, Bohm gave a brilliant and detailed exposition of a hidden-variables theory based on the familiar quantum formalism. This was a specific counterexample to the dogma that there could exist no causal version of quantum mechanics that was observationally equivalent to standard quantum mechanics. Given the presumed objectivity and impartiality of the scientific enterprise, one might expect that such an interpretation would be given serious consideration by the community of theoretical physicists. However, it was basically ignored, rather than either studied or rebutted. Just as what are often termed "external" factors had played a key role in establishing the Copenhagen hegemony, so they once again contributed to keeping this competitor from the field. That a generation of physicists had been educated in the Copenhagen dogma made it all the more difficult for Bohm's theory.

One finds no extensive examination of this new theory in the technical literature and Bohm did successfully respond to the few criticisms made on the basis of physics or of mathematics. In the main, sharp and quite negative responses came from men who were by then seen as the central, if dated, figures in the debates over the foundations of the Copenhagen interpretation. I illustrate this with the not-atypical reaction of one prominent member of this group.

Pauli had been one of the harshest critics of de Broglie's pilot-wave theory (of which Bohm's is a lineal descendant) at the 1927 Solvay Congress and he maintained the same attitude toward Bohm's theory. Interestingly enough, when Bohm sent him a copy of the paper in which Bohm showed Pauli's objections to the causal interpretation to be specious, Pauli never responded. In a letter to Born in 1953, Pauli referred to what he considered to be the retrograde efforts by Schrödinger, Bohm and Einstein and confidently predicted that the statistical character of the wave function and of the laws of nature would determine the style of the laws of physics for centuries to come. Any dream of a way back to the classical style of Newton and Maxwell seemed to him both hopeless and in bad taste. An extensive study of the correspondence between Pauli and the theoretical physicist Markus Fierz reveals that Pauli's interest in the philosophical possibilities opened up by uncertainty was one reason for his opposition to Bohm's attempt (and to hidden-variables theories generally) (Laurikainen 1988). Pauli worried that these ill-conceived efforts might, for ideological reasons, divert younger scientists from more fruitful (orthodox) avenues of research. His letter to Fierz in 1952 was quite sarcastic and contemptuous of Bohm's work. Fierz has offered the view that de Broglie was a Catholic and Bohm a Communist and that, to Pauli, this was the black and red alliance.[10] Pauli, so Fierz claims, felt that the de Broglie–Bohm theory was either prattle or an unnatural way of stating the same thing as standard quantum theory. Both Pauli and Fierz expressed strong, negative opinions in emotionally charged terms. For Pauli more was at stake that just physics.

Whereas materialism considers the spiritual as a mere byproduct of the material, Pauli especially understood the "epistemological lesson" of atomic physics as a signpost pointing in another direction. He, like Heisenberg, thought that quantum theory had clearly shown materialism to be unsatisfactory. Pauli was quite explicit in holding that it is wrong to imagine the existence of the material external world without including the observer and his psyche in that world view. The ideas of a detached observer and of realism were to be abandoned. He saw the Copenhagen interpretation as not being compatible with a materialistic world view. On the downside for the causal program from Pauli's perspective was the fact that psychological problems associated with observations were quite neglected by Bohm, as were the unconscious in his causal-deterministic universe. As a result, Pauli felt, Bohm saw no quantum paradoxes, but at the price of no free will either. He wanted no part of a deterministic, clock-work universe and believed that the trend of Western culture since the seventeenth century had been dangerous. The "irrational influences" that statistical causality implied were important for him. Pauli's concerns spilled over into the theological realm as well.

There were deep philosophical, moral and religious issues involved here. An attack on the canonical version of quantum theory resonated quite negatively and strongly with Pauli's beliefs about God and his views on epistemology. It was inconceivable to him that anything like a return to a "classical" world with causality and picturable, continuous processes in space-time was either possible or anything less than a disgusting loss of nerve and a return to darkness (Laurikainen 1988, 28–30). Partly because of ideological struggles involving scientists within the Soviet Union and partly because of Bohm's persecution by Senator Joseph McCarthy's crusade against un-American activities, Marxism and Communism also came to be associated with Bohm's interpretation by the early 1950s and that was certainly not conducive to an objective examination of his theory.

Even though in 1952 Bohm dispatched Pauli's specious earlier (1927) objection to a pilot-wave model and in 1964 Bell (inspired largely by Bohm's own 1952 model) did the same for von Newmann's celebrated "proof" of the impossibility of hidden-variables theories, the issue of the Copenhagen versus a causal interpretation remained effectively settled.

7. ALTERNATIVE APPROACHES – A RENEWED INTEREST

However, in the past decade or so an increasing sense of dissatisfaction has manifested itself about the inadequacies of the Copenhagen interpretation of quantum mechanics. A central interpretative problem underlying this unease with standard quantum mechanics has always been the measurement problem. Early on in the formal development of quantum mechanics, von Neumann distinguished between two types of time evolution for the state of a quantum system: the unitary time development governed by the Schrödinger equation and a very different collapse of the state vector upon measurement. Today it is generally recognized that von Neumann's "resolution" of the measurement problem either leads to an infinite regress or becomes essentially anthropocentric in nature. One of the older subsequent attempts at a resolution of the measurement problem, but wholly within the framework of Schrödinger evolution alone, is that based on the irreversible processes presumed necessary in measurements (Daneri, Loinger and Prosperi 1962; 1966). This program tries to account for the transition from an entangled pure state to an effective mixture as a practical constraint provided by what is actually doable experimentally. However, this neither gives a general resolution to an in-principle problem nor does it, even at the practical level, convincingly demonstrate that the technical conditions necessary to provide this "for all practical purposes" solution could never be circumvented by future experimental techniques. Within such a von Neumann (or standard-quantum mechanical) framework, the measurement problem remains in principle insoluble (Fine 1970; Brown 1986).

The general feeling that quantum mechanics as previously formulated is not equal to coping with such foundational issues has led to several more recent programs. These various research projects cover a broad spectrum. Let me progress from the more conservative to the more radical, as I briefly sample just some of the major contenders. Of course, such compact sketches and the illustrations of potential problems for each cannot do justice to these models and theories. The interested reader must consult the references cited for critical presentations and analyses. My purpose here is only to indicate the range of possibilities of theories alternative to the previous orthodoxy and the types of difficulties that each may face (i.e., none is manifestly better than BM on this score). Several essays in this volume address different aspects of these theories.

One of the origins of decoherence theory (Zurek 1981; 1982; 1993; Joos and Zeh 1985) was the realization that the environment is far more efficient than most people had appreciated in effectively reducing the interference among the terms of an entangled quantum state (e.g., by producing a superposition that has disjoint supports among its members) and perhaps even in selecting a preferred "pointer" basis. However, this is not really so much a theory as an important technical result about standard quantum mechanics applied to a system in contact with its environment. Even to the extent that it might illustrate how pure states do behave as mixtures in many practically realizable situations, it still leaves unresolved how it is that one of the possible terms in the mixture becomes actualized. The consistent histories theory (Griffiths 1984; 1993) similarly shows us that there is a lot more we can say about the possible "histories" of the evolution of a quantum system from its initial state to its final state than we had previously seen. Decoherence and consistent-histories ideas are mutually supportive (Gell-Mann and Hartle 1993; Zurek 1993). Again, though, this project may, finally, not give us any meaningful understanding of how just one particular history gets realized. At present it is unclear whether this consistent histories view can be made consistently complete (Goldstein and Page 1995; Dowker and Kent 1996).

A conceptually quite different interpretation, but still broadly within the standard Schrödinger evolution of the state vector, is the so-called modal interpretation (of which there are, in fact, several distinct versions) (van Fraassen 1981; 1991; Kochen 1985; Dieks 1989; 1994; Healey 1989; Bub 1992; 1994a). One early motivation was the technical result that the state ψ of a system consisting of two subsystems can be often uniquely[11] decomposed into a diagonal sum of direct products as

$$\psi = \sum_j c_j \, u_j \otimes w_j \tag{5}$$

where the $\{u_j\}$ and $\{w_j\}$ are orthonormal states for their respective subsystems. For certain modal versions of quantum mechanics, these particular eigenstates represent the actually existing, possessed quantities of the subsystems. In this or in some other fashion (such as the unique spectral resolution of the density operator for the system), one singles out a preferred observable (or a maximally consistent set of observables), the value of which is to be well defined at all times. A hope is that the standard measurement problem might be avoided (because the eigenvalue-eigenvector link between possessed values and the quantum-mechanical state of the system is now broken) by making a distinction between the physical state (which includes this special observable) and the usual dynamical state (which is governed by the schrödinger equation). Even aside from possible Kochen–Specker type contradictions for these preferred properties,[12] the ontological implications of a consistent time evolution for these preferred variables is yet to be fully worked out. Bohm's theory has even been viewed (admittedly somewhat anachronistically, from a historical perspective) as a type of modal interpretation in which the position variable is singled out (Vink 1993; Bub 1994a; Dickson 1995).

Probably the most dramatic interpretation of the standard formalism of quantum mechanics, and one of the oldest, is that of relative states, often termed many-worlds (Everett 1957; De Witt and Graham 1973). It avoids any problem of collapse by taking the unitary Schrödinger evolution of the state vector to be the complete description of the unfolding of the universe (which, on some readings, now contains many worlds). Aside from the profligacy of any literal interpretation of its ontology, it remains to be seen how this interpretation can adequately account for the observed quantum-mechanical statistics within any one branch of the universe (Albert and Loewer 1988; Kent 1990). There is a more recent variation on this theme, the many-minds interpretation (Albert and Loewer 1988; 1991; Albert 1992), that avoids this particular difficulty of Everett's theory.[13]

A very different undertaking countenances a modification of the formalism of quantum mechanics so that the collapse of the wave function becomes an objectively occurring process. This is the spontaneous localization (SL) program developed in various forms in recent years (Ghirardi, Rimini and Weber 1986; Pearle 1989; Ghirardi and Rimini 1990) (GRWP). One early criticism of the *continuous* versions of SL models (CSL) is that of the "tails"; that is, the fact that in any finite time the reduction to one localized term in the superposition is never complete (Albert 1992, 92–111; Karakostas 1994). However, such an objection can be countered by interpreting $|\psi|^2$ as a type of "matter" density, rather than as a probability density. These SL models are not in principle completely empirically equivalent to the standard formulation of quantum mechanics (i.e., to the unmodified Schrödinger equation). To date, none of these expected deviations have been able to be observed. Also, SL models have their own peculiar brand of nonlocality, one that may not fit well with relativity and with a reasonable criterion for objective reality. It remains to be shown that the GRWP approach to state-vector reduction can provide a fundamental resolution to the measurement problem. The basic motivation here seems to be to take seriously a unitary ontology in which there is just a wave (i.e., the "stuff" of the wave function ψ). Even though it departs from the exact form of the Schrödinger equation, it is very much in the spirit of Schrödinger's initial attempts at interpreting his wave function as the actual matter of the universe.

My brief summaries are not intended as an indictment against any of these diverse programs. Rather, they are meant to indicate that, while some of these are promising, each has its own set of potential difficulties. Which, if any, of these new directions in quantum mechanics will prove fruitful remains to be seen and constitutes the motivation for a large field of current research activity in the foundations of quantum mechanics.

As part of this general rekindling of interest in the foundational interpretive problems of quantum mechanics, the last decade has seen renewed work on Bohm's program[14] – among both theoretical physicists and philosophers of science concerned with finding a conceptually coherent framework within which to analyze and understand quantum phenomena. One of the great appeals of Bohm's theory is that it presents a definite, complete and functioning dynamics that can be studied in detail. Recent activities in Bohmian mechanics include a critical assessment of the roles of nonlocality and of probability in this theory, a clarification of its conceptual and mathematical framework and of the ramifications it has for our interpretation of the constraints provided by relativity, use of its graphic and realistic ontology to generate illustrative examples of great pedagogical value, technical applications of the theory in the relativistic domain, in field theory and in gravitation, and a reconsideration of the quantum mechanics of time measurements. There would be little point in my reviewing this work in any detail here since the majority of the essays in the present volume are devoted to developing and criticizing these and other aspects of this evolving theory. The reader will soon become aware from these papers that there are strong differences of opinion and even disagreements on technical matters among some of the proponents about various ways of fleshing out Bohm's general ideas. Rather than viewing this with alarm, I suggest that this is a sign of health in an area of growth and activity. This is not a body of dogma that brooks no dissent. David Bohm's seminal views on and his considerable technical contributions to quantum theory are finally receiving the long-overdue scrutiny they deserve to see whether or not they will prove to be consistent, empirically adequate and fruitful for further avenues of research.

University of Notre Dame

NOTES

1 For complete arguments and citations for many of the claims (especially historical ones) made here, see Cushing (1994).

2 The terminology "causal quantum theory program" has historically come to be used to

cover a number of theoretical attempts (such as stochastic mechanics) in which the common feature was not necessarily determinism, but rather the actual existence of entities (such as particles and trajectories) even for quantum phenomena. In my historical account here, the "causal program" is used in this wider sense.

[3] Although "Copenhagen" and "Bohm" are two different *theories*, the expressions "Copenhagen interpretation" and "Bohm interpretation" have become so commonly used in the literature that I often follow that practice.

[4] These include the usual Hilbert-space structure, a correspondence scheme between these mathematical objects and the physical states and observables they represent, rules for calculating expectation values of observables, and a projection postulate (either explicitly or effectively assumed) upon measurement.

[5] Dürr *et al.* (1992a) have emphasized that Bohmian mechanics is really a *first-order* theory characterized by Eqs. (1) and (3) together [rather than by Eqs. (1) through (4)]. However, I am here interested in the theory as Bohm himself presented it.

[6] The relevant state-dependent conditions on U were subtle enough to elude Einstein (1953) in his criticism of BM. See Holland (1993, Chapter 6).

[7] The divergent views of Einstein and of Bohr in their debate over quantum mechanics can be seen as rooted in their different styles of thinking (images versus words, respectively), rather than in broad cultural forces that were common to both (Kaiser 1994).

[8] This was the case with action at a distance in Newtonian gravitational theory.

[9] As Fine (1986, 98) discovered, Einstein unsuccessfully attempted a hidden-variables theory in the spring of 1927 (Cushing 1994, Chapter 8; Belousek 1996).

[10] Private communication (letter of 8/11/89).

[11] The caveat "often" here covers the possibility of degenerate states.

[12] Only certain modal interpretations may be susceptible to KS-type contradictions.

[13] In fact, Everett's original relative-state idea is arguably closer to the many-minds interpretation that it is to the many-worlds one with which it is usually associated.

[14] Antony Valentini has recently brought to my attention – and I quite agree – that Louis de Broglie (1928, 118–119) presented at the 1927 Solvay Congress a far more complete, many-particle pilot-wave theory (with a wave function on *configuration* space) than I indicated in my own book (Cushing 1994, 118–120, 149). For a detailed reconstruction that favors an essentially complete priority claim of de Broglie over Bohm on this issue, see Valentini (1996b). Valentini argues there for an interpretation different from mine on the nature and significance of the de Broglie-Pauli exchange (Cushing 1994, 119–120) at that conference.

DETLEF DÜRR, SHELDON GOLDSTEIN AND NINO ZANGHÌ

BOHMIAN MECHANICS AS THE FOUNDATION OF QUANTUM MECHANICS

In order to arrive at Bohmian mechanics from standard nonrelativistic quantum mechanics one need do almost nothing! One need only complete the usual quantum description in what is really the most obvious way: by simply including the positions of the particles of a quantum system as part of the state description of that system, allowing these positions to evolve in the most natural way. The entire quantum formalism, including the uncertainty principle and quantum randomness, emerges from an analysis of this evolution. This can be expressed succinctly – though in fact not succinctly enough – by declaring that the essential innovation of Bohmian mechanics is the insight that *particles move*!

1. BOHMIAN MECHANICS IS MINIMAL

Is it not clear from the smallness of the scintillation on the screen that we have to do with a particle? And is it not clear, from the diffraction and interference patterns, that the motion of the particle is directed by a wave? De Broglie showed in detail how the motion of a particle, passing through just one of two holes in screen, could be influenced by waves propagating through both holes. And so influenced that the particle does not go where the waves cancel out, but is attracted to where they cooperate. This idea seems to me so natural and simple, to resolve the wave-particle dilemma in such a clear and ordinary way, that it is a great mystery to me that it was so generally ignored. (Bell 1987a, 191)

According to orthodox quantum theory, the *complete* description of a system of particles is provided by its wave function. This statement is somewhat problematical: If "particles" is intended with its usual meaning – point-like entities whose most important feature is their position in space – the statement is clearly false, since the complete description would then have to include these positions; otherwise, the statement is, to be charitable, vague. Bohmian mechanics is the theory that emerges when we indeed insist that "particles" means particles.

According to Bohmian mechanics, the complete description or state of an N-particle system is provided by its wave function $\psi(q, t)$, where $q = (\mathbf{q}_1, \ldots, \mathbf{q}_N) \in \mathbb{R}^{3N}$, *and* its configuration $Q = (\mathbf{Q}_1, \ldots, \mathbf{Q}_N) \in \mathbb{R}^{3N}$, where the $\mathbf{Q}_k$ are the positions of the particles. The wave function, which evolves according to Schrödinger's equation,

$$(1) \qquad i\hbar \frac{\partial \psi}{\partial t} = H\psi,$$

choreographs the motion of the particles: these evolve according to the equation

J. T. Cushing et al. (eds.), Bohmian Mechanics and Quantum Theory: An Appraisal, 21–44.

$$\text{(2)} \qquad \frac{d\mathbf{Q}_k}{dt} = \frac{\hbar}{m_k} \frac{\mathrm{Im}(\psi^* \nabla_k \psi)}{\psi^* \psi} \mathbf{Q}_1, \ldots, \mathbf{Q}_N),$$

where $\nabla_k = \partial/\partial \mathbf{q}_k$. In Eq. (1), H is the usual nonrelativistic Schrödinger Hamiltonian; for spinless particles, it is of the form

$$\text{(3)} \qquad H = - \sum_{k=1}^{N} \frac{\hbar^2}{2m_k} \nabla_k^2 + V,$$

containing as parameters the masses $m_1 \ldots, m_N$ of the particles as well as the potential energy function V of the system. For an N-particle system of nonrelativistic particles, Eqs. (1) and (2) form a complete specification of the theory.[1] There is no need, and indeed no room, for any further *axioms*, describing either the behavior of other observables or the effects of measurement.

In view of what has so often been said – by most of the leading physicists of this century and in the strongest possible terms – about the radical implications of quantum theory, it is not easy to accept that Bohmian mechanics really works. However, in fact, it does: Bohmian mechanics accounts for all of the phenomena governed by nonrelativistic quantum mechanics, from spectral liens and quantum interference experiments to scattering theory and superconductivity. In particular, the usual measurement postulates of quantum theory, including collapse of the wave function and probabilities given by the absolute square of probability amplitudes, emerge as a consequence merely of the two equations of motion for Bohmian mechanics – Schrödinger's equation and the guiding equation – without the traditional invocation of a special and somewhat obscure status for observation.

It is important to bear in mind that regardless of which observable we choose to measure, the result of the measurement can be assumed to be given configurationally, say by some pointer orientation or by a pattern of ink marks on a piece of paper. Then the fact that Bohmian mechanics makes the same predictions as does orthodox quantum theory for the results of any experiment – for example, a measurement of momentum or of a spin component – at least assuming a random distribution for the configuration of the system and apparatus at the beginning of the experiment given by $|\psi(q)|^2$, is a more or less immediate consequence of (2). This is because the quantum continuity equation

$$\text{(4)} \qquad \frac{\partial |\psi(q, t)|^2}{\partial t} + \mathrm{div}\, J^{\psi}(q, t) = 0,$$

where

$$\text{(5)} \qquad J^{\psi}(q, t) = (\mathbf{J}_1^{\psi}(q, t), \ldots, \mathbf{J}_N^{\psi}(q, t)),$$

with

$$\text{(6)} \qquad \mathbf{J}_k^{\psi} = \frac{\hbar}{m_k} \mathrm{Im}(\psi^* \nabla_k \psi),$$

is the *quantum probability current*, an equation that is a simple consequence of Schrödinger's equation, becomes the classical continuity equation

$$\frac{\partial \rho}{\partial t} + \operatorname{div} \rho v = 0, \tag{7}$$

for the system $dQ/dt = v$ defined by (2) – the equation governing the evolution of a probability density ρ under the motion defined by the guiding equation (2) – when $\rho = |\psi|^2 = \psi^*\psi$, the *quantum equilibrium* distribution. In other words, if the probability density for the configuration satisfies $\rho(q, t_0) = |\psi(q, t_0)|^2$ at some time t_0, then the density to which this is carried by the motion (2) at any time t is also given by $\rho(q, t) = |\psi(q, t)|^2$. This is an extremely important property of Bohmian mechanics, one that expresses a certain compatibility between the two equations of motion defining the dynamics, a property which we call the *equivariance* of the probability distribution $|\psi|^2$. (It of course holds for any Bohmian system and not just the system-apparatus composite upon which we have been focusing.)

While the meaning and justification of the *quantum equilibrium hypothesis* that $\rho = |\psi|^2$ is a delicate matter, to which we shall later return, it is important to recognize at this point that, merely as a consequence of (2), Bohmian mechanics is a counterexample to all of the claims to the effect that a deterministic theory cannot account for quantum randomness in the familiar statistical mechanical way, as arising from averaging over ignorance: Bohmian mechanics is clearly a deterministic theory, and, as we have just explained, it does account for quantum randomness as arising from averaging over ignorance given by $|\psi(q)|^2$.

Note that Bohmian mechanics incorporates Schrödinger's equation into a rational theory, describing the motion of particles, merely by adding a single equation, the guiding equation (2), a first-order evolution equation for the configuration. In so doing it provides a precise role for the wave function in sharp contrast with its rather obscure status in orthodox quantum theory. Moreover, if we take Schrödinger's equation directly into account – as of course we should if we seek its rational completion – this additional equation emerges in an almost inevitable manner, indeed via several routes. Bell's preference is to observe that the probability current J^ψ and the probability density $\rho = \psi^*\psi$ would classically be related (as they would for any dynamics given by a first-order ordinary differential equation) by $J = \rho v$, obviously suggesting that

$$\frac{dQ}{dt} = v = \frac{J}{\rho}, \tag{8}$$

which is the guiding equation (2).

Bell's route to (2) makes it clear that it does not require great imagination to arrive at the guiding equation. However, it does not show that this equation is in any sense mathematically inevitable. Our own preference is to proceed in a somewhat different manner, avoiding any use, even in the motivation for the theory, of probabilistic notions, which are after all somewhat subtle, and see what symmetry considerations alone might suggest. Assume for simplicity that we are dealing with spinless particles. Then one finds (Dürr *et al.* 1992a) that, given Schrödinger's equation, the simplest choice, compatible with overall Galilean and time-reversal invariance, for an evolution equation for the configuration, the simplest way a suitable velocity vector can be extracted from the scalar field ψ, is given by

$$\text{(9)} \qquad \frac{dQ_k}{dt} = \frac{\hbar}{m_k} \operatorname{Im} \frac{\nabla_k \psi}{\psi},$$

which is of course equivalent to (2): The ∇ on the right-hand side is suggested by rotation invariance, the ψ in the denominator by homogeneity – i.e., by the fact that the wave function should be understood projectively, an understanding required for the Galilean invariance of Schrödinger's equation alone – and the "Im" by time-reversal invariance, since time-reversal is implemented on ψ by complex conjugation, again as demanded by Schrödinger's equation. The constant in front is precisely what is required for covariance under Galilean boosts.

2. BOHMIAN MECHANICS AND CLASSICAL PHYSICS

You will no doubt have noticed that the quantum potential, introduced and emphasized by Bohm (Bohm 1952a,b; Bohm and Hiley 1993) – but repeatedly dismissed, by omission, by Bell (Bell 1987a) – did not appear in our formulation of Bohmian mechanics. Bohm, in his seminal (and almost universally ignored!) 1952 hidden-variables paper (Bohm 1952a,b), wrote the wave function ψ in the polar form $\psi = Re^{iS/\hbar}$ where S is real and $R \geq 0$, and then rewrote Schrödinger's equation in terms of these new variables, obtaining a pair of coupled evolution equations: the continuity equation (7) for $\rho = R^2$, which suggests that ρ be interpreted as a probability density, and a modified Hamilton–Jacobi equation for S,

$$\text{(10)} \qquad \frac{\partial S}{\partial t} + H(\nabla S, q) + U = 0,$$

where $H = H(p, q)$ is the classical Hamiltonian function corresponding to (3), and

$$\text{(11)} \qquad U = -\sum_k \frac{\hbar^2}{2m_k} \frac{\nabla_k^2 V}{R}.$$

Noting that this equation differs from the usual classical Hamilton–Jacobi equation only by the appearance of an extra term, the *quantum potential* U, Bohm then used the equation to define particle trajectories just as is done for the classical Hamilton–Jacobi equation, that is, by identifying ∇S with *mv*, i.e., by

$$(12) \qquad \frac{d\mathbf{Q}_k}{dt} = \frac{\nabla_k S}{m_k},$$

which is equivalent to (9). The resulting motion is precisely what would have been obtained classically if the particles were acted upon by the force generated by the quantum potential in addition to the usual forces.

Bohm's rewriting of Schrödinger's equation via variables that seem interpretable in classical terms does not come without a cost. The most obvious cost in increased complexity: Schrödinger's equation is rather simple, not to mention linear, whereas the modified Hamilton–Jacobi equation is somewhat complicated, and highly nonlinear – and still requires the continuity equation for its closure. The quantum potential itself is neither simple nor natural [even to Bohm it has seemed "rather strange and arbitrary" (Bohm 1980, 80)] and it is not very satisfying to think of the quantum revolution as amounting to the insight that nature is classical after all, except that there is in nature what appears to be a rather ad hoc additional force term, the one arising from the quantum potential.

Moreover, the connection between classical mechanics and Bohmian mechanics that is suggested by the quantum potential is rather misleading. Bohmian mechanics is not simply classical mechanics with an additional force term. In Bohmian mechanics the velocities are not independent of positions, as they are classically, but are constrained by the guiding equation

$$(13) \qquad \mathbf{v}_k = \frac{\nabla_k S}{m_k}.$$

In classical Hamilton–Jacobi theory we also have this equation for the velocity, but there the Hamilton–Jacobi function S can be entirely eliminated and the description in terms of S simplified and reduced to a finite-dimensional description, with basic variables the positions and momenta of all the particles, given by Hamilton's or Newton's equations.

We wish to stress that since the dynamics for Bohmian mechanics is completely defined by Schrödinger's equation together with the guiding equation, there is neither need nor room for any further *axioms* involving the quantum potential! Thus the quantum potential should not be regarded as fundamental, and we should not allow it to obscure, as it all too easily tends to do, the most basic structure defining Bohmian mechanics.

We believe that the most serious flaw in the quantum potential formulation of Bohmian mechanics is that it gives a completely wrong impression of the lengths to which we must go in order to convert orthodox quantum theory into something more rational.[2] The quantum potential suggests, and indeed it has often been stated, that in order to transform Schrödinger's equation into a theory that can account, in what are often called "realistic" terms, for quantum phenomena, many of which are dramatically nonlocal, we must incorporate into the theory a quantum potential of a grossly nonlocal character.

We have already indicated why such sentiments are inadequate, but we would like to go further. Bohmian mechanics should be regarded as a first-order theory, in which it is the velocity, the rate of change of position, that is fundamental in that it is this quantity that is specified by the theory, directly and simply with the second-order (Newtonian) concepts of acceleration and force, work and energy playing no fundamental role. From our perspective the artificiality suggested by the quantum potential is the price one pays if one insists on casting a highly nonclassical theory into a classical mold.

This is not to say that these second-order concepts play *no* role in Bohmian mechanics; they are emergent notions, fundamental to the theory to which Bohmian mechanics converges int he "classical limit," namely, Newtonian mechanics. Moreover, in order most simply to see that Newtonian mechanics should be expected to emerge in this limit, it is convenient to transform the defining equations (1) and (2) of Bohmian mechanics into Bohm's Hamilton–Jacobi form. One then sees that the (size of the) quantum potential provides a rough measure of the deviation of Bohmian mechanics from its classical approximation.

It might be objected that mass is also a second-order concept, one that most definitely does play an important role in the very formulation of Bohmian mechanics. In this regard we would like to make several comments. First of all, the masses appear in the basic equations only in the combination $m_k/\hbar \equiv \mu_k$. Thus Eq. (2) could more efficiently be written as

$$\text{(14)} \qquad \frac{dQ_k}{dt} = \frac{1}{\mu_k}\frac{\mathrm{Im}(\psi^*\nabla_k\psi)}{\psi^*\psi},$$

and if we divide Schrödinger's equation by $\hbar$ it assumes the form

$$\text{(15)} \qquad i\frac{\partial\psi}{\partial t} = -\sum_{k=1}^{N}\frac{1}{2\mu_k}\nabla_k^2\psi + \hat{V}\psi,$$

with $\hat{V} = V/\hbar$. Thus it seems more appropriate to regard the naturalized masses μ_k, which in fact have the dimension of [time]/[length]^2, rather than the original masses m_k, as the fundamental parameters of the theory. Notice that if naturalized parameters (including also naturalized versions of the other coupling constants such as the naturalized electric charge $\hat{e} = e/\sqrt{\hbar}$) are used, Planck's constant $\hbar$ disappears form the formulation of this quantum theory. Where $\hbar$ remains is merely in the equations $m_k = \hbar\mu_k$ and $e^2 = \hbar\hat{e}^2$ relating the parameters – the masses and the charges – in the natural microscopic units with those in the natural units for the macroscopic scale, or, more precisely, for the theory, Newtonian mechanics, that emerges on this scale.

It might also be objected that notions such as *inertial* mass and the quantum potential are necessary if Bohmian mechanics is to provide us with any sort of *intuitive* explanation of quantum phenomena, i.e., explanation in familiar terms, presumably such as those involving only the concepts of classical mechanics. (See, for example, Baublitz and Shimony 1996.) It hardly seems necessary to remark, however, that physical explanation, even in a realistic framework, need not be in terms of classical physics.

Moreover, when classical physics was first propounded by Newton, this theory, invoking as it did action-at-a-distance, did not provide an explanation in familiar terms. Even less intuitive was Maxwell's electrodynamics, insofar as it depended upon the reality of the electromagnetic field. We should recall in this regard the lengths to which physicists, including Maxwell, were willing to go in trying to provide an intuitive explanation for this field as some sort of disturbance in a material substratum to be provided by the Ether. These attempts of course failed, but even had they not, the success would presumably have been accompanied by a rather drastic loss of mathematical simplicity. In the present century fundamental physics has moved sharply away from the search for such intuitive explanations in favor of explanations having an air of mathematical simplicity and naturalness, if not inevitability, and this has led to an astonishing amount of progress. It is particularly important to bear these remarks on intuitive explanation in mind when we come to the discussion in the next section of the status of quantum observables, especially spin.

The problem with orthodox quantum theory is not that it is unintuitive. Rather the problem is that

> . . . conventional formulations of quantum theory, and of quantum field theory in particular, are unprofessionally vague and ambiguous. Professional theoretical physicists ought to be able to do better. Bohm has shown us a way. (Bell 1987a, 173)

The problem, in other words, with orthodox quantum theory is not that it fails to be intuitively formulated, but rather that, with its incoherent babble about measurement, it is not even well formulated!

3. WHAT ABOUT QUANTUM OBSERVABLES?

We have argued that quantities such as mass do not have the same meaning in Bohmian mechanics as they do classically. This is not terribly surprising if we bear in mind that the meaning of theoretical entities is ultimately determined by their role in a theory, and thus when there is a drastic change of theory, a change in meaning is almost inevitable. We would now like to argue that with most observables, for example energy and momentum, something much more dramatic occurs: In the transition from classical mechanics they cease to remain properties at all. Observables, such as spin, that have no classical counterpart also should not be regarded as properties of the system. The best way to understand the status of these observables – and to better appreciate the minimality of Bohmian mechanics – is Bohr's way: what are called quantum observables obtain meaning *only* through their association with specific *experiments*. we believe that Bohr's point has not been taken to heart by most physicists, even those who regard themselves as advocates of the Copenhagen interpretation, and that the failure to appreciate this point nourishes a kind of naive realism about operators, an uncritical identification of operators with properties, that is the source of most, if not all, of the continuing confusion concerning the foundations of quantum mechanics.

Information about a system does not spontaneously pop into our heads, or into our (other) "measuring" instruments; rather, it is generated by an *experiment*: some physical interaction between the system of interest and these instruments, which together (if there is more than one) comprise the *apparatus* for the experiment. Moreover, this interaction is defined by, and must be analyzed in terms of, the physical theory governing the behavior of the composite formed by system and apparatus. If the apparatus is will designed, the experiment should somehow convey significant information about the system. However, we cannot hope to understand the significance of this "information" – for example, the nature of what it is, if anything, that has been measured – without some such theoretical analysis.

Whatever its significance, the information conveyed by the experiment is registered in the apparatus as an *output*, represented, say, by the orientation of a pointer. Moreover, when we speak of an experiment, we have in mind a fairly definite initial state of the apparatus, the ready state, one for which the apparatus should function as intended, and in particular one in which the pointer has some "null" orientation, say as in Figure 1.

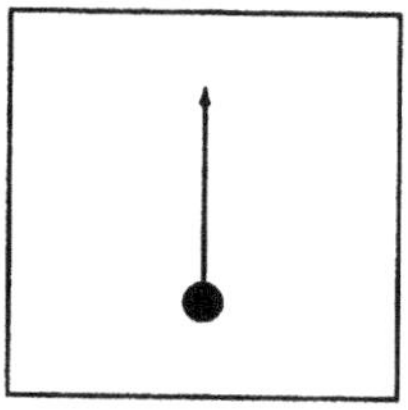

Fig. 1. Initial setting of apparatus.

For Bohmian mechanics we should expect in general that, as a consequence of the quantum equilibrium hypothesis, the justification of which we shall address in Section 4 and which we shall now simply take as an assumption, the outcome of the experiment – the final pointer orientation – will be random: Even if the system and apparatus initially have definite, known wave functions, so that the outcome is determined by the initial configuration of system and apparatus, this configuration is random, since the composite system is in quantum equilibrium, i.e., the distribution of this configuration is given by $|\Psi(x, y)|^2$, where Ψ is the wave function of the system-apparatus composite and x (respectively y) is the generic system (respectively apparatus) configuration. There are, however, special experiments whose outcomes are somewhat less random than we might have thought possible.

In fact, consider a *measurement-like* experiment, one which is *reproducible* in the sense that it will yield the same outcome as originally obtained if it is immediately repeated. (Note that this means that the apparatus must be immediately reset to its ready state, or a fresh apparatus must be employed, while the system is not tampered with so that its initial state for the repeated experiment is its final state produced by the first experiment.) Suppose that this experiment admits, i.e., that the apparatus is so designed that there are, only a finite (or countable) number of possible outcomes α,[3] for example, α = "left" and α = "right" as in Figure 2. The experiment also usually comes equipped with a *calibration* λ_α, an assignment of numerical values (or a vector of such values) to the various outcomes α.

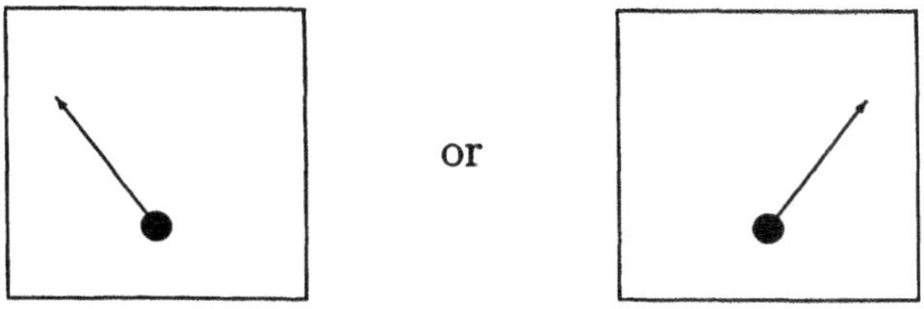

Fig. 2. Final apparatus readings.

It can be shown (Daumer *et al.* 1996), under further simplifying assumptions, that for such reproducible experiments there are special subspaces $\mathcal{H}_\alpha$ of the system Hilbert space $\mathcal{H}$ of (initial) wave functions, which are mutually orthogonal and span the entire system Hilbert space

$$(16)\qquad \mathcal{H} = \bigoplus_\alpha \mathcal{H}_\alpha,$$

such that if the system's wave function is initially in $\mathcal{H}_\alpha$, outcome α definitely occurs and the value λ_α is thus definitely obtained. It then follows that for a general initial system wave function

$$(17)\qquad \psi = \sum_\alpha \psi_\alpha \equiv \sum_\alpha P_{\mathcal{H}_\alpha}\psi,$$

where $P_{\mathcal{H}_\alpha}$ is the projection onto the subspace $\mathcal{H}_\alpha$, the outcome α is obtained with (the usual) probability[4]

$$(18)\qquad p_\alpha = \|P_{\mathcal{H}_\alpha}\psi\|^2.$$

In particular, the expected value obtained is

$$(19)\qquad \sum_\alpha p_\alpha\lambda_\alpha = \sum_\alpha \lambda_\alpha\|P_{\mathcal{H}_\alpha}\psi\|^2 = \langle\psi, A\psi\rangle,$$

where

$$(20) \qquad A = \sum_{\alpha} \lambda_\alpha P_{\mathcal{H}_\alpha},$$

and $\langle \cdot\,,\,\cdot \rangle$ is the usual inner product:

$$(21) \qquad \langle \psi, \phi \rangle = \int \psi^*(x)\phi(x)\mathrm{d}x.$$

What we wish to emphasize here is that, insofar as the statistics for the values which result form the experiment are concerned, the relevant data for the experiment are the collection $(\mathcal{H}_\alpha)$ of special subspaces, together with the corresponding calibration (λ_α), and *this data is compactly expressed and represented by the self-adjoint operator A, on the system Hilbert space* $\mathcal{H}$*, given by (20)*. Thus with a reproducible experiment $\mathscr{E}$ we naturally associate an operator $A = A_{\mathscr{E}}$,

$$(22) \qquad \mathscr{E} \rightarrow A,$$

a single mathematical object, defined on the system alone, in terms of which an efficient description of the possible results is achieved. If we wish we may speak of operators as observables, but if we do so it is important that we appreciate that in so speaking we merely refer to what we have just sketched: the role of operators in the description of certain experiments.[5]

In particular, so understood the notion of operator-as-observable in no way implies that anything is measured in the experiment, and certainly not the operator itself! In a general experiment no system property is being measured, even if the experiment happens to be measurement-like. Position measurements are of course an important exception. What in general is going on in obtaining outcome α is completely straightforward and in no way suggests, or assigns any substantive meaning to, statements to the effect that, prior to the experiment, observable A somehow had a value λ_α – whether this be in some determinate sense or in the sense of Heisenberg's "potentiality" or some other ill-defined fuzzy sense – which is revealed, or crystallized, by the experiment.[6]

Much of the preceding sketch of the emergence and role of operators as observables in Bohmian mechanics, including of course the von Neumann-type picture of "measurement" at which we arrive, applies as well to orthodox quantum theory.[7] In fact, it would appear that the argument against naive realism about operators provided by such an analysis has even greater force from an orthodox perspective: Given the initial wave function, at least in Bohmian mechanics the outcome of the particular experiment is determined by the initial configuration of system and apparatus, while for orthodox quantum theory there is nothing in the initial state which completely determines the outcome. Indeed, we find it rather surprising that most proponents of the von Neumann analysis of measurement, beginning with von Neumann, nonetheless seem to retain their naive realism about operators. Of course, this is presumably because more urgent matters –

the measurement problem and the suggestion of inconsistency and incoherence that it entails – soon force themselves upon one's attention. Moreover, such difficulties perhaps make it difficult to maintain much confidence about just what *should* be concluded from the "measurement" analysis, while in Bohmian mechanics, for which no such difficulties arise, what should be concluded is rather obvious.

It might be objected that we are claiming to arrive at the quantum formalism under somewhat unrealistic assumptions, such as, for example, reproducibility. (We note in this regard that many more experiments than those satisfying our assumptions can be associated with operators in exactly the manner we have described.) We agree. But this objection misses the point of the exercise. The quantum formalism itself is an idealization; when applicable at all, it is only as an approximation. Beyond illuminating the role of operators as ingredients in this formalism, our point was to indicate how naturally it emerges. In this regard we must emphasize that the following question arises for quantum orthodoxy, btu does not arise for Bohmian mechanics: For precisely which theory is the quantum formalism an idealization?

That the quantum formalism is merely an idealization, rarely directly relevant in practice, is quite clear. For example, in the real world the projection postulate – that when the measurement of an observable yields a specific value, the wave function of the system is replaced by its projection onto the corresponding eigenspace – is rarely satisfied. More important, a great many significant real-world experiments are simply not at all associated with operators in the usual way. Consider for example an electron with fairly general initial wave function, and surround the electron with a "photographic" plate, away from (the support of the wave function of) the electron, but not too far away. This set-up measures the position of "escape" of the electron from the region surrounded by the plate. Notice that since in general the time of escape is random, it is not at all clear which operator should correspond to the escape position – it should not be the Heisenberg position operator at a specific time, and a Heisenberg position operator at a random time has no meaning. In fact, there is presumably no such operator, so that for the experiment just described the probabilities for the possible results cannot be expressed in the form (18), and in fact are not given by the spectral measure for any operator.

Time measurements, for example escape times or decay times, are particularly embarrassing for the quantum formalism. This subject remains mired in controversy, with various research groups proposing their own favorite candidates for the "time operator" while paying little attention to the proposals of the other groups. For an analysis of time measurements within the framework of Bohmian mechanics, see Daumer *et al.* (1994) and Leavens (1996).

Because of such difficulties, it has been proposed (Davies 1976) that we should go beyond operators-as-observables, to "generalized obervables," described by mathematical objects even more abstract than operators. The basis of this generalization lies in the observation that, by the spectral theorem, the concept of self-adjoint operator is completely equivalent to that of (a normalized) projection-valued measure (PV), an orthogonal-projection-valued additive set function, on the value space $\mathbb{R}$. Since orthogonal projections are among the simplest examples of positive operators, a natural generalization of a "quantum observable" is provided by a (normalized) positive-operator-valued measure (POV). (When a POV is sandwiched by a wave function, as on the right-hand side of (19), it generates a probability distribution.)

It may seem that we would regard this development as a step in the wrong direction, since it supplies us with a new, much larger class of abstract mathematical entities about which to be naive realists. But for Bohmian mechanics POV's form an extremely natural class of objects to associate with experiments. In fact, consider a general experiment – beginning, say, at time 0 and ending at time t – with no assumptions about reproducibility or anything else. The experiment will define the following sequence of maps:

$$\psi \to \Psi = \psi \otimes \Phi_0 \to \Psi_t \to d\mu = \Psi_t^* \Psi_t dq \to \tilde{\mu} := \mu \circ F^{-1}.$$

Here ψ is the initial wave function of the system, and Φ_0 is the initial wave function of the apparatus; the latter is of course fixed, defined by the experiment. The second map corresponds to the time evolution arising from the interaction of the system and apparatus, which yields the wave function of the composite system after the experiment, with which we associate its quantum equilibrium distribution μ, the distribution of the configuration Q_t of the system and apparatus after the experiment. At the right we arrive at the probability distribution induced by a function F from the configuration space of the composite system to some value space, e.g., $\mathbb{R}$, or $\mathbb{R}^m$, or what have you: $\tilde{\mu}$ is the distribution of $F(Q_t)$. Here F could be completely general, but for application to the results of real-world experiments F might represent the "orientation of the apparatus pointer" or some coarse-graining thereof.

Notice that the composite map defined by this sequence, from wave functions to probability distributions on the value space, is "bilinear" or "quadratic," since the middle map, to the quantum equilibrium distribution, is obviously bilinear, while all the other maps are linear, all but the second trivially so. Now by elementary functional analysis, the notion of such a bilinear map is completely equivalent to that of a POV! Thus the emergence and role of POV's as "generalized observables" in Bohmian mechanics is merely an expression of the bilinearity of quantum equilibrium together with the linearity of Schrödinger's evolution. Thus the fact that with every experiment is associated a POV, which forms a compact expression of the statistics for the possible results, is a near mathematical triviality. It is therefore rather dubious that the occurrence of POV's as observables – the simplest case of which is that of PV's – can be regarded as suggesting any deep truths about reality or about epistemology.

The canonical example of a "quantum measurement" is provided by the Stern–Gerlach experiment. We wish to focus on this example here in order to make our previous considerations more concrete, as well as to present some further considerations about the "reality" of operators-as-observables. We wish in particular to comment on the status of spin. we shall therefore consider a Stern–Gerlach "measurement" of the spin of an electron, even though such an experiment is unphysical (Mott 1929), rather than of the internal angular momentum of a neutral atom.

We must first explain how to incorporate spin into Bohmian mechanics. This is very easy; we need do, in fact, almost nothing: Our derivation of Bohmian mechanics was based in part on rotation invariance, which requires in particular that rotations act on the value space of the wave function. The latter is rather inconspicuous for spinless particles – with complex-valued wave functions, what we have been considering up till now – since rotations then act in a trivial manner on the value space $\mathbb{C}$. The simplest nontrivial (projective) representation of the rotation group is the 2-dimensional, "spin $^1/_2$" representation; this representation leads to a Bohmian mechanics involving spinor-valued wave functions for a single particle and spinor-tensor-product-valued wave functions for many particles. Thus the wave function of a single spin $^1/_2$ particle has two components

$$(23)\qquad \psi(\mathbf{q}) = \begin{pmatrix} \psi_1(\mathbf{q}) \\ \psi_2(\mathbf{q}) \end{pmatrix},$$

which get mixed under rotations according to the action generated by the Pauli spin matrices $\boldsymbol{\sigma} = (\sigma_x, \sigma_y, \sigma_z)$, which may be taken to be

$$(24)\qquad \sigma_x = \begin{pmatrix} 0 & 1 \\ 1 & 0 \end{pmatrix}; \quad \sigma_y = \begin{pmatrix} 0 & -i \\ i & 0 \end{pmatrix}; \quad \sigma_z = \begin{pmatrix} 0 & 0 \\ 1 & -1 \end{pmatrix}.$$

Beyond the fact that the wave function now has a more abstract value space, nothing changes from our previous description; The wave function evolves via (1), where now the Hamiltonian H contains the Pauli term, for a single particle proportional to $\mathbf{B} \cdot \boldsymbol{\sigma}$, which represents the coupling between the "spin" and an external magnetic field $\mathbf{B}$. The configuration evolves according to (2), with the products of spinors now appearing there understood as spinor-inner-products.

Let's focus now on a Stern–Gerlach "measurement of $A = \sigma_z$." An inhomogeneous magnetic field is established in a neighborhood of the origin, by means of a suitable arrangement of magnets. This magnetic field is oriented more or less in the positive z-direction, and is increasing in this direction. We also assume that the arrangement is invariant under translations in the x-direction, i.e., that the geometry does not depend upon x-coordinate. An electron, with a fairly definite momentum, is directed towards the origin along the negative y-axis. Its passage through the inhomogeneous field generates a vertical deflection of its wave function away from the y-axis, which for Bohmian mechanics leads to a similar deflection of the electron's trajectory. if its wave function ψ were initially an eigenstate of σ_z of eigenvalue 1 (–1), i.e., if it were of the form[8]

$$(25) \qquad \psi = |\uparrow\rangle \otimes \phi_0 \quad (\psi = |\downarrow\rangle \otimes \phi_0),$$

where

$$(26) \qquad |\uparrow\rangle = \begin{pmatrix} 1 \\ 0 \end{pmatrix} \quad \text{and} \quad |\downarrow\rangle = \begin{pmatrix} 0 \\ 1 \end{pmatrix}.$$

then the deflection would be in the positive (negative) z-direction (by a rather definite angle). For a more general initial wave function, passage through the magnetic field will, by linearity, split the wave function into an upward-deflected piece (proportional to $|\uparrow\rangle$) and a downward-deflected piece (proportional to $|\downarrow\rangle$), with corresponding deflections of the possible trajectories.

The outcome is registered by detectors placed in the way of these two "beams." Thus of the four kinematically possible outcomes ("pointer positions") the occurrence of no detection defines the null output, simultaneous detection is irrelevant (since it does not occur if the experiment is performed one particle at a time), and the two relevant outcomes correspond to registration by either the upper of the lower detector. Thus the calibration for a measurement of σ_z is $\lambda_{up} = 1$ and $\lambda_{down} = -1$ (while for a measurement of the z-component of the spin angular momentum itself the calibration is the product of what we have just described by $\frac{1}{2}\hbar$).

Note that one can completely understand what's going on in this Stern–Gerlach experiment without invoking any additional property of the electron, e.g., its *actual* z-component of spin that is revealed in the experiment. For a general initial wave function there is no such property; what is more, the transparency of the analysis of this experiment makes it clear that there is nothing the least bit remarkable (or for that matter "nonclassical") about the *nonexistence* of this property. As we emphasized earlier, it is naive realism about operators, and the consequent failure to pay attention to the role of operators as observables, i.e., to precisely what we should mean when we speak of measuring operator-observables, that creates an impression of quantum peculiarity.

Bell has said that (for Bohmian mechanics) spin is not real. perhaps he should better have said: "*Even* spin is not real," not merely because of all observables, it is spin which is generally regarded as quantum mechanically most paradigmatic, but also because spin is treated in orthodox quantum theory very much like position, as a "degree of freedom" – a discrete index which supplements the continuous degrees of freedom corresponding to position – in the wave function. Be that as it may, his basic meaning is, we believe, this: Unlike position, spin is not *primitive*,[9] i.e., no *actual* discrete degrees of freedom, analogous to the *actual* positions of the particles, are added to the state description in order to deal with "particles with spin." Roughly speaking, spin is *merely* in the wave function. At the same time, as just said, "spin measurements" are completely clear, and merely reflect the way spinor wave functions are incorporated into a description of the motion of configurations.

It might be objected that while spin may not be primitive, so that the result of our "spin measurement" will not reflect any initial primitive property of the system, nonetheless this result *is* determined by the initial configuration of the system, i.e., by the position of our electron, together with its initial wave function, and as such – as a function $X_{\sigma_z}(\mathbf{q}, \psi)$ of the state of the system – it is some property of the system and in particular it is surely real. Concerning this, several comments.

The function $X_{\sigma_z}(\mathbf{q}, \psi)$, or better the property it represents, is (except for rather special choices of ψ) an extremely complicated function of its arguments; it is not "natural," not a "natural kind": It is not something in which, in its own right, we should be at all interested, apart from its relationship to the *result* of this particular experiment.

Be that as it may, it is not even possible to identify this function $X_{\sigma_z}(\mathbf{q}, \psi)$ with the measured spin component, since different experimental setups for "measuring the spin component" may lead to entirely different functions. In other words $X_{\sigma_z}(\mathbf{q}, \psi)$ is an abuse of notation, since the function X should be labeled, not by σ_z, but by the particular experiment for "measuring σ_z".

For example (Albert 1992, 153), if ψ and the magnetic field have sufficient reflection symmetry with respect to a plane between the poles of our SG magnet, and if the magnetic field is reversed, then the sign of what we have called $X_{\sigma_z}(\mathbf{q}, \psi)$ will be reversed: for both orientations of the magnetic field the electron cannot cross the plane of symmetry and hence if initially above respectively below the symmetry plane it remains above respectively below it. But when the field is reversed so must be the calibration, and what we have denoted by $X_{\sigma_z}(\mathbf{q}, \psi)$ changes sign with this change in experiment. (The change in experiment proposed by Albert is that "the *hardness box* is *flipped over*." However, with regard to spin this change will produce essentially no change in X at all. To obtain the reversal of sign, either the polarity or the geometry of the SG magnet must be reversed, but not both.)

In general X_A does not exist, i.e., $X_{\mathscr{E}}$, the result of the experiment ε, in general depends upon $\mathscr{E}$ and not just upon $A = A_{\mathscr{E}}$, the operator associated with $\mathscr{E}$. In foundations of quantum mechanics circles this situation is referred to as *contextuality*, but we believe that this terminology, while quite appropriate, somehow fails to convey with sufficient force the rather definitive character of what it entails: Properties which are merely contextual are not properties at all; they do not exist, and their failure to do so is in the strongest sense possible! We thus believe that contextuality reflects little more than the rather obvious observation that the result of an experiment should depend upon how it is performed!

4. THE QUANTUM EQUILIBRIUM HYPOTHESIS

The predictions of Bohmian mechanics for the results of a quantum experiment involving a system-apparatus composite having wave function ψ are precisely those of the quantum formalism, and moreover the quantum formalism of operators as observables emerges naturally and simply from Bohmian mechanics as the very expression of its empirical import, *provided* it is assumed that prior to the experiment the configuration of the system-apparatus composite is random, with distribution given by $\rho = |\psi|^2$. But how, in this deterministic theory, does randomness enter? What is special about $\rho = |\psi|^2$? What exactly does $\rho = |\psi|^2$ mean – to precisely which ensemble does this probability distribution refer? And why should $\rho = |\psi|^2$ be true?

We have already said that what is special about the quantum equilibrium distribution $\rho = |\psi|^2$ is that it is equivariant [see below Eq. (7)], a notion extending that of stationarity to the Bohmian dynamics (2), which is in general explicitly time-dependent. it is tempting when trying to justify the use of a particular "stationary" probability distribution μ for a dynamical system, such as the quantum equilibrium distribution for Bohmian mechanics, to argue that this distribution has a dynamical origin in the sense that even if the initial distribution μ_0 were different from μ, the dynamics generates a distribution μ_t which changes with time in such a way that μ_t approaches μ as t approaches ∞ (and that μ_t is approximately equal to μ for t of the order of a "relaxation time"). Such 'convergence to equilibrium' results – associated with the notions of 'mixing' and 'chaos' – are mathematically quite interesting. However, they are also usually very difficult to establish, even for rather simple and, indeed, artificially simplified dynamical systems. One of the nicest and earliest results along these lines, though for a rather special Bohmian model, is due to Bohm (Bohm 1953b).[10]

However, the justification of the quantum equilibrium hypothesis is a problem that by its very nature can be adequately addressed only on the universal level. To better appreciate this point, one should perhaps reflect upon the fact that the same thing is true for the related problem of understanding the origin of thermodynamic nonequilibrium (!) and irreversibility. as Feynman has said,

> Another delight of our subject of physics is that even simple and idealized things, like the ratchet and pawl, work only because they are part of the universe. The ratchet and pawl works in only one direction because it has some ultimate contact with the rest of the universe. . . . its one-way behavior is tied to the one-way behavior of the entire universe. (Feynman, Leighton and Sands 1963, 46–48)

An argument establishing the convergence to quantum equilibrium for local systems, if it is not part of an argument explaining universal quantum equilibrium, would leave open the possibility that conditions of local equilibrium would tend to be overwhelmed, on the occasions when they do briefly obtain, by interactions with an ambient universal nonequilibrium. In fact, this is precisely what does happen with thermodynamic equilibrium. In this regard, it is important to bear in mind that while we of course live in a universe that is not in universal thermodynamic equilibrium, a fact that is crucial to everything we experience, all available evidence supports universal quantum equilibrium. Were this not so, we should expect to be able to achieve violations of the quantum formalism – even for small systems. Indeed, we might expect the violations of universal quantum equilibrium to be as conspicuous as those of thermodynamic equilibrium.

Moreover, there are some crucial subtleties here, which we can begin to appreciate by first asking the question: Which systems should be governed by Bohmian mechanics? The systems which we normally consider are subsystems of a larger system – for example, the universe – whose behavior (the behavior of the whole) determines the behavior of its subsystems (the behavior of the parts). Thus for a Bohmian universe, it is only the universe itself which a priori – i.e., without further analysis – can be said to be governed by Bohmian mechanics.

So let's consider such a universe. Our first difficulty immediately emerges: In practice $\rho = |\psi|^2$ is applied to (small) subsystems. But only the universe has been assigned a wave function, which we shall denote by Ψ. What is meant then by the right hand side of $\rho = |\psi|^2$, i.e., by the wave function of a subsystem?

Fix an initial wave function Ψ_0 for this universe. Then since the Bohmian evolution is completely deterministic, once the initial configuration Q of this universe is also specified, all future events, including of course the results of measurements, are determined. Now let X be some subsystem variable – say the configuration of the subsystem at some time t – which we would like to be governed by $\rho = |\psi|^2$. How can this possibly be, when there is nothing at all random about X?

Of course, if we allow the initial universal configuration Q to be random, distributed according to the quantum equilibrium distribution $|\Psi_0(Q)|^2$, it follows from equivariance that the universal configuration Q_t at later times will also be random, with distribution given by $|\Psi_t|^2$, from which you might well imagine that it follows that any variable of interest, e.g., X, has the "right" distribution. But even if this were so (and it is), it would be devoid of physical significance! As Einstein has emphasized (Einstein 1953, 40; translated in Freistadt 1957, 40 (note)) "Nature as a whole can only be viewed as an individual system, existing only once, and not as a collection of systems."[11]

While Einstein's point is almost universally accepted among physicists, it is also very often ignored, even by the same physicists. We therefore elaborate: What possible physical significance can be assigned to an ensemble of universes, when we have but one universe at our disposal, the one in which we happen to reside? We cannot perform the *very same* experiment more than once. But we can perform many similar experiments, differing, however, at the very least, by location or time. In other words, insofar as the use of probability in physics is concerned, what is relevant is not sampling across an ensemble of universes, but sampling across space and time within a single universe. What is relevant is empirical distributions – actual relative frequencies for an ensemble of actual events.

At the expense belaboring the obvious, we stress that in order to understand why our universe should be expected to be in quantum equilibrium, it would not be *sufficient* to establish convergence to the universal quantum equilibrium *distribution*, even were it possible to do so. One simple consequence of our discussionis that proofs of convergence to equilibrium for the configuration of the universe would be of rather dubious physical significance: What good does it do to show that an initial distribution converges to some 'equilibrium distribution' if we can attach no relevant physical significance to the notion of a universe whose configuration is randomly distributed according to this distribution? In view of the implausibility of ever obtaining such a result, we are fortunate that it is also *unnecessary* (Dürr *et al.* 1992a,b), as we shall now explain.

Two problems must be addressed, that of the meaning of the wave function ψ of a subsystem and that of randomness. It turns out that once we come to grips with the first problem, the question of randomness almost answers itself. We obtain just what we want – that $\rho = |\psi|^2$ in the sense of empirical distributions; we find (Dürr *et al.* 1992a,b) that in a *typical* Bohmian universe an appearance of randomness emerges, precisely as described by the quantum formalism.

The term "typical" is used here in its mathematically precise sense: The conclusion holds for "almost every" universe, i.e., with the exception of a set of universes, or initial configurations, that is very small with respect to a certain natural measure, namely the universal quantum equilibrium distribution – the equivariant distribution for the universal Bohmian mechanics – on the set of all universes. It is important to realize that this guarantees that it holds for many particular universes – the overwhelming majority with respect to the only natural measure at hand – one of which might be ours.[12]

Before proceeding to a sketch of our analysis, we would like to give a simple example. Roughly speaking, what we wish to establish is analogous to the assertion, following from the law of large numbers, that the *relative frequency* of appearance of any particular digit in the decimal expansion of a *typical* number in the interval [0, 1] is $^1/_{10}$. In this statement two related notions appear: typicality, referring to an a priori measure, here the Lebesgue measure, and relative frequency, referring to structural patterns in an individual object.

It might be objected that unlike the Lebesgue measure on [0, 1], the universal quantum equilibrium measure will not in general be uniform. Concerning this, a comment: The uniform distribution – the Lebesgue measure on $\mathbb{R}^{3N}$ – has no special significance for the dynamical system defined by Bohmian mechanics. In particular, since the uniform distribution is not equivariant, typicality defined in terms of this distribution would depend critically on a somewhat arbitrary choice of initial time, which is clearly unacceptable. The sense of typicality defined by the universal quantum equilibrium measure is independent of any choice of initial time.

Given a subsystem, the x-system, with generic configuration x, we may write, for the generic configuration of the universe, $q = (x, y)$ where y is the generic configuration of the environment of the x-system. Similarly, we have $Q_t = (X_t, Y_t)$ for the actual configurations at time t. Clearly the simplest possibility for the wave function of the x-system, the simplest function of x which can sensibly be constructed from the actual state of the universe at time t (given by Q_t and Ψ_t), is

(27) $$\psi_t(x) = \Psi_t(x, Y_t),$$

the *conditional wave function* of the x-system at time t. This is almost all we need, almost but not quite.[13]

The conditional wave function is not quite the right notion for the *effective wave function* of a subsystem (see below; see also Dürr *et al.* 1992a,b), since it will not in general evolve according to Schrödinger's equation even when the system is isolated from its environment. However, whenever the effective wave function exists it agrees with the conditional wave function. Note, incidentally, that in an after-measurement situation, with a system-apparatus wave function as in note 4, we are confronted with the measurement problem if this wave function is the complete description of the composite system after the measurement, whereas for Bohmian mechanics, with the outcome of the measurement embodied in the configuration of the environment of the measured system, say in the orientation of a pointer on the apparatus, it is this configuration which, when inserted in (27), selects the term in the after-measurement macroscopic superposition that we speak of as defining the "collapsed" system wave function produced by the measurement. Moreover, if we reflect upon the structure of this superposition, we are directly led to the notion of effective wave function (Dürr *et al.* 1992a,b).

Suppose that

(28) $$\Psi_t(x, y) = \psi_t(x)\Phi_t(y) + \Psi_t^{\perp}(x, y),$$

where Φ_t and $\Psi_t^{\perp}$ have macroscopically disjoint y-supports. If

(29) $$Y_t \in \operatorname{supp} \Phi_t,$$

we say that ψ_t is the *effective wave function* of the x-system at time t. Note that it follows from (29) that $\Psi_t(x, Y_t) = \psi_t(x)\Phi_t(Y_t)$, so that the effective wave function is unambiguous, and indeed agrees with the conditional wave function up to an irrelevant constant factor.

We remark that it is the relative stability of the macroscopic disjointness employed in the definition of the effective wave function, arising from what are nowadays often called mechanisms of decoherence – the destruction of the coherent spreading of the wave function due to dissipation, the effectively irreversible flow of "phase information" into the (macroscopic) environment – which accounts for the fact that the effective wave function of a system obeys Schrödinger's equation for the system alone whenever this system is isolated. One of the best descriptions of the mechanisms of decoherence, though not the word itself, can be found in Bohm's 1952 "hidden variables" paper (Bohm 1952a,b). We wish to emphasize, however, that while decoherence plays a crucial role in the very formulation of the various interpretations of quantum theory loosely called decoherence theories, its role in Bohmian mechanics is of a quite different character: For Bohmian mechanics decoherence is purely phenomenological – it plays no role whatsoever in the formulation (or interpretation) of the theory itself.[14]

An immediate consequence (Dürr *et al.* 1992a,b) of (27) is the *fundamental conditional probability formula*:

$$(30) \qquad \mathbb{P}(X_t \in \mathrm{d}x \mid Y_t) = |\psi_t(x)|^2 \,\mathrm{d}x,$$

where $\mathbb{P}(\mathrm{d}Q) = |\Psi_0(Q)|^2 \,\mathrm{d}Q$.

Now suppose that at time t the x-system consists itself of many identical subsystems $x_1, \ldots, x_M$, each one having effective wave function ψ (with respect to coordinates relative to suitable frames). Then (Dürr *et al.* 1992a,b) the effective wave function of the x-system is the product wave function

$$(31) \qquad \psi_t(x) = \psi(x_1) \ldots \psi(x_M).$$

Note that it follows from (30) and (31) that the configurations of these subsystems are independent identically distributed random variables with respect to the initial universal quantum equilibrium distribution $\mathbb{P}$ conditioned on the environment of these subsystems. Thus the law of large numbers can be applied to conclude that the empirical distribution of the configurations $X_1, \ldots, X_M$ of the subsystems will typically be $|\psi(x)|^2$ – as demanded by the quantum formalism. For example, if $|\psi|^2$ assigns equal probability to the events "left" and "right," typically about half of our subsystems will have configurations belonging to "left" and half to "right." Moreover (Dürr *et al.* 1992a,b), this conclusion applies as well to a collection of systems at possibly different times as to the equal-time situation described here.[15]

It also follows (Dürr *et al.* 1992a,b) from the formula (30) that a typical universe embodies *absolute uncertainty*: the impossibility of obtaining more information about the present configuration of a system than what is expressed by the quantum equilibrium hypothesis. In this way, ironically, Bohmian mechanics may be regarded as providing a sharp foundation for and elucidation of Heisenberg's uncertainty principle.

5. WHAT IS A BOHMIAN THEORY?

Bohmian mechanics, the theory defined by Eqs. (1) and (2), is not Lorentz invariant, since (1) is a nonrelativistic equation, and, more importantly, since the right hand side of (2) involves the positions of the particles at a common (absolute) time. It is also frequently asserted that Bohmian mechanics cannot be made Lorentz invariant, by which it is presumably meant that no Bohmian theory – no theory that could be regarded somehow as a natural extension of Bohmian mechanics – can be found that is Lorentz invariant. The main reason for this belief is the manifest nonlocality of Bohmian mechanics (Bell 1987a). It must be stressed, however, that nonlocality has turned out to be a fact of nature: non-locality must be a feature of any physical theory accounting for the observed violations of Bell's inequality. (See Bell 1987a; Maudlin 1996; Squires 1996.)

A serious difficulty with the assertion that Bohmian mechanics cannot be made Lorentz invariant is that what it actually means is not at all clear, since this depends upon what is to be understood by a Bohmian theory. Concerning this there is surely no uniformity of opinion, but what we mean by a *Bohmian theory* is the following:

(1) A Bohmian theory should be based upon a clear ontology, the primitive ontology, corresponding roughly to Bell's local beables. This primitive ontology is what the theory is fundamentally about. For the nonrelativistic theory that we have been discussing, the primitive ontology is given by particles described by their positions, btu we see no compelling reason to insist upon this ontology for a relativistic extension of Bohmian mechanics.

Indeed, the most obvious ontology for a bosonic field theory is a field ontology, suggested by the fact that in standard quantum theory, it is the field configurations of a bosonic field theory that plays the role analogous to that of the particle configurations in the particle theory. However, we should not insist upon the field ontology either. Indeed, Bell (Bell 1987a, 173–180) has proposed a Bohmian model for a quantum field theory involving both bosonic and fermionic quantum fields in which the primitive ontology is associated only with fermions – with no local beables, neither fields nor particles, associated with the bosonic quantum fields. Squires (1996) has made a similar proposal.

While we insist that a Bohmian theory be based upon some clear ontology, we have no idea what the appropriate ontology for relativistic physics actually is.

(2) There should be a quantum state, a wave function, that evolves according to the unitary quantum evolution and whose role is to somehow generate the motion for the variables describing the primitive ontology.

(3) The predictions should agree (at least approximately) with those of orthodox quantum theory – at least to the extent that the latter are unambiguous.

This description of what a Bohmian theory should involve is admittedly vague, but greater precision would be inappropriate. But note that, vague as it is, this characterization clearly separates a Bohmian theory from an orthodox quantum theory as well as from the other leading alternatives to Copenhagen orthodoxy: The first condition is not satisfied by the decoherent or consistent histories (Griffiths 1984; Omnès 1988; Gell-Mann and Hartle 1993) formulations while with the spontaneous localization theories the second condition is deliberately abandoned (Ghirardi, Rimini and Weber 1986; Ghirardi, Pearle and Rimini 1990). With regard to the third condition, we are aware that it is not at all clear what should be meant by even an orthodox theory of quantum cosmology or gravity, let alone a Bohmian one. Nonetheless, this condition places strong constraints on the form of the guiding equation.

Furthermore, we do not wish to suggest here that the ultimate theory is likely to be a Bohmian theory, though we do think it very likely that if the ultimate theory is a quantum theory it will in fact be a Bohmian theory.

Understood in this way, a Bohmian theory is merely a quantum theory with a coherent ontology. If we believe that ours is a quantum world, does this seem like too much to demand? We see no reason why there can be no Lorentz invariant Bohmian theory. But if this should turn out to be impossible, it seems to us that we would be wiser to abandon Lorentz invariance before abandoning our demand for a coherent ontology.

ACKNOWLEDGMENTS

We are grateful to Karin Berndl, James Cushing, and Eugene Speer for valuable suggestions. This work was supported in part by NSF Grant No. DMS-9504556, by the DFG, and by the INFN.

Mathematisches Institut der Universität München

NOTES

[1] When a magnetic field is present, the gradients in the equations must be understood at the covariant derivatives involving the vector potential. If ψ is spinor-valued, the bilinear forms appearing in the numerator and denominator of (2) should be understood as spinor-inner-products. (See the discussion of spin in Section 3.) For indistinguishable particles, it follows from a careful analysis (Dürr *et al.* 1996b) of the natural configuration space, which will no longer be $\mathbb{R}^{3N}$, that when the wave function is represented in the usual way, as a function on $\mathbb{R}^{3N}$, it must be either symmetric or anti-symmetric under permutations of the labeled position variables. Note in this regard that according to orthodox quantum mechanics, the very notion of indistinguishable particles seems to be grounded on the nonexistence of particle trajectories. It is thus worth emphasizing that with Bohmian mechanics the

classification of particles as bosons or fermions emerges naturally from the very existence of trajectories.

2 The contortions required to deal with spin in the spirit of the quantum potential are particularly striking (Bohm and Hiley 1993; Holland 1993).

3 This is really no assumption at all, since the outcome should ultimately be converted to digital form, whatever its initial representation may be.

4 In the simplest such situation the unitary evolution for the wave function of the composite system carried the initial wave function $\Psi_i = \psi \otimes \Phi_0$ to the final wave function $\Psi_f = \sum_\alpha \psi_\alpha \otimes \Phi_0$ where Φ_0 is the ready apparatus wave function, and Φ_α is the apparatus wave function corresponding to outcome α. Then integrating $|\Psi_f|^2$ over supp Φ_α, we immediately arrive at (18).

5 Operators as observables also naturally convey information about the system's wave function after the experiment. For example, for an ideal measurement, when the outcome is α the wave function of the system after the experiment is (proportional to) $P_{\mathcal{H}_\alpha}\psi$. We shall touch briefly upon this collapse of the wave function, i.e., the projection postulate, in section 4, in connection with the notion of the effective wave function of a system.

6 Even speaking of the observable A as having value λ_α when the system's wave function is in $\mathcal{H}_\alpha$, i.e., when this wave function is an eigenstate of A of eigenvalue λ_α, insofar as it suggests that something peculiarly quantum is going on when the wave function is not an eigenstate whereas in fact there is nothing the least bit peculiar about the situation, perhaps does more harm than good.

7 It also applies to the spontaneous collapse models (Ghirardi, Rimini and Weber 1986; Ghirardi, Pearle and Rimini 1990), the interpretation of which (see, e.g., Albert 1992, 92–111) is often marred by naive realism about operators. See, however, Bell's presentation of GRW (Bell 1987a, 205) for an illuminating exception, as well as Ghirardi, Grassi and Benatti (1995) and Ghirardi and Grassi (1996).

8 Here we use the usual notation $\begin{pmatrix} a \\ b \end{pmatrix} \otimes \phi_0$ for $\begin{pmatrix} a\phi_0 \\ b\phi_0 \end{pmatrix}$.

9 We should probably distinguish two senses of "primitive": (i) the *strongly primitive* variables, which describe what the theory is fundamentally *about*, and (ii) the *weakly primitive* variables, the basic variables of the theory, those which define the complete state description. The latter may either in fact be strongly primitive, or, like the electromagnetic field in classical electrodynamics, they may be required in order to express the laws which govern the behavior of the strongly primitive variables in a simple and natural way. While this probably does not go far enough – we should further distinguish those weakly primitive variables which, like the velocity, are functions of the trajectory of the strongly primitive variables, and those, again like the electromagnetic field, which are not – these details are not relevant to our present purposes, so we shall ignore these distinctions.

10 As an illustration of the pitfalls in trying to establish convergence to quantum equilibrium, a recent attempt of Valentini (Valentini 1991a) is instructive. Valentini's argument is based on a "subquantum H-theorem," $d\bar{H}/dt \leq 0$, that is too weak to be of any relevance (since, for example, the inequality is not strict). The H-theorem is itself not correctly proven – it could not be since it is in general false. Even were the H-theorem true, correctly proven, and potentially relevant, the argument given would still be circular, since in proceeding from the H-theorem to the desired conclusion, Valentini finds it necessary to invoke "assumptions similar to those of classical statistical mechanics," namely that (Valentini 1992, 36) "the system is 'sufficiently chaotic'," which more or less amounts to assuming the very mixing which was to be derived.

11 For a rather explicit example of the failure to appreciate this point, see Albert (1992, 144): "And the statistical postulate . . . can be construed as stipulating something about the *initial conditions* of the universe; it can be construed (in the fairy-tale language, say) as stipulating that what God did when the universe was created was first to choose a wave function for it and sprinkle all of the particles into space in accordance with the quantum-mechanical probabilities, and then to leave everything alone, forever after, to evolve

deterministically. And note that just as the one-particle postulate can be derived . . . from the two-particle postulate, *all* of the more specialized statistical postulates will turn out to be similarly derivable from *this* one." (Note that an initial sprinkling in accordance with the quantum-mechanical probabilities need remain so only if "quantum-mechanical probabilities" is understood as referring to the quantum equilibrium distribution for the configuration of the entire universe rather than to empirical distributions for subsystems arising from this configuration. Note also that the analogy with the relationship between the one-particle and the two-particle postulates also requires that the universal "statistical postulate" be understood in this way.)

[12] It is important to realize that an appeal to typicality is unavoidable if we are to explain why the universe is at present in quantum equilibrium. This is because our analysis also demonstrates that there is a set B of initial configurations, a set of nonvanishing Lebesgue measure, that evolve to present configurations violating the quantum equilibrium hypothesis and hence the quantum formalism. This set cannot be wished away by any sort of mixing argument. Indeed, if, as is expected, mixing holds on the universal level, then this set B should be so convoluted as to be indescribable without a specific reference to the universal dynamics and hence cannot be dismissed as unphysical without circularity.

[13] For particles with spin, (27) should be replaced by $\Psi_t(x, Y_t) = \psi_t(x) \otimes \Phi_t(Y_t)$. In particular, for particles with spin, not every subsystem has a conditional wave function.

[14] However, decoherence is important for a serious discussion of the emergence of Newtonian mechanics as the description of the macroscopic regime for Bohmian mechanics, leading to the picture of a macroscopic Bohmian particle, in the classical regime, guided by a macroscopically well-localized wave packet with a macroscopically sharp momentum moving along a classical trajectory. It may, indeed, seem somewhat paradoxical that the gross features of our world should appear classical because of interaction with the environment and the resulting wave function entanglement (Joos and Zeh 1985; Gell-Mann and Hartle 1993), the characteristic quantum innovation (Schrödinger 1935).

[15] It should not be necessary to say that we do not claim to have established the impossibility – but rather the typicality – of quantum nonequilibrium. On the contrary, as we have suggested in Dürr *et al.* (1992a, 904), "the reader may wish to explore quantum nonequilibrium. What sort of behavior would emerge in a universe which is in quantum nonequilibrium?" Concerning this, we wish to note that despite what is suggested by the misuse of ensembles for the universe as a whole and the identification of the physical universal convergence of configurations to those characteristic of quantum equilibrium with the expected convergence of universal measures, of $P \rightarrow |\Psi|^2$, quantum equilibrium is not an attractor, and no "force" pushes the universal configuration to one of quantum equilibrium. Rather, any transition from quantum nonequilibrium to quantum equilibrium would be entropic and time-symmetric – driven indeed primarily by measure-theoretic effects, by the fact that the set of quantum equilibrium configurations is vastly larger than the set of configurations corresponding to quantum nonequilibrium – just as is the convergence to thermodynamic equilibrium. (For some speculations on the possible value of quantum nonequilibrium, see Valentini (1996a).)

ANTONY VALENTINI

PILOT-WAVE THEORY OF FIELDS, GRAVITATION AND COSMOLOGY*

All science is cosmology.
– Karl Popper

Our universe is characterised by: (i) The absence of practical instantaneous signals (signal-locality), and (ii) An all-pervading statistical 'noise' (uncertainty principle). Now, as Bell deduced, there is a fundamental nonlocality hidden behind quantum statistics. Relativity is not overtly violated, however. There seems to be a 'conspiracy' between relativity and quantum theory, whereby uncertainty noise prevents one from using subquantum nonlocality for practical signalling. Why should the nonlocality be *hidden* in this way? A physics whose coherence rests on such a peculiar conspiracy can hardly be regarded as fundamental.

We therefore suggest that the quantum-theoretical distribution $P = |\Psi|^2$ is not fundamental. It is merely an equilibrium state (analogous to the Maxwell distribution of molecular speeds), characterised by a statistical 'uncertainty noise' that *only just* manages to mask an underlying nonlocality. The universe today happens to be in this state, explaining the above conspiracy.

Plato, in *The Republic*, compared human beings to prisoners chained inside a cave since childhood, who grew up believing that the shadows moving on the wall in front of them were real things. Similarly, because their experience so far has been restricted to $P = |\Psi|^2$, physicists believe that this distribution – with its associated properties of locality and uncertainty – is fundamental. Further, they believe that the phenomenological language of the (approximately classical) macroscopic level is of fundamental significance. These erroneous beliefs are responsible for numerous paradoxical and spurious features of modern physics.

Bell showed by indirect reasoning that, behind the appearances of modern physics, the world is both nonlocal and contextual. These two clues suggest that our classical variables – from which quantum theory is constructed – are *not the correct set of elementary variables*. We use two arguments: (i) Since energy-momentum conservation is important and powerful only if it is a *local* principle, it would seem inappropriate that this principle should appear at the nonlocal subquantum level; and having then lost their raison d'être, the very concepts of 'energy' and 'momentum' should make no appearance. (ii) The fact that quantum-mechanical 'measurements' are contextual, i.e. are usually not true measurements (Bell 1987a) – in the sense that they do not necessarily yield information about some previously existing quantity belonging to the 'system' – clearly suggests that we are simply

J. T. Cushing et al. (eds.), Bohmian Mechanics and Quantum Theory: An Appraisal, 45–66.

using the wrong variables (just as the perplexities encountered in early theories of combustion indicated that chemical variables such as phlogiston were illusory – the paradoxical use of the word 'measurement' recalling the 'negative weight' once attributed to phlogiston). Since nonlocality already makes them suspect, 'energy' and 'momentum' are the obvious culprits.

We therefore seek a nonlocal theory based on new variables and a new dynamics, where 'energy' and 'momentum' make no appearance. The theory must account for the equilibrium $P = |\Psi|^2$ and for the associated phenomenology of quantum theory (which includes locality, uncertainty, and energy-momentum conservation).

1. ELEMENTS OF THE PILOT-WAVE THEORY

Such a theory may be obtained by suitably interpreting and developing ideas given long ago by de Broglie (1928) and Bohm (1952a,b). The quantum-mechanical wave function $\Psi(X, t)$ is interpreted as an objectively existing 'guiding field' (or pilot wave) in configuration space, which determines the *velocity* $\mathrm{d}X/\mathrm{d}t$ of the actual system configuration $X(t)$ at any time t. The guiding field satisfies the configuration-space Schrödinger equation (units $\hbar = 1$)

$$(1) \qquad \mathrm{i}\frac{\partial\Psi}{\partial t} = \hat{H}\Psi,$$

where $\hat{H}$ is the (configuration-space) Hamiltonian operator; while the velocity of the system is determined by

$$(2) \qquad \Pi = \frac{\partial S}{\partial X},$$

where Π is the canonical momentum and S is the phase of the guiding field (given locally by $S = \mathrm{Im}\ln\Psi$). The wave function evolves autonomously according to (1), generating an instantaneous velocity field throughout configuration space; and the system, wherever it happens to be, moves according to (2) with the local value of the velocity field. (A point X in configuration space may represent a collection of particle positions, the configuration of a field in space, or the geometry of 3-space itself. The symbol $\partial/\partial X$ then represents a generalised gradient. For nonrelativistic particles of mass m we have $\Pi = m\dot{X}$; for field theory Π is really the canonical momentum density and $\partial/\partial X$ a functional derivative.)

Historical Remark: The above (velocity-based) dynamics was originally proposed as a basis for quantum physics by de Broglie (1928, 118–119) – for a nonrelativistic many-body system with a wave function in *configuration* space (contrary to many erroneous claims made in the literature that de Broglie only had the one-body theory with a wave function in three-space). Bohm (1952a) revived de Broglie's theory, albeit in a pseudo-

Newtonian second-order form (based on acceleration determined by a 'quantum potential'), and showed how it could account for the general quantum theory of measurement (Bohm 1952b). The original first-order dynamics was popularised by Bell (1987a), who invariably used the historically accurate terminology 'de Broglie–Bohm theory' or 'pilot-wave theory' (the latter term having been originally used by de Broglie himself). Historically, a rather sophisticated precursor of this theory, based on relativistic phase waves, was put forward in de Broglie's seminal Doctoral Thesis of 1924. Subsequently de Broglie (rather like Maxwell) emphasised an underlying 'mechanical' picture: particles were assumed to be singularities of physical waves in space. In a paper of 1927, however, de Broglie recognised that one could abandon this scaffolding and that (like the electromagnetic theory) the resulting pilot-wave theory in configuration space could stand on its own feet. De Broglie's construction of this theory (with Schrödinger's assistance), and his subsequent abandonment of it in favour of the quantum formalism, is analysed elsewhere (Valentini 1996b).

The Schrödinger equation (1) and the de Broglie guidance condition (2) are, then, the fundamental equations of what might be termed *pilot-wave dynamics*. No reference whatever is made, or need be made, to acceleration, Newtonian 'force', 'energy' or 'momentum' (Dürr *et al.* 1992b, 8; Valentini 1992, 4, 8, 12–17, 34).[1] That the velocity is at all times given by the gradient of the phase of the wave function is quite sufficient to determine the motion of any system. *Causal thinking need not be based on 'forces' proportional to acceleration; the 'agency causing the motion' may just as well be an 'Aristotelian force' proportional to the velocity* (Valentini 1992, 12–17; 1996b). Thus one may adopt (2) as the fundamental law of motion for any system, with no need for further explanation. In this sense one may regard the concept of guiding field as fundamental and irreducible, like the electromagnetic field. Bohm's 'quantum potential' is, like the 'luminiferous ether', a concept best abandoned.

Since the guiding field determines velocity, two systems may reasonably be said to 'interact' whenever the velocity of one depends instantaneously on the configuration of the other. (Just as in Newtonian mechanics, where the law of motion determines acceleration, two systems are said to 'interact' whenever the acceleration of one depends instantaneously on the configuration of the other.) And from (2) it is clear that such interaction will occur *if and only if* the total wave function is nonseparable or 'entangled'. The essential physical significance of the law of motion (2) is therefore simply that *all 'interactions' are manifestations of EPR-entanglement and nothing but EPR-entanglement.* What might seem a mere change of terminology here is actually a radical shift of physical viewpoint, comparable to the abandonment of gravitational 'force' in general relativity. Even classical 'forces' may be seen, in the classical limit, as large-scale manifestations of nonseparability, of a form conceptually and physically identical to that seen in EPR-type experiments. (For classical field theory, the coupling between neighbouring field elements is a manifestation of entanglement.) Thus, instead of explaining nonlocal EPR-entanglement in terms of a pseudo-Newtonian 'quantum potential', as in Bohm's formulation, we take the opposite view: even classical 'forces' are manifestations of such entanglement.

The status of Ψ as a 'field' is radically different from that of ordinary fields. In particular, since Ψ exists in configuration space the notion of 'test particle' has no meaning. For if one introduces a 'test body', with the aim of probing the field Ψ associated with some system, then no matter how 'small' or otherwise physically insignificant the test body may be, the effect will be to *increase the dimension* of the configuration space: the body becomes compounded with the system, and one must now consider the total wave function Ψ_{total} for the new super-system. The same is true for the introduction of an 'apparatus'; and this remarkable feature is the ultimate source of contextuality. Nevertheless, a definite Ψ may be prepared, for instance by allowing a system to fall into its ground state. (The resulting Ψ may be *calculated*, assuming the truth of the Schrödinger equation.) And Ψ may be measured for a stationary state (Aharonov and Vaidman 1993).

The fundamental physics of pilot-wave dynamics may, however, be clearly perceived only by a 'subquantum demon' whose senses are directly receptive to subquantum trajectories. For such a demon, the law of motion (2) has a status similar to that of Newton's law for us. For instance, given the perceived velocity of a system, the demon would use (2) to deduce the configuration of a distant entangled system – just as we, given the acceleration of a particle, use Newton's law to deduce the position of a distant gravitating mass. In the pilot-wave theory, a *true measurement* must be of this form, defined in terms of the actual variables.

Subquantum H-Theorem and 'Heat Death'

An ensemble distribution P need have no a priori relation to $|\Psi|^2$. The equilibrium $P = |\Psi|^2$ is accounted for by means of a 'subquantum H-theorem' based on the configuration-space quantity

$$(3) \qquad H = \int dX\, P \ln\left(\frac{P}{|\Psi|^2}\right)$$

(where the 'subquantum entropy' $S = -H$). This is obtained from the classical $H_{\text{class}} = \int d\Omega\, p \ln p$ (on *phase* space) by replacing $p \to f \equiv P/|\Psi|^2$ and $d\Omega \to |\Psi|^2\, dX$, where p, $d\Omega$ and f, $|\Psi|^2\, dX$ are in each case preserved along trajectories.

In its coarse-graining version, the theorem describes the onset of 'mixing' of P and $|\Psi|^2$ (using the word 'mixing' in a non-technical sense). This is just the familiar Gibbs 'stirring' of two 'fluids', with densities P and $|\Psi|^2$, by the same (complicated) velocity field. An ensemble of complicated (and isolated) systems will in reasonable circumstances begin to evolve towards equilibrium, in the sense that P and $|\Psi|^2$ will tend to become indistinguishable on a coarse-grained level. The proof is analogous to that of the classical coarse-graining H-theorem due to the Ehrenfests. Assuming that at $t = 0$ P and $|\Psi|^2$ do not possess any (possibly 'conspiratorial') fine-grained microstructure – that is, that they are essentially constant over the coarse-graining cells – the theorem states that the coarse-grained H-function will at all times be less than or equal to its value at $t = 0$. (The possible equality sign is removed by assuming the initial velocity field $\dot{X}$ to satisfy $\text{var}(\dot{X} \cdot \nabla f) \neq 0$, where the variance is taken over a coarse-graining cell.) The coarse-grained H must decrease immediately after $t = 0$; and it is minimised by the coarse-grained $P = |\Psi|^2$. As in the classical case, $t = 0$ is a local maximum; the function decreases in both time directions. Further, the physically-plausible *assumption* must be made that there are 'appropriately complex' systems, widely realised in Nature, for which the initial steps towards equilibrium described by the H-theorem are subsequently brought to completion (to a good approximation).

Alternatively, one may develop an H-theorem based on external perturbations.[2] This implies that the subquantum and thermodynamic arrows of time must be aligned. Defining the direction of time in terms of the observed thermodynamic arrow, one may then rule out the possibility that the subquantum level happens to be evolving away from equilibrium (owing to 'conspiratorial' initial conditions).

Given equilibrium $P = |\Psi|^2$ for an ensemble of complicated systems (such as similar boxes of gas), it may be shown that extracting a subsystem (such as a particle) from each leads again to an equilibrium ensemble $\rho = |\psi|^2$.

Now it may also be shown that *nonlocal connections between distant entangled systems lead to practical instantaneous signals, at the statistical level, if and only if* $P \neq |\Psi|^2$. In equilibrium, such signals are precisely masked by uncertainty noise (where the uncertainty principle holds if and only if $P = |\Psi|^2$). The pilot-wave theory thus yields a concrete realisation of our view that signal-locality and uncertainty are related, and peculiar to equilibrium. (It seems likely that any realistic theory would exhibit a similar relationship.)

The universe presumably begins in a state $P \neq |\Psi|^2$, where instantaneous signalling would be possible. As one approaches the state $P = |\Psi|^2$ of maximum subquantum entropy, the possibility of instantaneous signalling fades away, and statistical uncertainty takes over. The equilibrium $P = |\Psi|^2$ may then be seen as a kind of 'subquantum heat death', in which the underlying nonlocal connections may no longer be used for practical signalling.[3] (In the classical heat death, thermal energy may no longer be used to do work.)

It may then be said that we are 'chained inside Plato's Cave', partly because our macroscopic functioning is essentially classical, but also because *a subquantum analogue of Boltzmann's heat death has actually happened in our universe*. The conspiracy between quantum theory and relativity is accounted for as a peculiarity of this subquantum heat death.

A positivist and verificationist epistemology has encouraged physicists to project human limitations onto Nature herself. The above picture suggests instead that present-day physicists are subject to experimental limitations that are not fundamental, but merely contingent.

Emergence of Quantum Theory and Classical Language

The human constitution is oblivious to the details of subquantum trajectories. We are unable to perform true measurements. Our constitution is based on the classical limit which arises, in accordance with Ehrenfest's theorem, from appropriately narrow wave packets. (Equilibrium is also required, to ensure that the actual configuration will almost certainly lie near the bulk of the narrow packet.) On encountering quantum phenomena, experimenters try to coordinate these in terms of their (supposedly fundamental) classical language. They adopt a 'measurement' scheme inspired by the language of classical physics. In a 'momentum measurement', say, one performs operations (such as scattering by an external field) that *classically* would constitute a measurement of momentum. It is *quite wrongly* assumed that these same operations yield the 'value of the momentum' even when performed on a nonclassical system. Instead the 'result' (for an ideal case) is equal to the eigenvalue of a linear operator $\hat{P}$ (acting on the final wave function guiding the system), which need have no bearing on any real property of the system prior to the 'measurement'.

Generally, if the wave function is an eigenfunction of an operator $\hat{Q}$ with eigenvalue q, then because our experimenter *believes* that he performs measurements correctly, he *says* that 'the observable has the value q'. But in the pilot-wave theory it is simply the case that $X(t)$ is guided by an eigenfunction ψ_q; it is usually in no sense true that 'the observable has the value q'.

How does our experimenter get away with saying that 'the observable has the value q'? This is formally possible because, owing to the linearity of the Schrödinger equation, there is a general mathematical correspondence between linear operators and classical variables. (The classical variable 'corresponding to' $\hat{Q}$ arises from guidance by a narrow packet whose $\hat{Q}$-spectrum is peaked around some value q: such a system may, phenomenologically, be *assigned* a classical variable with value q. This is no more peculiar than the emergence of 'gravitational force' in the Newtonian limit of general relativity.) This correspondence ensures that the von Neumann scheme works as a phenomenological book-keeping device. Solutions of the linear Schrödinger equation are conveniently classified in terms of a linear vector space; the analogy between operators and classical variables is useful as a heuristic mnemonic.

In the phenomenology of equilibrium, the trajectory plays a very minor role. It merely singles out the 'realised' branch of the wave function. Precise details of the trajectory disappear. Reference to the law of motion (2) may be dropped and, as shown in detail by Bohm (1952a,b), standard quantum theory emerges.

This practical success reassures the experimenter of the correctness of his statement that 'the observable has the value q'. However, the literal identification of eigenvalues with values of real physical quantities is *the* fundamental error of quantum 'measurement' theory. (The apparently innocent statement 'the observable has the value q' is the ultimate source of the 'schizophrenic' character of quantum theory: for a wave function with eigenvalue q is generally not an eigenfunction of some other 'incompatible' observable, which then seems to have *no definite value*.) And this basic error stems, as we have seen, from the mistaken belief that classical operations correctly measure nonclassical systems. As Einstein foresaw in 1926: 'Your theory will one day get you into hot water', because 'when it comes to observation, you behave as if everything can be left as it was, that is, as if you could use the old descriptive language' (Heisenberg 1971, 66).

Remark

Dürr *et al.* (1992a,b) suggest that, instead of appealing to a subquantum H-theorem, one may account for quantum statistics by adopting the 'natural measure' $|\Psi_0^{\mathrm{Univ}}|^2$ as a measure of 'typicality', or likelihood, for the initial configuration X_0^{Univ} of the universe (where Ψ_0^{Univ} is the initial universal wave function). If we consider the simple case of an ensemble of independent subsystems, this automatically leads to the equilibrium distribution $\rho_0 = |\psi_0|^2$ at $t = 0$ (and hence to $\rho = |\psi|^2$ at all future times). Now a different choice of measure, such as $|\Psi_0^{\mathrm{Univ}}|^4$, leads to an initial *dis*equilibrium distribution – in this case $\rho_0 \propto |\psi_0|^4$ (whose functional form will of course not be preserved in time). But Dürr *et al.* argue that because $|\Psi_0^{\mathrm{Univ}}|^2$ is the natural measure (singled out by the Schrödinger equation), and because disequilibrium states occupy a very small volume with respect to this natural measure, initial disequilibrium should be regarded as intrinsically very unlikely (though not impossible). In general, the quantum statistics observed today are 'explained' by showing that they occur, as empirical statistics across space and time, for initial configurations X_0^{Univ} that are 'typical' with respect to $|\Psi_0^{\mathrm{Univ}}|^2$.

The author disagrees with: (i) The suggestion that 'typicality' with respect to $|\Psi_0^{\mathrm{Univ}}|^2$ could be a satisfactory explanation of quantum statistics, and (ii) The claim that initial disequilibrium is intrinsically very unlikely.

Regarding (i): Equilibrium is indeed the 'natural' state, in the sense of being 'typical' with respect to the natural measure. But does this mean there is no need for a causal explanation of it? For example, most cosmologists feel that the observed homogeneity of the early universe requires a causal explanation, in terms of some homogenising process, and should not be put into the initial conditions by hand. And yet, one might argue, homogeneity is 'typical' with respect to the natural measure. Would 'typicality' be a satisfying explanation of the early homogeneity?

There are opposing philosophical standpoints here, concerning the nature of explanation, that cannot be resolved on purely logical grounds. In the author's view, it would be more satisfactory if quantum noise could be shown to arise for dynamical reasons – as suggested by the subquantum *H*-theorem.

Against the author's approach, the criticism might be raised that some notion of 'typicality' is in any case necessary to rule out 'exceptional' initial conditions that do not evolve towards equilibrium. But such initial conditions are ruled out by the observation of equilibrium today – for which we are seeking a causal explanation – while other initial (disequilibrium) conditions are not. One makes an assumption about the initial conditions in order to explain observations (as in classical statistical mechanics and, indeed, as in science in general). But one tries to derive the observed statistics as far as possible on dynamical grounds, for as wide a range of initial conditions as possible. In contrast, the use of 'typicality' (with respect to $|\Psi_0^{\text{Univ}}|^2$) to choose X_0^{Univ} amounts to inserting quantum noise into the initial conditions themselves.

Regarding (ii): Consider an isolated box of classical gas. What is the most likely state at some arbitrary instant, say $t = 0$? If we knew that, for $t < 0$, there had been an epoch of ergodic motion (on the energy surface), it would indeed be very likely that the gas is close to equilibrium at $t = 0$. (Though going back far enough in time, one would necessarily encounter a rare fluctuation – after which relaxation towards equilibrium, on the basis of some sort of *H*-theorem, must have occurred.) But if the box of gas was *created* at $t = 0$, there is no reason why the state should have been chosen randomly with respect to the dynamical measure. And if at $t = 5$ min the gas is found to be close to equilibrium (to within ε), any deduction about its initial state at $t = 0$ would have to consider *all possible* initial states consistent with the present. These would include all disequilibrium states that relax to within ε of equilibrium, within five minutes. And very little could be said about the relative likelihood of these initial states. (In principle one would even have to consider 'conspiratorial' initial states: perhaps the gas started out even closer to equilibrium, and is 'unravelling' towards a disequilibrium that becomes detectable, say, after one hour. For our universe we can rule out the analogous possibility in the subquantum case because, as mentioned, the subquantum and thermodynamic arrows of time must be aligned.)

Similarly, in the pilot-wave theory we know from experiment only that, today, $\rho \approx |\psi|^2$ to within some (small) ε.[4] The set of allowed initial states (consistent with experiment) includes not only equilibrium, but also all disequilibrium states that evolve to $\rho \approx |\psi|^2$ (to within ε) today – and we know that there are such states. Indeed, there are an uncountable number of distinct sets of allowed initial configurations X_0^{Univ}, of which the equilibrium set is just one.[5] Each of these sets is 'typical' (has measure unity) with respect to an appropriate measure, each is uncountable, none of them are in any way pathological, and they all agree with experiment. And there is no a priori mathematical measure of their relative likelihood. There is a 'natural measure', namely $|\Psi_0^{\mathrm{Univ}}|^2$, according to which the equilibrium set is overwhelmingly larger than its complement. But to use $|\Psi_0^{\mathrm{Univ}}|^2$ as a measure of intrinsic *likelihood*, for initial conditions, can only be a hypothesis. It is not required – or even strongly supported – by experiment. Use of $|\Psi_0^{\mathrm{Univ}}|^2$ effectively *defines* the other sets to be unlikely.

Thus, in the absence of a prior epoch of dynamical evolution, there is no justification for attaching an intrinsic mathematical likelihood to the equilibrium subset at $t = 0$. Indeed, no intrinsic mathematical likelihood can be attached to the complement of that subset either. We must turn to cosmology – and further experiment – if we wish to narrow down the range of possibilities, as discussed below.

On the other hand, in the absence of compelling evidence, one can try to motivate hypotheses about the beginning – such as the disequilibrium assumed by the author and the equilibrium assumed by Dürr *et al.* – on grounds of general physical plausibility (as opposed to measure-theoretic arguments). In the author's opinion, on general grounds it would be very peculiar (as well as disappointing) if Nature made no use at all of non-equilibrium phenomena. After all, we began by explaining the conspiracy between relativity and quantum theory by the hypothesis that these are equilibrium theories. It would be odd if this equilibrium were built into the universe from the beginning. Further, if quantum-mechanical uncertainty were indeed 'absolute', the details of the trajectories could never be seen and the law of motion (2) could never be tested. It seems doubtful that Nature should be constructed in such a 'malicious' manner (in Einstein's sense).

2. FIELD THEORY OF 'PARTICLES' AND OF SPIN

Field Theory of Massive 'Particles'

For a 'massive' real scalar field $\phi(x, t)$, the Schrödinger equation (1) for the wave functional $\Psi[\phi, t]$ reads (units $\hbar = c = 1$)

$$(4) \qquad i\frac{\partial\Psi}{\partial t} = \int d^3x \frac{1}{2}\left(-\frac{\delta^2}{\delta\phi^2} + (\nabla\phi)^2 + m^2\phi^2\right)\Psi,$$

while the equation of motion (2) reads

$$(5) \qquad \frac{\partial\phi(x, t)}{\partial t} = \frac{\delta S[\phi, t]}{\delta\phi(x)}.$$

These equations describe the evolution of a scalar field on Euclidean 3-space, in terms of a fixed time parameter t. (The scalar field was first treated by Bohm and Hiley (1984) using a 'super quantum potential'.) The field velocity $\dot{\phi}$ at the point x in space generally depends, for an entangled wave functional, on instantaneous values of the field at arbitrarily distant points $x' \neq x$. These nonlocal connections imply that our 3-space and our time are *absolute* (see below).

Restricting ourselves to equilibrium $P[\phi, t] = |\Psi[\phi, t]|^2$, we may use the formalism of standard quantum field theory. In the Schrödinger picture, the wave functional is just the Hilbert-space 'state vector' written in the basis $|\phi(x)\rangle$ of field configurations, i.e. $\Psi[\phi, t] = \langle\phi(x) \mid \Psi(t)\rangle$.

Consider now the state $|\Psi(t)\rangle = \sum_p c(p, t)|p\rangle$, which is a superposition of so-called 'single-particle states $|p\rangle$ of momentum p' (generated from the ground state $|0\rangle$ by the usual creation operator). In a low-energy approximation, this must somehow represent a nonrelativistic particle. But what does the 'particle' look like in terms of the field $\phi(x, t)$? Using the equilibrium probability $|\Psi[\phi, t]|^2$, it may be shown that the *most likely field* 'imitates' the nonrelativistic wave function. Calculating $\Psi[\phi, t]$ in a low-frequency ($\approx m$) approximation, one finds

$$(6) \qquad \Psi[\phi, t] \propto \exp\left(-\frac{m}{2}\int d^3x(\phi(x))^2\right)\int d^3x\ \phi(x)\psi(x, t).$$

The first (exponential) factor is just the ground-state wave functional $\Psi_0[\phi]$; the function $\psi(x, t)$ is the Fourier transform of $c(p, t)$, and may be shown to obey the nonrelativistic Schrödinger equation. Taking the modulus squared of (6), the equilibrium probability is found to be maximised for a field configuration given by

$$(7) \qquad \phi(x) = \mathrm{Re}\left[\frac{\psi(x, t)}{m\alpha(t)}\right],$$

where $\alpha(t) \equiv \int d^3x\ \phi(x)\psi(x, t)$. The most likely field is determined up to a sign. If ψ is real, the solution of (7) is just $\phi = \pm\psi/\sqrt{m}$.

Thus, the most likely field is essentially proportional to what we call the nonrelativistic wave function. If ψ is a localised packet of width Δx, the field will be a 'lump' of size Δx. Thus, while the actual field configuration lies somewhere within $\Psi[\phi, t]$, a 'particle' is *not* a point-like object lying somewhere within ψ. So while the pilot-wave theory of a nonrelativistic particle is useful as an illustrative model, it is in no sense the nonrelativistic limit of field theory, and is actually *incorrect*. Particle trajectories emerge only in the *classical* limit (of narrow ψ).

What happens during a 'position measurement'? The initial field may be completely delocalised. During the 'measurement' process the field is constrained to become concentrated in a small region around some point x, the 'measured position'. The value of x is determined by the initial field and apparatus variables (given the initial wave function and Hamiltonian). It has no simple relation to the initial field alone. Thus, in field theory, 'particle position' is a contextual variable.

Grassmann Field Theory of Spin

For fermions it is often claimed that, since the fields anticommute, the theory must be based on particle trajectories instead of on fields (Bohm, Hiley and Kaloyerou 1987). Such a theory is very inelegant: positrons are treated as 'holes' in the 'negative energy sea', and it is unsatisfactory to treat fermions as particles while bosons are treated as fields.

A field description may in fact be applied to fermions as well. In the path-integral formulation of quantum field theory, operators do not appear; for the fermion case one is obliged to make use of anticommuting c-numbers or Grassmann fields (which are an extension of the complex numbers). The situation is similar in pilot-wave field theory. We may represent fermions by an objective field of Grassmann numbers evolving in time, guided by a wave functional Ψ.

We consider the case of massive spin-1/2. This is usually described by a four-component Dirac field ψ. However, this is *not* appropriate for the pilot-wave theory. The momentum density conjugate to ψ is just $\pi = (i/2)\psi^\dagger$, so the equation of motion (2) would read $(i/2)\psi^\dagger = \delta S/\delta\psi$, which is absurd: for instead of telling us the rate of change of any given field, it restricts the allowed field. The origin of the difficulty is that the Lagrangian is linear in the time derivatives. We must therefore use the van der Waerden field, which is equivalent to the Dirac field, but whose Lagrangian is quadratic in the time derivatives (yielding a wave equation that is second-order in time).

The van der Waerden field is a *two*-component complex field ϕ_α (α = 1, 2) with Lagrangian density

$$\mathscr{L} = \dot{\phi}^*_\alpha\dot{\phi}_\alpha - (\sigma \cdot \nabla\phi)^*_\alpha(\sigma \cdot \nabla\phi)_\alpha - m^2\phi^*_\alpha\phi_\alpha,$$

where σ^i are Pauli spin matrices and α is summed over. By taking linear combinations of ϕ_α with its first space and time derivatives, one may construct a four-component field obeying the Dirac equation (Sakurai 1967, 91). As shown long ago (Feynman 1948; 1951; Feynman and Gell-Mann 1958; Brown 1958), the van der Waerden field is just as suitable as the Dirac field for the description of electrons and positrons.

In the pilot-wave theory we assume that the ϕ_α, ϕ^*_α are complex Grassmann fields. These anticommute among themselves; differentiation with respect to them must be defined 'from the left' and 'from the right' (Berezin 1966).

The Schrödinger equation (1) for $\Psi = \Psi[\phi_\alpha, \phi^*_\alpha, t]$ then reads

$$(8)\qquad \mathrm{i}\frac{\partial\Psi}{\partial t} = \int d^3x \left[-\frac{\overleftarrow{\delta}}{\delta\phi^*_\alpha}\left(\Psi\frac{\overleftarrow{\delta}}{\delta\phi_\alpha}\right) + (\sigma\cdot\nabla\phi)^*_\alpha(\sigma\cdot\nabla\phi)_\alpha\Psi + m^2\phi^*_\alpha\phi_\alpha\Psi \right],$$

while the equation of motion (2) reads

$$(9)\qquad \frac{\partial\phi^*_\alpha}{\partial t} = \frac{S\overleftarrow{\delta}}{\delta\phi_\alpha}, \qquad \frac{\partial\phi_\alpha}{\partial t} = \frac{\overleftarrow{\delta}S}{\delta\phi^*_\alpha}.$$

These equations define a pilot-wave theory of the massive spin-1/2 field. At any instant, there exists a definite Grassmann-valued field. ('Beables' need not be restricted to real or complex numbers.) As for the scalar case, localised 'particles' emerge as field 'lumps'. These may be thought of in terms of 'even' quantities, such as $\phi^*_\alpha\phi_\alpha$, which behave like complex numbers.

3. RELATIVISTIC INVARIANCE AND ELECTRODYNAMICS

Relativistic Invariance: An Equilibrium Symmetry

In pilot-wave field theory on flat spacetime, it is natural to assume that subquantum nonlocality singles out a preferred rest frame (Bohm and Hiley 1984). The nonlocality acts instantaneously across a true 3-space, defining an absolute simultaneity and a true time t. In equilibrium, nonlocality and the associated rest frame are invisible; the operator field equations of the Heisenberg picture show that relativistic invariance emerges.

Strictly speaking, what emerges in equilibrium is not Einstein's special relativity but Lorentz's earlier interpretation of the Lorentz transformations. The view of Lorentz – that one particular frame is at absolute rest – is perfectly consistent classically. A single frame with coordinates (x, t) is sufficient to describe all of physics – including the physical response of moving equipment. (See for example Bell (1987a).) One finds that absolute (uniform) motion of a physical reference frame – an experimenter with his equipment – 'cannot be detected', in the sense that the moving experimenter may consistently regard himself as being at rest, by making use of Lorentz-transformed coordinates and field variables. This 'Lorentz invariance' is a property of the (equilibrium) *dynamics*, not of space and time.

Relativistic spacetime is a construct of equilibrium observers who are unable to see nonlocality directly; it is, like 'quantum ontology', a misguided projection of contingent human limitations onto Nature herself.

Generally speaking, any maximum-entropy equilibrium state ought to show an especially high degree of symmetry. It is therefore perfectly reasonable that the state $P = |\Psi|^2$ of maximum subquantum entropy should display symmetries, such as Lorentz invariance, that the underlying physics does not possess. (This viewpoint has some affinity with broken symmetry in particle physics, where the actual vacuum state of the universe breaks the symmetry of the underlying theory; except that here the present state $P = |\Psi|^2$ of the universe, being a statistical equilibrium, has *more* symmetry than the underlying laws.)

In equilibrium, if a detector coupled to a scalar field is displaced through the vacuum by $+a$, its response will not change since the equilibrium distribution for the vacuum is translationally invariant, $|\Psi_0[\phi(x - a)]|^2 = |\Psi_0[\phi(x)]|^2$. While out of equilibrium its response will usually depend on its position (even though the space is flat), since an arbitrary distribution $P[\phi(x)]$ of field configurations (guided by the vacuum wave functional) is not translationally invariant.

Further, the response of a uniformly-accelerated detector (or the radiation emitted by a black hole) will be *thermal* only in equilibrium. There is a close relationship between quantum vacuum fluctuations and thermal fluctuations (Sciama *et al.* 1981), reinforcing our interpretation of $P = |\Psi|^2$ as a 'heat death'. Einstein's principle of equivalence is also an equilibrium symmetry only. A detector falling freely in a classical gravitational field perceives the same spectrum of quantum vacuum fluctuations as a similar free detector in Minkowski spacetime (Candelas and Sciama 1984). This result assumes $P = |\Psi|^2$ for each field vacuum and must break down if $P \neq |\Psi|^2$.

Absolute 3+1 Electrodynamics

Bohm (1952b) outlined a theory of the electromagnetic field based on the usual 4-potential A^μ written in the Coulomb gauge. The following approach is more satisfactory.

Classical electrodynamics may be regarded as a theory of a 3-vector potential A^i ($i = 1, 2, 3$) evolving in absolute space and time according to

$$(10) \qquad \nabla \cdot \dot{A} = -\rho, \qquad \ddot{A} + \nabla \times (\nabla \times A) = j,$$

where ρ and j are the charge and current densities. Defining $E = -\dot{A}$, $B = \nabla \times A$, one recovers Maxwell's equations. The theory is invariant under '3-gauge' transformations $A(x, t) \rightarrow A(x, t) + \nabla\lambda(x)$, where λ is *time-independent*. This 'absolute 3+1' theory is, mathematically speaking, just standard electrodynamics with the 4-vector potential A^μ written in the *temporal gauge*: the transformation $A^\mu \rightarrow A^\mu - \partial^\mu \int_0^t dt' \, A^0(x, t')$ sets A^0 equal to zero. However, we take the view that the absolute 3+1 theory is fundamental. Thus we do not 'set A^0 equal to zero'; A^0 simply does not exist.

As will become clear, A^0 is a 'Plato's Cave effect' – an unfortunate mathematical artifact introduced by misguided equilibrium observers who insist on fundamental Lorentz invariance. In covariant QED it leads to 'ghost particles' whose states have negative norm. In contrast, in our theory one automatically recovers (in equilibrium) QED in the *ghost-free* temporal gauge.

For the pure electromagnetic field, A^i is guided by a wave functional which we write as $\Psi = \Psi[A^i, t]$. In reality, however, Ψ is a function, not on the space of gauge-dependent fields A^i, but on the space of equivalence classes of fields connected by 3-gauge transformations. Thus, if one writes $\Psi = \Psi[A^i, t]$, the correct mathematical definition of Ψ requires that one append the gauge-invariance constraint

$$\nabla \cdot \left(\frac{\delta\Psi}{\delta A}\right) = 0. \tag{11}$$

This follows by putting $\delta\Psi = 0$ for an infinitesimal gauge change $\delta A = \nabla\lambda$. Having defined Ψ, the Schrödinger equation (1) reads

$$i\frac{\partial\Psi}{\partial t} = \int d^3x \frac{1}{2}\left(-\frac{\delta^2}{\delta A^2} + B^2\right)\Psi, \tag{12}$$

which preserves the constraint (11). The equation of motion (2) reads

$$\frac{\partial A(x, t)}{\partial t} = \frac{\delta S[A, t]}{\delta A(x)}. \tag{13}$$

Equations (11)–(13) define our subquantum theory of the electromagnetic field.

Restricting ourselves to equilibrium, the statistical predictions may be represented by operators. Transforming to the Heisenberg picture, (11) becomes

$$\nabla \cdot \hat{A}(x, t)|\Psi\rangle = 0. \tag{14}$$

(This is not a gauge condition, but a gauge-*invariance* condition; no particular 3-gauge has been specified.) One has equal-time commutation relations

$$[\hat{A}^i(x, t), \hat{\dot{A}}^j(x', t)] = i\delta^{ij}\delta^3(x - x'), \tag{15}$$

and the operator field equation

$$\hat{\ddot{A}} + \nabla \times (\nabla \times \hat{A}) = 0. \tag{16}$$

Taking the divergence of (16) implies $\nabla \cdot \hat{\ddot{A}} = 0$. Thus $\partial/\partial t(\nabla \cdot \hat{\dot{A}}|\Psi\rangle) = 0$ and the constraint (14) is preserved in time. From (14), $\nabla \cdot \hat{\dot{A}}$ is null on the physical subspace. However, $\nabla \cdot \hat{\dot{A}} \neq 0$ on the general Hilbert space; otherwise, taking the divergence of (15) would lead to an inconsistency.

Equations (14)–(16) are those of QED in the temporal gauge; equivalence to standard QED is therefore guaranteed (Leibbrandt 1987).

To exhibit the photon content, we split $\hat{A}$ into transverse and longitudinal parts, $\hat{A} = \hat{A}_T + \hat{A}_L$. From (16), $\hat{A}_T$ satisfies the usual wave equation and so may be expanded in terms of plane waves with two polarisation states; while $\hat{A}_L$ satisfies $\hat{\ddot{A}}_L = 0$ so that $\hat{A}_L = \hat{\alpha}(x)t + \hat{\beta}(x)$, where $\hat{\alpha}$, $\hat{\beta}$ are static and curl-free. The constraint (14) then reads $\nabla \cdot \hat{\alpha}(x)|\Psi\rangle = 0$, while of course $\nabla \cdot \hat{\alpha} \neq 0$. Thus $\hat{A}_L$ generates a static contribution $-\hat{\alpha}(x)$ to the total electric field operator $\hat{E} = -\dot{\hat{A}}$. However, $\nabla \cdot \hat{\alpha}(x)|\Psi\rangle = 0$ implies that $\hat{\alpha}$ is effectively zero on the physical subspace. (Nevertheless $\hat{\alpha} \neq 0$ is necessary to ensure $\nabla \cdot \dot{\hat{A}} \neq 0$; otherwise (15) would be violated.) The total Hamiltonian is the usual sum over photons with two polarisation states (coming from $\hat{A}_T$) *plus* a term $\hat{K} = (1/2)\int d^3x\hat{\alpha}^2$. From (14) one finds $\hat{K}|\Psi\rangle = 0$, so $\hat{K}$ has no effect on the energy levels.

Our approach is easily applied to interaction with a charged field (with local 3-gauge invariance). And it may be readily extended to non-Abelian gauge theories such as QCD: in equilibrium one obtains the standard theory written in the temporal gauge. Non-Abelian theories are greatly simplified in this gauge, where ghost particles do not appear in physical S-matrix elements (Leibbrandt 1987).

The general simplicity of the absolute 3+1 approach is a further indication that the fundamental level is indeed not Lorentz invariant.

4. GRAVITATION AND COSMOLOGY

Absolute 3+1 Gravitation

In standard quantum gravity, the arbitrariness of the spacetime slicing leads to the 'timeless' Wheeler–DeWitt equation $\hat{\mathcal{H}}\Psi = 0$, where $\hat{\mathcal{H}}$ is an appropriate Hamiltonian density operator in configuration space and Ψ is a function of the 3-geometry only. One might try to simply augment the standard theory by adding a de Broglie-Bohm 'trajectory' for the 3-metric, guided by the phase of Ψ (Vink 1992; Holland 1993). However, the slicing of spacetime is then arbitrary, as in standard general relativity. Subquantum nonlocality will propagate along arbitrary slices, which is surely inconsistent. Further, in the presence of a scalar field one will obtain, in the limit where spacetime is classical and flat, a scalar field theory with arbitrary slicing, in contradiction with our earlier assumption of an absolute frame. If there is an absolute space behind special relativity, the same must be true for general relativity.

We therefore propose that at the subquantum level there is an *absolute* curved 3-space whose evolution in absolute time is guided by a time-dependent wave function Ψ. Evolution of the geometry in one region generally depends instantaneously on the geometry at a distance; and this nonlocality distinguishes and maintains the absolute slicing of spacetime. However, as in the case of flat spacetime, the absolute space is (like the nonlocality itself) invisible at the quantum or classical level. (Any classical spacetime with at least one slicing into a family of spacelike 3-surfaces, labelled by a global time coordinate, may be considered to be generated by some as-yet-unknown theory of absolute curved space evolving in absolute time, by simply identifying one of the global times with absolute time. If the spacetime is globally hyperbolic, as we assume on general physical grounds, such a slicing always exists.)

Unfortunately, one does not know how classical spacetime ought to be sliced to recover the true underlying 3-space. Nevertheless our viewpoint is physically plausible, and overcomes the *conceptual* problems of standard quantum gravity.

In analogy with electrodynamics one might assume that the time components ${}^{(4)}g_{0\mu}$ of the 4-metric are, like A^0, spurious variables, and that these should be dropped by simply choosing 'temporal' coordinates such that ${}^{(4)}g_{0\mu} = \delta_{0\mu}$. The true 3-space would then correspond at the classical level to slicing with a constant lapse function $N = 1$. However, this cannot be correct, because for most spacetimes these coordinates very rapidly become singular, owing to the crossing of hypersurface-orthogonal geodesics. Volume elements shrink to a point and the 3-space itself becomes singular, even near the Sun (after about one hour). But this is no reflection on the general idea: there are other slicings that avoid this problem.

For example, York slicing chooses hypersurfaces on which K = constant, where K – the trace of the extrinsic curvature tensor – is itself the time label. For a closed universe satisfying reasonable conditions (such as the strong energy condition), York slicing exists and is unique (Marsden and Tipler 1980), with K extending from $-\infty$ (the 'big bang') to $+\infty$ (the 'big crunch'). Since K measures the local fractional contraction of 3-space volume (per unit of proper time normal to the hypersurface), a volume element may shrink to a point only as $K \to \infty$. *All* future singularities are consigned to $K = \infty$, making this slicing very convenient for numerical relativity. We may therefore assume that, in the classical limit, our absolute space coincides with slices K = constant; while our absolute time may be identified with York's 'cosmic time' $T = (4/3)K$. (Qadir and Wheeler (1985) have also proposed that this time is of cosmic significance, in the context of a possible change of the fundamental constants with time.) There is some evidence that York slicing may break down when the strong energy condition is violated, such as during inflation in the early universe (Goldwirth and Piran 1989). Even so, one could still regard York time as fundamental by assuming that classical general relativity breaks down in such conditions.

Despite its untenability, a subquantum gravity based on $N = 1$ is useful as an illustrative model. Writing $\Psi = \Psi[g_{ij}, t]$, one must append the well-known diffeomorphism-invariance condition

$$\left(\frac{\delta\Psi}{\delta g_{ij}}\right)_{|j} = 0, \tag{17}$$

which ensures that Ψ depends only on the geometry of 3-space (and not on the choice of spatial coordinates). Inserting $N^i = 0$, $N = 1$ into the usual Hamiltonian of canonical quantum gravity leads to the Schrödinger equation (units $16\pi G = 1$)

$$\mathrm{i}\frac{\partial\Psi}{\partial t} = \int \mathrm{d}^3x \left(-\frac{\delta}{\delta g_{ij}} G_{ijkl} \frac{\delta}{\delta g_{kl}} - g^{1/2}R\right)\Psi, \tag{18}$$

where the quantity in brackets is just the usual $\hat{\mathcal{H}}$ appearing in the Wheeler–DeWitt equation. Assuming a closed space, boundary terms are omitted. (Here G_{ijkl} is the superspace metric, $g = \det g_{ij}$ and R is the 3-scalar curvature. An explicit operator ordering has been chosen.) The constraint (17) is preserved in time by (18). Since the canonical momentum density p^{ij} is related to $\dot{g}_{ij}$ by $\dot{g}_{ij} = 2\, G_{ijkl}p^{kl}$, the equation of motion (2) reads

$$\text{(19)} \qquad \frac{\partial g_{ij}}{\partial t} = 2G_{ijkl}\frac{\delta S}{\delta g_{kl}}\,.$$

Equations (17)–(19) define our model of subquantum gravity.

The quantum level again arises for an equilibrium ensemble $P = |\Psi|^2$ of 3-geometries. It is well known that the operator equivalent of (17) implies ${}^{(4)}\hat{G}_{0j}\Psi = 0$ (on the physical subspace); while the dynamical operator equations arising from (18) (in the Heisenberg picture) are just ${}^{(4)}\hat{G}_{ij} = 0$. (Here, the Einstein 4-tensor $G_{\mu\nu}$ is written in our coordinates with ${}^{(4)}g_{0\mu} = \delta_{0\mu}$. As is well known, only its space-space components contain second time-derivatives of the 3-metric; and the classical dynamical equations $G_{ij} = 0$ preserve the constraints $G_{0\mu} = 0$ on the initial data.) A further condition ${}^{(4)}\hat{G}_{00}\Psi = 0$ arises in standard quantum gravity from the Wheeler–DeWitt equation $\hat{\mathcal{H}}\Psi = 0$. But there is no such constraint in our theory, where Ψ obeys the more general equation (18). Nevertheless, one may recover the complete Einstein equations in the classical limit by assuming Ψ to satisfy $\langle\hat{\mathcal{H}}\rangle \equiv \int Dg\ \Psi^*\hat{\mathcal{H}}\Psi = 0$, where Dg is an appropriate measure: for in the classical limit (where $\hat{\mathcal{H}} = (2g^{1/2})^{(4)}\hat{G}_{00}$) one then obtains the c-number equation ${}^{(4)}G_{00} = 0$. For a general Ψ, $\langle\hat{\mathcal{H}}\rangle$ will be some function $\langle\hat{\mathcal{H}}\rangle \equiv -g^{1/2}\tilde{\rho} \neq 0$, yielding ${}^{(4)}G_{00} = -(1/2)\tilde{\rho}$. On the right-hand-side of the classical Einstein equations we then have a term corresponding to an *effective* pressure-free 'energy' density (or 'dust'), of density $\tilde{\rho}$, whose only interaction with ordinary matter is via gravity. Perhaps $\tilde{\rho} = 0$ for the wave function of our universe. Otherwise, $\tilde{\rho} > 0$ might explain the apparent presence of 'cold dark matter' in space.

This effective 'energy' density in space arises because our theory contains an extra degree of freedom (per space point). The six functions g_{ij} minus three coordinate degrees of freedom imply three real degrees of freedom, whereas according to the usual count the gravitational field has only two. In standard quantum gravity it is thought that some function of the g_{ij} must be extracted to play the role of time. Here, however, we already have a time, and the extra degree of freedom shows up at the classical level as an arbitrary function $\tilde{\rho}_0(x)$ at $t = 0$.

This model based on $N = 1$ may be readily applied to a (minisuperspace) homogeneous Friedmann universe. However, as we have noted, in general it leads to an evolution for the absolute 3-space that is, in the classical limit, unacceptably singular. Nevertheless, the model suffices to show that pilot-wave quantum gravity is able to overcome the conceptual difficulties of the standard approach: (i) In the standard approach the physical meaning of equal-time commutation relations is obscure; whereas here one may straightforwardly contemplate such relations on our absolute 3-space. (ii) Since we have a Schrödinger equation, rather than a 'timeless' Wheeler–DeWitt equation, $|\Psi|^2$ may (in equilibrium) be identified with a probability density in the usual way. (iii) Time is usually said to emerge phenomenologically, via quantum clocks correlated with some internal degree of freedom of the gravitational field. This procedure is physically obscure; the choice of which degree of freedom to identify with time is highly arbitrary (corresponding to the arbitrariness of time in general relativity), and it is not clear that different choices lead to equivalent theories. Here we have a unique time, which is physically well-defined at the nonlocal subquantum level. In the classical limit it coincides with one of the classical time functions (as is clear from the way in which the classical Einstein equations emerge), this probably being the one that is cosmologically preferred (see below).

The flaw in the $N = 1$ model may be repaired by adopting a better guess as to how the true 3-space is related to classical spacetime. York slicing may be the answer. Classically, on surfaces T = constant the unconstrained configuration degrees of freedom may be identified with a *conformal* 3-geometry (York 1972). This has metric $\gamma_{ij} \equiv g^{-1/3}g_{ij}$ and non-vanishing Hamiltonian $H = \int d^3x\, g^{1/2}$ (the volume of the universe). The usual constraint $\mathcal{H} = 0$ may be solved formally to yield $g^{1/2} = g^{1/2}[\gamma_{ij}, \pi^{ij}, T]$, where π^{ij} is 'almost' canonically-conjugate to γ_{ij} (Choquet-Bruhat and York 1980). Unfortunately, this expression is defined only implicitly, as a solution of the Lichnerowicz equation. Proceeding nevertheless, one may formally write down a Schrödinger equation

$$i\frac{\partial \Psi}{\partial T} = \int d^3x\, \hat{g}^{1/2}(\hat{\gamma}_{ij}, \hat{\pi}_{ij}, T)\Psi. \tag{20}$$

Here, $\Psi = \Psi[\gamma_{ij}, T]$ is subject to an appropriate constraint ensuring that it is really a function of the coordinate-independent conformal 3-geometry (and of time T). Since $\det \gamma_{ij} = 1$, the six functions γ_{ij} minus three coordinate degrees of freedom correctly yield two degrees of freedom per space point (in contrast with the $N = 1$ model). The time-dependent Hamiltonian is equal, in the classical limit, to the volume of Riemannian 3-space, but is unfortunately not known as an explicit function of the new variables. Thus, while one may suppose the main physical features of the theory to be correct, we are as yet unable to write down a complete mathematical realisation of it. Some other basis must be sought for deriving the gravitational Hamiltonian.

Classically, conformal 3-geometry is usually seen as a convenient tool for obtaining an initial Riemannian 3-geometry satisfying the constraints. But at the subquantum level, the evolving entity may actually *be* the conformal geometry. Perhaps Riemannian geometry emerges only in the classical limit, by *defining* a 'volume' equal to the classical Hamiltonian. This hypothesis is very attractive in the nonlocal, holistic pilot-wave theory, since a conformal geometry is 'distanceless', having local shape but no size. Indeed, the uniqueness of York slicing for a closed universe and its capacity to ensure that gravitational collapse occurs *simultaneously* throughout the universe (Qadir and Wheeler 1985) suggest that there may be a relationship between conformal 3-geometry and fundamental nonlocality. A deeper understanding of this might enable us to derive an explicit expression for the gravitational Hamiltonian, or perhaps a deeper theory of gravitation altogether, on the basis of some as-yet-unknown physical principle.

Subquantum Cosmology. Disequilibrium in the Early Universe

Our view that the universe is now in a state of 'subquantum heat death' suggests that there may have been disequilibrium $P \neq |\Psi|^2$ in the remote past, perhaps near the big bang. 'Evidence' for this is provided by the 'horizon problem' – the puzzling homogeneity of the early universe (deduced from the isotropy of the microwave background) over distances much larger than the size of causally-connected domains – and by the (closely-related) problem of the perturbations away from homogeneity, on similar scales across space, that seem to be needed to seed galaxy formation. It is widely believed that these problems can only be solved either by inflation (for which no satisfactory model exists), or by some sort of nonlocality that abolishes causal horizons in the very early universe (quantum fluctuations in the causal structure of spacetime often being invoked). But such nonlocality is easily – and very naturally – provided by the pilot-wave theory, locality being violated if $P \neq |\Psi|^2$. The problems of principle arising from causal horizons in classical cosmology simply disappear if initial disequilibrium is assumed. Further, the absolute 3-space along which subquantum nonlocality acts (leading to the early homogeneity) will necessarily coincide with the observed rest-frame defined by the uniform microwave background. We therefore predict that, if the nonlocality ever becomes directly observable, it will be found to propagate along the hypersurface defined by the observed cosmological rest frame. (Note that York slicing coincides with cosmological rest for a Friedmann model.)

Since a finite (proper) time has elapsed since the 'beginning', relaxation to equilibrium cannot have taken place to arbitrary accuracy. Even today, some residual disequilibrium $P \neq |\Psi|^2$ must be present on a sufficiently small scale – at which quantum theory will break down. A priori, independently of particular cosmological models, residual disequilibrium on the Planck scale $l_P \sim 10^{-33}$ cm seems likely. As a simple (if unrealistic) example, for a hypothetical universe consisting of $\sim 10^{80}$ Hydrogen atoms in the ground-state, the actual distribution of electron positions with respect to their nuclei (in the nonrelativistic particle model) might be $\propto e^{-2r/a_0} \cos^2(r/l_P)$, instead of the usual $\propto e^{-2r/a_0}$ (where a_0 is the Bohr radius).[6]

Given an initial state, sufficiently far from equilibrium that the ensuing nonlocality is sufficient to abolish causal horizons at early times, one may in principle calculate forwards to the present and predict the scale of residual disequilibrium. In practice such calculations are very difficult. The subquantum H-theorem enables one to define a typical time scale over which significant relaxation occurs, given an initial wave function and coarse-graining length. On this basis one may try to estimate the scale of residual disequilibrium soon after the big bang for, say, a model Friedmann universe filled with a scalar field. Such models are widely used in standard quantum cosmology, though how realistic they are is hard to say. At very early times, the disequilibrium scale might even be much smaller than l_P. However, if inflation does occur, single-particle wave functions will be enormously expanded, and this might magnify any residual disequilibrium to *macroscopic* scales (see below). Subsequent evolution might then conceivably leave us with disequilibrium on scales considerably larger than l_P.

Experimental Search for Residual Disequilibrium

In this view quantum noise is, in effect, a remnant of the big bang. Just as tiny variations in temperature have recently been detected for the microwave background so, we suggest, residual deviations from quantum noise might be found.

Experimental detection of such deviations, on scales $\sim l_p$ (say), might seem out of the question. However, some cosmological 'particles' decoupled early in the history of the universe and have since travelled over cosmological distances essentially freely. Neutrinos decoupled very early (within a few seconds of the big bang), while photons continued to interact until recombination. The (quantum-mechanical) mean free path of cosmological photons is comparable to the size of the observable universe, while that of cosmological neutrinos is *very* much larger. This essentially free propagation entails an enormous spreading, on an intergalactic scale, of their wave packets (which almost never undergo 'reduction'). For these particles, the residual disequilibrium scale (at the time of decoupling) will subsequently 'expand' to accessible dimensions.

To see this, consider an ensemble of nonrelativistic particles guided by a freely-spreading wave function, whose initial spread Δx_0 becomes magnified to $\Delta x(t)$ in time t. As the wave packet spreads, the particles are carried along with it. Particles that were initially separated by a distance δx_0 are later separated by (roughly) $\delta x(t) \approx [\Delta x(t)/\Delta x_0]\delta x_0$. It is easy to see that this results in an expansion, by a factor $\sim\Delta x(t)/\Delta x_0$, of the scale on which disequilibrium is present. (Field theory may be shown to yield the same results.)

There is no conflict with the subquantum H-theorem, which only describes the behaviour of P *relative to* $|\Psi|^2$ (for complex systems). If

$|\Psi|^2$ itself expands, the absolute scale at which disequilibrium is present may increase.

As an illustrative example, consider a particle generated by an atomic transition in deep space, some one billion light years ($\sim 10^{27}$ cm) distant from Earth. Assume for simplicity that its wave function expands in a roughly spherically-symmetric manner. (This assumption is not essential to our argument.) On emission, the wave function will have a spread of, say, $\Delta x_0 \sim 10^{-6}$ cm. For an ensemble, some of the particles will eventually be detected on Earth. Just prior to detection the wave function – assuming it to have freely evolved – will have an enormous spread $\Delta x(t) \sim 10^{27}$ cm. The wave packet will therefore have been magnified in scale by a factor $\Delta x(t)/\Delta x_0 \sim 10^{33}$. Any disequilibrium initially present at, say, $l_P \sim 10^{-33}$ cm will now be present on a scale ~1 cm, easily accessible, in principle, to experiment.[7] In particular, if some of the incoming particles should impinge on a macroscopic two-slit screen, *the usual interference pattern will be blurred.*

Interaction with the intergalactic and galactic media is utterly negligible for neutrinos. For photons, the cumulative dynamical effect of tiny scattering terms in the wave function could be large; possibly, this will not scramble the free expansion of the disequilibrium scale. At the Planck scale, no system is free; this might prevent the expansion of disequilibrium from scales $\sim l_P$. If so, one must hope that the residual disequilibrium scale is considerably larger than l_P.

Since cosmological neutrinos are undetectable at present, we suggest that quantum-mechanical experiments (such as single-particle interference) be performed with cosmological microwave photons. Alternatively, one might use 'weakly interacting massive particles' – the hypothetical components of dark matter – which are currently being searched for experimentally. (Although their wave functions are unknown, incoming particles may be subjected to a state preparation, or simply treated as plane waves.) Deviations from precisely quantum noise might be found. Even setting aside the hypotheses made here, a test of quantum theory for such enormously expanded wave packets would be of intrinsic interest.

ACKNOWLEDGEMENTS

It is a pleasure to thank Prof. Marcello Cini for his kind support and Prof. James T. Cushing for his interest in this work.

University of Rome 'La Sapienza'

NOTES

* This article is a brief outline of recent work by the author on the de Broglie–Bohm theory. For further details the reader is referred to Valentini (1991a,b; 1992; 1996b).

1 The abstract *canonical* momentum makes an appearance in (2), in order to relate $\partial S/\partial X$ to the velocity; but the usual physical momentum plays no role.

2 Bohm (1953b) showed that external perturbations lead to $\rho \to |\psi|^2$ for the special case of an ensemble of two-level molecules subjected to random collisions, where Bohm (1952b, 186) had already suggested that $\rho = |\psi|^2$ might arise in a manner analogous to thermal equilibrium in classical statistical physics. In neither case was a general *H*-theorem developed, however. Instead, Bohm and Vigier (1954) introduced subquantum 'fluid fluctuations' that drive $\rho \to |\psi|^2$ for a general system.

3 The approximate classicality of the experimenter also plays a role here, ensuring that he cannot 'see inside' wave packets.

4 In the context of *quantum* theory, the assumption that $\rho = |\psi|^2$ exactly is necessary for self-consistency, and so is supported by the validity of quantum theory generally. But in the context of the pilot-wave theory, disequilibrium ($\varepsilon \neq 0$) is perfectly possible, so there is *no* reason to suppose from its empirical smallness that ε is actually zero.

5 This is clear even in the absence of rigorous results on relaxation. For independent subsystems there will necessarily exist, simply by continuity, some $\delta > 0$ such that all smooth (fine-grained) distributions satisfying $\rho_0 \approx |\psi_0|^2$ to within δ must evolve to $\rho \approx |\psi|^2$ to within ε today. And more generally, continuity guarantees that if X_0^{Univ} should be 'typical' with respect to any measure that is to sufficient accuracy $\approx |\Psi_0^{\mathrm{Univ}}|^2$, consistency with experiment will be obtained.

6 For 10^{80} electrons in a configuration volume a_0^3, one has $\sim 10^5$ electrons per volume l_P^3, so an approximately smooth distribution of positions is well defined on the scale l_P. Disequilibrium on this scale is therefore also well defined.

7 Similarly, if the volume of the universe inflates by, say, $\sim 10^{10^4}$ (Linde 1982), the resulting wave function expansion for free particles might magnify disequilibrium to macroscopic scales.

LUCIEN HARDY

CONTEXTUALITY IN BOHMIAN MECHANICS

1. INTRODUCTION

The Kochen–Specker theorem shows that noncontextual hidden variable interpretations of quantum theory are impossible. This does not mean, however, that hidden variable theories are not possible. In fact, the Bohm model is just such a theory. We show by considering an example involving interferometers how the Bohm model is contextual, thus circumventing the Kochen–Specker theorem. We will find that the result of a measurement of an operator in Bohmian mechanics is not just dependent on the context of other measurements that are also being performed at the same time, but also dependent on the way in which a particular measurement is performed.

2. PRELIMINARY DISCUSSION

In quantum theory the state of a system is described by a state vector in a Hilbert space. It is sometimes stated that it is a postulate of quantum mechanics that the state vector provides a complete description of a system. Actually this is an unnecessary postulate since it introduces an ontological notion of completeness that has no bearing on the application of quantum theory to make predictions. A more accurate postulate would be that no more information about the state of a system than is contained in the state vector is experimentally accessible. Given such a postulate one is then free to suppose either that the state vector is in fact a complete description or that it is not. The second possibility amounts to supposing that there are additional variables usually called hidden variables that further specify the state of a system. It is very often argued that Ockham's razor should be applied here to rule against introducing such hidden variables, but this rather depends on how one interprets the objective of applying Ockham's razor. Indeed, there must be a balance between the application of Ockham's razor and the need to provide explanation. There are perhaps three reasons why one might want to introduce hidden variables into quantum mechanics: (1) To restore determinism (2) To provide a clear ontology (3) To solve the measurement problem. determinism, although the most obvious reason to introduce hidden variables, is probably not the most important. It would be good to have a clear ontology (that is, a picture of what is happening in quantum systems). However, the measurement problem is by far the most important problem in quantum mechanics. It is because of the measurement problem that we are forced to admit that

J. T. Cushing et al. (eds.), Bohmian Mechanics and Quantum Theory: An Appraisal, 67–76.

the standard formulation of quantum mechanics must eventually be superseded and this provides the strongest reason to consider hidden variables.

Unfortunately, the debate about hidden variable approaches to quantum theory was suppressed for a long time since many people believed that von Neumann had proven that hidden variable theories are impossible for quantum mechanics. Even now many physicists not working on foundations of quantum mechanics continue to believe this. Von Neumann's proof was shown by Bell (1966) to depend on over restrictive (and indeed evidently false) assumptions. We will not discuss it here. A more important theorem often cited to show the impossibility of hidden variable interpretations of quantum theory is due to Kochen and Specker (1967; see also Bell 1966). In fact, this theorem only shows the impossibility of a restricted class of hidden variable theories.

As anyone reading this book will be aware, the question of whether hidden variable interpretations of quantum mechanics are possible was solved by Bohm in 1952. Bohm constructed an explicit hidden variable interpretation that reproduces all the predictions of quantum mechanics. So that's it. Hidden variable interpretations are possible. End of story.

Rather than end the story here, however, we will take up the question of how Bohm's model circumvents the Kochen–Specker theorem. That it must is clear since Bohm's interpretation reproduces all the predictions of quantum mechanics. There two reasons for taking up this question. First, that we will learn something about measurements in the Bohm model. Second, that we will re-enforce that the Kochen–Specker theorem applies only to a restricted class of hidden variable models to which the Bohm model does not belong. Comprehensive recent discussions of how Bohm's model circumvents the Kochen–Specker theorem have already been given (Dewdney 1992; Pagonis and Clifton 1995). Their discussion is in the context of spin, as was the original Kochen–Specker theorem. From the point view of the Bohm model, however, spin does not constitute a very clear example since it can be difficult to picture the way in which the spin is influencing the dynamics. (To further complicate matters, there are two different versions of spin in Bohmian mechanics: Bohm *et al.* 1955; Bell 1966.) However, the Kochen–Specker theorem applies to any system whose state vector lives in a three or more dimensional Hilbert space and so we are not restricted to spin systems. It has recently been suggested by Reck *et al.* (1994) that interferometers be used to run the Kochen–Specker argument. we will take up this suggestion here since it is very simple to apply the Bohm interpretation to interferometers (e.g. Dewdney 1985).

3. THE KOCHEN-SPECKER THEOREM

Consider a system described by a state vector in a three-dimensional Hilbert space and consider a projection operator $P_d = |d\rangle\langle d|$. If one were to make a measurement corresponding to this operator, then he would get the result

0 or 1 corresponding to the two eigenvalues of P_d. The idea behind the Kochen–Specker theorem is that, if there are hidden variables λ, then they would determine the outcome of this measurement. (We are now assuming that the quantum state $|\psi\rangle$ of the system is included with the hidden variables λ so that the latter provide a complete description of the system.) Thus there exists a function $P_d(\lambda)$ which determines the result of the measurement if it is made. This function takes the value 0 or 1 depending on the values taken by the hidden variables. This function has a value even if the corresponding measurement is not made, but instead some other measurement is made. An assumption has crept in at this stage. It has been assumed that the value of $P_d(\lambda)$ is independent of the context of the measurement. It is possible that other measurements could be made at the same time corresponding to operators that commute with P_d. Let $(|d\rangle, |e\rangle, |f\rangle)$ and $(|d\rangle, |e'\rangle, |f'\rangle)$ be two triads of orthonormal state vectors. It is possible to measure the three projection operators corresponding to the first triad or the second triad. In both cases P_d is being measured. By assuming that the result of the P_d measurement is determined by a function $P_d(\lambda)$ we assume that it takes the same value whether it is measured in conjunction with P_e and P_f or $P_{e'}$ and $P_{f'}$. We will see that it is exactly this assumption, called noncontextuality, that is violated by the Bohm model.

For any given λ, the values taken by the result functions must be consistent with one another. This requires that if $|d\rangle$, $|e\rangle$ and $|f\rangle$ are orthonormal, then only one of $P_d(\lambda)$, $P_e(\lambda)$ and $P_f(\lambda)$ can be equal to 1, the other two being equal to 0. It can be shown that this constraint leads to a contradiction. For the details of this proof the reader is referred to Redhead (1989, 119). The conclusion is that it is not possible to assign values corresponding to projection operators if these values are not allowed to depend on the measurement context.

4. USING INTERFEROMETERS TO MAKE MEASUREMENTS

Consider a system in which a particle can be in a superposition of being in a wave packet propagating along one of three paths a, b, c (which we shall visualize as running parallel to one another) with corresponding states $|a\rangle$, $|b\rangle$, $|c\rangle$. The system is thus effectively described by a state vector in a 3-dimensional Hilbert space. We may wish to measure a set of projection operators P_d, P_e and P_f, where $|d\rangle$, $|e\rangle$ and $|f\rangle$ are orthonormal states and therefore related to $|a\rangle$, $|b\rangle$ and $|c\rangle$ by a unitary transformation. It has been shown by Reck *et al.* (1994) that *any* such unitary transformation (not just for a three dimensional system, but for the general N dimensional case) can be realized using beam splitters, mirrors and phase shifters such that afterwards the three paths correspond to the states $|d\rangle$, $|e\rangle$ and $|f\rangle$ (see Figure 1, for example). Then, by placing detectors into the paths, one is able to measure P_d, P_e, P_f. Since any triad of commuting projection operators can be measured in this way, this type of system is rich enough to run a

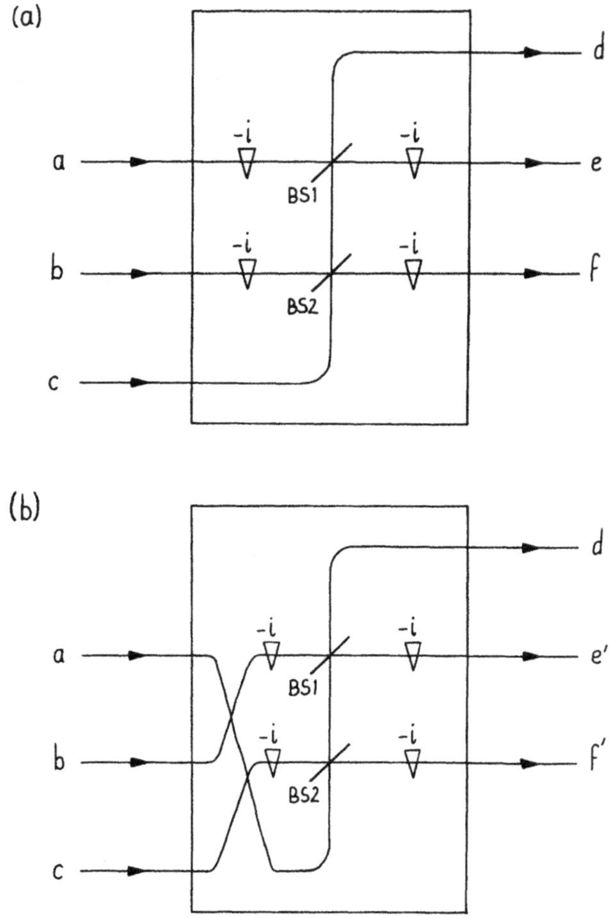

Fig. 1. Interferometric arrangements to perform two different measurements with one projector in common. Note that in (b) the paths do not actually intersect in three dimensions at those places without a beamsplitter.

Kochen–Specker type argument. Rather than consider all the configurations that would be necessary to do this, we will just consider the measurement of P_d in two different contexts as discussed above.

Figure 1 shows two such configurations. The phase shifters are shown with phase factors (if ϕ is the phase shift, then $e^{i\phi}$ is the corresponding phase factor). On passing through a phase shifter with phase factor $e^{i\phi}$ the state $|a\rangle$ would become $e^{i\phi}|a\rangle$. If a beam splitter with reflectivity R and transmittance $T = 1 - R$ has input paths u and v and output paths s and t where u is transmitted into s then on evolving through the beam splitter (see Figure 2), the states $|u\rangle$ and $|v\rangle$ become

(1) $|u\rangle \to \sqrt{T}|s\rangle + i\sqrt{R}|t\rangle,$

(2) $|v\rangle \to i\sqrt{R}|s\rangle + \sqrt{T}|t\rangle.$

The phase factor i is picked up on reflection from the beam splitter. Note that there are other correct ways of choosing the phase factors in these transformations. The only condition that must be satisfied is that $|u\rangle$ and $|v\rangle$ evolve into states that are orthonormal since they are themselves orthonormal (so that the corresponding transformation matrix is unitary). We will ignore phase factors associated with propagation along the paths (the path lengths can always be chosen so that these are equal to 1) and also we will assume that all the path lengths are chosen such that wave packets coming into any given beam splitter arrive at the same time. The paths appear to cross at some places in Figure 1 (and also in Figure 4) where there is no beam splitter, but we will assume that they so not actually cross (as can easily be arranged in three dimensions). This is important in the bohm model since if they did cross the particle trajectories would be effected introducing additional unnecessary complications. In the transformations (1) the term on the right evolves into the term on the left. In what follows we will equate the two sides for simplicity so that $|u\rangle = \sqrt{T}|s\rangle + i\sqrt{R}|t\rangle$ (strictly it is the time evolved state $|u(t)\rangle$ which is equal to the right hand side).

Now consider the apparatus in Figure 1(a). Employing the above considerations, it is easily shown that

(3) $|d\rangle = \frac{1}{\sqrt{3}}(|a\rangle + |b\rangle + |c\rangle),$

(4) $|e\rangle = \frac{1}{\sqrt{6}}(-2|a\rangle + |b\rangle + |c\rangle),$

(5) $|f\rangle = \frac{1}{\sqrt{2}}(-|b\rangle + |c\rangle).$

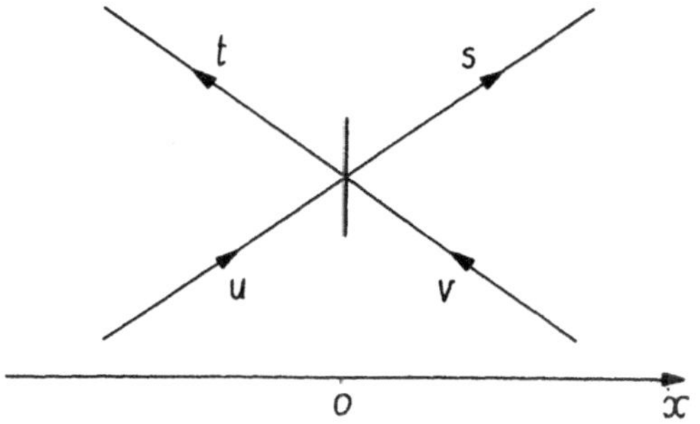

Fig. 2. Input and output paths from a beam splitter.

With detectors placed in the outcoming paths, this apparatus measures the three projection operators P_d, P_e and P_f. For the apparatus in Figure 1(b) it can be shown that

$$|d\rangle = \frac{1}{\sqrt{3}}(|a\rangle + |b\rangle + |c\rangle), \tag{6}$$

$$|e'\rangle = \frac{1}{\sqrt{6}}(|a\rangle - 2|b\rangle + |c\rangle), \tag{7}$$

$$|f'\rangle = \frac{1}{\sqrt{2}}(|a\rangle - |c\rangle). \tag{8}$$

With detectors in the outcoming paths, this apparatus measures the three projection operators P_d, $P_{e'}$ and $P_{f'}$. Hence we have an example of the type of situation discussed above where the operator P_d can be measured in two different contexts.

5. INTERFEROMETERS IN BOHMIAN MECHANICS

It is very easy to draw rough diagrams showing Bohmian particle trajectories through interferometers made up of beam splitters and phase shifters. This is facilitated by the following two rules in Bohmian mechanics (we will consider a one-particle system moving in one dimension with coordinate x):

(i) If the probability density $\rho(x)$ for an ensemble of systems satisfy $\rho(x) = |\psi(x)|^2$ at some initial time t_0, then it will continue to satisfy this relationship for all later times.

(ii) If the trajectories corresponding to different members of an ensemble (with the same wavefunction) are drawn on the same spacetime diagram, then no two trajectories can cross.

The reason for rule (ii) this is that the velocity of a particle is given by an expression of the form

$$\dot{x} = v_\psi(x, t). \tag{9}$$

If two trajectories were to cross, then the function v_ψ would have to be double valued, which is not permitted. When there are more degrees of freedom, then rule (ii) applies to a diagram of the position in configuration space verses time. The form of the function v_ψ must be chosen so that rule (i) is satisfied (if Galilean invariance and time symmetry are imposed then rule (i) is sufficient to determine uniquely the form of the function v_ψ (Dürr *et al.* 1992a; Squires 1996).

Now consider how to apply these rules to a particle impinging on a beam splitter. The beam splitter is shown in Figure 2. There are three space dimensions, but only the x coordinate marked plays a non-trivial role. Hence we can study this system by considering only this coordinate.[1] The beam splitter is placed at $x = 0$. First consider the case where the initial state is a wave packet $|u\rangle$ impinging on a 50:50 beam splitter. The initial position of the particle is somewhere in the wave packet. Figure 3(a) shows various trajectories for various initial positions in the wave packet. Note that if the particle is initially in the front of the wave packet it will be transmitted and if it is initially in the back of the wave packet it will be reflected. This must be the case otherwise rule (ii) would be violated. Note further that the proportions being transmitted and reflected are T and R in accordance with rule (i). Also note that those particles that are transmitted (reflected) spread out to fill the whole of the transmitted (reflected) wave packet, again in accordance with rule (i). Another example is shown in Figure 3(b). In this example, the initial state is some superposition like $1/\sqrt{2}(|u\rangle + e^{i\theta}|v\rangle)$ where θ is chosen such that there is a greater amplitude squared in the outgoing s path than the t path. Applying rules (i) and (ii) it is again easy to sketch some trajectories.

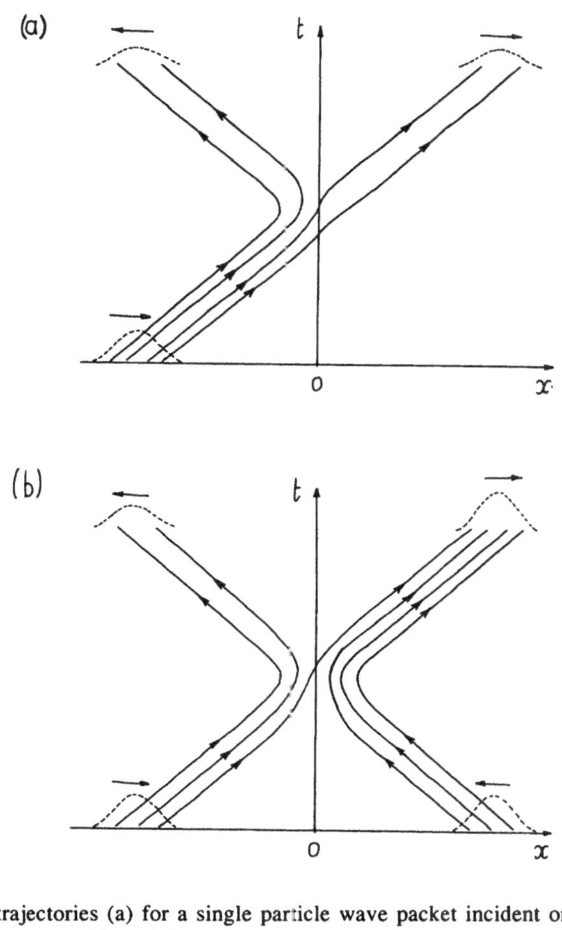

Fig. 3. Bohm trajectories (a) for a single particle wave packet incident on one input and (b) for a supperposition of single particle wave packets incident on both inputs.

It is also relatively straightforward to sketch particle trajectories for multiparticle systems, although one should remember that rule (ii) must then be modified to apply to configurations pace. For our purposes it will be sufficient to stick with one particle.

6. AN EXAMPLE OF CONTEXTUALITY IN BOHMIAN MECHANICS

One of the peculiarities of the Kochen–Specker theorem, as opposed to the Einstein–Podolsky–Rosen argument or Bell's theorem, is that it does not require a particular initial state to be specified. Nevertheless, if we are to discuss an example, then it is necessary to specify an initial state. Our system consists of a particle in one of three paths *a*, *b* or *c*. In Bohmian mechanics it is necessary to specify both the quantum state of the system and the position of the particle. We will take the initial state to be $|a\rangle$ and the position of the particle to be in the back third proportion of this wave packet in path *a* (this proportion refers to the integrated value of $|\psi(x)|^2$). We will call this state λ_1. With the system in this state it is possible to perform either of the measurements in Figure 1. Consider Figure 1(a) first. When the particle reaches *BS*1 it will reflected into path *d* since it is in the back third of the wave packet and the beam splitter has reflectivity $R = 1/3$. This will cause the detector in path *d* to fire and hence $P_d(\lambda_1) = 1$. Now consider Figure 1(b). When the particle reaches *BS*2, it will be reflected into path f' since it is in the back third of the wave packet and the beam splitter has reflectivity $R = 1/2$. This will cause the detector in path f' to fire so that we now have $P_d(\lambda_1) = 0$. Hence, we see with this simple example that the result of the P_d measurement is dependent on the measurement context. In this way the Kochen–Specker argument is circumvented.

7. EVEN MORE CONTEXTUALITY

We gave an example above that shows how the result of a measurement can depend on what else is being measured. However, there is an even more basic type of contextuality in Bohmian mechanics. The result of a measurement of a given operator can depend on the method used to measure it. We will give a very simple example. We take our system to consist of a particle in path *a* or *b* so that our Hilbert space is two dimensional (in which the Kochen–Specker argument could not be run). Now we wish to measure the operators P_c and P_d where

$$|a\rangle = |d\rangle \qquad |b\rangle = |c\rangle \tag{10}$$

Figure 4 shows two ways of doing this. In Figure 4(a) the path *a* runs directly into path *d* and path *b* runs directly into *c*. As the paths do not interfere it is clear that regardless of the initial quantum state of the system a particle starting in path *a* would end up in path *d* and likewise a particle starting in path *b* will end up in path *c*. (Recall that the although the paths

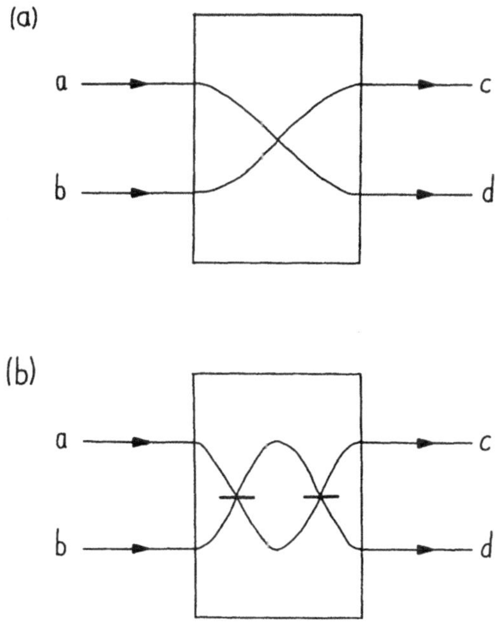

Fig. 4. Two interferometric arrangements that perform the same measurement. Note that in (a) the paths do not actually intersect in three dimensions.

appear to cross in the figure, in three dimensions there is no intersection.) Now consider Figure 4(b). A Mach–Zehnder type interferometer connects the paths. Using the transformations (1) it is easily shown that the state $|a\rangle$ evolves into $|d\rangle$ and the state $|b\rangle$ evolves into $|c\rangle$. Thus, this apparatus measures the same operator as the apparatus in Figure 4(a). If the initial state of the system is $|a\rangle$ and the particle is in path a, then by rule (i) it will end up in path d as expected. However, if the initial state of the system is $1/\sqrt{2}(|a\rangle + |b\rangle)$ and the particle is in path a, then it will end up in path c as we will now show. After evolving past the first beam splitter, this state will lead to equal amplitudes for the two internal paths (as can be seen by applying (1)). Hence, by rules (i) and (ii), it must be reflected (since if it were transmitted another trajectory in an ensemble satisfying $\rho = |\psi|^2$ initially but starting in path b would have to be transmitted to preserve the balance to satisfy rule (i), but these two trajectories would have to cross which is forbidden by rule (ii)). At the second beam splitter the same thing happens. There are equal amplitudes for path c and d. Hence, the particle must be reflected again ending up in path c. This means that even though the system is in the same initial state (both its quantum state and the particle position), and even though the same operator is being measured, different outcomes are possible since different measurement procedures are used. This shows that the result of a measurement in the

Bohm model is dependant not only on what else is being measured but also on the method used to make the measurement (see also the spin example given in Albert 1992).

8. CONCLUSIONS

We have shown how Bohmian mechanics can circumvent the Kochen–Specker argument against hidden variable theories. The fact that the Bohm model can reproduce all the predictions of quantum mechanics makes it clear that there can be no proof that hidden variable interpretations of quantum mechanics are impossible – the Bohm model will always serve as a counter example. An interesting question arises here: Does there exist any probabilistic theory that cannot be given a hidden variable interpretation? If an example could be found, it would show that von Neumann and others were not trying in vain. However, it seems more likely that probabilistic theories can always be supplemented by hidden variables to yield a deterministic theory. A related and perhaps more important issue is the question of symmetries. The probabilistic theory may have certain symmetries (for example quantum mechanics is Lorentz invariant at this level). However, on being supplemented by hidden variables, these symmetries may be broken. The Bohm model has a preferred frame of reference that is not evident at the level of the probabilistic quantum predictions. In fact, the Kochen–Specker theorem may be viewed in this way. The probability of getting $P_d = 1$ for a given quantum state is independent of the measurement context. This contextuality only becomes evident when hidden variables are introduced.

The above examples of measurements with interferometers tell us something about the Bohm model. Namely, that trying to associate elements of reality with operators in the Bohm model can lead to confusion. The basic elements of reality are the positions of the particles along with the quantum state vector. The outcome of a measurement of a given operator can depend on the way the measurement is carried out. It does not make sense to allocate elements of reality to operators without also specifying the full experimental arrangement (see also Dürr *et al.* 1996a).

University of Durham

NOTE

[1] In fact, it is not obvious that one can consider only this coordinate. However, if we assume that the three-dimensional wave-packets are spherically symmetric, then symmetry arguments can be employed to show that the essential motion is along the coordinate x. Alternatively, the paths here can be thought of as 1-dimensional channels. Then there is only one coordinate.

KARIN BERNDL

GLOBAL EXISTENCE AND UNIQUENESS OF BOHMIAN TRAJECTORIES

1. INTRODUCTION

Consider the equations of motion of Bohmian mechanics for a system of N particles with masses $m_1, \ldots, m_N$ moving in physical space $\mathbb{R}^3$: the wave function ψ evolves according to Schrödinger's equation

$$(1) \qquad i\hbar \frac{\partial \psi_t(q)}{\partial t} = \left(- \sum_{k=1}^{N} \frac{\hbar^2}{2m_k} \Delta_k + V(q) \right) \psi_t(q),$$

and the configuration $Q = (\mathbf{Q}_1, \ldots, \mathbf{Q}_N) \in \mathbb{R}^{3N}$, with $\mathbf{Q}_k \in \mathbb{R}^3$ denoting the position of the k-th particle, evolves according to Bohm's equation

$$(2) \qquad \frac{dQ_t}{dt} = v^{\psi_t}(Q_t),$$

where the velocity field $v^{\psi} = (\mathbf{v}_1^{\psi}, \ldots, \mathbf{v}_N^{\psi})$ is determined by the wave function ψ

$$(3) \qquad \mathbf{v}_k^{\psi}(q) = \frac{\hbar}{m_k} \operatorname{Im} \frac{\nabla_k \psi(q)}{\psi(q)}.$$

(Bohm 1952a). This velocity field is regular on the subset of $\mathbb{R}^{3N}$ where $\psi \neq 0$ and is differentiable. The question arises as to what happens if the configuration Q_t, when moving in accordance with (2) along integral curves of v^{ψ_t}, reaches at some time τ a singularity of v^{ψ}, for example a node of the wave function (i.e., a point where $\psi = 0$). In general, the event of reaching a singularity of v^{ψ} corresponds to a singularity of the motion: the velocity dQ_t/dt becoming infinite, Q_t being discontinuous, or even "exploding" (i.e., reaching infinity as $t \to \tau$). In some cases it may be possible to continue the Bohmian trajectory through a singularity of the velocity field, but in any case, Eq. (2) is not defined at singularities of v^{ψ}, and the theory would have to be supplemented by a suitable prescription for how to extend the motion through a singularity, or which trajectories to avoid.

We should therefore consider the following problem – the *initial value problem in Bohmian mechanics: For given initial values ψ_0 and Q_0 at a time t_0 (we shall put $t_0 = 0$), do there exist global unique solutions (ψ_t, Q_t) of* (1, 2) *such that $\psi_{t_0} = \psi_0$ and $Q_{t_0} = Q_0$?* A positive answer to this question for all or at least for a suitable majority of initial conditions, or for a class of "physically relevant" initial conditions, is certainly important: the limits of global existence of solutions may hint at limits of validity of the theory.

J. T. Cushing et al. (eds.), Bohmian Mechanics and Quantum Theory: An Appraisal, 77–86.

Let us first give a simple explicit example where indeed some trajectories reach nodes of the wave function. Consider the one-dimensional harmonic oscillator (with $\hbar = m = \omega = 1$) with potential $V(q) = q^2/2$, and take as the wave function a superposition of the ground state and the second excited state:

$$\psi_t(q) = e^{-q^2/2}(1 + (1 - 2q^2)e^{-2it})e^{-it/2}. \tag{4}$$

This wave function leads to a velocity field v^ψ which is an odd function of q and periodic in time with period π. The Bohmian motion is therefore invariant under reflection $q \to -q$ and periodic in t with period π: If $Q(t)$ is a solution of (2), then $-Q(t)$ is also a solution of (2), and $Q(t + \pi) = Q(t)$. The wave function ψ has nodes at $(q, t) = (0, (n + 1/2)\pi)$ and at $(q, t) = (\pm 1, n\pi)$ for all integers n. There are three trajectories which periodically run into nodes of ψ (see Figure 1). One of them, the constant $Q_t = 0$ for $t \neq (n + 1/2)\pi$, is a solution of (2) which runs with velocity 0 at times $t \to (n + 1/2)\pi$ into nodes of ψ. The other two "node-crossing" trajectories (which are reflection images of each other) are singular at the nodes: for example, in the vicinity of the node at $q = 1$, $t = 0$ the trajectory running into it has the form $Q_t = (3t^2/4)^{1/3} + 1$ (i.e. it has infinite velocity at $t = 0$).

In this example the trajectories may be continued through the nodes in an obvious and consistent way. This is an artifact of the low dimensionality. In fact, in d = 1 dimension, Q_t satisfies

$$\int_{-\infty}^{Q_t(q_0)} |\psi_t(q)|^2 dq = \int_{-\infty}^{q_0} |\psi_0(q)|^2 dq, \tag{5}$$

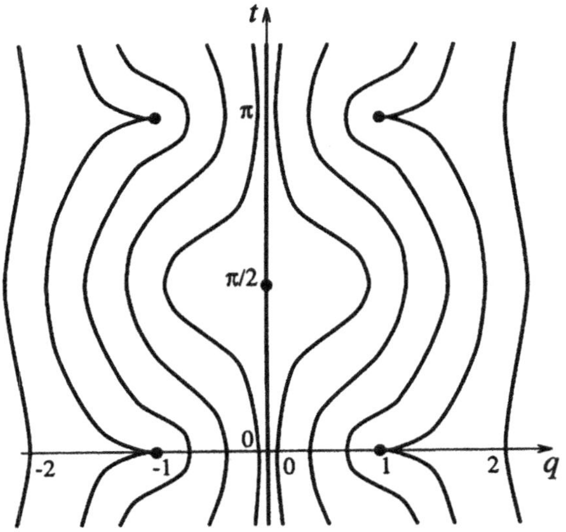

Fig. 1. Sketch of the course of Bohmian trajectories for the wave function (4). Nodes of the wave function are marked by dots.

employing the equivariance of the $|\psi_t|^2$-measure (i.e. that the density $\rho_0 = |\psi_0|^2$ evolves under the Bohmian dynamics to $\rho_t = |\psi_t|^2$ (Dürr, *et al.* 1992a; see also Section 4) and the property that trajectories do not cross in configuration-space-time. Clearly, much less regularity of ψ is needed for this definition of the motion of the Bohmian configuration – in particular, nodes of the wave function are no problem – and, if both definitions (2) and (5) are possible, they agree. However, a generalization of (5) to the physically most interesting case of d = $3N$-dimensional configuration space is not known. We shall turn now to the question of the existence of global unique solutions of the equations of motion of Bohmian mechanics, (1) and (2).

2. THE SCHRÖDINGER EQUATION

In Bohmian mechanics, the evolution of the wave function ψ according to Schrödinger's equation (1) is independent of the evolution of the actual particle configuration Q according to the guiding equation (2), while for the integration of (2) we need ψ. Therefore, when solving the initial value problem, we may consider Schrödinger's equation first.

The linear partial differential equation (1) is usually discussed in the frame-work of the Hilbert space $\mathcal{H} = L^2(\mathbb{R}^{3N})$ of square integrable functions, by viewing the Hamiltonian $\tilde{H} = -\sum_{k=1}^{N} (\hbar^2/2m_k)\, \Delta_k + V(q)$ as the generator of the unitary group $U_t = e^{-iHt/\hbar}$ (Kato 1951; Reed and Simon 1975; Simon 1977). Usually, boundary conditions on the wave function have to be specified in order to get a unique time evolution $(U_t)_{t\in\mathbb{R}}$. In more technical terms, one has to select a self-adjoint extension H of the partial differential operator $\tilde{H}$, which is a priori defined only on sufficiently smooth functions. As different self-adjoint extensions generate different time evolutions of the wave function, the choice of the right self-adjoint extension is a matter of physics. We shall comment on this in Section 4. Given $(U_t)_{t\in\mathbb{R}}$, for any initial $\psi_0 \in \mathcal{H}$ we have a global unique wave function $\psi_t = U_t\psi_0$. This wave function, however, in general is not a genuine solution of Schrödinger's equation: Generic $\psi_0 \in \mathcal{H}$ are not smooth functions, and ψ_t will not be differentiable, so that the differential equation (1) cannot be discussed. The standard procedure now is to forget about Schrödinger's equation, and argue that all one is interested in is a unitary time evolution – or, even more abstractly, a unitary representation of the time translation group. This is, however, not sufficient for our approach. From the point of view of Bohmian mechanics, the wave function ψ is a smooth field on configuration-space-time solving Schrödinger's equation. To obtain this, we have to put suitable conditions on the initial wave function ψ_0. It turns out that a suitable subspace of $\mathcal{H}$ is given by the so-called C^∞-vectors of the Hamiltonian H, $C^\infty(H) = \bigcap_{n=1}^{\infty} \mathcal{D}(H^n)$. $\mathcal{D}(H^n)$ denotes the domain of the n-th power of the Hamiltonian, i.e. the set of wave functions for which the expectation value of the $2n$-th power of H, the "energy," is finite, $\int_{\mathbb{R}^{3N}}(H^n\psi)^2\, dq = \langle\psi \mid H^{2n}\psi\rangle < \infty$.[1] $C^\infty(H)$ is a dense subset of $\mathcal{H}$, and invariant under the time evolution U_t. Eigenfunctions and wave packets are special C^∞-vectors, so all wave functions usually considered in physics are included. It should not at all be regarded as a defect of Bohmian mechanics that it is not defined for all $\psi_0 \in \mathcal{H}$. From the point of view of Bohmian mechanics, the Hilbert space $L^2(\mathbb{R}^{3N})$ is not the state space of the wave function, but a useful tool for the analysis of the theory. (In this context, see also Dürr *et al.* (1996a) and Daumer (1996).)

We have: *For any* $\psi_0 \in C^\infty(H)$, $\psi_t = U_t\psi_0 = e^{-itH/\hbar}\psi_0$ *is a global smooth solution of Schrödinger's equation on* $\Omega \times \mathbb{R}$, where Ω denotes the set where the potential is smooth. This result cannot be genuinely strengthened: the wave function cannot be expected to be regular at points where the potential is singular. For instance, the ground state eigenfunction of the Coulomb potential $V(q) = 1/|q|$, $e^{-|q|}$, is not differentiable at the singularity of the potential at $q = 0$. Important examples of potentials are the N-particle Coulomb interaction for particles with charges e_i

$$V_{\text{Coulomb}}(\mathbf{q}_1, \ldots, \mathbf{q}_N) = \sum_{i=1}^{N} \sum_{j=i+1}^{N} \frac{e_i e_j}{|\mathbf{q}_i - \mathbf{q}_j|},$$

with $\Omega_{\text{Coulomb}} = \mathbb{R}^{3N}\backslash\bigcup_{i=1}^{N}\bigcup_{j=i+1}^{N}\{\mathbf{q}_i = \mathbf{q}_j\}$, and the N-electron atom

$$V_{\text{atom}}(\mathbf{q}_1, \ldots, \mathbf{q}_N) = \sum_{i=1}^{N} \frac{e_n e_i}{|\mathbf{q}_i|} + \sum_{i=1}^{N} \sum_{j=i+1}^{N} \frac{e_i e_j}{|\mathbf{q}_i - \mathbf{q}_j|},$$

with $\Omega_{\text{atom}} = \mathbb{R}^{3N}\backslash(\bigcup_{i=1}^{N}\{\mathbf{q}_i = 0\} \cup \bigcup_{i=1}^{N}\bigcup_{j=i+1}^{N}\{\mathbf{q}_i = \mathbf{q}_j\})$, in the approximation that the nucleus with charge e_n is at rest at the origin, acting like an external Coulomb field.

3. THE BOHMIAN TRAJECTORIES

With the wave function $\psi \in C^\infty(\Omega \times \mathbb{R})$, the velocity field v^ψ (3) can be formed on the set where the wave function $\psi \neq 0$. We shall call the set where the velocity field is regular $\mathcal{G} = (\Omega \times \mathbb{R})\backslash\mathcal{N}$, where $\mathcal{N}$ is the (space-time) set of nodes of the wave function $\mathcal{N} = \{(q, t) \in \Omega \times \mathbb{R}: \psi_t(q) = 0\}$. On this set,the first order ordinary differential equation (2) is locally integrable. If we extend the solution as far as possible, we obtain for each initial value $q_0 \in \mathcal{G}_0$ (i.e., $(q_0, 0) \in \mathcal{G}$) a *maximal solution* $Q(t; q_0)$ on a maximal time interval of existence $(\tau^-(q_0), \tau^+(q_0))$. The solution is called *global* if $\tau^- = -\infty$ and $\tau^+ = +\infty$. In the introduction we have given an example showing that we cannot expect to have global solutions for *all* initial values q_0. But what indeed holds true is that for *almost all* initial configurations we have global unique solutions. "Almost all" here is with respect to the natural – the equivariant – measure P^{ψ_0} on configuration space $\mathbb{R}^{3N}$, that is the measure with the density $|\psi_0|^2$. We have proved that *for a large class of potentials, including the N-particle Coulomb interaction with arbitrary masses and charges, as well as arbitrary positive potentials,* $P^{\psi_0}(\tau^+ = +\infty$

and $\tau^- = -\infty) = 1$. In other words, for almost any initial point Q_0, the solution of the guiding equation (2) is global (i.e., nodes of the wave function or other singularities of the velocity field will not be reached in finite time) and the solution does not "explode" (i.e., reach infinity in finite time). (Cf. the example of Figure 1: the set of initial configurations for which the solution of (2) is not global consists of 3 points – certainly as set of P^{Ψ_0}-measure 0 – while for all the other initial values the solution is global.)

We shall comment on the proof – which is in fact quite intuitive – in the next section. (For the details, see Berndl, Dürr, Goldstein, Peruzzi and Zanghì 1995.) Despite the relative simplicity of the proof, the generality of the result is rather surprising. The analogous problem in Newtonian mechanics – global existence and uniqueness of solutions of the N-body problem for Newtonian gravity – is a classical problem of mathematical physics which has been investigated with a great variety of methods (Moser 1973; Diacu 1992; 1993), but for which the analogous result – global existence of solutions for Lebesgue-almost all initial values in the N-particle phase space $\mathbb{R}^{6N}$ – has not yet been established. In addition to the possibility of collision singularities, the N-body problem with $N > 3$ yields marvelous scenarios of so-called pseudocollisions, where some particles, while oscillating wildly, reach infinity in finite time. Examples of such catastrophies have been constructed by Mather and McGehee (1975),[2] by Gerver (1991), and by Xia (1992). For special situations, such as for example an almost planar solar system with weakly eccentric planets, the KAM theorem furnishes among other things the existence of global solutions for a set of initial values of positive Lebesgue measure in phase space (cf., for example, Arnold 1963; 1989). For the general case of the N-body problem with $N \geq 5$, it is still open whether for almost all initial values the solutions of Newtonian gravity are global or whether pseudocollisions occur on a set of positive measure.

The explanatory power of Newtonian mechanics rests largely on the analysis of specific solutions of the equations of motion, such as, for example, the planetary motion, or the gyroscope motion – physical situations where the bodies are modelled by fixed mass densities. The more ambitious program of Newtonian mechanics for a system composed of a huge number of point particles or rigid balls, for example the statistical gas theory, is difficult for two reasons. Firstly, to get global existence of solutions for almost all initial values – certainly a prerequisite of a statistical analysis – one has to deal, for example, with the problem of singular interactions in the case of point particles or, in the case of rigid balls, multiple collisions. Secondly, more interesting than equilibrium properties and much more difficult to analyse are the physical effects occurring in the transition from nonequilibrium to equilibrium.

In Bohmian mechanics, actual trajectories are interesting in some cases. However, to determine the equilibrium properties they are not needed. And in contrast to the situation in classical physics, we have the strongest empirical evidence that our world is in quantum equilibrium (Dürr *et al.* 1992a,b). Our theorem of global existence of Bohmian trajectories is thus exactly what is necessary and sufficient for a complete analysis of Bohmian mechanics. It provides the rigorous basis for the derivation of the quantum formalism as well as scattering theory from an equilibrium analysis of Bohmian mechanics. (These topics are described in Dürr *et al.* (1996a) and in Daumer (1996).) In this sense, what we have proved is just that $|\psi|^2$ *is an equivariant density for the Bohmian particle motion,*

$$\rho_0 = |\psi_0|^2 \Rightarrow \rho_t = |\psi_t|^2 \qquad \text{for all } t \in \mathbb{R}. \tag{6}$$

4. THE QUANTUM FLUX

The preceding sentence may have confused readers: Why isn't the equivariance of the $|\psi_t|^2$-measure clear without an intricate examination of the global existence and uniqueness of solutions of Bohmian mechanics? Why doesn't (6) follow immediately from a comparison of the continuity equation for an ensemble of configurations moving with velocity v^ψ and having a density $\rho_t(q)$

$$\frac{\partial}{\partial t}\rho_t(q) + \sum_{k=1}^{N} \mathrm{div}_k(\mathbf{v}_k^{\psi_t}(q)\rho_t(q)) = 0, \tag{7}$$

with the "quantum continuity equation"

$$\frac{\partial}{\partial t}|\psi_t(q)|^2 + \sum_{k=1}^{N} \mathrm{div}_k\, \mathbf{j}_k^{\psi_t}(q) = 0, \tag{8}$$

noting that the quantum probability current $j^\psi = (\mathbf{j}_1^\psi, \ldots, \mathbf{j}_N^\psi)$ is given by

$$\mathbf{j}_k^\psi = \mathbf{v}_k^\psi\,|\psi|^2 = \frac{\hbar}{m_k}\,\mathrm{Im}(\psi^*\nabla_k\psi)?$$

The answer is that the continuity equation (7) expresses how an ensemble density ρ_0 evolves under the deterministic evolution of the trajectories, and holds therefore on just that set which is covered by the integral curves of v^{ψ_t}. We shall denote this set – the image set of the maximal solution Q – by $\mathcal{I}$:

$$\mathcal{I} = \{(q, t) \in \mathcal{G}: \exists q_0 \in \mathcal{G}_0 \text{ with } t \in (\tau^-(q_0), \tau^+(q_0)) \text{ and } q = Q_t(q_0)\}.$$

Equation (8), on the other hand, is an identity for every ψ_t which solves Schrödinger's equation. Thus (6) holds on $\mathcal{I}$: $\rho_t(q) = |\psi_t(q)|^2$ for $(q, t) \in \mathcal{I}$. By putting $\rho_t(q) = 0$ on the set which is not reached by the Bohmian trajectories $(\Omega \times \mathbb{R})\backslash\mathcal{I}$, we obtain $\rho_t(q) \le |\psi_t(q)|^2$ for $(q, t) \in \Omega \times \mathbb{R}$. This insight is fundamental to the proof of global existence.

The second fundamental insight comes from the consideration of the configuration-space-time flux $J(q, t) = (v^{\psi_t}(q)\rho_t(q), \rho_t(q))$. To establish $P^{\psi_0}(\tau^+ = +\infty \text{ and } \tau^- = -\infty) = 1$ (i.e., that for P^{ψ_0}-almost all initial configurations q_0 the maximal solution is global), we have to show that the solution will reach singularities of the velocity field or infinity in finite time at most for a set of initial configurations of P^{ψ_0}-measure zero. The probability of such a "bad event" is estimated with the help of the flux $J(q, t)$: For any hypersurface Σ in configuration-space-time with local normal vector field $n(q, t)$ and surface area element $d\sigma$,

$$\int_\Sigma |J(q, t) \cdot n(q, t)| d\sigma,$$

is the expected number of crossings of the hypersurface Σ by the Bohmian trajectory and hence a bound for the probability of crossing Σ. We obtain

$$P^{\psi_0}\,(Q_t \text{ crosses } \Sigma) \leq \int_\Sigma |J(q, t) \cdot n(q, t)| d\sigma \leq \int_\Sigma |J^{\psi_t}(q) \cdot n(q, t)| d\sigma, \tag{9}$$

with the quantum flux $J^{\psi_t}(q) = (j^{\psi_t}(q), |\psi_t(q)|^2)$.

Consider now neighborhoods around the singularities of the velocity field: $\mathcal{N}^\varepsilon$, a (configuration-space-time) neighborhood of thickness ε around the set of nodes $\mathcal{N}$ of the wave function, $\mathcal{S}^\delta$, a (configuration space) neighborhood of thickness δ around the set of singularities of the potential $\mathcal{S} = \partial\Omega$, and $\mathcal{K}^r$, a sphere in configuration space of radius r to control escape to infinity. $\mathcal{G}^{\varepsilon\delta r}$ denotes the set of "ε-δ-r-good" points in configuration-space-time: $\mathcal{G}^{\varepsilon\delta r} = ((\mathcal{K}^r \setminus \mathcal{S}^\delta) \times \mathbb{R}) \setminus \mathcal{N}^\varepsilon$ (see Figure 2).

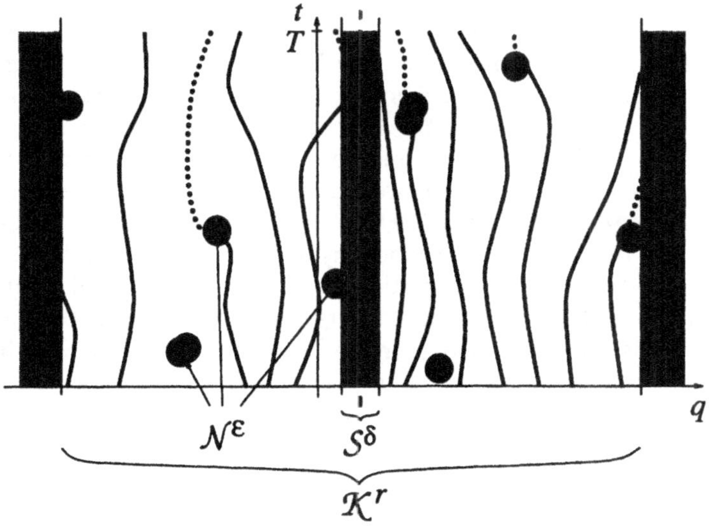

Fig. 2. Bohmian trajectories run in the white area $\mathcal{G}^{\varepsilon\delta r} = ((\mathcal{K}^r \setminus \mathcal{S}^\delta) \times \mathbb{R}) \setminus \mathcal{N}^\varepsilon$. The set of singularities of the potential $\mathcal{S} \times \mathbb{R}$ is dashed, nodes of the wave function are marked by x. Trajectories having crossed $\partial\mathcal{G}^{\varepsilon\delta r}$ are dotted. By letting the grey neighborhoods $\mathcal{N}^\varepsilon$ and $\mathcal{S}^\delta$ around the singularities of the velocity field shrink and the sphere $\mathcal{K}^r$ blow up, the set of dotted trajectories shrinks to a set of measure zero.

A Bohmian trajectory approaching a singularity of the velocity field or infinity first has to cross the boundary of $\mathcal{G}^{\varepsilon\delta r}$. From (9), we obtain the following bound for the probability of a "bad event" in the time interval $[0, T]$: for all $0 < T, \varepsilon, \delta, r < \infty$

$$(10) \qquad P^{\psi_0}(\tau^+ < T) \leq P^{\psi_0}(\mathcal{G}_0 \setminus \mathcal{G}_0^{\varepsilon\delta r}) + \int_{\partial\mathcal{G}^{\varepsilon\delta r} \cap (\mathbb{R}^{3N} \times [0, T])} |J^{\psi_t}(q) \cdot n(q, t)| \, d\sigma$$
$$\leq P^{\psi_0}(\mathcal{G}_0 \setminus \mathcal{G}_0^{\varepsilon\delta r}) + \mathbf{N}(\varepsilon, \delta, r) + \mathbf{S}(\delta) + \mathbf{I}(r),$$

with

$$\mathbf{N}(\varepsilon, \delta, r) = \int_{\partial\mathcal{N}^\varepsilon \cap ((\mathcal{K}^r \setminus \mathcal{S}^\delta) \times [0, T])} |J^{\psi_t}(q) \cdot n(q, t)| \, d\sigma,$$

$$\mathbf{S}(\delta) = \int_{(\partial\mathcal{S}^\delta \cap \Omega) \times [0, T]} |J^{\psi_t}(q) \cdot n(q, t)| \, d\sigma,$$

$$\mathbf{I}(r) = \int_{(\partial\mathcal{K}^r \cap \Omega) \times [0, T]} |J^{\psi_t}(q) \cdot n(q, t)| \, d\sigma.$$

By proving that for appropriate choices of sequences $\varepsilon \to 0$, $\delta \to 0$, and $r \to \infty$ the right hand side of (10) gets arbitrarily small we show that for all $T > 0$, $P^{\psi_0}(\tau^+ < T) = 0$, and thus $P^{\psi_0}(\tau^+ < \infty) = 0$. From time reversal invariance we obtain that also $P^{\psi_0}(\tau^- > -\infty) = 0$, and that altogether the solutions of Bohm's equation (2) with the velocity field (3) are global for almost all initial configurations.

Heuristically it is rather immediate that the flux integrals **N**, **S**, and **I** get arbitrarily small as $\varepsilon \to 0$, $\delta \to 0$, and $r \to \infty$. For **N**, observe that the flux $J^\psi = 0$ at the nodes of the wave function. From the continuity of ψ, J^ψ is small in the vicinity of $\mathcal{N}$. Furthermore, the nodal set $\mathcal{N}$ itself is generically small: since ψ is a complex function, the set where $\psi_t(q) = 0$, i.e. $\mathrm{Re}\, \psi_t(q) = 0$ and $\mathrm{Im}\, \psi_t(q) = 0$, is generically of codimension 2 in configuration-space-time. Thus the area of $\partial\mathcal{N}^\varepsilon$ should be small. Of the set $\mathcal{S}$ of singularities of the potential V, we assume in Berndl, Dürr, Goldstein, Peruzzi and Zanghì (1995) that it be contained in a union of a finite number of $(3N-3)$-dimensional hyperplanes, as is certainly the case for the N-particle Coulomb interaction $V = V_{\mathrm{Coulomb}}$ and the N-electron atom $V = V_{\mathrm{atom}}$. The area of $\partial\mathcal{S}^\delta$ is therefore very small. Moreover, in certain cases it is required as a boundary condition for self-adjointness of the Hamiltonian that the flux into the singularities of V vanishes, see below. That the flux to infinity is small can be derived from the fact that the quantum flux $J^{\psi_t}(q)$ tends rapidly to 0 as $|q| \to \infty$, which follows from the square integrability of ψ and $\nabla\psi$ (i.e., from the normalizability of the $|\psi|^2$-distribution and finite "kinetic energy"). These conditions are automatically fulfilled for the considered class of potentials for $\psi_0 \in C^\infty(H)$, which is just what we required for the existence of a global classical solution of Schrödinger's equation.

We are thus lead back to the question of the existence of global classical solutions of Schrödinger's equation! In fact, the connection between the global existence of Bohmian trajectories and global solutions of Schrödinger's equation is most remarkable. The clue is the quantum flux J^{Ψ}, which in Bohmian mechanics has the interpretation of a flux of particles moving along deterministic trajectories with velocity v^{Ψ}: The condition that there is no flux into the critical points ensures firstly, as explained above, that the Bohmian configuration will not reach the critical points and thus exists globally. Secondly, it provides suitable boundary conditions for the domain of the Hamiltonian $\mathcal{D}(H)$ such that the Hamiltonian will be self-adjoint on $\mathcal{D}(H)$ and thus Schrödinger's equation has global unique solutions as explained in Section 2. This connection is realized by considering that from integration by parts

$$\int_M \psi^*(H\psi)\mathrm{d}q - \int_M (H\psi^*)\psi\mathrm{d}q = -i\hbar \int_{\partial M} j^{\psi} \cdot n\mathrm{d}s,$$

for $M = \mathcal{K}\setminus\mathcal{S}^{\delta}$, and thus by the self-adjointness of the Hamiltonian

$$\lim_{\delta \to 0,\, r \to \infty} \left(\int_{\partial\mathcal{S}^{\delta} \cap \mathcal{K}^{r}} j^{\psi} \cdot n\mathrm{d}s + \int_{\partial\mathcal{K}^{r}\setminus\mathcal{S}^{\delta}} j^{\psi} \cdot n\mathrm{d}s \right) = 0.$$

This is only slightly weaker than the vanishing of $\mathbf{S}(\delta)$ and $\mathbf{I}(r)$ in the limit $\delta \to 0$, $r \to \infty$, which is part of our sufficient condition for global existence of Bohmian trajectories. Moreover, in situations where the self-adjoint extension of $\tilde{H}$ (cf. Section 2) is not unique, the particle picture of Bohmian mechanics supplies an interpretation of the different possible boundary conditions yielding different time evolutions of the wave function and thus a basis for the choice of one over the others. (For more on these points, see Berndl, Dürr, Goldstein, Peruzzi and Zanghì (1995).)

In this way, the point of view of Bohmian mechanics provides genuine understanding to the mathematics around the self-adjointness of Schrödinger Hamiltonians! Usually, the self-adjointness of the Hamiltonian – via its equivalence to the existence of a unitary group – is motivated by the conservation of $|\psi|^2$-probability. Probability of what? The standard answer – the probability of *finding* a particle in a certain region – is justified by Bohmian mechanics: A particle is found in a certain region because, in fact, it's there. By incorporating the positions of the particles into the theory, and thus by interpreting $|\psi|^2$ as a probability density of particles being and the quantum flux J^{ψ} as a flux of particles moving, Bohmian mechanics can be regarded as providing the foundation for all intuitive reasoning in quantum mechanics.

Mathematisches Institut der Universität München

NOTES

[1] If only the expectation of H^2, the squared energy, is finite, $\psi_0 \in \mathfrak{D}(H)$, "Schrödinger's equation holds in the L^2-sense" (i.e. $i \lim_{h \to 0}(\psi_{t+h} - \psi_t)/h = H\psi_t$, where the equality and convergence of the limit holds with respect to the Hilbert space norm and not pointwise as expressed by Schrödinger's equation).

[2] In this (one-dimensional) example the system explodes only after infinitely many two-particle collisions; thus it does not describe a genuine pseudocollision.

MARTIN DAUMER

SCATTERING THEORY FROM A BOHMIAN PERSPECTIVE

Quantum mechanical scattering theory is a subject with a long and winding history. We shall pick out some of the most important concepts and ideas of scattering theory and look at them from the perspective of Bohmian mechanics: Bohmian mechanics, having real particle trajectories, provides an excellent basis for analyzing scattering phenomena.

1. A VERY BRIEF HISTORICAL SKETCH

We begin with a quote taken from Born (1926), shortly after Heisenberg (1925) had invented matrix mechanics and Schrödinger (1926) his wave mechanics:

> Neither of these two conceptions appear satisfactory to me. I should like to attempt here to give a third interpretation and to test its utility on collision processes. In this attempt, I adhere to an observation of Einstein on the relationship of wave field and light quanta; he said, for example, that the waves are present only to show the corpuscular light quanta the way, and he spoke in the sense of a "ghost field." This determines the probability that a light quantum . . . takes a certain path; . . . And here it is obvious to regard the de Broglie–Schrödinger waves as the ghost field or, better, "guiding field."
>
> I should therefore like to investigate experimentally the following idea: the guiding field, represented by a scalar function ψ of the coordinates of all the particles involved and the time, propagates in accordance with Schrödinger's differential equation. . . . The paths of these corpuscles are determined only to the extent that the laws of energy and momentum restrict them; otherwise, only a probability for a certain path is found.
>
> * * *
>
> Closing remarks: On the basis of the above discussion, I should like to put forward the opinion that quantum mechanics permits not only the formulation and solution of the problem of stationary states, but also that of transition processes. In these circumstances Schrödinger's version appears to do justice to the facts in by far the easiest manner; moreover, it permits the retention of the conventional ideas of space and time in which events take place in a completely normal manner. On the other hand, the proposed theory does not correspond to the requirement of the causal determinacy of the individual event. In my preliminary communication I stressed this indeterminacy quite particularly, since it appears to me in best agreement with the practice of the experimenter. But it is natural for him who will not be satisfied with this to remain unconverted and to assume that there are other parameters, not given in the theory, that determine the individual event. In classical mechanics these are the "phases" of the motion, i.e. the coordinates of the particles at a given instant. It appears to me *a priori* improbable that quantities corresponding to these phases can easily be introduced into the new theory, but Mr. Frenkel has told me that this may perhaps be the case. However this may be, this possibility would not alter anything relating the practical indeterminacy of collision processes, since it is in fact impossible to give the values of the phases; it must in fact lead to the same formulae as the "phaseless" theory proposed here. (Born 1926; Ludwig 1968, 207–208, 224–225)

Schrödinger (1926) had shown with his wave mechanics how to obtain

J. T. Cushing et al. (eds.), Bohmian Mechanics and Quantum Theory: An Appraisal, 87–97.

the discrete energy levels of the hydrogen atom, by seeking stationary square-integrable solutions of the Schrödinger equation (in units $\hbar = m = 1$)

$$(1) \qquad i\frac{\partial\psi_t(\mathbf{x})}{\partial t} = -\frac{1}{2}\,\Delta\psi_t(\mathbf{x}) + V(\mathbf{x})\psi_t(\mathbf{x}).$$

What are called stationary states are solutions of the form $\psi_t(\mathbf{x}) = e^{-iEt}\psi(\mathbf{x})$, where $\psi(\mathbf{x})$ obeys the stationary Schrödinger equation

$$(2) \qquad -\frac{1}{2}\,\Delta\psi(\mathbf{x}) + V(\mathbf{x})\psi(\mathbf{x}) = E\psi(\mathbf{x}),$$

which has square-integrable solutions only for certain discrete energies $E_n < 0$, in agreement with the experimental results. The physical meaning of the wave function was however unclear.

A description of a time-dependent scattering process was soon given by Max Born (1926), who explored the hypothesis that the wave function might be a "guiding field" for the motion of the electron. As a consequence of this hypothesis, Born is led in his paper to the "statistical interpretation" of the wave function: $\rho_t(\mathbf{x}) = |\psi_t(\mathbf{x})|^2$ is the probability density for a particle to be at point $\mathbf{x}$ at time t. It follows from (1) that there is a conserved flux corresponding to this density, the quantum flux $\mathbf{j}^{\psi_t}(\mathbf{x}) := \mathrm{Im}(\psi_t^*(\mathbf{x})\nabla\psi_t(\mathbf{x}))$, which obeys the continuity equation

$$(3) \qquad \frac{\partial|\psi_t(\mathbf{x})|^2}{\partial t} + \nabla\cdot\mathbf{j}^{\psi_t}(\mathbf{x}) = 0.$$

Born interpreted the quantum flux as a probability current of particles. His basic Ansatz, sometimes called "naive" scattering theory (Reed and Simon 1979, 355), is to seek non-normalized solutions of the stationary Schrödinger equation (2) for *positive* energies $E = k^2/2$, which have the long-distance behavior

$$(4) \qquad \psi(\mathbf{x})^{x\to\infty}\; e^{i\mathbf{k}\cdot\mathbf{x}} + f(\theta,\phi)\frac{e^{ikx}}{x}.$$

$e^{i\mathbf{k}\cdot\mathbf{x}}$ is interpreted as the incoming plane wave and $f(\theta,\phi)^{e^{ikx}}/_x$ as the outgoing spherical wave with angular dependent density. The flux corresponding to the incoming wave is $\mathbf{k}$, such that the number of crossings per unit time and a unit surface orthogonal to $\mathbf{k}$ is $k := |\mathbf{k}|$. The flux corresponding to the spherical wave is $^{\mathbf{x}}/_{x^3}k|f(\theta,\phi)|^2$ and is obviously purely radial. The number of crossings of the surface element $x^2 d\Omega$ of a distant sphere in a direction specified by the angles θ, ϕ per unit time divided by the number of incoming particles per unit time and unit surface is called "differential cross section," and one finds

$$(5) \qquad \frac{d\sigma}{d\Omega} = |f(\theta,\phi)|^2.$$

This description of a scattering process is however not convincing for simple reasons: there is no hint in the equations what the particles do which are responsible for the flux and how their motion is related to the wave function such that $\rho_t(\mathbf{x}) = |\psi_t(\mathbf{x})|^2$ holds for all times. Moreover, the picture is entirely time-independent although a scattering process is certainly a process in space and time; stuff moves. The arguments leading to the formula (5) for the cross section "wouldn't convince an educated first grader" (Goldberger, cited in Simon 1971, 94).

From a physical point of view it might have seemed natural that a time-dependent justification of Born's time-independent method was developed, involving a detailed analysis of the behavior of the wave function and its corresponding flux, but scattering theory proceeded along a different direction. (For the reasons why Born later abandoned the idea of a guiding field see, e.g., Beller 1990.)

Heisenberg aimed at casting all in-principle-measurable quantities, such as energy levels and the cross section, into a single abstract object: the unitary *S*-matrix (see Cushing (1990) for a history of this program), or better, writing $S = e^{i\eta}$, into the self-adjoint "phase matrix" η (Heisenberg 1946). *S* should be the map from freely evolving "in"-states, which are controllable by an experimenter, to the freely evolving "out"-states, whose properties can be measured. Motivated by this idea Møller introduced the concept of "wave operators" (Møller 1945), the building blocks of the *S*-matrix: one expects that in a scattering process any wave function has a simple asymptotic behavior for large negative and positive times, where it should evolve almost freely with e^{-iH_0t} ($H_0 := -{}^1\!/_2\Delta$). Hence there should be states ϕ_+ and ϕ_- such that

$$(6) \qquad \lim_{t\to\pm\infty} \|e^{-iHt}\psi - e^{-iH_0t}\phi_\pm\|_2 = 0,$$

where $\|\cdot\|_2$ is the norm in L^2. If the wave operators

$$(7) \qquad \Omega_\mp = \text{s-}\lim_{t\to\pm\infty} e^{iHt}e^{-iH_0t},$$

("s-lim" denotes the strong limit) exist, then, if we now imagine that $\phi_\pm$ are given, obviously $\psi := \Omega_\mp\phi_\pm$ obeys (6). (The strange sign convention is a tradition; see Reed and Simon 1979, 17.)

The operators $\Omega_\pm$ should be unitary (we assume here that there are no bound states) such that $\Omega_\pm^\dagger = \Omega_\pm^{-1}$. The *S*-matrix can now be defined as $S = \Omega_-^\dagger\Omega_+$, because it maps freely evolving "in"-states onto "out"-states.

There has been a lot of work on the precise definition and properties of wave operators. We should mention here the concept of modified wave operators which was introduced (Dollard 1964) in order to bypass the problem, that the usual wave operators don't exist for long range potentials such as the Coulomb potential. The program of "asymptotic completeness," which aims at proving certain additional properties of the wave operators and the spectrum of the Hamiltonian, kept mathematicians and physicists busy for a long time, until recently this problem could be solved in great generality (Dereziński 1993).

Interestingly enough, these achievements would not have been possible without the "renaissance" of geometrical ideas (Ruelle 1969; Enss 1978), namely by realizing that the wave packets evolve in space and time and are far away from the scattering center most of the time, instead of working with very abstract and complicated methods in momentum space (Faddeev 1963), where this simple geometrical picture easily gets lost.

Most of this work on the S-matrix, wave operators and asymptotic completeness is however not much concerned with the original question of scattering theory, the justification of the formula for the cross section (5) of the "naive" scattering theory, which was commonly used to do the actual numerical calculations. It is true that formulas for the differential cross section have been suggested from an analysis of $\langle \mathbf{k}'|S|\mathbf{k}\rangle$, which was interpreted as the "probability density" to find the momentum eigenstate $|\mathbf{k}'\rangle$ in the final state $S|\mathbf{k}\rangle$ (Lippmann and Schwinger 1950 used "adiabatic switching," Gell-Mann and Goldberger 1953 Abelian limits), but these arguments were not much more convincing than the original argument to arrive at (5) (see also Reed and Simon 1979, 356).

An exception is the work of Ikebe (1960), who rigorously established the physicist's notation of expansions in continuum eigenfunctions (for "Ikebe"-potentials) and linked wave operators with solutions of the "Lippmann–Schwinger equation." The time-dependent wave function (in physicists notation formally $\langle \mathbf{x}|e^{-iHt}|\psi\rangle = \int d^3k e^{-i(k^2t/2)}\langle \mathbf{x}|\tilde{\mathbf{k}}\rangle\langle \tilde{\mathbf{k}}|\psi\rangle$) may be written as

$$(8)\qquad \psi_t(\mathbf{x}) = (e^{-iHt}\psi)(\mathbf{x}) = (2\pi)^{-\frac{3}{2}}\int d^3k e^{-i\frac{k^2t}{2}}\,\phi(\mathbf{x},\mathbf{k})\psi^{\#}(\mathbf{k}),$$

where the "generalized eigenfunctions" $\phi(\mathbf{x}, \mathbf{k})$ are solutions of the Lippmann–Schwinger equation

$$(9)\qquad \phi(\mathbf{x},\mathbf{k}) = e^{i\mathbf{k}\cdot\mathbf{x}} - \frac{1}{2\pi}\int d^3y\,\frac{e^{-ik|\mathbf{x}-\mathbf{y}|}}{|\mathbf{x}-\mathbf{y}|}\,V(\mathbf{y})\phi(\mathbf{y},\mathbf{k}).$$

Here $\psi^{\#}(\mathbf{k}) := (2\pi)^{-3/2}\int d^3x\phi^*(\mathbf{x},\mathbf{k})\psi(\mathbf{x})$ is the "generalized Fourier transform" of ψ and is connected with the wave operators by $(\widehat{\Omega_-^{\dagger}\psi})(\mathbf{k}) = \psi^{\#}(\mathbf{k})$, where ^ denotes the usual Fourier transform. (There is another set of eigenfunctions corresponding to Ω_+ which are in fact the ones most frequently used.) The solutions of the Lippmann–Schwinger equation are also solutions of the stationary Schrödinger equation with the asymptotic behavior as in (4) (the ones for Ω_+) such that the differential cross section may be read off their long distance asymptotics as in (5). At least one could now find the formula (5) of the "naive" scattering theory somewhere in an appropriate expansion of the time-dependent wave function.

Another exception is the work of Dollard (1969). Dollard suggested using the probability to find a particle in the far future in a given cone

$C \subset \mathbb{R}^3$ as a natural time-dependent *definition* of the cross section. Dollard's "scattering-into-cones-theorem" relates this probability to the wave operators:

$$(10) \qquad \lim_{t\to\infty} \int_C d^3x |\psi_t(\mathbf{x})|^2 = \int_C d^3\mathbf{v} |\Omega_-^\dagger \psi(\mathbf{v})|^2.$$

The scattering-into-cones-theorem has come to be regarded as the fundamental result from which the differential cross section ought to be derived (e.g., Reed and Simon 1979, 356; Enss and Simon 1980).

Dollard's approach was however criticized by Combes *et al.* (1975). They observe that the experimental relevance of the scattering-into-cones-theorem rests on the connection of the probability of finding the particle in the far future in a cone with the probability that the particle has, at some time, crossed a given distant surface subtended by the cone. Heuristically, the last probability should be given by integrating the quantum-mechanical flux over the total time interval and this surface. (The flux is often used that way in textbooks). Combes *et al.* hence conjecture the "flux-across-surfaces-theorem"

$$(11) \qquad \lim_{R\to\infty} \int_0^\infty dt \int_{C\cap\partial B_R} \mathbf{j}^{\psi_t} \cdot \mathbf{n} d\sigma = \int_C d^3\mathbf{v} |\Omega_-^\dagger \psi(\mathbf{v})|^2,$$

where B_R is the ball with radius R and outward normal $\mathbf{n}$. There exists no proof of this theorem. Even the "free flux-across-surfaces-theorem" for freely evolving ψ_t,

$$(12) \qquad \lim_{R\to\infty} \int_0^\infty dt \int_{C\cap\partial B_R} \mathbf{j}^{\psi_t} \cdot \mathbf{n} d\sigma = \int_C d^3\mathbf{v} |\hat{\psi}(\mathbf{v})|^2,$$

which should be physically good enough, because the scattered wave packet is expected to move almost freely after the scattering is essentially complete, has not been proven.

This is certainly strange because the physical importance of (11) and (12) for scattering theory is obvious and the mathematical problem does not seem to be too hard. Perhaps it was the vagueness in the meaning attached to the flux which is responsible for this matter of fact. For example, the authors try to reformulate the problem in operator language and are faced with the problem that for general L^2-functions the current across a given surface may well be infinite. Instead of using smooth functions they use "smeared-out" surfaces and therefore fail to find a proof of the original theorem. Or, for example, the argue that

> At large distances the scattering part of the wave function contains outgoing particles only. Therefore the particles cannot describe loops there and the flux can be measured by the interposition of counters on ∂B_R. (Combes *et al.* 1975, 368)

and seem to have in mind a picture very similar to Born's original proposal (cf. the Born quote), also without giving a precise guiding law for the trajectories which would allow, for example, to check the "no-loop-conjecture."

Next we want to show how the ideas of Combes *et al.* arise naturally by analyzing a scattering process in the framework of Bohmian mechanics (Bell 1987a; Bohm 1952a,b; Bohm and Hiley 1993; Dürr *et al.* 1992a; 1996a; Holland 1993): in this theory *particles move* along trajectories determined by the quantum flux, controlling the expected number of particles crossings of surfaces. We will sketch the main ideas of the proof of (12) (for the complete proof see Daumer 1995) and indicate the extension to the interacting case. We shall see that the flux-across-surfaces-theorem in Bohmian mechanics is a relation between the flux across a distant surface and the asymptotic probability of outward crossings of the trajectories of this surface – obviously the quantity of interest for the scattering analysis of *any* mechanical theory of point particles.

2. BOHMIAN MECHANICS

Bohmian mechanics does what not only Born found "a priori improbable," namely it shows that the introduction of additional parameters into the theory, represented by the Schrödinger equation (1), which "determine the individual event," is easily possible (see the essay of Dürr *et al.* 1996a): the integral curves of the velocity field

$$\mathbf{v}^{\psi_t}(\mathbf{x}) = \frac{\mathbf{j}^{\psi_t}}{\rho_t}(\mathbf{x}) = \operatorname{Im}\frac{\nabla\psi_t}{\psi_t}(\mathbf{x}), \tag{13}$$

which are solutions of

$$\frac{\mathrm{d}}{\mathrm{d}t}\mathbf{x}(t) = \mathbf{v}^{\psi_t}(\mathbf{x}(t)), \tag{14}$$

together with an initial position $\mathbf{x}_0$ determine the trajectory of the particle. (For a proof of the global existence of the solutions for general N-particle systems and a large class of potentials see Berndl 1996.) The initial position is distributed according to the quantum equilibrium probability $\mathbb{P}^\psi$ (ψ is normalized) with density $\rho = |\psi|^2$ (for a justification of "quantum equilibrium" see Dürr *et al.* 1992a).

Thus, in Bohmian mechanics a particle moves along a trajectory guided by the particle's wave function. Hence, given ψ_t, the solutions $\mathbf{x}(t, \mathbf{x}_0)$ of equation (14) are random trajectories, where the randomness comes from the $\mathbb{P}^\psi$-distributed random initial position $\mathbf{x}_0$, ψ being the initial wave function.

Consider now a region $G \subset \mathbb{R}^3$ and let $N^{\Sigma,\Delta}$ be the number of crossings of $\mathbf{x}(t)$ of subsets $\Sigma \subset \partial G$ in time intervals $\Delta \subset [0, \infty)$. Splitting $N^{\Sigma,\Delta} =: N_+^{\Sigma,\Delta} + N_-^{\Sigma,\Delta}$, where $N_+^{\Sigma,\Delta}$ denotes the number of outward crossings and $N_-^{\Sigma,\Delta}$ the number of backward crossings of Σ in Δ, we define for the number of "signed crossings" $N_s^{\Sigma,\Delta} := N_+^{\Sigma,\Delta} - N_-^{\Sigma,\Delta}$. By the very meaning of the probability flux it is rather clear (and it can easily be computed, Berndl 1995), that the expectated value of these numbers of crossings in quantum equilibrium is given by integrals of the current, namely

$$\mathbb{E}^{\Psi}(N^{\Sigma,\Delta}) = \int_{\Delta} \mathrm{d}t \int_{\Sigma} |\mathbf{j}^{\Psi_t} \cdot \mathbf{n}| \mathrm{d}\sigma \left(= \int \mathrm{d}^3\mathbf{x} |\psi(\mathbf{x})|^2 N^{\Sigma,\Delta}(\mathbf{x})\right) \tag{15}$$

$$\mathbb{E}^{\Psi}(N_s^{\Sigma,\Delta}) = \int_{\Delta} \mathrm{d}t \int_{\Sigma} \mathbf{j}^{\Psi_t} \cdot \mathbf{n} \mathrm{d}\sigma. \tag{16}$$

(This relation between the current and the expected number of crossings is also one of the fundamental insights used in the proof of global existence of solutions.)

3. SCATTERING ANALYSIS OF BOHMIAN MECHANICS

We want to analyze the *scattering regime* of Bohmian mechanics, i.e. the asymptotic behavior of the distribution of crossings of the trajectories traversing some distant surface surrounding the scattering center (see also Daumer 1995).

As surfaces we choose, for the sake of simplicity, spheres and we fix the notation illustrated in Figure 1. (We may imagine the surface surrounding the scattering center or, more generally, simply an area in which the particle happens to be.)

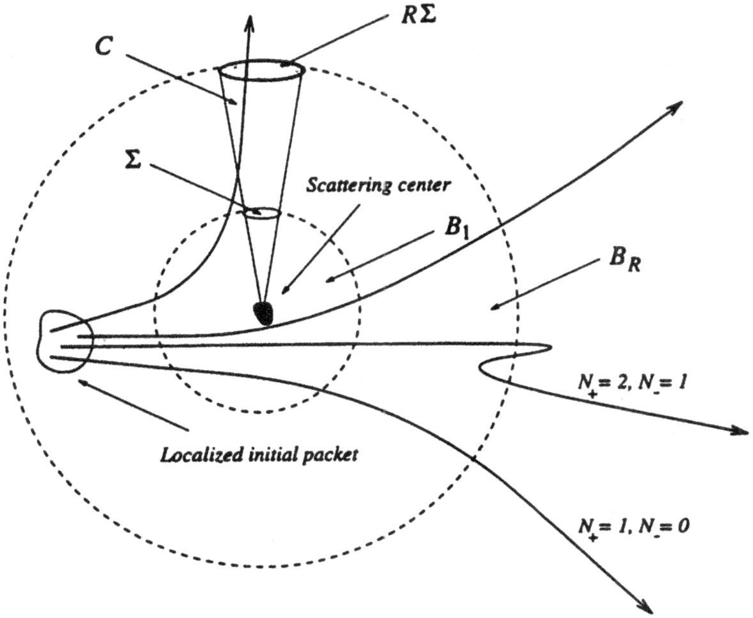

Fig. 1. Possible trajectories of a particle starting in the wave packet localized in the ball B_R.

We consider the random variables (functions of the paths) *first exit time* from B_R

(17) $t_e := \inf\{t \geq 0 | \mathbf{x}(t) \notin B_R\}$,

and the corresponding *exit position*

(18) $\mathbf{x}_e = \mathbf{x}(t_e)$.

Upon solving (14) and (1) the statistical distributions for t_e and $\mathbf{x}_e$ can of course be calculated (see, e.g., Leavens 1990; 1996). In general we should expect this to be a very hard task, but it turns out that if ∂B_R is at most crossed once by every trajectory – this is what we expect to happen asymptotically in the scattering regime – a very simple formula involving the current obtains.

The probability of the exit positions $\mathbb{P}^\Psi(\mathbf{x}_e \in R\Sigma)$ should become "independent" of R for large R such that we may focus on the map $\sigma^\Psi : \mathcal{B}(S^2) \to \mathbb{R}^+$ defined by

$$(19) \qquad \sigma^\Psi(\Sigma) := \lim_{R \to \infty} \mathbb{P}^\Psi(\mathbf{x}_e \in R\Sigma) = \lim_{R \to \infty} \mathbb{P}^\Psi\left(\frac{\mathbf{x}_e}{x_e} \in \Sigma\right),$$

where $\mathbf{x}_e/x_e$ is the exit direction, which we expect to be a probability measure on the unit sphere (if eventually all trajectories go off to infinity, as they should.) This measure gives us the asymptotic probability of outward crossings of a distant surface, certainly the quantity of interest for the scattering analysis of any mechanical theory of point particles and it seems appropriate to *define* σ^Ψ in (19) as the cross section measure (see also the last section).

How can we find a handy expression for this probability?

With formula (15) we have already a formula for the expected number of crossings and the expected number of signed crossings of the surface $R\Sigma$ in the time interval Δ. For large R the sphere ∂B_R should be crossed at most once, from the inside to the outside, such that the number of crossings equals the number of signed crossings, both being either 0 or 1, such that furthermore their expectation value equals the probability, that the particle has crossed the surface $R\Sigma$ at some time.

Hence, if there are asymptotically no backward crossings, i.e. if $\lim_{R \to \infty} \mathbb{E}^\Psi(N_-^{\partial B_R, [0, \infty)}) = 0$, we find for the asymptotic *probability* that a trajectory crosses the surface $R\Sigma$ from the inside to the outside

$$(20) \qquad \sigma^\Psi(\Sigma) = \lim_{R \to \infty} \mathbb{P}^\Psi(\mathbf{x}_e \in R\Sigma) = \lim_{R \to \infty} \int_0^\infty \mathrm{d}t \int_{R\Sigma} \mathbf{j}^{\Psi_t} \cdot \mathbf{n} \mathrm{d}\sigma.$$

This is very nice result, because it connects our (natural) definition of the cross section (19) with the quantity considered in the flux-across-surfaces-theorem (11).

Up to now the discussion leading to formula (20) has been completely general concerning the time evolution. Let us now process (20) further, taking the simplest case, namely free evolution. Our goal is to find a formula where the limit is taken.

The flux will contribute to the integral in (20) only for large times, because the packet has to travel a long time until it reaches the distant sphere ∂B_R such that we may use the long-time asymptotics of the free evolution. We use the well-known formula (Reed and Simon 1975, 59)

$$(21) \qquad \psi_t(\mathbf{x}) = (e^{-iH_0 t}\psi)(\mathbf{x}) = \int d^3y \, \frac{e^{i(|\mathbf{x}-\mathbf{y}|^2/2t)}}{(2\pi i t)^{3/2}} \, \psi(\mathbf{y}),$$

and obtain with the splitting

$$(22) \qquad \psi_t(\mathbf{x}) = \frac{e^{i(x^2/2t)}}{(it)^{3/2}} \, \hat{\psi}\left(\frac{\mathbf{x}}{t}\right) + \frac{e^{i(x^2/2t)}}{(it)^{3/2}} \int \frac{d^3y}{(2\pi)^{3/2}} \, e^{-i\frac{\mathbf{x}\cdot\mathbf{y}}{t}} (e^{i\frac{y^2}{2t}} - 1)\psi(\mathbf{y}),$$

neglecting the second term, as $t \to \infty$

$$(23) \qquad \psi_t(\mathbf{x}) \sim (it)^{-\frac{3}{2}} e^{i\frac{x^2}{2t}} \hat{\psi}\left(\frac{\mathbf{x}}{t}\right).$$

This asymptotics for scattering theory has since long been realized as important (e.g., Brenig and Haag 1959; Dollard 1969, who proved that the asymptotics (23) holds in the L_2 sense). From (23) we find for $t \to \infty$

$$(24) \qquad \mathbf{j}^{\psi_t}(\mathbf{x}) = \operatorname{Im}\psi_t^*(\mathbf{x})\nabla\psi_t(\mathbf{x}) \approx \frac{\mathbf{x}}{t} t^{-3} |\hat{\psi}\left(\frac{\mathbf{x}}{t}\right)|^2,$$

and note that the current and hence also the velocity is strictly radical for large times, i.e. parallel to the outward normal $\mathbf{n}$ of ∂B_R, reflecting our expectation that the expected value of backward crossings of ∂B_R vanishes as $R \to \infty$.

We now return to (20). Using the approximation (24) and substituting $\mathbf{v} := \mathbf{x}/t$ we arrive at

$$(25) \qquad \int_0^\infty dt \int_{R\Sigma} \mathbf{j}^{\psi_t} \cdot \mathbf{n} d\sigma \approx \int_0^\infty dt \int_{R\Sigma} t^{-3} |\hat{\psi}\left(\frac{\mathbf{x}}{t}\right)|^2 \frac{\mathbf{x}}{t} \cdot \mathbf{n}(\mathbf{x}) d\sigma$$

$$= \int_0^\infty dv v^2 \int_C d\Omega |\hat{\psi}(\mathbf{v})|^2 = \int_C d^3\mathbf{v} |\hat{\psi}(\mathbf{v})|^2.$$

This heuristic argument for the free flux-across-surfaces-theorem (12) is so simple and intuitive that it is indeed strange that it does not appear in any primer on scattering theory!

Now let us turn to the interacting case. Using Ikebe's eigenfunction expansion (8) and the relation of the generalized eigenfunctions with the wave operators we find that

$$(26) \qquad \psi_t(\mathbf{x}) = (2\pi)^{-\frac{3}{2}} \int d^3k e^{-ik^2 t/2} \phi(\mathbf{x}, \mathbf{k}) \widehat{\Omega_-^\dagger \psi}(\mathbf{k}).$$

The Lippmann–Schwinger equation (9) for $\phi(\mathbf{x}, \mathbf{k})$ allows us to split off the free evolution of $\Omega_-^{\hat{\dagger}}\psi$ and we obtain

$$\psi_t(\mathbf{x}) = e^{-iH_0t}\Omega_-^\dagger\psi \tag{27}$$

$$- (2\pi)^{-\frac{3}{2}} \int d^3k e^{-ik^2t/2}\, \Omega_-^{\hat{\dagger}}\psi(\mathbf{k}) \left(\frac{1}{2\pi}\int d^3y \frac{e^{-ik|\mathbf{x}-\mathbf{y}|}}{|\mathbf{x}-\mathbf{y}|} V(\mathbf{y})\phi(\mathbf{y},\mathbf{k})\right).$$

Now observe that the first term immediately gives with (25) the desired result

$$\lim_{R\to\infty} \int_0^\infty dt \int_{C\cap\partial B_R} \mathbf{j}^{\psi_t}\cdot \mathbf{n}d\sigma = \int_C d^3\mathbf{v}|\Omega_-^{\hat{\dagger}}\psi(\mathbf{v})|^2. \tag{28}$$

The other terms which appear in the flux should not contribute for the following reason: they all contain the phase factor $e^{-i(k^2t/2)-ikx}$, where we used $e^{-ik|\mathbf{x}-\mathbf{y}|} \approx e^{-ikx}$ for x far away from the range of the potential. This factor is rapidly oscillating for large x and t such that the k-integrals should decay fast enough in x and t to give no contribution to the flux across surfaces.

4. MORALS

Let us recollect what we have achieved from our Bohmian perspective.

The analysis of the scattering regime of Bohmian mechanics suggests a natural *definition* of the cross section measure as the asymptotic probability distribution of the exist positions, formula (19). This is very similar to the definition of the cross section in classical mechanics (e.g., Reed and Simon 1979, 15). The random distribution of impact parameters used in classical mechanics to define the cross section measure corresponds to the $|\psi|^2$ distribution of the initial positions of the particle. The "individual event" (cf. the Born quote at the beginning), that is the deflection of one particle in a certain direction, is indeed determined by the initial position. However, the initial positions are randomly distributed according to $|\psi|^2$ such that there is no way to control the individual event. What is relevant for a scattering experiment with a given wave function is the statistical distribution of these individual events and the flux-across-surfaces-theorem provides us with a formula for the asymptotic distribution.

Let us now examine how this formula may be connected with the usual operator formalism of quantum mechanics. The first step has already been done, by writing formula (19) in terms of wave operators instead of using the generalized eigenfunctions directly. (Existence of the wave operators and asymptotic completeness may appear as a by-product, once an eigenfunction expansion has been established (Green and Lanford 1960).) Note that formula (19) may be rewritten as

$$\sigma(\Sigma) = \int_C d^3\mathbf{v}|\Omega_-^{\hat{\dagger}}\psi(\mathbf{v})|^2 = (\psi, \Omega_-\mathcal{F}^{-1}P_C\mathcal{F}\Omega_-^\dagger\psi), \tag{29}$$

where P_C denotes the projection operator on the cone C and $\mathcal{F}$ is the unitary operator of the fourier transformation. The map $\Sigma \to \Omega_{-}\mathcal{F}^{-1}P_C\mathcal{F}\Omega_{-}^{\dagger}$ is an explicit example of what is called a projection operator valued measure which corresponds to a unique self-adjoint operator by the spectral theorem. This particular example of escape statistics exemplifies the general situation, namely that operators as "observables" appear merely as computational tools in the phenomenology of certain types of experiments – those for which the statistics of the result are governed by a projection operator valued measure.

Now we come to the end of the closing remarks of the Born quote at the beginning of this essay. Is it true that any deterministic completion of quantum mechanics must lead to the same formulae for the cross section as Born's formula?

Certainly not! Further assumptions would be required, e.g., the initial packet must be close to a plane wave, initially far away from the scattering center, in order to have a chance to arrive from formula (11) (or from Dollard's formula (10)) at Born's formula (5).

This is, however, not the strongest point we can make here, because Born's formula is not really taken seriously as the fundamental formula for the cross section nowadays; but Dollard's formula certainly is.

But what is the quantum-mechanical prediction if we go further and do not want to make such strong idealizations, e.g. if we happen to place de tectors around the scattering center which are *not* that far away? What are the predictions for the times and positions at which the detectors click? Dollard's formula doesn't apply here, because it is an asymptotic formula.

Maybe – after some reflection – one comes up with what seems a very natural candidate for the joint density of exit position and exit time, namely $\mathbf{j}^{\Psi_t}(\mathbf{x}) \cdot \mathbf{n}(\mathbf{x})d\sigma dt$. But further scrutinizing this answer within quantum mechanics should leave one uneasy. After all, this is a formula – a prediction – which is not at all a quantum-mechanical prediction of the common type: $\mathbf{j}^{\Psi_t}(\mathbf{x}) \cdot \mathbf{n}(\mathbf{x})$ can well be negative and thus the formula makes sense only for (very) particular wave functions, for which $\mathbf{j}^{\Psi_t}(\mathbf{x}) \cdot \mathbf{n}(\mathbf{x})$ is positive. But what is then the right formula for all the other wave functions? Why is there such a simply formula for some wave functions?

In Bohmian mechanics there is no need to feel uneasy. $\mathbb{P}^{\Psi}(\mathbf{x}_e, t_e) \in (d\sigma, dt))$ is the probability that the particle exits at $\mathbf{x}_e$ at time t_e – and that can be calculated in the usual way for *any* wave function. For some wave functions it turns out (Daumer *et al.* 1994) to be indeed given by

$$\text{(30)} \qquad \mathbb{P}^{\Psi}((\mathbf{x}_e, t_e) \in (d\sigma, dt)) = \mathbf{j}^{\Psi_t}(\mathbf{x}) \cdot \mathbf{n}(\mathbf{x})d\sigma dt,$$

for others it isn't. That's alright, isn't it?

Mathematisches Institut der Universität München

PETER R. HOLLAND

IS QUANTUM MECHANICS UNIVERSAL?

1. INTRODUCTION

The domain of validity of quantum mechanics is an unsolved problem of contemporary physics. The extraordinary success of the quantum treatments of mechanical and optical systems has led to a belief in the universality of the quantum description. That is, given suitable computational prowess, it is widely considered that there is no natural process to which the quantum theory cannot be successfully applied, up to and including the Universe.

Appliers of quantum mechanics to the Universe assign to it a wave function depending on all the variables associated with all the things the Universe comprises (assumed to be reducible to particles and fields). But it is questionable on empirical and theoretical grounds whether this is a meaningful concept. In particular, it is not established that the things the Universe is made from actually are quantum-mechanical in character. Are all the valid results of other sciences that describe the structure and evolution of matter really instances of an enveloping quantum description?

Obvious potential exceptions are the evolutionary and emergent processes characteristic of biological systems which seem to imply notions of causality not explicitly appearing in current physical theories. Bohr (1958b) propounded a non-reductionist view that physics does not underpin biology. Rather, he felt that biological and physical explanations are complementary in the sense that both are necessary for a complete account of the behaviour of living systems, but that the conditions under which biological and physical concepts may be applied are mutually exclusive. For example, the act of empirically investigating a bio-system using physical techniques will deprive it of the very quality that characterises it as biological, namely its 'life'.

In Bohr's formulation it would not make sense to apply quantum mechanics to a biological subject. Just this was attempted by Schrödinger (1944) who applied the theory directly to genetic processes by attributing to them well-defined quantum states. It seems this is necessary as key aspects of the process will be missed if the molecules are treated as just classical objects (Bohm 1969). It has been suggested that DNA may engage in behaviour akin to a quantum measurement process (Percival 1991; Home and Chattopadhyay 1996), but as yet there is no general method for analysing systems which somehow combine both quantum and classical characteristics. Indeed, there is evidence that the process of self-replication is incompatible with quantum laws (Wigner 1961; Wootters and Zurek 1982; for further discussion see Landsberg 1984). In a wider context it

J. T. Cushing et al. (eds.), Bohmian Mechanics and Quantum Theory: An Appraisal, 99–110.

has been argued that if one desires a physical explanation of the processes that seem to characterise 'consciousness', new physics beyond conventional quantum mechanics will be required (e.g., Penrose 1995).

These are open problems. The aim of this essay is not to address them but to argue that there are grounds for doubting the universality of quantum mechanics at a more elementary level, without going to complex life forms. In the more restricted arena of a Universe embracing purely mechanical phenomena, it is a problem to show how the macroscopic behaviour of matter, which we know is adequately described by classical mechanics, can emerge if the Universe is basically a quantum system. As we shall see, the de Broglie–Bohm picture implies such derivation is not generally possible because the formal structure of quantum mechanics is not rich enough to embrace the full variety of potential classical motions. This result, one of the most important new insights of the de Broglie–Bohm model, casts doubt on the assumption that quantum mechanics is a universal physical theory within the domain of phenomena to which physics is traditionally applied.

2. BOHR'S VIEW OF CORRESPONDENCE

The relation between new and superseded physical theories is a subtle and complex problem. Physicists have tended to work with the simple criterion that a new theory shall contain an old theory as a special case in those situations where the latter is valid. The new theory is then construed as 'better' because it corrects the errors of the old and extends the predictive power of science into previously unexplored domains. Along this line, a common textbook view in quantum mechanics is that classical mechanics is deducible in certain specified circumstances, for example when Planck's constant $\hbar$ is small relative to some other variable quantity having the dimension of action, and that this provides an unproblematic description of the relation between the two physical theories.

This notion of one theory containing another as a special case was not Bohr's view of the correspondence principle. Bohr's rendering was as follows (Bohr 1934, 85; for further discussion see Hanson 1958, 149–157, and Feyerabend 1978, 276). There is a domain, such as that of high quantum numbers, where quantum-mechanical quantities, such as energy, will approximately coincide with the corresponding classical quantities when a classical description is made of the same system. Bohr asserts the existence of a limiting domain of *numerical* coincidence, but warns that this limit in no way implies the gradual emergence of the classical theory, for in the limit the quantum postulate still maintains its significance. In short, numerical coincidence does not imply conceptual identity. The word 'energy' carries different connotations in the two theories and this difference in meaning is not diminished in the mathematical limit. Rather, the existence of this limit allows us to apply classical-mechanical pictures but these are independently invoked from a separate theoretical realm and not deduced from quantum mechanics. The system remains quantum mechanical.

This disjunction between the quantum and classical languages is in accord with Bohr's analysis of measurement. It will be recalled that for Bohr the quantum algorithm can be physically interpreted only with the aid of classical concepts. It is assumed that there is a level at which the latter are approximately correct and the approximately valid classical laws may be applied alongside the quantum laws. In the context of measurement Bohr (1948) characterises this domain in terms of the relative heaviness of the apparatus; when the observing device is sufficiently massive relative to the atomic objects of investigation, the fine quantal structure of the matter making up the device can be neglected and classical concepts brought in.

Thus, according to Bohr, the quantum and classical theories are conceptually independent yet connected by a numerical consistency condition, and both are necessary for a complete account of microphenomena. One (quantum) is needed to make predictions pertaining to the novel properties of matter, while the other (classical) provides the means to connect the predictions with empirical outcomes. This connection is possible because there is a realm in which we can view matter in two ways (classical and quantum), using mechanical notions or not.

This is a sophisticated world view whose subtlety is not captured by textbook criteria such as $\hbar \to 0$. It implies that more than one representative of a group of incompatible theories may be applied to the same class of physical processes, at least approximately. Philosophers of science have been sensitive to this possibility, but it does not often appear in physicists' accounts of the problem. We have recalled Bohr's attitude to correspondence because it displays a surprising concordance with the implications of the de Broglie–Bohm theory, as we shall see.

3. PROBLEMS WITH BOHR

We have seen that for Bohr all matter is essentially quantum mechanical but there is a domain where it is legitimate to introduce mechanical pictures alongside the quantum description. The basic technical difficulty with Bohr in this context is that he does not give any generally valid criterion to judge when we can expect the quantum and classical formalisms will approximately coincide so that we can apply the classical concepts. Which elements of the two descriptions should be compared to decide this issue? As we shall see presently, there is no basis to the claim that high quantum numbers, large mass or $\hbar \to 0$ are sufficient conditions to generate classical behaviour, except in very special cases.

It may be true that we obtain asymptotic agreement between observable quantities such as the energy levels and the classical energy for a hydrogen atom for high quantum numbers, but is this in itself sufficient to claim we have entered the domain of applicability of the classical description? Suppose we wish to describe the classical system corresponding to the quantum system in its quasi-classical limit in the Hamilton–Jacobi language, but that in the limit the classical Hamilton–Jacobi equation is not retrieved from the Schrödinger equation, as it must be if we are to achieve consistency between the quantum and classical descriptions. What justification is there for using 'classical pictures' in this case?

The lack of mathematical precision in Bohr's formulation is reflected in other conventional treatments. Solvers of the classical-limit problem who claim success when agreement is reached between just one pair of observable quantities such as energy levels, probability densities or scattering cross-sections have not finished the job. Numerical coincidence in one observable does not guarantee empirical equivalence for other observables, although this would be expected if the classical theory is applicable. For example, the quasi-classical, high-quantum-number energy may be associated with a probability density that is classically non-realisable (this happens for real stationary states). And, in any case, the mathematical coincidence of selected functions misses the physical point. The core of the classical description is a substantive corpuscle guided by forces in accordance with Newton's laws of motion. Recovering segments of the classical formalism does not constitute a derivation of the classical theory or a demonstration of consistency with the quantum theory, unless the classical law of motion is found (or some other expression of it such as appears in the Hamilton-Jacobi theory).

4. NEW CORRESPONDENCE PRINCIPLE

We may elucidate the difficulty with the mathematical aspect of Bohr's correspondence principle in a simple way as follows. Consider the Schrödinger equation for a single body of mass m moving in a potential V:

$$i\hbar \frac{\partial \psi}{\partial t} = \left(-\frac{\hbar^2}{2m}\nabla^2 + V\right)\psi. \tag{1}$$

Writing the wave function as $\psi(x, t) = R\exp(iS/\hbar)$, we may decompose (1) into two real equations coupling the real functions R and S:

$$\frac{\partial S}{\partial t} + \frac{1}{2m}(\nabla S)^2 + Q + V = 0 \tag{2}$$

$$\frac{\partial R^2}{\partial t} + \nabla \cdot \left(\frac{R^2 \nabla S}{m}\right) = 0. \tag{3}$$

Here $Q = (-\hbar^2/2m)\nabla^2 R/R$ is what is termed in the de Broglie–Bohm theory the 'quantum potential' (but for the present we just consider 'pure' quantum theory and do not introduce the trajectory concept).

It is often asserted (e.g., Messiah 1962, 223) that when $\hbar \to 0$, $Q \to 0$ and we recover from (2) the classical Hamilton–Jacobi equation. Taking this in combination with the conservation law (3) then yields, it is claimed, an ensemble of non-interacting classical particles whose statistical distribution is given by $R^2 = |\psi|^2$. This is stated to establish the conditions for the emergence of the 'classical limit of quantum mechanics'. In this limiting domain the law of motion of a mass point,

$$\dot{x} = \frac{\nabla S}{m}, \tag{4}$$

must be added to the equations (2) and (3).

Following our remarks in §3, this 'derivation' is open to the following four objections:

(a) The parameter dependence of Q is a delicate affair. In general Q does not become negligible when $\hbar \to 0$ (or when other parameters on which it depends take on special values) because R is a function of $\hbar$ (and the other parameters). Hence the classical limit of quantum mechanics is not mathematically characterised by the condition $\hbar \to 0$.

(b) Granted that there are at least some solutions of the Schrödinger equation for which $Q \to 0$ in some spacetime region or through a variation of parameters, it is not generally the case that we recover *generic* classical motions in the potential V.

(c) In this limiting case of the Schrödinger equation quantum mechanics remains valid and so the usual Born interpretation that $|\psi|^2$ represents the probability of measurement outcomes remains intact. This concept is logically distinct from the classical interpretation of R^2 in Eq. (3) which pertains to the likely actual location of the particle. In a domain of validity of quantum mechanics, a new interpretation of $|\psi|^2$ has been introduced that is not logically derivable from Born's interpretation.

(d) For the classical limit to be meaningful a corpuscle obeying the law (4) must be added to a limiting case of quantum mechanics. As discussed in §5, this entity cannot be derived from the wave function.

As described elsewhere in this volume, a natural solution to the problems (c) and (d) is obtained in the de Broglie–Bohm formulation which adopts the law (4) in the exact quantum limit. In particular, Born's interpretation follows as a particular instance of the assumption that $|\psi|^2$ describes the distribution of actual (pre-measurement) positions. This approach also provides a natural framework for comprehending the implications of (a) and (b) (see §§6,7). When it exists, the classical limit is obtained through the new correspondence principle $Q \to 0$. What results mathematically is a single-valued ensemble of classical orbits whose position distribution fixes its momentum distribution. This differs from traditional statements of correspondence such as $\hbar \to 0$ by focusing on the role of the quantum state.

The limit $\hbar \to 0$ may be a means of achieving the classical regime but it does not *characterise* it.

5. IDENTIFICATION OF THE PARTICLE WITH A PACKET

Sometimes attempts are made to derive classical behaviour from quantum mechanics by using the wave function as a model of the particle. This is legitimate insofar as the potential or time scale are chosen such that a wave function that is initially a localised packet retains its integrity. But such conditions are very exceptional; for most physical potentials of interest the wave function spreads and this expansion is significant in the lifetime of the process. It is hard to see how this behaviour can be reconciled with the empirical fact that particles are localised entities. Indeed the particle may be found far from the peak of the packet. An example is radioactive decay where the wave may take many centuries to leak out of the source but the emitted particle may be detected after only a few minutes. The wave is not itself giving a physical representation of the particle. Attempts to use other attributes of the packet, such as associating its *mean* motion with a classical orbit via Ehrenfest's theorem, are also problematic (Holland 1993, 254).

These conclusions remain valid even in cases where the spreading is negligible. For the wave function can be coherently split into a set of disjoint wave functions which are also approximately non-spreading and these may separate to great distances. Each of these sub-waves is associated with the same mass, charge, etc. as the original wave. It does not make sense to suppose that an initially localised particle may split into another set of 'particles', all having the same mass, charge, etc.

To conclude, the classical conception of the state of a physical system, the position and momentum of a corpuscle, is not contained in the quantal conception, i.e., the wave function, as a special case. They are unconnected concepts.

6. QUANTUM SYSTEMS WITH NO CLASSICAL ANALOGUE

Whether it is possible to obtain a negligible quantum potential for a given external classical potential depends on the choice of quantum state. Examples of the various possibilities have been given elsewhere (Holland 1993, chapter 6). For example, for real stationary states it is impossible to achieve $Q \to 0$, whatever the parameter values. Indeed, it rapidly becomes clear when one examines concrete cases that this correspondence limit is exceptional. Fortunately, there do exist examples where exact classical behaviour for the ensemble can be obtained; for the harmonic oscillator in a coherent state, the limit where the width of the packet is much smaller than the amplitude of oscillation does yield the classical oscillator. Lesser requirements, such as $Q \to 0$ in a restricted spacetime region or for a subensemble, can sometimes be satisfied, but for many quantum systems no hint of classical motion can be found. Correspondingly, no agreement will be found in observable quantities such as probabilities.

7. CLASSICAL SYSTEMS WITH NO QUANTUM ANALOGUE

We have just seen that classical mechanics does not generally emerge from the quantum description when limiting processes are applied to specific wave functions. Of course, we may be ingenious enough to select a quantum state that does imply the classical domain in some limit. Here we show that, even if this were possible (and it has never been demonstrated to be so), the classical behaviour that will be found can only be a subset of that which is allowed by classical laws. That is, there exist physically realisable classical systems that can never be reached as the limit of some quantum systems. Not all of classical mechanics where it is correct can be a special case of quantum mechanics.

If classical mechanics really is a limiting case of quantum mechanics, we should assign a quantal wave function to a classical ensemble. Being a limiting case of an exact quantal wave function, this 'classical wave function' will retain the usual mathematical properties of ψ, especially that it is a single-valued function of position with a conserved amplitude. But the property of single-valuedness, which implies that at each moment at most one trajectory may pass through each space point, is not a characteristic of generic classical ensembles. The quantum wave function is not rich enough to embrace such ensembles and we cannot therefore expect to solve typical statistical problems in classical mechanics using this tool.

A simple example that illustrates why the wave function is not a useful concept in classical physics is reflection from a one-dimensional infinite potential barrier. The barrier occupies the space $x > 0$ so the particle moves in the region $x < 0$. Let us consider first the behaviour of a classical object, say a ball, elastically reflected at the wall (Holland 1993, §2.6). An ensemble of possible trajectories having the same momentum but a range of initial positions is shown in Figure 1. It will be noted that each space point

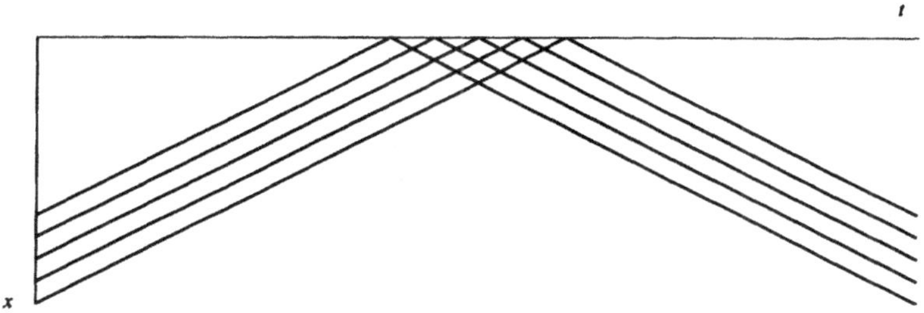

Fig. 1. Reflection of an ensemble of classical trajectories at an infinite barrier (along the t-axis).

supports two trajectories, one incoming (momentum p_0) and one outgoing (momentum $-p_0$). For a statistical ensemble the initial phase space distribution function is:

$$f_0(x, p) = \rho_0(x)\delta(p - p_0), \quad x < 0, \tag{5}$$

and zero for $x > 0$. At time t solution of the Liouville equation in the domain $x < 0$ yields the normalised function

$$f(x, p, t) = \rho_0(x - p_0t/m)\delta(p - p_0) + \rho_0(-x - p_0t/m)\delta(p + p_0). \tag{6}$$

Classical wave functions of the quantum type, that is, ones that have conserved amplitudes and whose phases obey the classical Hamilton–Jacobi equation, cannot reproduce this statistical distribution. To see this we associate the incident beam with an incoming wave function

$$\psi_i(x, t) = R_0(x - p_0t/m)e^{i(p_0x - Et)/\hbar}, \quad x < 0, \tag{7}$$

where $R_0(x) = +\sqrt{\rho_0(x)}$ and $E = p_0^2/2m$, and the reflected beam with the wave function

$$\psi_r(x, t) = R_0(-x - p_0t/m)e^{i - (p_0x - Et)/\hbar}, \quad x < 0. \tag{8}$$

Introducing the classical density matrix

$$\rho(x, x', t) = \psi_i(x)\psi_i^*(x') + \psi_r(x)\psi_r^*(x'), \tag{9}$$

the mean density in the domain $x < 0$ will be

$$\rho(x, x) = |\psi_i(x)|^2 + |\psi_r(x)|^2 \tag{10}$$

$$= \rho_0(x - p_0t/m) + \rho_0(-x - p_0t/m), \tag{11}$$

and the (local) mean momentum at each point is

$$\left(\frac{1}{\rho(x, x)}\right) \mathrm{Re} - i\hbar \left.\frac{\partial\rho(x, x')}{\partial x}\right|_{x = x'} = \frac{p_0(|\psi_i(x)|^2 - |\psi_r(x)|^2)}{|\psi_i(x)|^2 + |\psi_r(x)|^2} \tag{12}$$

$$= \frac{p_0[\rho_0(x - p_0t/m) - \rho_0(-x - p_0t/m)]}{\rho_0(x - p_0t/m) + \rho_0(-x - p_0t/m)}. \tag{13}$$

Formulas (11) and (13) are the correct expressions implied by projecting the phase space distribution (6) into the (one-dimensional) configuration space.

But, in order that the 'wave functions' (7) and (8) reproduce the correct distributions, they cannot at the same time be quantal wave functions. Although the phases satisfy the classical Hamilton–Jacobi equation and generate the correct motions via (4), the amplitudes do not obey the conservation law (3). Hence, the functions (7) and (8) do not correspond to the limit of Schrödinger wave functions in this potential.

The true Schrödigner wave function implies a qualitatively distinct ensemble of particle motions and there is no limit in which these reduce to the classical behaviour exhibited in Figure 1. Suppose the quantum theory is applied to the same system as above, i.e., the same mass and initial wave function $\psi_0(x) = \psi_i(x, 0)$ which we take to be a packet localised far from the wall. In this case the statistical problem is solved by *linearly* superposing the incident and reflected wave functions (as opposed to non-linearly combining them as in (9)). To compare as closely as possible with the classical case, we choose the parameters on which the packet depends and a time scale for the whole process so that we may neglect the spreading of the packet. The approximate quantal wave function in the domain $x < 0$ which satisfies the boundary condition $\psi(0, t) = 0$ for all t is then

$$(14) \qquad \psi(x, t) = (1/\sqrt{2})[\psi_i(x, t) - \psi_r(x, t)], \qquad x < 0.$$

This function exhibits interference effects in the region where ψ_i and ψ_r overlap. To plot the trajectories we need to specify the initial distribution $\rho_0(x)$ which we choose to be a Gaussian (the trajectories for the exact Gaussian wave function are given by Leavens and Aers 1993). For our purposes we require only a schematic illustration that depicts the essential characteristics of the ensemble. To this end we choose the parameters so that the interference terms may be neglected, as well as the spreading. The orbits are plotted in Figure 2. Although in this approximation there are no interference terms in the domain of overlap, there is still a finite quantum potential causing reflection of the paths.

The quantal trajectories differ individually from their classical counterparts in the same potential (Figure 1) and this implies that globally the ensembles are distinct. The segments of each trajectory far from the wall ($x \to -\infty$, where there is negligible overlap of the waves) coincide with classical rectilinear motion. But, as each particle approaches the barrier, it is turned away and, unlike the classical orbit, never reaches it. The particles in the leading edge of the packet spend longer in the vicinity of the barrier (in the exact case the individual quantum orbits differ in other ways, such as curving around nodes). This behaviour is a manifestation at the individual level of the single-valuedness of the wave function, for it implies that globally the quantal paths never cross.

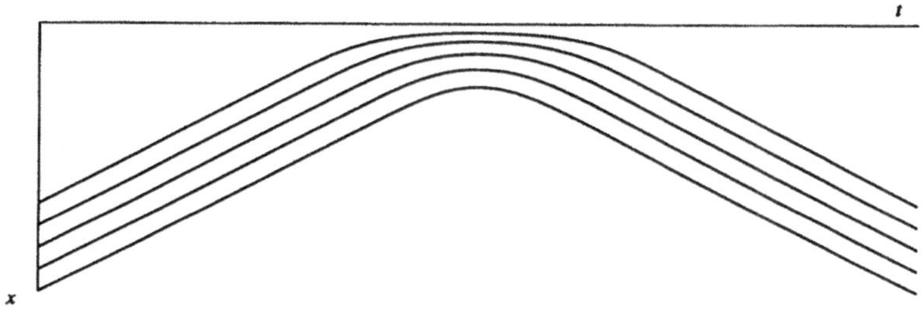

Fig. 2. Reflection of an ensemble of quantal trajectories having the same initial conditions as in Figure 1.

If quantum mechanics is the more basic theory, how can the single-valued pattern of Figure 2 turn into the multivalued pattern Figure 1 as a limiting case? Clearly, it cannot. The connection between the two patterns is that the quantal structure reproduces in this limit the local *mean* motion derived from the classical set of orbits. This is easy to see if we compute the momentum ∇S for the wave function (14) in the case of negligible interference terms. The result is the expression (12) for the classical local mean momentum field.

In demonstrating that quantum wave functions cannot account for generic classical behaviour, we have called attention to the role of the single-valuedness requirement in quantum mechanics, one manifestation of which is that the ensemble is globally restricted. Note, though, that the key issue here is not the absence of quantum multivalued trajectory fields *per se*. Rather, the congruent character of the ensemble associated with a pure quantum state is significant because it is an expression of the irreducible deviation of an individual quantum orbit from the corresponding classical orbit (having the same initial position and momentum and moving in the same classical potential). We can after all obtain multivalued quantum patterns by substituting a statistical mixture for the pure state (Holland 1993, 61, 227). But the use of a mixed state does not alleviate the basic difficulty because it does not change the way the quantum paths are calculated. Each orbit in the total ensemble is determined by one of the pure states making up the mixture, using the usual de Broglie–Bohm formula (4) for that pure state. So, the problem of getting a quantum particle to travel along a classical orbit remains the same, *viz.*, how to obtain a vanishing quantum potential for the pure state. As we have seen, a wave function cannot imply generic classical motions and so neither can a density matrix. Although the latter implies multivalued sets of orbits, it will not in general reproduce a classical ensemble.

From the perspective of this example (and that of many similar ones), we see that (the apparently correct part of) classical dynamics allows a different set of possible ensemble structures to that implied by quantum dynamics. Whatever the statistical state, quantum mechanics does not generate the full variety of valid classical motions. In the case of quantum mixtures the total ensemble is built from orbits each of which is an element of a single-valued family. In contrast, a single classical orbit will usually belong to several such families (classical Hamilton–Jacobi functions are typically multivalued). This point is illuminated by comparing the quantum and classical phase spaces in terms of their respective Lagrangian manifolds (for a clear introduction to this concept in the classical case see Littlejohn 1992).

8. CONCLUSION

The appearance of $\hbar$ in quantum mechanics and its absence in classical mechanics does not mean that the latter emerges from the former as a special case when $\hbar \rightarrow 0$. There are many other formal and informal elements distinguishing the two theories. The formal insight of the de Broglie–Bohm model is that quantum mechanics is at once a generalisation of classical mechanics (appearance of $\hbar$, linear superposition and the quantum potential) and a specialisation (to single-valued ensembles). Informally, the two theories possess some common descriptive elements (such as the orbit of a mass point), but they diverge fundamentally in their conception of physical states. For the de Broglie–Bohm theory an essential component of a system is a physical field mathematically described by Schrödinger's wave function. If in some limit we achieve a vanishing quantum potential and hence the laws describing a (single-valued) classical ensemble, the wave function retains its reality; we are just in a domain where it is not manifest. This concept differs fundamentally from a genuine classical wave function which describes only our knowledge of a situation, and not the actual situation.

From a quite different perspective we thus arrive at a conception of the quantum/classical interface similar to Bohr's. We have two theories of motion which are both necessary to account for physical phenomena. These theories agree in their description of motion for a certain class of phenomena (this class is different from and apparently smaller than that claimed by Bohr). But even for that class, they remain conceptually distinct. In short, quantum and classical mechanics are *different* physical theories with only limited numerical and conceptual overlap.

Unlike Bohr, this conclusion has been reached by analysing the dynamics of objective material objects rather than through an abstract analysis of their measurement. Bohr made his interpretation of quantum mechanics depend on classical iconography and this requires him to assert the existence of a limiting domain where the classical images can be applied. But he has no general method to decide when this iconography is valid. The advantage of the de Broglie–Bohm approach is that its interpretation of quantum mechanics is self-contained and does not require reference to other theories. That classical dynamics is not generally a special case has then no implications for the validity of the interpretation. Rather, this result represents an insight into the relation between the two theories which is not obtained in other approaches.

How far are these conclusions artefacts of the de Broglie–Bohm model which might go away in a different formulation of quantum mechanics? One way to gauge this is to drop the assumption that a material particle travels along the integral curves of the quantum-mechanical current. Then we are back to orthodox quantum mechanics and the particle tracks may be interpreted as attributes of the statistically-interpreted wave function, namely, flow liens of probability. Within orthodox quantum theory we may compare these flow lines with the analogous classical probability flow lines obtained by projecting the phase space flow into configuration space. In §7 we presented an example where (after several approximations were made) agreement was found between these two sets of lines. But in general, even at this statistical level, the two sets of flow lines differ fundamentally (for a worked example, a particle in a box, see Holland 1993, §6.5) (and the underlying particle orbits which generate them even more so). This suggests the problem is intrinsic to the quantum formalism and not an artefact of one particular interpretation.

It seems premature to embark on a study of the Universe conceived as a single quantum-mechanical entity when, even in the restricted realm of mechanical phenomena, quantum mechanics does not provide a universal description. One of the problems is that there is no natural limitation in the quantum theory telling us the scale or type of system to which it potentially applies. An implication of the above discussion is that, if we do desire a consistent universal theory in the mechanical sphere, then this should embrace the quantum and classical theories as special cases of a more general structure rather than trying to treat one as an instance of the other. In the context of the theory discussed in this paper, we need to develop a generalised phase space theory which includes as limiting cases both the multivalued classical Hamilton–Jacobi waves *and* the linear superposition principle.

Yet perhaps we should not always demand the mutual consistency of all branches of science, for we might exclude too early a theory from which we can still learn. The chaotic phenomena of classical mechanics, which have led to a renaissance of that subject, were discovered independently of quantum mechanics, not through studying its limiting cases. The notion that in any epoch there must be one dominant and universal theory may be an unwarranted imposition on science. As Whitehead observed, a clash of doctrines is not a disaster; it is an opportunity.

University of the West of England

C. RICHARD LEAVENS

THE "TUNNELING-TIME PROBLEM" FOR ELECTRONS

1. INTRODUCTION

The very concept of the motion of an electron during the time interval between its preparation in a given state and the subsequent "collapse" of the time-evolved state during a measurement is widely regarded as a meaningless one within conventional quantum mechanics. Despite this, a large number of theoretical papers, including several reviews (e.g., Hauge and Støvneng 1989; Landauer and Martin 1994), have been devoted to various characteristic times associated with the motion of a particle interacting with a potential barrier. Most of the approaches involve, at least implicitly, some relatively straightforward extension of the fundamental postulates of standard quantum theory. These will be referred to as "conventional" approaches to distinguish them from those based on alternatives to quantum theory, such as Bohmian mechanics. In this paper, the approach based on Bohm's theory is compared to several "conventional" methods that do not involve such radical departures from orthodoxy. In section 2, the characteristic times of interest are defined in words and the underlying reason for the difficulty in translating these words into unique, universally accepted mathematical expressions discussed. In Section 3, Bohmian mechanics is applied to the derivation of expressions for the mean dwell, transmission, reflection and arrival times. The systematic projector approach of Brouard *et al.* (1994) is considered in Section 4 where it is shown that none of the infinite number of possibilities for the mean transmission time that are generated by this method can be equal to the (unique) result from Bohmian mechanics. In Section 5, "conventional" probability current approaches are criticized from the point of view of Bohm's theory. In Section 6, the quantum clock of Salecker and Wigner (1958) is applied to the calculation of the mean and mean-square dwell time. The results for the free particle case (no barrier) are particularly illuminating. Concluding remarks are made in Section 7.

2. DEFINING THE PROBLEM

The often asked question "How long does it take to tunnel?" is at the same time both too vague (What precisely is meant by "it"?) and too specific (Why restrict one's attention to tunneling, i.e. propagation through a classically forbidden region?). The characteristic times of interest in this paper will be precisely defined (at least from the point of view of Bohmian mechanics) and no restriction will be made to the tunneling regime.

J. T. Cushing et al. (eds.), Bohmian Mechanics and Quantum Theory: An Appraisal, 111–129.

Consider an ensemble of a very large number of identically prepared single-particle one-dimensional scattering experiments. In each, a Schrödinger electron with the same initial wave function $\psi(z, t = 0)$ is incident from the left on the potential barrier $U(z) = V(z)\Theta(z)\Theta(d - z)$ assumed to be zero outside the range $0 \leq z \leq d$ [$\Theta(z) = 1$ for $z > 0$ and 0 for $z < 0$]. The mean transmission (reflection) time $\tau_{T(R)}(z_1, z_2)$ is defined as the average time spent in the region $z_1 \leq z \leq z_2$ subsequent to $t = 0$ by those electrons that are ultimately transmitted (reflected). The mean dwell time $\tau_D(z_1, z_2)$ is the average time spent by electrons between z_1 and z_2 irrespective of their ultimate fate. It is assumed throughout that the initial wave packet is normalized to unity and is localized far enough to the left of the barrier that the integrated probability density $|\psi(z, 0)|^2$ from z = Minimum[z_1, 0] to ∞ is negligibly small compared to the transmission probability $|T|^2$, usually by a factor of 10^{-4} or smaller.

The above three characteristic times are intrinsic quantities in the sense that there is no reference to a measuring apparatus in their definitions (other than the implied particle detector well outside the region of interest to determine whether or not an incident electron is eventually transmitted). Because of the position-momentum uncertainty relation it is clear that these intrinsic quantities cannot in general be measured directly by placing stop-watch type particle detectors at z_1 and z_2 (Cushing 1995a). The difficulty with calculating them using the standard formulation of quantum mechanics stems from the fact that there are no universally accepted hermitian operators $\hat{t}_X(z_1, z_2)$ associated with the time spent by a particle of type $X(= T, R$ or $D)$ in the region $[z_1, z_2]$.[1] As regards the mean arrival time $\tau_A(b)$ at the point $z = b$, Allcock (1969) has proven that there does not exist a complete set of measurement eigenstates associated with the time of arrival of a particle at a spatial point, not even for a free particle.

A frequently heard objection to the notion of transmission and reflection times is that they imply the existence of microscopically well-defined particle trajectories, a concept which is expressly forbidden in orthodox quantum mechanics on the grounds that it is impossible, even in principle, to observe such trajectories because of the position-momentum uncertainty relation. Such trajectories, although unobservable,[2] are of central importance in Bohm's causal alternative to quantum mechanics (Bohm 1952a,b; Bohm, Hiley and Kaloyerou 1987; Bell 1987a; Bohm and Hiley 1993; Holland 1993; Dürr *et al.* 1992a; Valentini 1991a,b; Albert 1994; Cushing 1994). In the next section, Bohm's theory is used to derive expressions for the characteristic times introduced above.

3. BOHM TRAJECTORY APPROACH

In Bohmian mechanics, tailored to the problem of interest here, it is postulated that an electron is an actual particle that is always accompanied by a field which probes the potential $U(z)$ and guides the particle's motion accordingly so that it has a well defined position $z(t)$ and velocity $v(t)$ at each instant of time. It is further postulated that the guiding field is the solution $\psi(z, t)$ of the time-dependent Schrödinger equation and that the particle's equation of motion is $v(t) \equiv dz(t)/dt = v(z, t)\,|_{z=z(t)}$ where the velocity field $v(z, t)$ is given by

$$(1) \qquad v(z, t) = \frac{J(z, t)}{|\psi(z, t)|^2},$$

with $J(z, t) \equiv (\hbar/m)\mathrm{Im}[\psi^*(z, t)\partial\psi(z, t)/\partial z]$ the probability current density. It follows from these postulates that $|\psi(z, t)|^2\, dz$ is, as assumed by Bohm, the probability of the electron *being* between z and $z + dz$ at time t (Dürr *et al.* 1992a; Valentini 1991a).

Now, given the initial wave function $\psi(z, t = 0)$ and position $z^{(0)} \equiv z(t = 0)$ of an electron, its subsequent motion is uniquely determined by simultaneous integration of the time-dependent Schrödinger equation for $\psi(z, t)$ and the equation of motion for $z(t)$ to obtain the Bohm trajectory $z(z^{(0)}, t)$. Uncertainty enters only through the probability distribution $|\psi(z^{(0)}, 0)|^2$ for the unknown initial position $z^{(0)}$ of the particle. Hence, the ensemble average of a particle property f, say, can be obtained by determining the value $f(z^{(0)})$ of the property for a particle following the trajectory $z(z^{(0)}, t)$ and then averaging over all $z^{(0)}$:

$$(2) \qquad \langle f(z^{(0)})\rangle_{\mathrm{Bohm}} \equiv \int_{-\infty}^{\infty} dz^{(0)} |\psi(z^{(0)}, 0)|^2 f(z^{(0)}).$$

This simple prescription is now applied to the various characteristic times defined above.

For a particle that is at $z = z^{(0)}$ at $t = 0$ the time that it spends thereafter in the region $[z_1, z_2]$ is given by the classical stop-watch expression

$$(3) \qquad t(z_1, z_2; z^{(0)}) = \int_0^{\infty} dt \int_{z_1}^{z_2} dz\, \delta[z - z(z^{(0)}, t)].$$

The mean dwell time is then given by

$$(4) \qquad \tau_D(z_1, z_2) = \langle t(z_1, z_2; z^{(0)})\rangle_{\mathrm{Bohm}} = \int_{-\infty}^{\infty} dz^{(0)} |\psi(z^{(0)}, 0)|^2\, t(z_1, z_2; z^{(0)})$$
$$= \int_0^{\infty} dt \int_{z_1}^{z_2} dz \int_{-\infty}^{\infty} dz^{(0)} |\psi(z^{(0)}, 0)|^2\, \delta[z - z(z^{(0)}, t)].$$

The innermost integral is just the probability distribution of particle positions at time t, i.e. $|\psi(z, t)|^2$, leading immediately to the widely accepted expression

$$(5) \qquad \tau_D(z_1, z_2) = \int_0^{\infty} dt \int_{z_1}^{z_2} dz\, |\psi(z, t)|^2.$$

An important property of Bohm trajectories $z(z^{(0)}, t)$ with different $z^{(0)}$ (but the same $\psi(z,0)$!) is that they do not intersect each other: if $z_i^{(0)} \neq z_j^{(0)}$, then $z(z_i^{(0)}, t) \neq z(z_j^{(0)}, t)$ for any t (Leavens and Aers 1993, 111; Holland 1993, 85). This fact allows the probability density $|\psi(z, t)|^2$ in

(5) to be decomposed into components $|\psi(z,t)|^2\Theta[z - z_c(t)]$ and $|\psi(z,t)|^2\Theta[z_c(t) - z]$ associated with transmission and reflection, respectively. The bifurcation curve $z_c(t)$ separating transmitted trajectories from reflected ones is the trajectory $z(z_c^{(0)}, t)$ given implicitly by

$$(6) \qquad |T|^2 = \int_{z_c(t)}^{\infty} \mathrm{d}z \, |\psi(z, t)|^2$$

where $|T|^2$ is the transmission probability (the reflection probability, $1 - |T|^2$, is denoted by $|R|^2$). Inserting $1 \equiv \Theta[z - z_c(t)] + \Theta[z_c(t) - z]$ into the integrand of (5) immediately gives

$$(7) \qquad \tau_D(z_1, z_2) = |T|^2 \, \tau_T(z_1, z_2) + |R|^2 \, \tau_R(z_1, z_2),$$

$$(8a) \qquad \tau_T(z_1, z_2) = \frac{1}{|T|^2} \int_0^{\infty} \mathrm{d}t \int_{z_1}^{z_2} \mathrm{d}z |\psi(z, t)|^2 \Theta[z - z_c(t)],$$

$$(8b) \qquad \tau_R(z_1, z_2) = \frac{1}{|R|^2} \int_0^{\infty} \mathrm{d}t \int_{z_1}^{z_2} \mathrm{d}z |\psi(z, t)|^2 \Theta[z_c(t) - z].$$

Within Bohmian mechanics, the average characteristic times $\tau_{X=T,R,D}(z_1, z_2)$ are uniquely defined and each is obviously a real, non-negative, additive[3] quantity. Including the effect of some proposed "measurement" device on the particle trajectories could very well lead to significant departures of the calculated characteristic times from the above intrinsic values but cannot lead to complex-valued or negative results. Hence, if the pointer positions of an ensemble of such devices indicate a negative result for one or more of the characteristic times τ_X then, *from the point of view of Bohmian mechanics*, the devices cannot be properly timing either the intrinsic or perturbed motion of the particles.

Probability distributions $P_X[t_X(z_1, z_2)]$ for the various times can be computed by a straightforward histogram method involving the calculation of a very large number of trajectories $z(z^{(0)}, t)$ and for each adding the weight $|\psi(z^{(0)}, 0)|^2$ to the channel determined by $t_X(z_1, z_2; z^{(0)})$. Several examples are given by Leavens and Aers (1993, 120–132). A much more efficient method, based on the probability current density $J(z, t)$, has been developed (McKinnon and Leavens 1995).

Since transmitted electrons are associated with the leading $|T|^2$ part of $|\psi(z, t)|^2$ it is clear that (8a) with $(z_1, z_2) = (0, d)$ will not in general agree with the mean transmission time $\tau_T(0, d)$ derived on the assumption that the peak (or centroid) of the incident wave packet evolves into the peak (or centroid) of the transmitted wave packet (Hauge, Falck and Fjeldly 1987). As is well-known (Hartman 1962, 3431) the latter approach can lead to superluminal (i.e. $d/\tau_T(0, d) > c$) and even negative values of $\tau_T(0, d)$.

For stationary-state scattering of "incident" electrons with precisely defined wave number $k > 0$ and energy $E \equiv \hbar^2k^2/2m$, expression (5) for the mean dwell time becomes (Leavens and Aers 1990, 60)

(9) $$\tau_D(k;\, z_1, z_2) = \frac{1}{J_{k,i}} \int_{z_1}^{z_2} dz |\psi_k(z)|^2,$$

where the stationary-state wave function $\psi_k(z)\exp(-iEt/\hbar)$ is normalized so that the incident probability current density $J_{k,i}$ associated with the plane wave $\exp(ikz)$ is $\hbar k/m$. Spiller *et al.* (1990) postulated that for $|T(k)|^2 > 0$ the corresponding mean transmission time is given by

(10) $$\tau_T(k;\, z_1, z_2) = \int_{z_1}^{z_2} \frac{dz}{v_k(z)},$$

where $v_k(z) \equiv J_k / |\psi_k(z)|^2$ with $J_k \equiv |T(k)|^2 J_{k,i}$ the stationary-state probability current density. Since $v_k(z)$ is independent of $z^{(0)}$ it follows that all the transmitted trajectories are parallel to each other and hence that $P_T[t_T(k;\, z_1, z_2)] = \delta[t_T(k;\, z_1, z_2) - \tau_T(k;\, z_1, z_2)]$ for $z_1 > -\infty$. It follows immediately from (9) and (10) that $\tau_T(k;\, z_1, z_2) = |T(k)|^{-2}\tau_D(k;\, z_1, z_2)$ and then, from (7), that $|R(k)|^2\tau_R(k;\, z_1, z_2) = 0$. Hence for $0 < |R(k)|^2 < 1$ it follows that $\tau_R(k;\, z_1, z_2) = 0$ and that $P_R[t_R(k;\, z_1, z_2)] = \delta[t_R(k;\, z_1, z_2)]$ for $z_1 > -\infty$. That, in the stationary-state limit, reflected electrons never enter the region $[z_1, z_2]$ can be counter-intuitive if one's intuition is based on pictures of the time evolution of wave packets. The author accordingly tends to regard the result as a peculiarity not of Bohmian mechanics but of the stationary-state limit itself, namely the assumed coherence of a single-particle state over all of space and time from $t = -\infty$.

For an example of the behaviour of $\tau_{X=T,R,D}(0, d)$, consider an initial gaussian wave function

(11) $$\psi(z, t = 0) = [2\pi(\Delta z)^2]^{-1/4} \exp\left[-\left(\frac{z - z_0}{2\Delta z}\right)^2 + ik_0 z\right],$$

with $E^{(0)} \equiv \hbar^2 k_0^2/2m = 5eV$ and $z_0 \cong -4.70\Delta z$ incident on a rectangular barrier $V_0\Theta(z)\Theta(d - z)$ with $V_0 = 2E_0$ and $d = 2.5$Å. Figure 1 shows the dependence of $\tau_D(0, d)$, $|T|^2\tau_T(0, d)$ and $|R|^2\tau_R(0, d)$ on the spread in wave number $\Delta k = 1/2\Delta z$ (Leavens *et al.* 1995). The parameters were chosen to illustrate the approach to the effective stationary-state regime $\Delta k/k_0 \lesssim 2^{1/2}e^{-2k_0 d}$ (for $E_0 = V_0/2$ and $|T(k_0)|^2 \ll 1$) in which $|R|^2\tau_R(0, d) \ll \tau_D(0, d)$ and $\tau_T(k_0;\, 0, d)$ is a good approximation to $\tau_T(0, d)$.

Now consider the distribution of arrival times at a point $z = b \geq d$ on the far side of the barrier *assuming that there are no reentrant Bohm trajectories through this point.* For $z^{(0)}$ in the range $[z_c^{(0)}, b]$ let $t(b;\, z^{(0)})$ denote the time at which an electron following the trajectory $z(z^{(0)}, t)$ reaches $z = b$. Trajectories with $z^{(0)} < z_c^{(0)}$ are reflected ones and, in keeping with the above assumption, never reach $z = b$. For these $t(b;\, z^{(0)})$ is a meaningless quantity (never is not equated with $+\infty$ in our analysis) and trajectories with $z^{(0)} < z_c^{(0)} \equiv z_c(0)$ are excluded from the outset. Those with $z^{(0)} \geq b$ should also be excluded but since the initial state is prepared so that trajectories with $z^{(0)} > b \geq d$ are of negligible importance, this fact will simply be

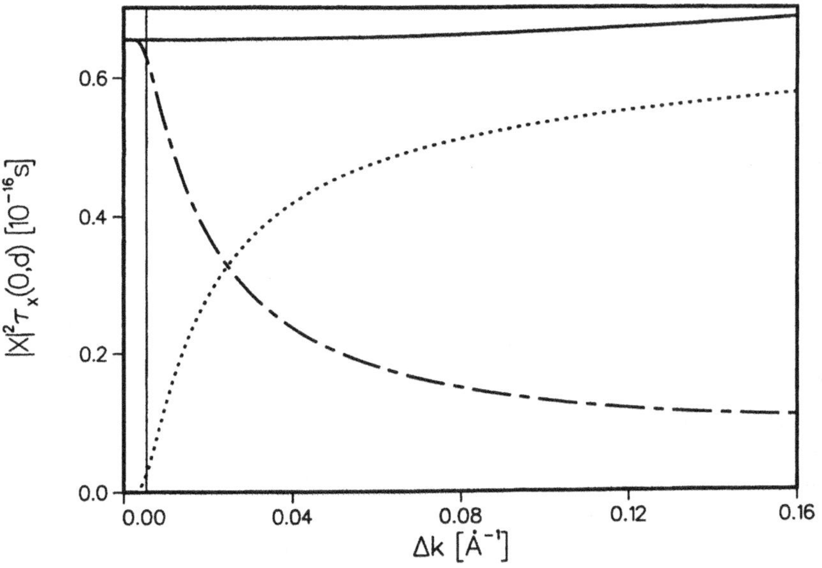

Fig. 1. Dependence on Δk of the transmission (— - —) and reflection (----------) contributions, $|T|^2\tau_T(0, d)$ and $|R|^2\tau_R(0, d)$ respectively, to the mean dwell time (———). The value of Δk below which (10) is expected to be good approximation is indicated by the vertical line. The parameters of the initial wave function and of the scattering potential are given in the text.

ignored. The distribution of arrival times at $z = b$ is then given by

$$(12) \qquad P_A[t(b)] = \left(\int_{z_c^{(0)}}^{\infty} dz^{(0)} |\psi(z^{(0)}, 0)|^2\right)^{-1} \int_{z_c^{(0)}}^{\infty} dz^{(0)} |\psi(z^{(0)}, 0)|^2 \delta[t(b) - t(b; z^{(0)})].$$

From (6), the first factor is $|T|^{-2}$. Now

$$(13) \qquad \delta[z(z^{(0)}, t) - b] = \frac{\delta[t - t(b; z^{(0)})]}{|dz(z^{(0)}, t)/dt|} = \frac{\delta[t - t(b; z^{(0)})]}{|v[z(z^{(0)}, t), t]|}\ .$$

There is only one term because of the assumption that for $z_c^{(0)} \leq z^{(0)} \leq b$ the trajectory $z(z^{(0)}, t)$ passes through $z = b$ once and only once for $t > 0$. Hence,

$$(14) \qquad P_A[t(b)] = \frac{|v[b, t(b)]|}{|T|^2} \int_{z_c^{(0)}}^{\infty} dz^{(0)} |\psi(z^{(0)}, 0)|^2 \delta[z(z^{(0)}, t(b)) - b]$$
$$= \frac{v[b, t(b)]\ |\psi[b, t(b)]|^2}{|T|^2} = \frac{J[b, t(b)]}{|T|^2}\ ,$$

using (1). The absolute value sign has been dropped because $v[b, t(b)]$ can not be negative when there are no reentrant trajectories through

$z = b > d$. When this restrictive assumption is removed one obtains after a more lengthy analysis

$$P[t(b)] = \frac{|J[b, t(b)]|}{\int_0^\infty dt\, |J(b, t)|}. \tag{15}$$

The mean arrival time $\tau_A(b)$ is obtained by integrating $t(b)P_A[t(b)]$ over $t(b)$ from 0 to ∞. When (14) is used for $P_A[t(b)]$ one immediately obtains the result derived by Dumont and Marchioro (1993) using "conventional" quantum mechanics. That their result is obtained using (14) rather than (15) suggests that their derivation contains an implicit assumption corresponding to the above one of no reentrant trajectories through $z = b$. Such an assumption can be difficult to articulate without the concept of *actual* particle trajectories. This concept is not an *explicit* ingredient of the "conventional" approaches considered in the next three sections. For each method, the reader should judge for himself whether or not it slips into the interpretation of the mathematics as an implicit assumption.

4. THE SYSTEMATIC PROJECTOR APPROACH

Brouard *et al.* (BSM) (1994) have developed a systematic approach to the determination of a class of possible expressions for τ_T and τ_R. It is based on the following projections:

$$\hat{D}(z_1, z_2)\psi(z, t) \equiv \Theta(z - z_1)\Theta(z_2 - z)\psi(z, t), \tag{16}$$

$$\hat{T}\psi(z, t) \equiv \psi_T(z, t), \qquad \hat{R}\psi(z, t) \equiv \psi_R(z, t). \tag{17}$$

The operator $\hat{D}(z_1, z_2)$ associated with the question "Is the particle between z_1 and z_2?" projects from the wave function $\psi(z, t)$ at any time t the part that is "located" in that region. The operator $\hat{T}$ associated with the question "Will the particle ultimately be transmitted?" projects from $\psi(z, t)$ the part which will have only positive wave number components at $t = +\infty$, i.e. the part $\psi_T(z, t)$ associated with transmission. $\hat{R} \equiv \hat{1} - \hat{T}$ projects from $\psi(z, t)$ the part $\psi_R(z, t)$ associated with reflection. For $|T|^2 < 1$, $\psi_T(z, t = 0)$ has two components, one to the left and one to the right of the barrier. For an opaque barrier the latter component completely dominates the former. [Figures 10 and 11 of Leavens (1995b) show snapshots of $\psi_T(z, t)$ and $\psi_R(z, t)$, for the case of an initial gaussian wave packet incident on a rectangular barrier, at a sequence of times t spanning the duration of the scattering process.]

The probability of *finding* that an electron is in the region $z_1 \leq z \leq z_2$ at time t is obtained by integrating $\psi^*(z, t)\, \hat{D}(z_1, z_2)\psi(z, t)$ over all z. This is standard. Integration of the resulting expression over t from 0 to ∞ immediately gives the right-hand-side of (5) for the mean dwell time $\tau_D(z_1, z_2)$. Experimentally, the latter integration would apparently require either strong (i.e. instantaneous, wave function collapsing) $\hat{D}$ measurements on a very large number of ensembles, one for each time t in the summation, or temporally extended weak (i.e. non wave function collapsing) measurements on a single ensemble. Neither approach would likely be widely accepted at the present time as part of conventional quantum mechanics.

Now, consider the following two resolutions of $\hat{D}$ into two parts, one associated only with transmission and one only with reflection:

$$\text{(18)} \qquad \hat{D} = \begin{cases} \hat{1}\hat{D} = (\hat{T} + \hat{R})\hat{D} = \hat{T}\hat{D} + \hat{R}\hat{D}, \\ \hat{D}\hat{1} = \hat{D}(\hat{T} + \hat{R}) = \hat{D}\hat{T} + \hat{D}\hat{R}. \end{cases}$$

Since $\hat{T}$ and $\hat{D}$ do not commute with each other there is no *a priori* reason for interpreting either the spatial integral over all z of $\psi^*(z, t)\hat{D}(z_1, z_2)\hat{T}\psi(z, t)$ or of $\psi^*(z, t)\hat{T}\hat{D}(z_1, z_2)\psi(z, t)$ as the probability of finding that a "to be transmitted" electron is in the region $z_1 \le z \le z_2$ at time t. BSM acknowledged that the proper response to this dilemma could very well be to regard the concept of transmission time as a meaningless one. However, they adopted the positive response of investigating the properties of the mean transmission times associated with different combinations of $\hat{T}$ and $\hat{D}$ such as $\hat{T}\hat{D}$, $\hat{D}\hat{T}$ and the (hermitian) symmetrized operator $(\hat{T}\hat{D})_{\text{symm}} = (\hat{T}\hat{D} + \hat{D}\hat{T})/2$. Since $\hat{D}$ is a projection operator ($\hat{D}^2 = \hat{D}$) this approach can be continued indefinitely by systematically inserting one or more factors of $\hat{1} \equiv \hat{T} + \hat{R}$ into $\hat{D}^n$ with $n = 1, 2, \ldots$ and leads to an infinite number of resolutions of $\hat{D}$ and an infinite number of possible mean transmission, reflection and interference times:

$$\text{(19)} \qquad \hat{D} = F(\hat{T}, \hat{D}) + F(\hat{R}, \hat{D}) + G(\hat{T}, \hat{R}, \hat{D}),$$

$$\text{(20a)} \qquad \tau_T^{F(T,D)}(z_1, z_2) = \frac{1}{|T|^2}\int_0^{\infty}\mathrm{d}t \int_{-\infty}^{\infty}\mathrm{d}z\ \psi^*(z, t)F(\hat{T}, \hat{D})\psi(z, t),$$

$$\text{(20b)} \qquad \tau_R^{F(R,D)}(z_1, z_2) = \frac{1}{|R|^2}\int_0^{\infty}\mathrm{d}t \int_{-\infty}^{\infty}\mathrm{d}z\ \psi^*(z, t)F(\hat{R}, \hat{D})\psi(z, t),$$

$$\text{(20c)} \qquad \tau_{T-R}^{G(T,R,D)}(z_1, z_2) = \int_0^{\infty}\mathrm{d}t \int_{-\infty}^{\infty}\mathrm{d}z\ \psi^*(z, t)G(\hat{T}, \hat{R}, \hat{D})\psi(z, t),$$

satisfying

$$\text{(21)} \qquad \tau_D(z_1, z_2) = |T|^2\tau_T^{F(T,D)}(z_1, z_2) + |R|^2\tau_R^{F(R,D)}(z_1, z_2) + \tau_{T-R}^{G(T,R,D)}(z_1, z_2).$$

BSM showed that some of the simplest resolutions of $\hat{D}$ lead to mean transmission and reflection times that are identical to ones obtained by a variety of well-known approaches. For example,

$$\text{(22a)} \qquad \tau_T^{TD}(z_1, z_2) = \frac{1}{|T|^2}\int_0^{\infty}\mathrm{d}t \int_{z_1}^{z_2}\mathrm{d}z\ \psi_T^*(z, t)\psi(z, t),$$

$$\text{(22b)} \qquad \tau_R^{RD}(z_1, z_2) = \frac{1}{|R|^2}\int_0^{\infty}\mathrm{d}t \int_{z_1}^{z_2}\mathrm{d}z\ \psi_R^*(z, t)\psi(z, t),$$

are equal to the complex-valued mean transmission and reflection times derived by Sokolovski and co-workers (Sokolovski and Baskin 1987; Sokolovski and Connor 1990) using the (appropriately weighted) Feynman paths connecting the space-time points ($z < 0$, $t = 0$) and ($z' > d$, t_∞) for transmission or ($z' < 0$, t_∞) for reflection to decompose the mean dwell time. (The time t_∞ must be large enough that the scattering process is essentially completed.) There is no interference term in this case and the sum rule (7) is satisfied exactly.

BSM also showed that $\tau_T^{(TD)\mathrm{symm}} = \mathrm{Re}[\tau_T^{TD}]$ and $\tau_R^{(RD)\mathrm{symm}} = \mathrm{Re}[\tau_R^{RD}]$ which are identical to the mean transmission and reflection times derived by Rybachenko (1967) using the Larmor precession of an electron's spin in a magnetic field as a clock (Baz' 1967).[4] They are also identical to the results of Iannaccone and Pellegrini (1994) based on a different Feynman path integral approach, of Leavens and McKinnon (1994) using the Salecker–Wigner (1958) clock, and of Steinberg (1995a; 1995b) based on the weak measurement theory of Aharonov and his coworkers (Aharonov *et al.* 1988; Aharonov and Vaidman 1990). Steinberg's work is especially noteworthy for two reasons: (1) Giving the post-selected state $\psi_T(z, t_\infty)$ or $\psi_R(z, t_\infty)$ the same status as the initially prepared state $\psi(z, 0)$ automatically leads within weak measurement theory to the unique choice $F(\hat{T}, \hat{D}) = \hat{T}\hat{D}$, $F(\hat{R}, \hat{D}) = \hat{R}\hat{D}$, $G(\hat{T}, \hat{R}, \hat{D}) = 0$. (2) The theory reveals clearly the physical meaning of the real *and* imaginary parts of (22). The real part of τ_T^{TD} is the average elapsed time indicated by the pointers on those stop-watches belonging to the subensemble for which the scattered electron is transmitted; the imaginary part is proportional to the average momentum shift $\langle \Delta P_{\mathrm{pointer}} \rangle_T$ of these pointers arising from the assumed particle-pointer interaction $\lambda t_\infty^{-1} \Theta(t)\Theta(t_\infty - t)\hat{D}(z_1, z_2)\hat{P}_{\mathrm{pointer}}$. This average momentum shift goes to zero in the limit of vanishingly small precision in the initial position of the pointer.

As a final example, the decomposition $\hat{D} = \hat{1}\hat{D}\hat{1} = (\hat{T} + \hat{R})\hat{D}(\hat{T} + \hat{R}) = \hat{T}\hat{D}\hat{T} + \hat{R}\hat{D}\hat{R} + \hat{T}\hat{D}\hat{R} + \hat{R}\hat{D}\hat{T}$ leads to the mean transmission, reflection and interference times introduced by Muga *et al.* (1992) and van Tiggelen *et al.* (1993).

An important and useful property of the mean transmission and reflection times generated by the systematic approach of BSM (1994) is expressed by

$$\text{(23a)} \quad \tau_T^{F(T,D)}(z_1, z_2) = \frac{1}{|T|^2} \int_0^\infty \frac{dk}{2\pi} |\varnothing(k)|^2 \, |T(k)|^2 \, \tau_T^{F(T,D)}(k; z_1, z_2),$$

$$\text{(23b)} \quad \tau_R^{F(R,D)}(z_1, z_2) = \frac{1}{|R|^2} \int_0^\infty \frac{dk}{2\pi} |\varnothing(k)|^2 \, |R(k)|^2 \, \tau_R^{F(R,D)}(k; z_1, z_2),$$

where $\varnothing(k)$ is the Fourier transform of $\psi(z, 0)$, $|T(k)|^2$ and $|R(k)|^2$ are the transmission and reflection probabilities, and $\tau_T^{F(T,D)}(k; z_1, z_2)$ and $\tau_R^{F(R,D)}(k; z_1, z_2)$ the mean transmission and reflection times for the stationary-state scattering problem. The derivation of (23a) and (23b) is critically dependent on the fact that the integrands of (22a) and (22b) are bilinear in $\psi(z, t)$. This is not the case for the integrands of (8a) and (8b) for the corresponding Bohm trajectory results where $\psi(z, t)$ enters non-bilinearly because $z_c(t)$ defined by (6) is itself an implicit functional of $\psi(z, t)$. Hence, equations of the form (23a) and (23b) do not apply to the average Bohm trajectory transmission and reflection times.[5] This means that there is no conceivable choice of F for which $\tau^F_{T(R)}$ is equal to expression (8a(b)) for $\tau_{T(R)}$ for all initial wave packets and potential barriers. This has been confirmed (Leavens 1995a, 91) by an explicit counter example to the conjecture that such a choice exists.

Now consider the special case $F(T, D) = (TD)_{\text{symm}}$ and $F(R, D) = (RD)_{\text{symm}}$ for the stationary-state scattering of electrons by a rectangular barrier. It is well-known (Rybachenko 1967) that $\tau_T^{(TD)\text{symm}}(k;\, 0,\, d)$ is essentially independent of d in the opaque barrier regime $\kappa d \gg 1$, with $\kappa \equiv [2m(V_0 - E)]^{1/2}/\hbar$, where $|T(k)|^2 \propto \exp(-2\kappa d)$. Calculations of $\tau_T^{(TD)\text{symm}}(k;\, 0,\, z)$ for such a barrier imply that transmitted electrons move through the interior ($\kappa^{-1} \lesssim z \lesssim d - \kappa^{-1}$) of the barrier exceedingly quickly and that superluminal speeds occur at the center $z = d/2$ for quite ordinary barrier parameters (e.g. $d \gtrsim 6\text{Å}$ for $V_0 = 2E = 10eV$) (Leavens and Aers 1990, 66).[6] An even more surprising result (Leavens and McKinnon 1994, 16) is obtained by considering reflected electrons in the region $[z_1 = d,\, z_2 = z]$ on the far side of the barrier. For the special case $V_0 = 2E$ it is easy to show that $\tau_R^{(RD)\text{symm}}(k;\, d,\, z) \le 0$. There is no doubt that the stop-watch pointers for the reflected electron subensemble would, on average, be displaced backwards if the corresponding weak $\hat{D}$ measurement were to be carried out on each system. However, in the author's opinion this should not be interpreted to mean that a reflected electron, either as an actual particle or even as encoded information, can spend a negative amount of time in the region $[d, z]$.

For the non-stationary case, Steinberg (1995b) has suggested that the complex-valued quantities $\psi_T^*(z, t)\psi(z, t)$ and $\psi_R^*(z, t)\psi(z, t)$ appearing in (22) as a result of the decomposition $|\psi|^2 = (\psi_T + \psi_R)^*\psi$ be considered as "elements of reality". For $|T|^2 < 1$, both $\psi_T^*(z, 0)\psi(z, 0)$ and $\psi_R^*(z, 0)\psi(z, 0)$ are very much more strongly localized on the left side of the barrier than are $\psi_T(z, 0)$ and $\psi_R(z, 0)$. Figure 2 shows three snapshots of $\text{Re}[\psi_R^*(z, t)\psi(z, t)]$ for z inside and immediately to the right of a thin rectangular barrier with $d = 1.5\text{Å}$ and $V_0 = 2E_0 = 10\, eV$ (the centroid and width of the initial incident gaussian are $z_0 = -38.9\text{Å}$ and $\Delta z = 6.25\text{Å}$). The oscillations about zero for $z \ge d$ are of decreasing amplitude with the first one negative. This is the origin of the negative values of $\tau_R^{(RD)\text{symm}}(d, z) = \text{Re}[\tau_R^{RD}(d, z)]$ obtained from (22b).

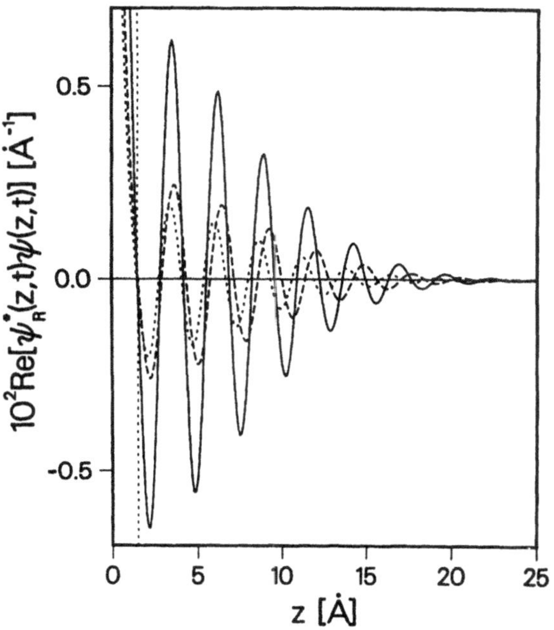

Fig. 2. The "element of reality" $\mathrm{Re}[\psi_R^*(z, t)\psi(z, t)]$ for $z > 0$ and $t = nt_0/4$ [$t_0 \equiv -z_0/(\hbar k_0/m)$] with $n = 3$ (----------), 4 (———) and 5 (———). The parameters are given in the text.

5. "CONVENTIONAL' PROBABILITY CURRENT DENSITY APPROACH

It has been postulated (Olkhovsky and Recami 1992; Muga, Brouard and Sala 1992) that the mean transmission and reflection times for the barrier region can be expressed in terms of the probability current density $J(z, t)$ as

$$\text{(24a)} \quad \tau_T^J(0, d) = \frac{\int_0^\infty \mathrm{d}t\, t\, J_+(d, t)}{\int_0^\infty \mathrm{d}t\, J_+(d, t)} - \frac{\int_0^\infty \mathrm{d}t\, t\, J_+(0, t)}{\int_0^\infty \mathrm{d}t\, J_+(0, t)},$$

$$\text{(24b)} \quad \tau_R^J(0, d) = \frac{\int_0^\infty \mathrm{d}t\, t\, J_-(0, t)}{\int_0^\infty \mathrm{d}t\, J_-(0, t)} - \frac{\int_0^\infty \mathrm{d}t\, t\, J_+(0, t)}{\int_0^\infty \mathrm{d}t\, J_+(0, t)}.$$

The first term on the right side of (24a) is interpreted as the mean exit time of particles from the barrier through its right edge at $z = d$ and the first term on the right side of (24b) as the mean exit time through its left edge at $z = 0$. The second term on the right side of both expressions is interpreted as the mean time at which incident particles enter the barrier.

The right-going and left-going components $J_+(z, t)$ and $J_-(z, t)$ respectively, of the probability current density are not uniquely defined within conventional quantum mechanics and many would accordingly regard them as ill-defined or meaningless quantities. Olkhovsky and Recami (1992) postulated that

$$(25) \qquad J_\pm(z, t) = J(z, t)\,\Theta\,[\pm J(z, t)],$$

while Muga *et al.* (1992) assumed that

$$(26) \qquad J_+(z, t) = \int_0^\infty \mathrm{d}p \left(\frac{p}{m}\right) f_W(z, p, t),$$

$$J_-(z, t) = \int_{-\infty}^0 \mathrm{d}p \left(\frac{p}{m}\right) f_W(z, p, t),$$

where $p \equiv \hbar k/m$ is free particle momentum and $f_W(z, p, t)$ is the Wigner function. In both cases $J(z, t) = J_+(z, t) + J_-(z, t)$ with no interference term.

The last term of (24a) is identical to the last term of (24b). Olkhovsky and Recami and also Muga *et al.* considered this a strong point in favour of the approach because it reflects the fact that within conventional quantum mechanics one cannot separate $J_+(0, t)$ into "to be transmitted" and "to be reflected" components $[J_+(0, t)]_T$ and $[J_+(0, t)]_R$. Olkhovsky *et al.* (1994) argue that this non-separability arises from quantum interference terms in $J_+(0, t)$. [Straightforward substitution of the wave function decomposition $\psi = \psi_T + \psi_R$ of the projector approach into the expression for J leads to $J = J_T + J_R + J_{T-R}$ with a $T - R$ interference term J_{T-R} that is nonzero in general.] However, the last term of (24a) and (24b), being a quotient of two integrals involving $J_+(0, t)$, is of the form $[A_T + A_R + A_{T-R}]/[B_T + B_R + B_{T-R}]$ and hence is in general non-separable even if the interference terms A_{T-R} and B_{T-R} are zero. This means that (24a) and (24b) do not apply in general even to the corresponding classical scattering problem as is clearly shown by the simple example constructed by Delgado *et al.* (1995).

It is instructive to look at (24a) and (24b) from the point of view of Bohmian mechanics. Although innumerable possibilities for the decomposition $J = J_+ + J_-$ have been suggested (Leavens 1995a, 92), Bohm's theory allows only one of these, namely the decomposition (25) postulated by Olkhovsky and Recami. This follows readily from the fact that within Bohmian mechanics only a single particle trajectory contributes to $J(z, t) \equiv |\psi(z, t)|^2\, v(z, t)$ at each space-time point (z, t) because of the non-intersection property. Moreover, the existence of deterministic non-intersecting trajectories also leads to a unique decomposition of $J(z,t)$ into components associated only with transmission or reflection:

$$(27) \qquad J(z, t) \equiv [J(z, t)]_T + [J(z, t)]_R$$

with

$$(28) \qquad [J(z, t)]_T = J(z, t)\Theta[z - z_c(t)], \qquad [J(z, t)]_R = J(z, t)\Theta[z_c(t) - z].$$

Now, the right-hand-side of expression (5) for the dwell time can be written in terms of $J(z, t)$ by multiplying the continuity equation $\partial|\psi(z, t)|^2/\partial t + \partial J(z, t)/\partial z = 0$ by t and then integrating over t from 0 to infinity and over z from z_1 to z_2 to obtain (Jaworsky and Wardlaw 1988)

$$\tau_D(z_1, z_2) = \int_0^\infty \mathrm{d}t\; t[J(z_2, t) - J(z_1, t)]. \tag{29}$$

Substitution of (27) into (29) immediately gives the Bohm trajectory mean transmission and reflection times in the form[7]

$$\tau_T(z_1, z_2) = \frac{1}{|T|^2}\int_0^\infty \mathrm{d}t\; t\{[J(z_2, t)]_T - [J(z_1, t)]_T\}, \tag{30a}$$

$$\tau_R(z_1, z_2) = \frac{1}{|R|^2}\int_0^\infty \mathrm{d}t\; t\{[J(z_2, t)]_R - [J(z_1, t)]_R\}. \tag{30b}$$

Within Bohmian mechanics the mean time at which "to be transmitted" electrons enter the barrier is not equal to the mean time at which "to be reflected" electrons do. For example, for the case in which both transmitted and some reflected trajectories enter the barrier (i.e. for $\Delta k/k_0$ not too small), the simplest situation involves two times t_T and t_R with $t_T < t_R$ such that only transmitted trajectories enter the barrier for $0 \le t \le t_T$, only reflected trajectories enter for $t_T \le t \le t_R$, and none enter for $t > t_R$. Clearly, for this case the mean time at which a transmitted particle enters the barrier is less than the mean time of entrance for all particles that enter the barrier. This means that the subtracted term in (24a) is too large and can lead to anomalously small and even negative values of $\tau_T^J(0, d)$ (Leavens 1993; Delgado *et al.* 1995).

6. THE QUANTUM CLOCK OF SALECKER AND WIGNER

The quantum clock devised by Salecker and Wigner (1958) was applied by Peres (1980) to a time-of-flight measurement of the velocity of a free nonrelativistic particle of well-defined energy E. The relativistic version was treated subsequently by Davies (1986). Leavens and McKinnon (1994) applied the quantum clock approach, suitably modified, to a calculation of the characteristic times $\tau_X(k; z_1, z_2)$ for both nonrelativistic and relativistic electrons. In this section this work is reviewed and then extended in an attempt to calculate the mean-square dwell time.

Following Peres, consider a quantum clock with an odd number $N \equiv 2j + 1$ of states which are conveniently represented by the alternative sets of orthonormal basis functions

$$u_m(\theta) = (2\pi)^{-1/2} e^{im\theta} \qquad (m = -j, \ldots, 0, \ldots, j) \tag{31}$$

and

$$v_n(\theta) = N^{-1/2} \sum_{m=-j}^{j} e^{-i2\pi nm/N} u_m(\theta) \qquad (n = 0, \ldots, j, \ldots, N-1) \tag{32}$$

with $0 \leq \theta < 2\pi$. The $u_m(\theta)$ are eigenfunctions of the clock Hamiltonian

$$(33) \qquad \hat{H}_c \equiv -i\hbar\omega\frac{\partial}{\partial\theta} \qquad \left(\omega \equiv \frac{2\pi}{N\tau}\right)$$

with eigenvalues $m\hbar\omega$ while the $v_n(\theta)$ are eigenfunctions of the clock-time operator

$$(34) \qquad \hat{T}_c \equiv \sum_{n'=0}^{N-1} n'\tau\hat{P}_{n'} \qquad [\text{where } \hat{P}_{n'}v_n(\theta) \equiv \delta_{n',n}v_n(\theta)],$$

with eigenvalues $n\tau$. The expectation value of $\hat{H}_c$ in the state $v_n(\theta)$ is zero. For large N the basis function $v_n(\theta)$ is sharply peaked at $\theta = 2\pi n/N$ with an uncertainty of $\pm\pi/N$, corresponding to the time $n\tau$ with an uncertainty of $\pm\tau/2$. In addition, the $v_n(\theta)$ have the important digital clock-like property

$$(35) \qquad e^{-i\hat{H}_c\tau/\hbar}v_n(\theta) = v_{n+1(\text{modulo } N)}(\theta).$$

To determine the time spent by a free $[V(z) = 0]$ particle of well-defined energy $E \equiv \hbar^2k^2/2m$ in the region $[z_1, z_2]$, Peres considered a stationary-state scattering experiment described by the Hamiltonian

$$(36) \qquad \hat{H} = -\frac{\hbar^2}{2m}\frac{\partial^2}{\partial z^2} + \Theta(z - z_1)\Theta(z_2 - z)\hat{H}_c.$$

The associated classical picture is of a stop-watch that runs only when the otherwise free particle is in the region of interest. Solution of the Schrödinger equation for the "final" state $\psi_f(z, \theta)\exp[-iEt_\infty/\hbar]$ that "evolves" from the assumed "initial" state $\psi_i(z, \theta) = A_k\exp(ikz)v_0(\theta)$ and calculation of the expectation value of the elapsed time, $[\langle\psi_f^*(z, \theta)\hat{T}_c\psi_f(z, \theta)\rangle - \langle\psi_i^*(z, \theta)\hat{T}_c\psi_i(z, \theta)\rangle]$, recorded by the ensemble of quantum clocks gives the expected result $\tau_T(z_1, z_2) = (z_2 - z_1)/(\hbar k/m)$ provided that the clocked region $[z_1, z_2]$ is very much longer than the de Broglie wavelength $\lambda = 2\pi/k$ of the free particle. This criterion is obtained by demanding that the conditions for negligible perturbation of the particle by the clock, $E \gg j\hbar\omega$, and for good time resolution, $\tau_T(z_1, z_2) \gg \tau$, are met simultaneously.

Now add the scattering potential $V(z)\Theta(z)\Theta(d - z)$ to the Hamiltonian (36) and consider the application of the quantum clock method to the calculation of the characteristic times $\tau_X(k; z_1, z_2)$. First consider the important special case of the mean dwell time $\tau_D(k; 0, d)$ for an opaque ($\kappa d \gg 1$) rectangular barrier of height V_0. A necessary condition for negligible perturbation of the particle-barrier scattering process of interest by the particle-clock interaction is $V_0 - E \gg j\hbar\omega$ ($|T|^2$ for the undisturbed system depends exponentially on $(V_0 - E)^{1/2}$). Writing this in terms of the expected result $\tau_D(k; 0, d) = \hbar V_0^{-1}E^{1/2}(V_0 - E)^{-1/2}$ obtained from (9) assuming $\kappa d \gg 1$, one obtains after a little algebra $\tau_D(k; 0, d) \ll \tau/4$. This condition is completely incompatible with the criterion $\tau_D(k; 0, d) \gg \tau$ for good time resolution. Leavens and McKinnon (1994) by-passed the latter condition by using the actual elapsed time $t - t_i$ to calibrate the mean elapsed time $[\langle\varphi^*(\theta, t)\hat{T}_c\varphi(\theta, t)\rangle - \langle\varphi^*(\theta, t_i)\hat{T}_c\varphi(\theta, t_i)\rangle]$ calculated for an ensemble of *freely running* quantum clocks $[\hat{H} = \hat{H}_c]$, each prepared in the same initial state $\varphi(\theta, t_i) \equiv \exp(-i\hat{H}_c t_i/\hbar)v_0(\theta)$. This calibration is then applied to the quantity $[\langle\psi_f^*(z, \theta)\hat{T}_c\psi_f(z, \theta)\rangle - \langle\psi_i^*(z, \theta)\hat{T}_c\psi_i(z, \theta)\rangle]$, with $\psi_i(z, \theta) = A_k\exp(ikz)\varphi(\theta, t_i)$, calculated for the clocked particle-(arbitrary) barrier scattering process of interest. Since $\tau \equiv 2\pi/N\omega$ has now lost its meaning as the time resolution of the (calibrated) clock there is no advantage to having $N\omega$ very large. In particular, one can choose the smallest reasonable value for N, i.e. $N = 3$, and let the corresponding particle-clock interaction energy $\hbar\omega$ become arbitrarily small. For $\tau < t_i < 2\tau$ this procedure gives precisely the expected result (9) for the mean dwell time $\tau_D(k; z_1, z_2)$.[8] With the use of a distant ideal particle detector, the complete set of clock readings can be sorted into two subsets, one for transmitted and one for reflected particles, without perturbing the scattering process under investigation. The corresponding (calibrated) quantum clock results for the mean transmission and reflection times $\tau_T(k; z_1, z_2)$ and $\tau_R(k; z_1, z_2)$ are identical to the weak measurement results of Steinberg (1995a; 1995b). Leavens and McKinnon (1994, 17) extended their (calibrated) Salecker–Wigner clock analysis to Dirac electrons but found that the superluminal speeds and negative reflection times discussed in Section 4 persisted. Within Bohmian mechanics, on the other hand, the natural extension of the equation of motion (1) to Dirac electrons (Holland 1993, 503–518) gives a particle velocity that cannot exceed c in magnitude.

For an ensemble of freely running $N = 3$ quantum clocks each prepared in the state $\varphi(\theta, t_i = \tau) = v_1(\theta)$ it is straightforward to show that

$$(37) \qquad \langle\varphi^*(\theta, t)\,(\hat{T}_c - \tau)^2\varphi(\theta, t)\rangle \cong (8\pi^2/27)\,(t - \tau)^2, \qquad (|t - \tau| \ll \tau).$$

If one interprets the expectation value $\lim_{\omega\to 0}\langle\psi_f^*(z, \theta)(\hat{T}_c - \tau)^2\psi_f(z, \theta)\rangle$ multiplied by the calibration factor $(8\pi^2/27)^{-1}$ from (37) as the mean-square dwell time $\overline{t_D^2(k; z_1, z_2)}$ then one obtains precisely the Feynman path integral result derived by Iannaccone and Pellegrini (1994, 16551) and suggested by Schulman and Ziolkowski (1989, 270–277).[9] For the special case of a symmetrical $[V(d/2 - z') = V(d/2 + z')]$ opaque barrier one obtains $\overline{t_D^2(k; z_1, z_2)} = [\overline{t_D(k; z_1, z_2)}]^2 \equiv [\tau_D(k; 0, d)]^2$ to an excellent approximation. Unless one invokes negative probabilities, this leads to the surprising conclusion that virtually every reflected electron in the ensemble spends almost exactly the same amount of time inside such a barrier. Now consider the special case of a *free* particle $[V(z) = 0]$. In this limit one obtains $\overline{t_D^2(k; z_1, z_2)} = [\overline{t_D(k; z_1, z_2)}]^2\,\{1 + \sin^2[k(z_2 - z_1)]/[k(z_2 - z_1)]^2\}$ where $\overline{t_D(k; z_1, z_2)} = m(z_2 - z_1)/\hbar k$. This means that for arbitrary z_1 every electron in the ensemble spends precisely the same length of time $m(z_2 - z_1)/\hbar k$ in the region $[z_1, z_2]$ if and only if $k(z_2 - z_1) = \infty$ or $n\pi$ with $n = 1, 2, 3, \ldots$. Because of these bizarre results, the author cannot accept the Salecker–

Wigner clock result for the mean-square dwell time as the second moment of the assumed distribution of intrinsic dwell times. This conclusion is completely consistent with the often repeated claim of Bell (1987a), Bohm and Hiley (1993) and Dürr *et al.* (1996a) that, with the exception of position, one does not in general measure pre-existing values of particle properties.

The first two moments of the dwell time derived by Iannaccone and Pellegrini (1994) are identical to the real parts of the corresponding complex-valued moments derived by Sokolovski and Connor (1991). If this correspondence between the results of the two path integral methods extends to higher moments, then one encounters obvious inconsistencies. For example, for an opaque symmetrical barrier not only is $\overline{t_D^3(k;\,0,\,d)}$ not close to $[\overline{t_D(k;\,0,\,d)}]^3$, as implied by the results for the first two moments, but it can be negative! Similarly, for the free particle case $\overline{t_D^3(k;\,z_1,\,z_2)}$ should be equal to $[m(z_2 - z_1)/\hbar k]^3$ whenever $k(z_2 - z_1) = n\pi$ with $n = 1, 2, 3, \ldots$ but this is found not to be the case.

For completeness, it should be recalled that the Bohm trajectory result for the distribution of intrinsic dwell times in the stationary-state limit is

$$P_D[t_D(k;\,z_1,\,z_2)] = |T(k)|^2\delta\{t_D(k;\,z_1,\,z_2) - \tau_T(k;\,z_1,\,z_2)] + |R(k)|^2\delta\{t_D(k;\,z_1,\,z_2) - \tau_R(k;\,z_1,\,z_2)], \tag{38}$$

for $z_1 > -\infty$ and *arbitrary* barrier potential $V(z)$. For $|T(k)|^2 > 0$, $\tau_R(k;\,z_1,\,z_2) = 0$.

7. CONCLUDING REMARKS

It is well known that Bohmian mechanics leads to exactly the same experimental predictions as conventional quantum theory whenever the predictions of the latter are unambiguous. The concepts of transmission and reflection times for quantum particles are not meaningful within standard text-book quantum mechanics, strictly applied, or within some modern reinterpretations such as those based on consistent histories (Griffiths 1984; Gell-Mann and Hartle 1990; Omnès 1992; Yamada 1996). Despite this, there is a large and rapidly expanding theoretical literature devoted to calculation of these quantities. Several of the approaches that have been proposed, including the one based on weak measurement theory, are contained in the systematic projector approach which leads in principle, via rather straightforward extension of conventional quantum mechanics, to an unlimited number of possible expressions for the mean transmission and reflection times. None of these "conventional" possibilities for τ_T and τ_R can be equal to the results that emerge uniquely and almost trivially from Bohmian mechanics. An essential difference between the projector and Bohm trajectory approaches is that the former is based on the wave-like decomposition $\psi = \psi_T + \psi_R$ while the latter is based on the particle-like decomposition $|\psi|^2 = [|\psi|^2]_T + [|\psi|^2]_R$. The "conventional" probability current density approach is based solely on the ill-defined $J = J_+ + J_-$ decomposition of J into right-going and left-going components. This approach is conventional only in that no attempt is made to decompose J into transmitted and reflected components J_T and J_R. Although particle-like decompositions $J = J_+ + J_-$ and $J = J_T + J_R$ are both uniquely defined within Bohmian mechanics, only the latter is needed to calculate τ_T and τ_R from J.

Since expression (5) for the mean dwell time is well established, it appears from the above considerations that it is the ambiguity (within "conventional" quantum mechanics) in the decomposition into transmitted and reflected components that is at the heart of the "tunneling time problem". Weak measurement theory removes this ambiguity but leads to expressions for τ_T and τ_R that many, including the author, find unacceptable because they can lead to superluminal mean transmission speeds and negative mean reflection times. The (calibrated) Salecker–Wigner quantum clock approach throws further light on the problem. The calibration procedure allows the particle-clock interaction to be made arbitrarily weak and, perhaps not surprisingly, it gives precisely the weak measurement results for τ_D, τ_T and τ_R. The approach is readily extended to Dirac electrons and shows that the superluminal and negative times mentioned above are not artifacts of using the non-relativistic Schrödinger equation. The approach can also be extended to calculate the mean-square dwell time and gives precisely an expression obtained using Feynman paths. The resulting behaviour is bizarre in both the zero barrier and opaque (symmetric) barrier limits. This behaviour cannot be blamed on the particular transmission-reflection decomposition used because none was, of course, needed. Moreover, since peculiar behaviour is obtained even for free electrons, it seems that the word "tunneling" in "tunneling-time problem" is a red herring. The underlying problem is simply that conventional quantum mechanics is not a theory for electron motion (Holland 1993).

Is it consistent for the author to regard the quantum clock analysis leading to (5) as a proper derivation of the mean dwell time, rather than a mathematical fluke, and simultaneously to reject the corresponding results for the mean transmission and reflection times and the mean-square dwell time as being unphysical? It is consistent if there is something fundamentally different about the latter three quantities. There *is* within Bohmian mechanics: the mean dwell time can be computed without needing to calculate any particle trajectories; the other three quantities cannot [e.g., computation of the mean transmission and reflection times using (8a) and (8b) requires the trajectory $z_c(t)$].

From the above it is concluded that the basic difference between the Bohm trajectory and the "conventional" approaches considered is that Bohmian mechanics consistently treats a "particle" as an actual particle[10] at all times (Dürr *et al.* 1992a, 843), not just at those instants when an ideal (i.e. strong) measurement is made of its position. Aside from those instants, the "conventional" approaches are stuck with waves and the resulting ambiguities associated with timing an extended entity, especially one that spreads and splits into two or more components.[11] In the Bohm trajectory approach it is the presence of an actual particle in a given region that is being clocked. If the advocates of one of the "conventional" approaches considered in this paper should make the same claim, then the two approaches are clearly incompatible. If, on the other hand, it is some wave-like aspect of the electron that is being clocked then there is the possibility that the two approaches are complementary. In this case, it should be stated as clearly as possible just which aspect of the electron, if any, other than its statistical effect on the pointer of a clock, is being considered and whether or not the presence of this aspect in the region of interest is actually being timed. If this time can be negative, then a clear explanation should be given of what a negative time of presence in a region of space actually means. In the author's opinion it is not good enough to pass it off as just another magical feature of quantum mechanics. This is particularly so because a "conventional" approach as defined here necessarily involves some as yet unjustified extension of standard text-book quantum mechanics. If, however, the mean transmission and reflection times tell us nothing more about the electrons than what average shifts they impart to pointers in the respective subensembles, then this involves completely standard but unmagical quantum mechanics. There is now nothing mysterious about negative pointer shifts but does the apparatus really merit being called a clock when it is not actually timing anything?

ACKNOWLEDGEMENTS

The author has benefited from numerous discussions with his NRC colleagues G. C. Aers and W. R. McKinnon and also with J. T. Cushing (Notre Dame), S. Goldstein (Rutgers), G. Iannaccone (Pisa) and J. G. Muga (La Laguna).

National Research Council of Canada

NOTES

[1] For the special case of the dwell time ($X = D$) there is widespread agreement on the correct expression for the mean value τ_D and one can construct a hermitian operator that produces this result; but it is not at all clear that the nth power of this operator will produce the correct nth moment of the dwell time for $n > 1$ [Daumer *et al.* 1994].

[2] A Bohm trajectory can be retrodicted from a precise measurement of the time of arrival at $z = b$, if $\psi(z, 0)$ is known for all z.

[3] $\tau_x(z_1, z_3) = \tau_x(z_1, z_2) + \tau_x(z_2, z_3)$ with $z_1 \leq z_2 \leq z_3$.

[4] The quantities $-\mathrm{Im}[\tau_T^{TD}]$ and $-\mathrm{Im}[\tau_R^{RD}]$ are identical to the so-called mean spin-rotation transmission and reflection times introduced by Büttiker (1983); $|\tau_T^{TD}|$ and $|\tau_R^{RD}|$ are equal to the well-known Büttiker–Landauer (1982) expressions for τ_T and τ_R.

[5] This is readily confirmed for the reflection case by recalling that within Bohmian mechanics $\tau_R(k;\, z_1,\, z_2) = 0$ for $0 < |R(k)|^2 < 1$ but $\tau_R(z_1,\, z_2)$ is in general not equal to zero (see Figure 1) as would be the case if (23b) applied to these quantities.

[6] It is readily shown that in the tunneling regime $V_0 > E$, the Bohm trajectory result for $\tau_T(k;\, 0,\, d)$ *exceeds* the free particle ($V_0 = 0$) result $md/\hbar k$ for all $d > 0$.

[7] These expressions were suggested to the author by S. Goldstein.

[8] For $\tau < t_i < 2\tau$, the initial clock state is not an eigenstate of $\hat{T}_c$. Hence, a strong measurement of the initial mean clock time would collapse the wave function for each ensemble member to one of the eigenstates of $\hat{T}_c$, thus aborting the desired initial state. Any such attempt to verify the initial mean clock time must be done on a different (but identically prepared) ensemble from the one used to determine the final mean clock time.

[9] $\overline{t_D^2(k;\, z_1,\, z_2)} = |T(k)|^2\, \overline{t_T^2(k;\, z_1,\, z_2)} + |R(k)|^2\, \overline{t_R^2(k;\, z_1,\, z_2)}$ relates the mean-square dwell time to the mean-square transmission and reflection times actually considered in these references.

[10] In the author's opinion, Feynman paths, unlike Bohm trajectories, are to be interpreted only as useful mathematical constructs, not as actually realizable particle trajectories.

[11] That the Bohm trajectory and projector approaches both lead to (5) for the mean dwell time is consistent with the fact that $|\psi(z,\, t)|^2$ can be interpreted both as a probability distribution for the position of a particle and as the local intensity of a wave.

EUAN J. SQUIRES

LOCAL BOHMIAN MECHANICS

A derivation of the Bohm model is given and some general observations are made about it. A local modification of the model is introduced and, for a simple experiment, its properties are studied.

1. THE BOHM MODE, AS A SIMPLE REALISTIC QUANTUM THEORY

Non-relativistic quantum mechanics, as it is presented in almost all text-books, is a theory which is either incorrect or incomplete, even within the domain where non-relativistic approximations are adequate. It apparently gives the correct predictions for the outcome of measurements, but nowhere it does it contain those outcomes, or allow any description of the processes whereby they are produced. By far the simplest method of curing this problem is that introduced by David Bohm in 1952. Here quantum theory is correct, but it is incomplete. The complete theory has, in addition to the wave function, trajectories for individual particles exactly as in classical mechanics. Indeed, as we discuss in the next section, it is possible to regard the Bohm model as being an extension of classical mechanics.

We shall first consider the simple logical steps which allow us to derive the Bohm model directly from the rules of orthodox quantum theory. The starting point here is to note that all observations are in reality observations of *position*. We deduce the results of measurements of other quantities by an observation of position (consider for example the measurement of a spin projection by a Stern–Gerlach device). Exactly *why* this is so is an interesting question itself (Squires 1990a). Next we recall that quantum theory gives statistical predictions. Thus we require a model in which objects at all times have positions, and which gives the correct statistical distribution of these positions.

To see what this means we suppose that we have N particles with positions $\mathbf{x}_i(t)$, where t is the time. We can represent these by the vector $\mathbf{X(t)}$ in the $3N$ dimensional configuration space. If we denote the probability distribution of the positions at time t by $\rho(\mathbf{x}, t)$, then the condition that $\rho(\mathbf{x}, t + \mathrm{d}t)$ gives the probability distribution at $t + \mathrm{d}t$ is clearly

$$(1.1) \qquad \rho(\mathbf{X}, t)\mathrm{d}^{3N}\mathbf{X} = \rho(\mathbf{X} + \dot{\mathbf{X}}\mathrm{d}t, t + \mathrm{d}t)\mathrm{d}^{3N}\mathbf{X}(t + \mathrm{d}t),$$

which leads to the standard continuity equation

$$(1.2) \qquad \nabla \cdot (\rho\dot{\mathbf{X}}) \equiv \sum_i \nabla_i \cdot (\rho\dot{\mathbf{x}}_i) = -\frac{\partial\rho}{\partial t},$$

where $\dot{\mathbf{X}}(\mathbf{X}, t)$ is the vector field giving the velocity of a particle at $\mathbf{X}$.

J. T. Cushing et al. (eds.), Bohmian Mechanics and Quantum Theory: An Appraisal, 131–140.

The r.h.s. of Eq. (1.2) can be evaluated if we write the density in terms of a wave function evolving according to the Schrödinger equation:

$$(1.3) \quad \rho = \Psi^*\Psi,$$

with

$$(1.4) \quad i\hbar\dot{\Psi} = \left(-\frac{\hbar^2}{2}\sum_i \frac{1}{m_i}\frac{\partial^2}{\partial x_i^2} + V\right)\Psi.$$

A simple calculation yields

$$(1.5) \quad \frac{\partial\rho}{\partial t} = \frac{i\hbar}{2}\sum_i \frac{1}{m_i}\nabla_i \cdot [\Psi^*\Delta_i\Psi - \Psi\Delta_i\Psi^*]$$

$$(1.6) \quad = -\sum_i \nabla_i \cdot \frac{1}{m_i}\Re(\Psi^*\mathbf{p}_i\Psi),$$

where $\mathbf{p}_i$ is the momentum operator for the ith particle,

$$(1.7) \quad \mathbf{p}_i = -i\hbar\nabla_i.$$

Comparing Eqs. (1.6) and (1.2), we see that

$$(1.8) \quad \dot{\mathbf{x}}_i = \frac{1}{m_i}\Re\left(\frac{\mathbf{p}_i\Psi}{\Psi}\right) + \left(\frac{1}{\rho}\right)\mathbf{c}_i,$$

where the $\mathbf{c}_i$ are arbitrary vectors which satisfy

$$(1.9) \quad \sum_i \nabla_i \cdot \mathbf{c}_i = 0.$$

To obtain the Bohm expression we take the simplest form (i.e., zero) for the $\mathbf{c}_i$. This can be justified essentially on the grounds of simplicity, together with the fact that any arbitrary vectors $\mathbf{c}$, not dependent on Ψ, would break rotational invariance and would give the very unphysical result that the velocity would go to infinity as the density went to zero. It is also the obvious choice if we use the standard form for the probability current, which is similarly underdetermined.

Before proceeding we note, however, that there is a simple case where the neglect of the final term is rather less "natural". This is when Ψ, and hence ρ, is independent of time. Then the r.h.s. of Eq. (1.2) is zero, which would suggest zero velocity as the natural solution. The fact that the Bohm model need not give zero velocity in such a situation may be significant in quantum cosmology (Valentini 1992; Vink 1992; Squires 1992; 1995). Here, according to the Wheeler–deWitt equation, the wave function of the universe (which is the only wave function that actually *exists*!) is independent of time. This is a consequence of the fact that the theory must be invariant under reparameterisation of time. For any real solution of this equation, the straightforward generalisation of Bohmian mechanics to quantum cosmology predicts zero velocities (i.e., a universe in which nothing ever moves). Presumably this is not a good prediction! There is of course an analogous prediction in the microscopic world where for example the model predicts that an electron in the ground state of a hydrogen atom does not move. In this case, however, the result is not a problem because we know that predictions for the results of *measurements* of the electron velocity, which will be related to positions of certain probes, will be correct. There is no similar escape in the cosmological case – a stationary universe is simply a stationary universe! Thus it is essential to select a (non-trivially) complex solution of the Wheeler–deWitt equation and to use the fact that such a wave function can give non-zero velocities, even if the wave function itself is constant.

Equations (1.4) and (1.8), with the $\mathbf{c}_i$ equal to zero, completely define the theory. Provided that in any experiment the initial distribution of positions agrees with that given by the quantum rule at the initial time, then they will do so at the end of the experiment, which then guarantees that the model will always agree exactly with the predictions of quantum theory. Note, especially, that the model automatically avoids the hidden-variable "no-go" theorems (Bell 1966; Hardy 1996; Pagonis and Clifton 1995); in other words, the form of Eq. (1.8) ensures that there is the necessary contextuality of measurements (see remarks below Eq. 3.2).

To find an analogue of Newton's second law of motion we put

$$(1.10)\quad \ddot{\mathbf{x}}_i = \frac{\partial \dot{\mathbf{x}}_i}{\partial t} + \sum_j \frac{d\mathbf{x}_j}{dt} \cdot \nabla_j \dot{\mathbf{x}}_i$$

$$(1.11)\quad = \frac{1}{m_i} \Re \left[-i\hbar \frac{\partial}{\partial t} \left(\frac{\nabla_i \Psi}{\Psi} \right) + \sum_j \dot{\mathbf{x}}_j \cdot \nabla_j \left(\frac{-i\hbar \nabla_i \Psi}{\Psi} \right) \right],$$

where we have used Eqs. (1.7) and (1.8). After a little rearrangement this becomes

$$(1.12)\quad m_i \ddot{\mathbf{x}}_i = -\nabla_i [V + Q],$$

which is Newton's second law with the potential, $V + Q$, given by

$$(1.13)\quad V + Q = \Re \left(\frac{H\Psi}{\Psi} \right) + \left(\frac{1}{2} \right) m_j \dot{\mathbf{x}}_j^2,$$

where H is the hamiltonian. We can separate out the "quantum potential", Q, by using Eq. 1.4. This gives

$$(1.14)\quad \Re \sum_j \frac{1}{2m_j} \left(\frac{\mathbf{p}_j^2 \Psi}{\Psi} - m_j \dot{\mathbf{x}}_j^2 \right).$$

This equation reveals some interesting features of the model. For example, if we replace the operator $\mathbf{p}$ by its classical value $m\dot{\mathbf{x}}$, the quantum potential becomes zero, so in this sense the Bohm equation for the velocity may be said to contain Newton's second law. Nevertheless, although we expect that the quantum potential should be a small quantum correction, it clearly exactly cancels (up to an irrelevant constant) the "classical" potential in the case when the state is an energy eigenstate with a constant phase, as occurs in particular for a lowest-energy bound state.

It is important to note, however, that this step of introducing the potential is not necessary. Unlike Newtonian mechanics, Bohmian mechanics gives an equation for the velocities, not for the accelerations. The initial conditions for a Bohmian universe are not positions and velocities, but positions (together of course with the wave function). Even the positions are not free, but have to satisfy the constraint that, at some initial time, they will give probability distributions consistent with the Born rule (see Dürr *et al.* 1992a, for a detailed discussion).

2. COMPARISON WITH CLASSICAL MECHANICS

Before proceeding it is of interest to see how the Bohm model relates to other models. In my opinion the model looks much more convincing if we emphasise its similarity to *classical* mechanics rather than to (orthodox) *quantum* mechanics. The ontology of the model is that of classical mechanics; it has real particles, which at all times have positions. The law describing how the particles move has the same form as in classical mechanics; the trajectories are defined by Newton's second law of motion.

On the other hand, there is none of the indeterminism or special role of observations, which are characteristic of quantum theory.

There are, of course, differences between Bohmian mechanics and Newtonian mechanics, although to some extent these can be regarded as additions to the latter rather than as changes. One such difference is that, as we have noted, the second-order equation of motion can be integrated to give a first-order equation, without any need for additional boundary conditions. To appreciate the significance of this point we might imagine a world in which time is discrete. Then it is clear that in Newtonian mechanics, the equations plus conditions at one time do not completely determine the state at future times. In the Bohm model, however, they do.

The other difference is that there is an additional "quantum" force. Of course, classical physics is easily able to cope with new forces, but this particular addition is not as innocent as it sounds. The quantum force is unlike all the other forces in nature because it is not derivable from the positions of the particles. We can compare it with, for example, gravity. The force of gravity acting on a given particle is a unique function of the positions of the other particles, which act as sources in the Poisson equation. This, of course, is only true if we neglect the "complementary function" (i.e., solution of the Laplace equation), a procedure which we normally justify by imposing some sort of boundary condition at infinity. The quantum potential, however, is derived from the Schrödinger equation in which there are no sources, so the analogue of the neglected term is here everything. Also the actual quantum force is independent of the *magnitude* of the field from which it derived (see Eq. (1.14)).

These differences strongly suggest that there is some underlying theory, which is not quantum theory and not classical mechanics, but which combines (and explains?) certain aspects of both. It is one of the great merits of the Bohm model that, in addition to giving a proper, respectable, explanation of all quantum phenomena, it encourages speculation about such an underlying theory, and even about possible theories which give results different to those of standard quantum theory. An example is discussed below.

3. RELATIVISTIC INVARIANCE AND THE BOHM MODEL

The Bohm model exposes the non-locality which has long been recognised as one of the significant features of the difference between quantum mechanics and classical physics. Bell's theorem demonstrated that this non-locality is not peculiar to the particular form of realistic model used by Bohm. Any "completion" of quantum theory which is consistent with all its predictions must be non-local. Since the experimental tests (e.g., Aspect *et al.* 1981; 1982a; 1982b) seem to agree with quantum theory, most physicists have come to accept this non-locality in some form or other. Although the discussions are normally carried out in a non-relativistic context, it is clear that agreement with the predictions of quantum theory strongly suggests that a realistic model should be non-Lorentz invariant. In particular, it should require that there exists a preferred frame (Hardy 1992; Hardy and Squires 1992).

The need for such a frame is of course evident in the Bohm model because the expression for the velocity of one particle (Eq. (1.8)) requires knowledge of the position of all the others *at the same time*, which clearly is not a frame-independent concept. However, the fact that the model so clearly reveals the non-locality means that it also shows how it might be removed, at the cost of course of a failure to agree at all times with the predictions of quantum theory. The idea is suggested by the analogy with a classical potential given in Section 2. Any classical potential (e.g., the electrostatic potential) is also defined in the configuration space of the particles, and requires simultaneous positions. However, we know how this is dealt with in a proper relativistic treatment: we use the "retarded" potential in which the positions are determined on the backward light cone.

It is possible to do something similar in the Bohm model (Squires 1993; Mackman and Squires 1995) and to replace Eq. 1.8 by

$$\text{(3.1)} \qquad \dot{\mathbf{x}}_i(t_i) = \frac{1}{m_i} \Re \left(\frac{\mathbf{p}_j \Psi(\mathbf{x}_1(t_1), (\mathbf{x}_2(t_2) \ldots)}{\Psi(\mathbf{x}_1(t_1), (\mathbf{x}_2(t_2) \ldots)} \right),$$

where

$$\text{(3.2)} \qquad t_k = t_i - \frac{|(\mathbf{x}_i(t_i) - (\mathbf{x}_k(t_k)|}{c}.$$

In Eq. (3.1) we have ignored the explicit time dependence of the wave function, and it is not clear how we should treat this. Part of the problem is that we are working within the framework of non-relativistic quantum theory. At the fundamental level, we could perhaps take refuge in the fact noted above that the actual wave function of the universe is constant. Further work is needed here, but for the moment we shall ignore the problem. In the example discussed below an unambiguous procedure suggests itself.

It is clear that Eq. (3.1) goes some way towards removing the obvious non-locality from the Bohm model, and it is important to study the nature of its inevitable disagreement with quantum theory predictions. In principle this is possible because the model allows explicit calculations.

Consider, for example, an experiment in one space dimension in which a photon is emitted from an origin in the form of two wave-packets, of equal size, one travelling in the positive direction and one in the negative. We suppose that there are photon detectors at positions l and $-l$ and that the purpose of the experiment is to determine in which direction the photon is observed to travel (i.e., which detector actually "sees" the photon). The difficulty with this type of discussion is how we model a real photon detector. One simple possibility (see Squires 1993; and Squires and Mackman 1994) is to take the detectors to be free particles, initially in zero-momentum gaussian wave-packets, which receive a momentum p when they detect a photon of momentum p. Thus, just after the photon is emitted, the wave function of the system is given by

$$(3.3) \qquad |\Psi\rangle = 2^{-1/2}[\phi_L(y) + \phi_R(y)]\psi_L(x_L)\psi_R(x_R),$$

where $\phi_{L,R}(y)$ is the part of the photon wave function moving to the L, R respectively, and $\psi_{L,R}$ are the two detector states given by

$$(3.4) \qquad \psi_{L,R}(x_{L,R}) = \left(\frac{a}{\pi}\right)^{\frac{1}{4}} \exp\left[-\frac{1}{2}a(x_{L,R} \pm l)^2\right].$$

At a later time, after the photon wave packets have interacted with the detectors, the wave function has the form:

$$(3.5) \qquad |\Psi\rangle = 2^{-1/2}[\phi_L(y)\psi_L^{-p}(x_L)\psi_R(x_R) + \phi_R(y)\psi_L(x_L)\psi_R^{p}(x_R)],$$

where the $\psi^{\pm p}$ now represent moving wave packets, e.g.,

$$(3.6) \qquad \psi^{-p}(x_L) = \left(\frac{a}{\pi}\right)^{\frac{1}{4}} \exp\left[i\left(x_L p + \frac{p^2 t}{2m}\right) - \frac{1}{2}a\left(x_L + l + \frac{pt}{m}\right)^2\right].$$

Note that in the last expression we have neglected the quantum evolution of the free detector state.

Now we recall that in a spin measurement, as for example in the experiments of Aspect *et al.*, the actual outcome (i.e., the value of the spin that is observed) is determined in the Bohm model by the value(s) of the, so-called, hidden variable(s) in the detector. To study something analogous to this we suppose that there are no photon trajectories (this is the case in at least some versions of the Bohm model). Then, as we shall see below, the measurement outcome, which we refer to as the position of the photon, again depends upon the values of the hidden variables of the detectors i.e., the positions of the particles).

First, it is necessary to modify the standard Bohm formula for particle velocities by integrating over the positions of those particles without trajectories (Bell 1981a; Squires and Mackman 1994). Thus, in general, Eq. (1.8) must be replaced by

$$(3.7) \qquad \dot{\mathbf{x}}_i = \Re\left(\frac{\int \mathrm{d}^3\mathbf{y}\Psi^*\mathbf{p}_i\Psi}{m_i\int \mathrm{d}^3\mathbf{y}\Psi^*\Psi}\right),$$

In our experiment this leads to the result

$$(3.8) \qquad m\dot{x}_L = \Re\left(\frac{|\psi_R|^2\psi_L^{-p*}p_L^{op}\psi_L^{-p} + |\psi_R^p|^2\psi_L^*p_L^{op}\psi_L}{|\psi_R|^2|\psi_L^{-p}|^2 + |\psi_R^p|^2|\psi_L|^2}\right),$$

and a similar equation for $\dot{x}_R$, where m is the mass of the detector particle. Here we have assumed that there is no overlap between the right and left moving photon wave packets. This will be approximately true for any reasonable definition of the space-time surface implicit in Eq. (3.7).

We first use the non-retarded, and hence non-local, Bohm model. Then, inserting the previous wave functions into Eq. (3.8), we find

$$(3.9) \qquad \dot{u} = (1 + \exp[2t(v - u)])^{-1},$$

and

$$(3.10) \qquad \dot{v} = (1 + \exp[2t(u - v)])^{-1},$$

where we have simplified the notation by sing units in which $a = p/m = 1$ and by defining

$$(3.11) \qquad u = x_R - l,$$

and

$$(3.12) \qquad v = -(x_L + l).$$

Adding (3.9) and (3.10) we obtain $\dot{u} + \dot{v} = 1$, hence

$$(3.13) \qquad u + v = t + u_0 + v_0,$$

where u_0 and v_0 are the values of u and v at the time when the interaction occurs, taken to be $t = 0$. If we substitute (3.13) into (3.9) and (3.10) we find

$$(3.14) \qquad \dot{u} = (1 + \exp[-2t(2u - \mathrm{t} - u_0 - v_0)])^{-1},$$

and

$$(3.15) \qquad \dot{v} = (1 + \exp[-2t(2v - t - u_0 - v_0)])^{-1}.$$

We compare these results with what we obtain with only one detector, say the one at the left. Then the Bohm equation would give

$$(3.16)\quad \dot{v} = (1 + \exp[-t(2v - t)])^{-1}.$$

Clearly the small t behaviour is $v \simeq {}^1\!/_2 t + v_0$, leading to

$$(3.17)\quad \dot{v} \simeq (1 + \exp[-2tv_0])^{-1},$$

for small t. It follows that this detector will record the photon (in the sense that the detector particle will continue to move, with velocity approaching 1), if $v_0 > 0$, but not if $v_0 < 0$. Incidently, we can here see the fact that an initial distribution agreeing with quantum theory will give the correct quantum theoretic outcome: such an initial distribution will have the v_0 equally distributed between positive and negative values, leading to half the photons being detected in the left detector, as required.

If we treat Eq. (3.15) in a similar way we find

$$(3.18)\quad \dot{v} \simeq (1 + \exp[-2t(v_0 - u_0)])^{-1}.$$

Hence in this case the condition that the left detector records (fails to record) the photon is that $v_0 - u_0$ is positive (negative). Clearly the opposite is true for the right detector, so (as required) one, and only one, detector will see the photon. Again an initial distribution agreeing with quantum theory will have $v_0 - u_0$ positive and negative with equal frequency, so giving the expected output results. We note also that this example reveals the contextuality of this version of the Bohm model: comparison of (3.17) with (3.18) shows that the result emerging from the right detector, say, is affected by the presence of the left detector.

The differential equations (3.14) and (3.15) can in fact be solved directly to give, for example,

$$(3.19)\quad \int_{\frac{-(u_0 - v_0)}{2}}^{\frac{(u_0 - v_0)}{2}} dy e^{-2y^2} = \int_{t-\left(u - \frac{-(u_0 + v_0)}{2}\right)}^{u - \frac{(u_0 + v_0)}{2}} dy e^{-2y^2}.$$

Clearly, if $(u_0 - v_0)$ is positive, then $(u - t)$ must remain constant as t becomes large; on the other hand if it is negative, then u becomes constant and small for large t. This confirms that we will obtain the expected measurement outcomes.

We must now consider what happens in this experiment if we use the retarded Bohm model. Since we have well-localised wave-packets we can solve the problem noted below Eq. 3.1 by evaluating the wave-packet from Eq. 3.6 at the appropriate retarded time as well as retarded position. We define T, the time for light to travel from one detector to the other, according to

$$(3.20)\quad T = \frac{2l}{c}.$$

Then, for $t < T$, the two detectors will behave as if the other one was not present. Hence, for $t < T$,

$$(3.21)\quad \dot{u}(t) = (1 + \exp[-t(2u(t) - t)])^{-1}$$

and

$$(3.22)\quad \dot{v}(t) = (1 + \exp[-t(2v(t) - t)])^{-1},$$

where, for reasons which will be immediately evident, we have explicitly written the time arguments.

Thus, up to $t = T$, the detectors behave independently and record the presence, or otherwise, of the photon strictly according to their own initial position. Clearly this means that in some cases "wrong" results are occurring (i.e., both, or neither, detector is seeing the photon). However, at $t = T$, the situation changes because information about the presence of the other detector becomes available. Thus, for $t > T$,

$$(3.23)\quad \dot{u}(t) = (1 + \exp[-2u(t)t + 2v(t - T) \cdot (t - T) + T \cdot (2t - T)])^{-1}$$

and

$$(3.24)\quad \dot{v}(t) = (1 + \exp[-2v(t)t + 2u(t - T) \cdot (t - T) + T \cdot (2t - T))^{-1}.$$

Note that, as expected, these equations agree with Eqs. (3.9) and (3.10) if T is put equal to zero.

I have not been able to solve these equations analytically, but it is clear that they give the expected results (Squires 1993). In particular, if T is sufficiently small, and $|v_0 - u_0|$ sufficiently large, then again, one, and only one, detector will record the photon. There are, however, circumstances in which both, or neither, detector will see the photon. This would correspond to a "wrong" result, in the sense that the predictions of orthodox quantum theory would be violated.

More precisely, the condition that there will be a significant number of "wrong" results (zero or two photons), is that

$$(3.25)\quad T \geq |v_0 - u_0|_{typical}.$$

With the units restored this means

$$(3.26)\quad \frac{l}{c} \geq \frac{m}{pa^{1/2}}.$$

If we now assume that the detector acquires all the initial momentum of the detected photon, then $p = \hbar/\lambda$ so the condition for wrong results becomes

$$(3.27)\quad \frac{l\hbar}{\mathrm{med}\lambda} \geq 1,$$

where $d \sim a^{-1/2}$ is the spatial spread of the initial wave function of the detector particle. In fact numerical solutions of Eqs. (3.23) and (3.24) show that when

$$\frac{l\hbar}{med\lambda} = 1, \tag{3.28}$$

then about one in ten events give wrong results (H. Movahhedian, private communication).

In a typical experiment the separation l is only a few metres, so if for m we take the mass of a macroscopic pointer, it is clear that the condition in Eq. (3.27) is not satisfied. On the other hand, if we suppose the detection comes about by the photon being absorbed by an electron, then for an optical photon, the LHS of (3.27) is of the order of $10^{-5}/d$, with d measured in metres. If the electron is initially confined to within an atomic distance, then clearly this is much greater than 1, so there will be many events in which both, or neither, electron records the photon. Of course, we would certainly not directly observe a single electron, and it would be essential here to use a device which was genuinely responsive to the electron trajectory (this is not always a trivial issue – see, for example, Englert *et al.* 1992; Dewdney *et al.* 1993).

The tentative conclusion of this analysis is that it is unlikely that deviations from the quantum theory results, which would arise in our retarded model, would have been seen in any experiments that have been performed. Nevertheless, a better analysis of the actual experiments is required, and such an analysis could well reveal the possibility of realistic tests for retarded effects in future, carefully designed experiments. Such experiments would need the largest possible values of l, detectors where the effective "mass" of the detector is as small as possible, and of course efficient detectors.

University of Durham

YAKIR AHARONOV AND LEV VAIDMAN

ABOUT POSITION MEASUREMENTS WHICH DO NOT SHOW THE BOHMIAN PARTICLE POSITION

This work is inspired by our discussions with David Bohm in different periods in Tel-Aviv, London, and South Carolina. We were excited by the results we obtained in the framework of two-state vector formalism about weak measurements (Aharonov and Vaidman 1990) and we were trying to understand these results using Bohm's causal interpretation (Bohm 1952a,b). Both Bohm's theory and the two-state vector formalism yield the same predictions for the results of experiments as the standard quantum theory. Therefore, clearly, there cannot be a technical contradiction between these two approaches. However, it is a legitimate question to ask how close or how different are the concepts of the two theories. Friction between the basic concepts might lead to a direction for the modification of quantum theory; the search for a useful modification was the goal of David Bohm.

I (L.V.) have a vivid memory of an excursion the day after the conference in South Carolina honoring 30 years of the Aharonov–Bohm effect when I, David and Saral Bohm were riding in a carriage in the old streets of Charleston. I remember discussing with David my passion for the many-worlds interpretation (Everett 1957), and then, what I find most attractive in the Bohm theory: that it is unambiguous, deterministic and complete. I see it as the best candidate for the final theory of the world for a physicist (such as Y.A.) who does not want to accept the existence of many worlds. David said to me that what I liked in his theory did not have much value for him. He had a strong belief that a man cannot find the final theory of the world. All that we can do is to look for a better and better approximation to the correct theory which is intrinsically unattainable to us. His vision was that the causal interpretation should suggest a way for generalization to the next-level theory which also, by no means, will be the final theory of the world.

We believe that the observations we make in this work are somewhat disturbing for a physicist who wants to see in the causal interpretation the final word about the world, but they are not a real threat for the theory from the perspective of Bohm. Apart from discussing weak measurements, we will consider recently proposed "delayed observation" measurements (Englert *et al.* 1992) which exhibit similar features. Note, however, that the difficulties we see follow mostly form a particular approach to the Bohm theory we adopt, in which only the Bohmian particle corresponds to the "reality" which we experience, while the wave function is just a pilot wave which governs the motion of the particle. By the position of the "Bohmian particle" we mean what is frequently called actual position of the particle in the Bohm theory. For a composite system consisting of many

J. T. Cushing et al. (eds.), Bohmian Mechanics and Quantum Theory: An Appraisal, 141–154.

particles we shall, somewhat abusing the language, refer to the point in configuration space formed from the coordinates of all the Bohmian particles as the "Bohmian particle."

We consider the following two principles desirable for a causal interpretation. In Bohmian mechanics in most cases they are indeed valid. However, as we will show, there are situations in which they are not.

(I) *A procedure which we usually consider as a good measurement of position should yield the position of the Bohmian particle.*

(II) *An empty wave (the one without the Bohmian particle inside) should not yield observable effects on other particles.*

The motivation for the first principle is clear. The second principle we find desirable because otherwise the Bohmian picture becomes very complex. In the Bohm theory there is no collapse of the wave function, so the total wave incorporates all the complexity of the many worlds of the Everett interpretation. We hoped that the Bohm theory could avoid it. The Bohm theory, as we understand it, says that when we observe a person whose Schrödinger wave is split into a superposition of two macroscopically different waves, we actually see only the component of the superposition which has the Bohmian particle inside. We thought that this is true also when we "observe" a single particle – by means of any type of interaction that we might speak of as detecting it.

Let us start with the following example. The wave function of the particle consists of two identical wave packets running in opposite directions (see Figure 1). For simplicity we will consider a rectangular shape and assume

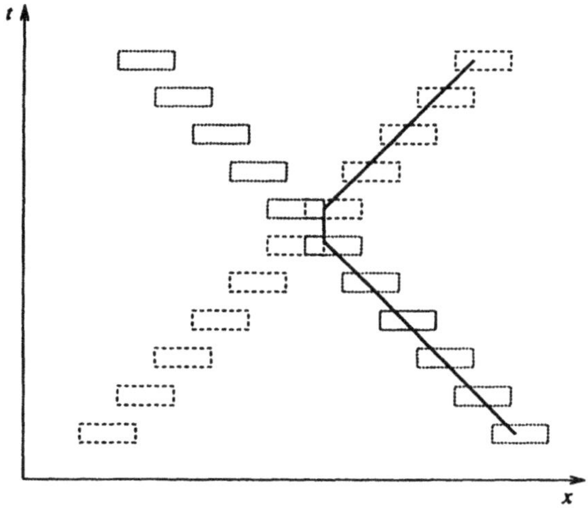

Fig. 1. Space-time diagram for the wave and the Bohm coordinate of the particle. The initial state is the superposition of two identical wave packets moving one towards the other. The Bohmian particle coordinate located initially in the right wave packet. Contrary to the naive expectation the Bohmian particle makes a turn in the region where no (Hamiltonian type) interaction takes place.

that the spread of the wave packets during the process can be neglected. In the beginning the Bohmian particle is inside the right wave packet. The Bohm theory yields an extremely simple prescription for finding the evolution of the system. Until the overlap, the Bohmian particle runs together with the right wave packet. At the moment it reaches the area of the overlap, it stops (the currents of the two wave packets cancel each other in the calculation of the Bohmian particle velocity). Since at that moment the left wave packet has its whole length to go over the particle, while the right wave packet has only a part of its length, the particle will end up inside the left wave packet and will run with it to the right.

One of us (L.V.) used to view the Bohm interpretation as the most elegant way of pointing out one of the many worlds of the Everett interpretation as "real". This example shows that the Bohmian world might be different from any of the Everett worlds. Indeed, no turns take place in Everett worlds in this example.

Now let us add a robust position measurement on the left side of the wave packet (the empty wave; see Figure 2). By a robust measurement we mean a von Neumann-type interaction between the particle and the pointer variable of the measuring device. We can model it by the interaction Hamiltonian

(1) $$H = g(t)P\Pi_V,$$

where P is the conjugate momentum to the pointer variable Q which is, say, the spatial coordinate of the pointer. Π_V is a projection operator on the volume V. The left wave packet of the particle passes the volume V during a certain time t. The coupling constant g is chosen in such a way that during that time the wave function of the pointer is shifted by a distance much larger than its spread. We consider the initial state of the particle and the position of its Bohm coordinate to be the same as in the previous case. Of course, initially the Bohmian particle of the pointer is somewhere inside its initial wave packet. It is very easy to see the evolution, according to the causal interpretation, also in this situation. The wave function of the particle and the system become entangled; the right wave packet is entangled with no shift of the pointer while the left wave packet is entangled with the shifted wave packet of the pointer. The Bohmian particle which started inside the right wave packet runs with it all the way to the left, and the Bohmian particle of the pointer does not move. Indeed, when the wave packets of the particle overlap, there is no overlap for the wave function of the composite system (particle and the pointer) and therefore the left wave packet does not change the motion of the Bohmian particle moving inside the right wave packet (see Figure 2(a)). (If, instead, initially the Bohmian particle was inside the left wave packet, then it will run with this wave packet all the way to the right, while the Bohmian particle of the pointer will move inside the shifted wave packet of the pointer outside its initial location; see Figure 2(b).)

In this example we can see the two principles at work. The position

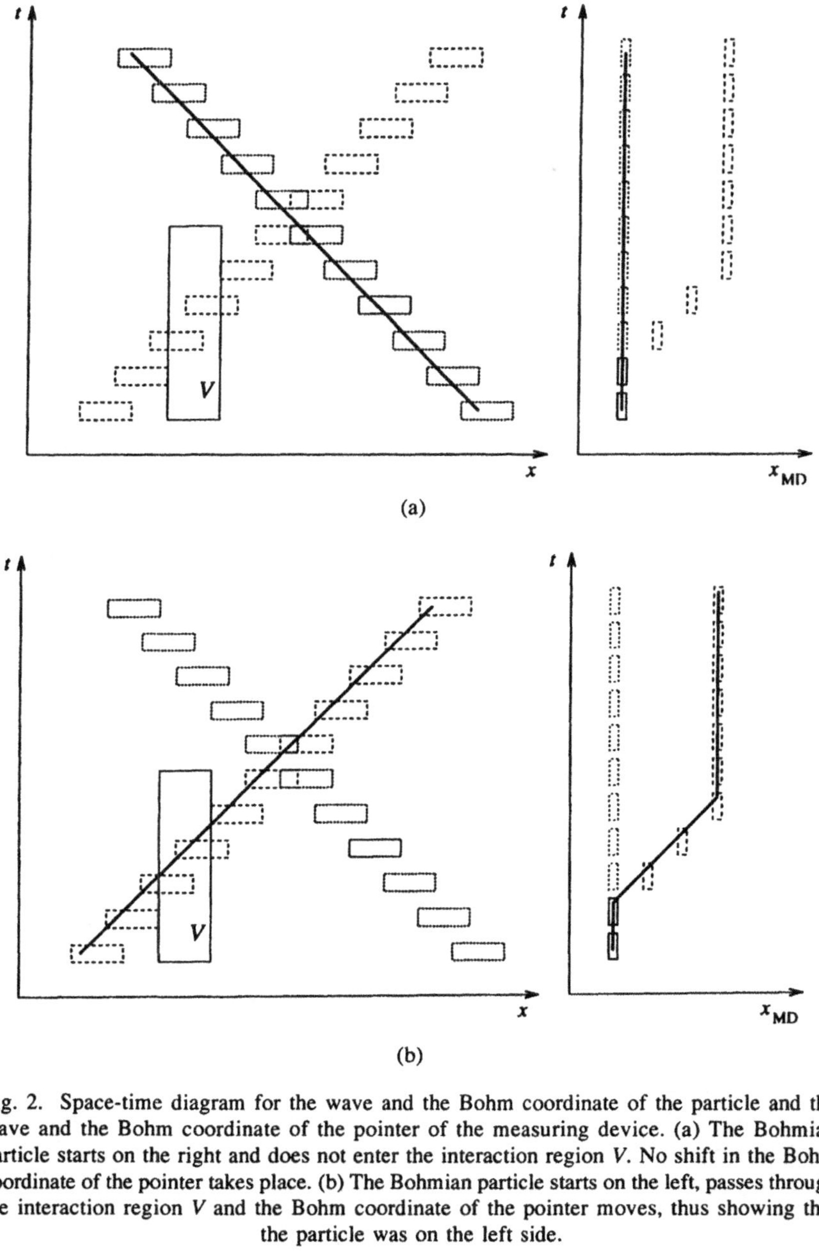

Fig. 2. Space-time diagram for the wave and the Bohm coordinate of the particle and the wave and the Bohm coordinate of the pointer of the measuring device. (a) The Bohmian particle starts on the right and does not enter the interaction region *V*. No shift in the Bohm coordinate of the pointer takes place. (b) The Bohmian particle starts on the left, passes through the interaction region *V* and the Bohm coordinate of the pointer moves, thus showing that the particle was on the left side.

measurement shows that the article is on the left if and only if the Bohmian particle was on the left. And, also, observable action on the measuring device occurs only when the Bohmian particle is at the location of the measuring device.

Let us turn now to an example in which there is some difficulty with the causal interpretation. We will consider again the same situation but, instead of a robust measurement, we will discuss "weak measurement". This is a standard measurement with a weakened coupling. The difficulty appears when we consider the pre- and post-selected ensemble. The particles are all pre-selected in the initial state which is the superposition of the two wave packets as described above. At the end we observe the location of the particle and consider only the cases in which the particle is found in the right side. Contrary to the example above, no *a priori* assumption is made about the initial Bohm coordinates. Since the final measurement is considered to be robust, we assume that the final Bohm coordinate is in the right side too.

For simplicity, instead of taking the usual "weak measurement", we will consider now a simple model which, in this case, exhibits similar features. We will assume that the wave function of the pointer has a rectangular form and that its spread during the process can be neglected. The coupling constant $g(t)$ is taken to be very small such that the shift caused by the particle passing through the volume V is equal to 10% of the width of the wave packet (see Figure 3). In this situation one measurement yields usually (in 90% of the cases) no information, but the outcomes of a number of results obtained on a large ensemble of identical pre- and post-selected systems will yield a clear answer: the particle is on the left. Indeed, the statistical outcomes are identical to those obtained from measurements on a pre-selected ensemble with a particle placed on the left side. In order to fulfill the first principle we need that in such a situation we can claim with high probability that the Bohmian particle was also on the left. This is, however, not so.

Assuming uniform distribution of the Bohmian particles we can easily see that in 90% of the cases, in spite of the results of the measurements, the Bohmian particle was not there. Indeed, if the Bohmian particle of the pointer was anywhere in the last 90% of its wave packet (i.e., not close to the beginning edge) and the Bohm coordinate of the particle was on the right, then it will behave in the same way as in the first example: the Bohmian particle will move together with the right wave packet, stops when it enters the region of the overlap of the two wave packets until it is taken to the right by the left wave packet (see Figure 3(a)). For Bohmian particle starting on the left side the situation is similar. In 90% cases, when the pointer Bohmian particle is in the beginning part of the wave function, the Bohmian particle makes the turn and, only when the Bohm's pointer coordinate is in the last 10% percent of the wave function, does it go straight to the right (see Figure 3(b)). Thus, the 90% of the Bohmian particles which ended at the right side, started from the right, and only 10% started from the left. The apparent difficulty is that the outcomes of the measurements, if interpreted in a natural way, correspond to all particles being on the left side, while only 10% of the Bohmian particles were there.

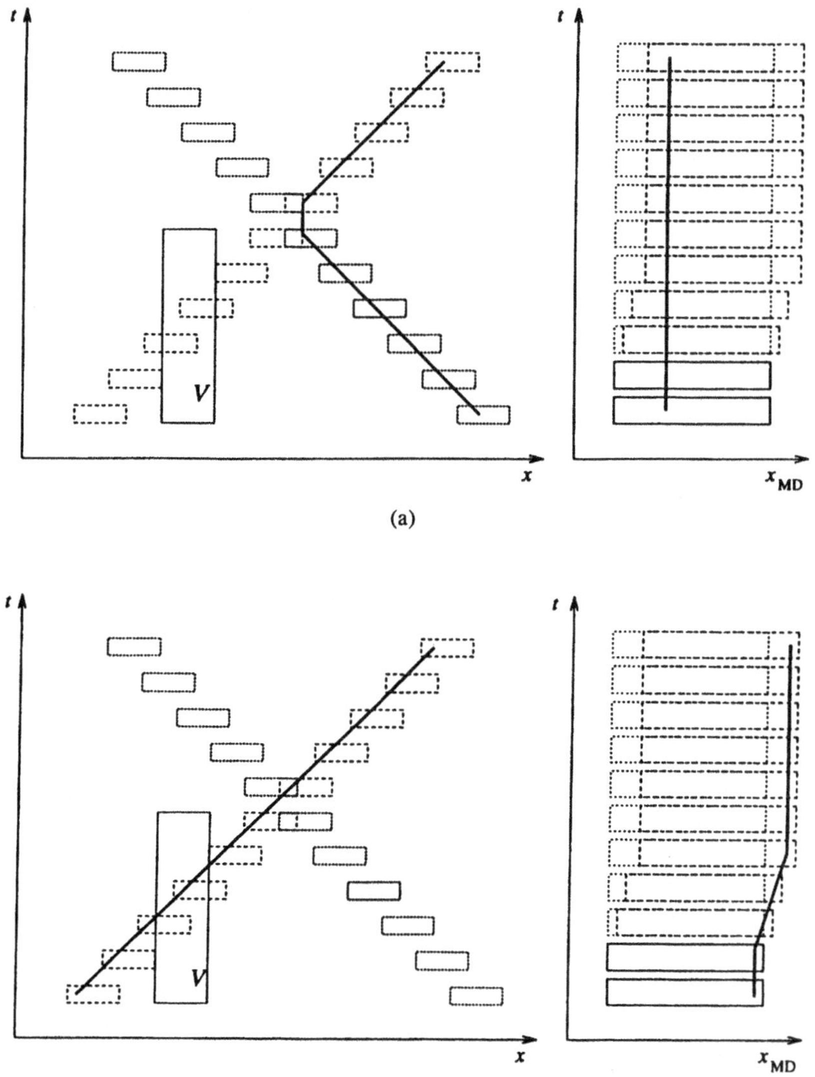

Fig. 3. Space-time diagram for the "weak measurement". Only the cases when the particle ends up on the right are considered. (a) The Bohm coordinate of the particle starts on the right and the Bohm coordinate of the pointer starts anywhere in the right 90% of the wave. (b) The Bohm coordinate of the particle starts on the left and the Bohm coordinate of the pointer starts anywhere in the right 10% of the wave. The final distribution of the Bohm coordinates of the measuring device cooresponds (according to the usual, pre-select only, situation) to the particles which were initially on the left. Indeed, the Bohm coordinate of the pointer cannot be found at the end in the first (left) 10% of the initial wave. Nevertheless, 90% of the Bohmian particles started on the right.

It is important to emphasize that we have difficulty only with the first principle. The change in the measuring device in all cases can be seen as a direct influence of the nonempty waves. Although we had just 10% of particles on the left side, for all of them the Bohmian particle of the pointer was at the far end, such that after the shift due to the interaction, they fill the missing part of the spatial distribution of the pointer relative to the distribution corresponding to no particle on the left.

The difficulty which we have in this example can be seen only on an ensemble of pre- and post-selected systems. We may try to consider the whole ensemble of N particles as a single system, but then the difficulty will not appear. Indeed, in this case we have to consider a single measuring device. If we take the same measuring device which is coupled, one after the other, to all particles, then the picture becomes different. Straightforward analysis shows that for $N \gg 10$ most of Bohmian particles start from the left and pass the volume V.

In order to explain our next example, which is a weak measurement performed on a single (pre- and post-selected) system, we will start with a brief review of the two-state vector formalism. In 1964 Aharonov, Bergmann and Lebowitz considered measurements performed on a quantum system between two other measurements, results of which were given. They proposed describing the quantum system between two measurements by using two states: the usual one, evolving towards the future from the time of the first measurement, and a second state evolving backwards in time, from the time of the second measurement. If a system has been prepared at time t_1 in a state $|\Psi_1\rangle$ and is found at time t_2 in a state $|\Psi_2\rangle$, then at time t, $t_1 < t < t_2$, the system is described by

$$(2) \qquad \langle\Psi_2| e^{i\int_{t_2}^{t} H dt} \qquad \text{and} \qquad e^{-i\int_{t_1}^{t} H dt} |\Psi_2\rangle.$$

For simplicity, we shall consider the free Hamiltonian to be zero; then, the system at time t is described by the two states $\langle\Psi_2|$ and $|\Psi_1\rangle$. In order to obtain such a system, we prepare an ensemble of systems in the state $|\Psi_1\rangle$, perform a measurement of the desired variable using separate measuring devices for each system in the ensemble, and perform the post-selection measurement. If the outcome of the post-selection was not the desired result, we discard the system and the corresponding measuring device. We look only at measuring devices corresponding to the systems post-selected in the state $\langle\Psi_2|$.

The basic concept of the two-state approach, the weak value of a physical variable A in the time interval between pre-selection of the state $|\Psi_1\rangle$ and post-selection of the state $|\Psi_2\rangle$, is given by (Aharonov and Vaidman 1990)

$$(3) \qquad A_w \equiv \frac{\langle\Psi_2|A|\Psi_1\rangle}{\langle\Psi_2|\Psi_1\rangle} .$$

Let us show briefly how weak values emerge from a measuring procedure with a sufficiently weak coupling. We consider a sequence of measurements: a pre-selection of $|\Psi_1\rangle$, a (weak) measurement interaction of the form of Eq. (1), and a post-selection measurement finding the state $|\Psi_2\rangle$. The state of the measuring device (which was initially in a Gaussian state) after this sequence is given (up to normalization) by

$$(4) \qquad \Phi(Q) = \langle\Psi_2|e^{-iPA}|\Psi_1\rangle e^{-Q^2/2\Delta^2}.$$

In the P-representation we can rewrite it as

$$(5) \qquad \tilde{\Phi}(P) = \langle\Psi_2|\Psi_1\rangle e^{-iA_wP}\, e^{-\Delta^2P^2/2} + \langle\Psi_2|\Psi_1\rangle \sum_{n=2}^{\infty} \frac{(iP)^n}{n!}\,[(A^n)_w - (A_w)^n]e^{-\Delta^2P^2/2}.$$

If Δ is sufficiently large, we can neglect the second term of (5) when we Fourier transform back to the Q-representation. Large Δ corresponds to weak measurement in the sense that the interaction Hamiltonian (1) is small. Thus, in the limit of weak measurement, the final state of the measuring device (in the Q-representation) is

$$(6) \qquad \Phi(Q) = e^{-(Q - A_w)^2/2\Delta^2}.$$

This state represents a measuring device pointing to the weak value, A_w.

Although we have showed this result for a specific von Neumann model of measurements, the result is completely general: any coupling of a pre- and post-selected system to a variable A, provided the coupling is sufficiently weak, results in effective coupling to A_w. Since Δ has to be large, the weak coupling between a single system and the measuring device will not, in most cases, lead to a distinguishable shift of the pointer variable, but collecting the results of measurements on an ensemble of pre- and post-selected systems will yield the weak values of a measured variable to any desired precision.

As an example, consider a spin-1/2 particle prepared with the spin state pointing in the x direction and found later with the spin in the y direction. We consider weak measurement of the spin component in the ξ direction which is the bisector of $\hat{x}$ and $\hat{y}$, i.e., $\sigma_\xi = (\sigma_x + \sigma_y)/\sqrt{2}$. Thus, $|\Psi_1\rangle = |\uparrow_x\rangle$, $|\Psi_2\rangle = |\uparrow_y\rangle$, and the weak value of σ_ξ in this case is:

$$(7) \qquad (\sigma_\xi)_w = \frac{\langle\uparrow_y|\sigma_\xi|\uparrow_x\rangle}{\langle\uparrow_y|\uparrow_x\rangle} = \frac{1}{\sqrt{2}}\frac{\langle\uparrow_y|(\sigma_x + \sigma_y)|\uparrow_x\rangle}{\langle\uparrow_y|\uparrow_x\rangle} = \sqrt{2}.$$

This value is, of course, "forbidden" in the standard interpretation where a spin component can obtain the (eigen)values ±1 only. If the angle between the spin directions of the initial and final states is close to 180°, then the weak measurement in the direction of the bisector can be even much larger (Aharonov *et al.* 1988).

Discussing spin-measurements in the framework of the Bohm theory requires additional care. The most elegant way to analyze spin in causal interpretation is not to attribute to it some kind of a Bohm pointer (e.g., Dewdney *et al.* 1988), but to consider the spin as a property of the quantum wave. Then the probability one-half in the result of a spin z measurement of a spin prepared in, say, the $\sigma_x = 1$ state corresponds to uniform distribution of the Bohmian particles inside the wave packet. If the Bohmian particle is in the upper half of the wave, the Stern–Gerlach measurement will end up with the particle going up; otherwise it will go down. This is a manifestation of *contextuality* of quantum measurement: the fact that the particle goes up is interpreted as spin "up" or spin "down" depending on the direction of the gradient of the magnetic field in the Stern–Gerlach device (see, for example, Albert 1992, 153–155).

Bohm's picture nicely explains the peculiar result of a weak measurement of the spin component. Due to the coupling of the weak measurement, the Bohm coordinates of the pointer variable (which is the coordinate of the particle) are not shifted by the "forbidden" value $(\sigma_\xi)_w$. The interaction causes correlations such that the post-selection measurement will pick the Bohmian particle from the appropriate region. In the case of a large value of $(\sigma_\xi)_w$, the post-selected Bohmian particles are coming from the "tail" of the wave function. We will not present here the derivation of this result, but we will give a hint how it works. Assume that the initial state of the particle is a Gaussian with the spread Δ. Then, to calculate the motion of the Bohmian particles in the post-selection measurement we have to consider a mixture of two, almost identical, slightly shifted Gaussians. The particles in a certain region go towards a certain (post-selection) direction if the "weight" of one packet is significantly larger than that of the other. But this, for very small shift, happens far away in the tail: $e^{-(x-\delta)^2/\Delta^2}/e^{-x^2/\Delta^2} \gg \Delta^2/\delta$.

The example we want to consider here is less interesting for practical applications, but more striking conceptually. It is a weak measurement performed on a single system with a particular (very rare) post-selection which does lead to a distinguishable shift of the pointer. Such was the example considered in the first work on weak measurements (Aharonov *et al.* 1987). In this work a single system of a large spin N is considered. The system is pre-selected in the state $|\Psi_1\rangle = |S_x = N\rangle$ and post-selected in the state $|\Psi_2\rangle = |S_y = N\rangle$. At an intermediate time the spin component S_ξ is weakly measured and again the "forbidden" value $\sqrt{2}N$ is obtained. The uncertainty has to be only slightly larger than $\sqrt{N}$. The probability distribution of the results is centered around $\sqrt{2}N$, and for large N it lies clearly outside the range of the eigenvalues, $(-N, N)$. More generally, we can consider pre-selection and post-selection in the directions such that the angle with the bisector $\hat{\xi}$ is θ. Then, the weak value of S_ξ is $(S_\xi)_w = N/\cos\theta$ and it can be much larger than N.

Our purpose here is to consider a position measurement performed on a single system. This is, however, different from the "position" measurements we discussed above. In previous examples we measured a projection operator on a certain volume. Now we consider a measurement of the operator of position. Our example is similar to the spin-N particle experiment discussed above. We consider a massive particle which was prepared in a state $|\Psi_1\rangle$ and was found at time t_2 in a state $|\Psi_2\rangle$. The states $|\Psi_1\rangle$ and $|\Psi_2\rangle$ are the superpositions of very well localized wave packets around locations between -1 and 1. The states, up to normalization, are

$$(8) \qquad |\Psi_1\rangle = \sum_{i=0}^{N} \left| x = \frac{N-2i}{N} \right\rangle, \qquad |\Psi_2\rangle = \sum_{i=0}^{N} c_i \left| x = \frac{N-2i}{N} \right\rangle,$$

where $c_i = (-\tan^2 \theta/2)^i/i!(N-i)!$. In this case the weak measurement (the standard measurement of position with precision of the order $\sqrt{N}$) yields a number around the weak value of x which is $x_w = 1/\cos\theta$. Mathematically, this is identical to the superposition of small forces which is equivalent to a large force, as discussed by Aharonov, Anandan, Popescu and Vaidman (1990). The details can be found there.

If we choose θ and N such that $\sqrt{N} \gg 1/\cos\theta$, we may view this experiment as a good measurement of position: we can repeat the procedure several times (including the pre-selection, the position measurement and the post-selection) and we will obtain values around x_w with relatively small statistical spread. Nevertheless, the outcome of this kind of position measurement has no relation to the Bohmian particle position; the latter can have only values between -1 and 1. It is not that there is any real contradiction, or that the causal interpretation cannot explain the outcome of the measurement. If we consider the Bohmian particle position of the pointer variable, it will be around the value x_w. Again, it was not shifted by the coupling of weak measurement, but "picked" from the tail in the post-selection process, and this is due to the peculiar wave function correlation created by the whole process. The unfortunate feature is that again, the first principle is not filled in this specific case: the procedure which is very close to the usual position measurement yields consistently results which have no relation to the Bohmian particle position.

Another example where the Bohmian particle position does not help to understand the result of measurement is the *protective measurement* (Aharonov and Vaidman 1993) of the Schrödinger wave function. If the only "reality" is the Bohmian particle position and the Schrödinger wave is just a pilot wave which governs the motion of the particle, then the wave should not be observable directly. Let us briefly show how, in certain cases, we can observe the complete Schrödinger wave of a single particle, in spite of the fact that the Bohmian particle position represents almost no features of the wave.

As an example of a simple protective measurement, let us consider a particle in a discrete nondegenerate energy eigenstate $\Psi(x)$. The standard von Neumann procedure for measuring the value of the projection operator, Π_V, on the volume V involves an interaction Hamiltonian (1). In the protective measurements $g(t) = 1/T$ for most of the time T and it goes to zero gradually before and after the period T. For $g(t)$ smooth enough we obtain an adiabatic process in which the particle cannot make a transition from one energy eigenstate to another, and, in the limit $T \rightarrow \infty$, the interaction Hamiltonian does not change the energy eigenstate. For any given value of P, the energy of the eigenstate shifts by an infinitesimal amount given by first order perturbation theory: $\delta E = \langle H_{int} \rangle = \langle \Pi_V \rangle P/T$. The corresponding time evolution $e^{-iP\langle \Pi_V \rangle}$ shifts the pointer by the average value $\langle \Pi_V \rangle$. By measuring the averages of a sufficiently large number of projection operators of different regions, the density of the full Schrödinger wave can be reconstructed to any desired precision.

Again, we would like naively to think that the projection operator Π_V should yield 1, if the Bohmian particle is inside V and zero if it is outside V. The measurement, however yields $|\Psi(x)|^2$ irrespectively of where the Bohmian particle is. If the system is in an energy eigenstate, then the Bohmian particle does not move, so, in fact, for measurements of almost all projection operators the Bohmian particle is outside the volume during the whole period of the measurement, and for one projection operator it is always inside. Thus, the second principle is broken here: the measuring devices measuring $|\Psi(x)|^2$ in the "empty" volumes yield non-null outcomes.

More precisely, the measuring interaction changes the state of the system and the Bohmian particle starts moving. However, in the adiabatic limit its velocity is very small. Certainly, the bohmian particle does not visit every volume many times during the measurement in such a way that the average time inside the volume V is proportional to $|\Psi(x)|^2$ – a possible naive explanation based on Bohm's ontology.

The breakdown of the second principle (as well as the first one) can be seen much more clearly in an experiment suggested recently by Englert *et al.* (1992). In order to explain their idea, we consider again our first example in which the initial state of the particle is the superposition of two identical wave packets moving in opposite directions; the Bohmian particle initially is inside the right wave packet and we measure on the left side the existence of the particle. Our measurement is robust in the sense that the measuring device evolves into an orthogonal state if the particle is in the left side, but the density of the wave function of the measuring device is not changed significantly during the time of motion of the particle. Englert *et al.* considered a spin flip which is a very clean example, but, an example which fits better the spirit of the present work is a standard von Neumann interaction which makes a shift of the wave function of the pointer in the momentum representation. Then the Hamiltonian is $H = g(t)Q\Pi_V$, instead of (1). In fact, the realization of the von Neumann measurement in the laboratory is of this kind: in the Stern–Gerlach experiment the pointer is the particle itself and it gets a shift in momentum due to the interaction. Only later, the shift in momentum is translated into the shift in position. We consider the case that the shift in momentum will lead to the shift in position comparable with the width of the spatial wave function of the pointer only much later, such that at the time of the overlap of the wave packets of the particle, the two wave packets of the pointer, the one shifted in momentum which correlated to the left wave packet of the particle and the other, undisturbed, which correlated to the right wave packet, will essentially coincide. In this case, the behavior of the Bohmian particle will be as in the case that no measurement was performed at all (see Figure 4). The Bohmian particle will stop in the area of the overlap and will finally change the direction of its motion. The particle started on the right will end up on the right side and will never reach the left side. The Bohmian particle of the pointer, however, will follow the wave packet which got the shift in the momentum: it eventually will move far away from its original place. This corresponds to what we usually consider the result of the measurement saying that the particle is on the left side. Therefore, it goes against the first principle. But it also goes against the second principle.

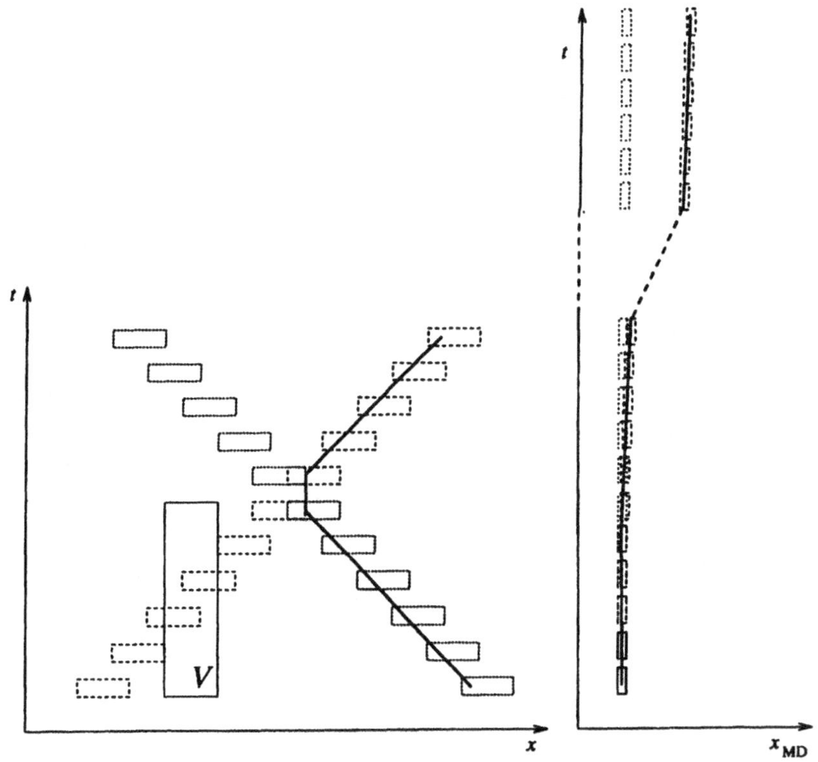

Fig. 4. Space-time diagram for the "delayed measurement". The Bohmian particle does not pass through the interaction region V, but, nevertheless, the Bohm coordinate of the measuring device moves. Note that it starts moving not immediately after the interaction in the region V, but only after the overlap of the two wave packets of the particle and that the Bohm coordinate of the pointer moves out of the location of its initial wave packet only at much later time.

The Bohmian particle was never on the left side. There was only an "empty" wave there. Nevertheless, it causes the action of the measuring device located on the left side; the Bohmian particle of the pointer moves because of the existence of the particle.

It is instructive to see how, exactly, the empty wave causes the change in the measuring device. At the time of the interaction between the particle and the measuring device, when the empty wave passes the region V, no change in the motion of the Bohmian particle of the pointer takes place. Only after the overlap of the empty and nonempty waves of the particle, the Bohmian particle of the pointer starts moving. Note, that if we "look" on the problem in configuration space, then no surprising behavior is observed. The empty wave in configuration space does not influence the motion of the Bohmian particle in configuration space until the particle is "inside" the wave and the wave ceases to be empty. The difficulty is seen only in the ordinary three dimensional space. It is even more dramatic when our particle moves inside a special bubble chamber. The bubbles created due to the passage of the particle are developed slowly enough such that during the time of the motion of the particle the density of the spatial wave function of each bubble does not change significantly. Then, what we will see as a trace of bubbles is the particle moving from the left to the right, while the Bohmian particle, in fact, will move from the right side, stop, and come back to the right. Note, however, that the picture we see via appearance of the bubbles is time delayed. In a "fast" bubble chamber, in which we see a real time motion, the Bohmian particle moves together with the trace of bubbles.

The examples considered in this work do not show that the Bohm's causal interpretation is inconsistent. It shows that Bohmian trajectories behave not as we would expect from a classical type model. Before seeing the examples discussed in this work we could think otherwise. In fact we did, and we were very reluctant to accept that the Bohmian picture is different, in some cases, from a naive picture based on the outcomes of the experiments; we worked hard, but in vain, searching for an error in our and the Englert *et al.* arguments. If we follow David Bohm, viewing his theory as an alternative formalism which should lead us to fruitful generalizations and modifications, the difficulties we have discussed can be considered on the positive side as showing a direction for constructing a better theory. However, in order to put our work in a proper perspective, we have to note that the proponents of the Bohm theory do not see the phenomena we described here as difficulties of the theory; see, for example, Dürr, Fusseder, Goldstein and Zanghi (1993). The fact that we see these difficulties follows form our particular approach to the Bohm theory in which the wave is not considered to be a "reality". Thus, our difficulties are not surprising in the light of the words of Bell (1987a, 128): "*No one can understand this theory until he is willing to think of* ψ *as a real objective field rather than just a 'probability amplitude'.*" Although we are certainly sympathetic to the approach in which ψ is real (see Aharonov and Vaidman 1993), one of us (L.V.) does not see it fit well into the framework of the Bohm theory, since in his view, the main purpose of the Bohm theory is to avoid the reality of the many worlds that he believes is incorporated in the wave function of the Universe (Vaidman 1993).

We are grateful to Sheldon Goldstein for illuminating correspondence which clarified and sharpened enormously this work. This research was supported in part by grant 614/95 of the Basic Research Foundation (administered by the Israel Academy of Sciences and Humanities).

Tel-Aviv University

P. N. KALOYEROU

AN ONTOLOGICAL INTERPRETATION OF BOSON FIELDS

Although there are an increasing number of physicists who are dissatisfied with Bohr's complementarity, complementarity nevertheless remains an essential element of the commonly accepted interpretation of the quantum theory. Complementarity can be criticised on many grounds, but the main motivating element for me to seek an alternative interpretation is the following. I take Bohr's complementarity to deny the possibility of a description of underlying physical reality, or equivalently, to deny that a single well-defined model consistent with the quantum theory is possible.[1]

In 1952 Bohm introduced just the single well-defined model considered impossible by complementarity (Bohm 1952a; 1952b). Because of similarities (though there are important differences) with de Broglie's pilot wave theory (a simplification of his theory of the double solution), the model has come to be called the Bohm–de Broglie causal interpretation (de Broglie 1956). The model consists of objectively existing particles, guided by the wave function $\psi = Re^{(i/\hbar)S}$, where R and S are regarded as objectively existing fields which codetermine one another. Mass is localised at the position of the particle, and dynamical variables such as position, energy, and momentum are attributed precise values at every instant of time.

Here we consider an extension of the causal interpretation to relativistic boson fields. We emphasise that we adopt a view that classical ideas being close to the way we think are useful to aid thought and will be used as far as possible. We reject that oft quoted view that the newness of the quantum theory makes heretic a return to older discredited classical ideas.

Because of the different statistics satisfied by bosons and fermions, it is not clear whether the ontology of these entities should be the same or different. A criterion that has been introduced by Bohm, regarded as preliminary by the present author, is that where the classical limit of the equation of motion of the field is a wave equation, then the entity can be consistently regarded as an objectively existing field, but where the classical limit is a particle equation, then the entity must be regarded as an objectively existing particle. The former is the case for bosons, such as the electromagnetic field and the mesons, and the latter for fermions. The problem with this criteria is that the field ontology of bosons is in direct conflict with that of fermions when it is recalled that some bosons are fermion composites (e.g., mesons are quark-antiquark pairs) and quarks are fermions. It seems likely instead that fermions and bosons should have the same ontology.

An early attempt at a relativistic causal interpretation was a particle

J. T. Cushing et al. (eds.), Bohmian Mechanics and Quantum Theory: An Appraisal, 155–167.

ontology for massive, spin-zero bosons described by the Klein–Gordon equation (Kaloyerou 1985; 1995). To resolve the negative probability difficulties, the Klein–Gordon equation was modified by the introduction of an evolution parameter (as an alternative to its reinterpretation as a field equation, which leads to a field ontology). This approach, however, leads to a model which suffers from a number of pathologies which make the model untenable. In the absence of any reasonable alternative approach, we conclude that a consistent particle ontology of massive, spin-zero bosons is probably not possible. It seems likely therefore, that in order to resolve the difficulty specified in the previous paragraph, fermions will have to be treated as fields. Such a programme faces the immediate problem that fermions obey Fermi-Dirac statistics and hence the Pauli exclusion principle. In this case, it is difficult to see how to build anything resembling a classical field from fermions.

The current situation for fermions is that Bohm and Hiley have proposed a model for fermions as objectively existing particles through an interpretation of the Dirac equation (Bohm and Hiley 1989; 1993). Valentini, on the other hand, has proposed that the Dirac equation be rejected in favour of the mathematical equivalent van der Waerden equation and that fermions be treated as an objectively existing field of Grassmann numbers (Valentini 1992). Valentini's approach is promising, but before it can be accepted, the question of how two apparently equivalent equations yield two opposing ontologies must be addressed.

The causal interpretation of boson fields was first suggested by Bohm in the appendix of his 1952 article (Bohm 1952b). There, the electromagnetic field was treated together with the photoelectric and Compton effects. The extension to other bosonic fields follows along very similar lines. Specifically, the extension to the complex Klein–Gordon equation, describing spin-zero charged massive bosons, was given by the present author with Bohm (the treatment of the real Klein–Gordon equation describing spin-zero uncharged massive bosons is virtually identical) (Kaloyerou 1985); and to massless scalar particles by the present author, Bohm and Hiley (this unrealistic case was chosen for its simplicity) (Bohm, Hiley and Kaloyerou 1987). A detailed treatment of the electromagnetic field was given by the present author (Kaloyerou 1994).

Using the electromagnetic field as a typical example, we may proceed to outline the principles, emphasising that despite mass and charge the interpretation follows through in exactly the same way for the Klein–Gordon fields. Essentially, we will give a summary of Kaloyerou (1994).

Second quantization in usual quantum field theory is effected by imposing non-zero equal-time commutation relations on the field functions and their conjugate momenta. The causal interpretation begins by noticing that the procedure is exactly equivalent to introducing what we shall call a field Schrödinger equation.[2]

For the electromagnetic field the appropriate (equal-time) commutation relations are[3]

$$[\Pi_i(\mathbf{x}, t), A_j(\mathbf{x}', t)] = -i\hbar\delta^{tr}_{ij}(\mathbf{x} - \mathbf{x}'),$$

where δ^{tr}_{ij} is the transverse delta function, A_j are the components of the vector potential, and Π_j their corresponding conjugate momenta. We adopt the radiation (Coulomb, or transverse) gauge where $\nabla \cdot \mathbf{A} = 0$ and the scalar potential $\phi = 0$, and use Heavyside units.

The field Schrödinger equation is

$$\int \mathcal{H}(\mathbf{A}(\mathbf{x}', t), \Pi(\mathbf{x}', t))\Phi[(\mathbf{A}(\mathbf{x}, t), t)]d\mathbf{x}' = i\hbar \frac{\partial\Phi[(\mathbf{A}(\mathbf{x}, t)]}{\partial t},$$

where $\mathcal{H}$ is the Hamiltonian density (corresponding to the Heisenberg–Pauli Lagrangian density) given by

$$\mathcal{H} = \frac{1}{2}(\mathbf{E}^2 + \mathbf{B}^2) = \frac{1}{2}[c^2\Pi^2 + (\nabla \times \mathbf{A})^2].$$

E and **B** are the electric and magnetic field strengths, respectively.

With the field Schrödinger equation established, we may proceed in complete analogy with Bohm's non-relativistic causal interpretation. By making the substitution $\Phi = R[\mathbf{A}(\mathbf{x}, t), t]e^{(i/\hbar)S[\mathbf{A}(\mathbf{x}, t), t]}$ into the field Schrödinger equation, and separating real and imaginary terms, we obtain a continuity equation

$$\frac{\partial R^2}{\partial t} + c^2 \int \frac{\delta}{\delta \mathbf{A}'}\left(R^2 \frac{\delta S}{\delta \mathbf{A}'}\right) d\mathbf{x}' = 0,$$

and a Hamilton–Jacobi equation

$$-\frac{\partial S}{\partial t} = \frac{1}{2}\int \left[\left(\frac{\delta S}{\delta \mathbf{A}'}\right)^2 c^2 + (\nabla \times \mathbf{A}')^2 + \left(-\frac{\hbar^2 c^2}{R}\frac{\delta^2 R}{\delta \mathbf{A}'^2}\right)\right] d\mathbf{x}' = 0,$$

The Hamilton–Jacobi equation differs from the classical Hamilton–Jacobi equation by the extra term

$$Q = -\frac{1}{2}\int \frac{\hbar^2 c^2}{R}\frac{\delta^2 R}{\delta \mathbf{A}'^2} d\mathbf{x}',$$

which we call the field quantum potential.

By analogy with classical Hamilton–Jacobi theory for a field, the total energy and conjugate momentum are given by

$$E = -\frac{\partial S}{\partial t}, \qquad \Pi = \frac{\delta S[\mathbf{A}, t]}{\delta \mathbf{A}'}.$$

The expectation value of an operator $\hat{\mathrm{O}}$ is given by

$$\bar{\mathrm{O}} = \int \Phi^* \hat{\mathrm{O}} \Phi \mathcal{D}\mathbf{A}(\mathbf{x}, t).$$

where $\mathcal{D}\mathbf{A}(\mathbf{x}, t)$ denotes functional integration.

In order that our formalism be mathematically equivalent to the usual one, we must make the following three assumptions (analogous to the assumptions of the nonrelativistic causal interpretation):

1. The wave functional $\Phi[\mathbf{A}, t]$ satisfies the field-Schrödinger equation.
2. The momentum conjugate to $\mathbf{A}$ is given by $\Pi = \delta S[\mathbf{A}]/\delta\mathbf{A}'$.
3. For any given state of the field, the initial field configuration is not precisely known, but is given with a certain probability derived from the probability density $P[\mathbf{A}] = |\Psi[\mathbf{A}]|^2$.

We see from the above that the state of the field is represented by a wave functional $\Phi[\mathbf{A}(\mathbf{x}, t)]$. A functional is a rule which assigns a number to each function of some class. Only those functionals which are square integrable, and hence can be normalised, are regarded as physically acceptable:

$$\int |\Phi|^2 \mathcal{D}\mathbf{A} = 1.$$

In this case $|\Phi|^2$ can be interpreted as the probability density for a given $\mathbf{A}(\mathbf{x}, t)$ (i.e. for a given field configuration). The continuity equation expresses the conservation of probability in function space.

To obtain an ontology it is then most natural to postulate that the wave functional is the probability amplitude for finding an objectively existing field $\mathbf{A}(\mathbf{x}, t)$. The probability element enters in that we cannot know the initial $\mathbf{A}(\mathbf{x}, t = 0)$, but can only determine it with some probability. The initial field configuration is thus regarded as a hidden variable. Compare this to the wave function of the nonrelativistic theory which gives the probability of localised detection, for which it is natural to postulate that it is the probability amplitude for finding an objectively existing particle, and hence leading to a particle ontology.

The fact that the wave functional cannot be interpreted as a probability for localised detection is already a strong indication that a particle ontology for the electromagnetic field is not possible. For further reasons against the possibility of a particle ontology of the electromagnetic field see Bohm, Hiley and Kaloyerou (1987, 366).

To develop the ontology, we further postulate that the terms $E = -(\partial S/\partial t)$, and $\Pi = \delta S/\delta\mathbf{A}$ in the Hamilton–Jacobi equation represent objectively existing values of the total energy and conjugate momentum, respectively. By substituting $\Pi = \delta S/\delta\mathbf{A}$ into the classical formulae for the momentum density, angular momentum density, and spin for the electromagnetic field we obtain what can be regarded as objectively existing values of these quantities. The argument of the integral in the Hamilton–Jacobi equation can be viewed as an energy density. The total energy, momentum, angular momentum, and spin can be obtained be integrating the corresponding densities over all space. We emphasise that the values so defined are pre-measurement values and will not equal the values obtained upon measurement, except in the special case of eigenstates.

The argument of the integral in the expression of the field quantum potential may be viewed as the field quantum potential density and is also considered to be an objectively existing field. In the sense that if the field quantum potential term were absent from the Hamilton–Jacobi equation we would simply have a classical field, we can say that the field quantum potential gives rise to the quantum properties of the field. However, it must be remembered that many aspects of the quantum behaviour of the field can also be calculated by use of the S-functional. This fact has prompted some authors, incorrectly in the author's view, to abandon the quantum potential altogether, particularly with regard to the nonrelativistic case.

It appears from the above that we can maintain an ontology of the electromagnetic field (and of bosons generally) similar to that of classical fields. There is, however, a crucial difference. In general the quantum field possesses an inherent nonlocality in the sense that the development in time of the field at one spatial point depends on the field at all other points. This implies that a local interaction at one point in the field affects the field at all other points.

Because of the nonlocal nature of the field, what we have called densities above cannot be densities in the classical sense. The change with time of a classical density in some infinitesimal volume depends only on the flow in or out of the infinitesimal volume, but the change with time of our "quantum densities" in some infinitesimal volume will in general depend nonlocally on the values of the density at all points outside of the infinitesimal volume.

The nonlocality in the field may be seen a number of ways, but it is implied immediately by the fact that the state of the field is described by a functional. One way to see this is to use the method of finite differences, employed extensively by Euler in the study variational problems, by which a functional can be approximated by a function of many variables. Applying the method of finite differences to the Hamilton–Jacobi equation, and especially to the continuity equation, helps to bring out the meaning of these equations but we shall not do this here explicitly.

The method of finite differences proceeds by dividing space into small cells of volume $\Delta\mathbf{x}_i$, where each cell is labelled by the index i. In each cell the field is described by the time dependent function $\mathbf{A}_i(t) \equiv \mathbf{A}(\mathbf{x}_i, t)$. The space vectors $\mathbf{x}_i$ are regarded not as variables but as labels marking the position of each cell in space. The $\mathbf{A}_i(t)$ can be regarded as field coordinates analogous to the generalised coordinates of a mechanical system, where i denotes the degrees of freedom. With this scheme the wave functional can be approximated as a function of many variables

$$\Phi[\mathbf{A}, t] \rightarrow \Phi\,(\mathbf{A}_1, \mathbf{A}_2, \ldots, t) \equiv \Phi(\mathbf{A}_i, t).$$

Written this way the $\Phi(\mathbf{A}_i, t)$ is analogous to the many-body wave function

$\Phi(\mathbf{x}_1, \mathbf{x}_2, \ldots, t)$ and the nonlocality is made explicit since Φ is seen to depend on the $\mathbf{A}_i$'s in all the cells.

Another way to see the nonlocal nature of a quantum field is through the equation of motion of the field function (i.e., the wave equation). The wave equation for the electromagnetic field can be obtained by taking the functional derivative of the Hamilton–Jacobi equation:

$$\nabla^2\mathbf{A} - \frac{1}{c^2}\frac{\partial^2\mathbf{A}}{\partial t^2} = \frac{\delta Q'}{\delta\mathbf{A}}.$$

In general, the functional derivative of the quantum potential $\delta Q'/\delta\mathbf{A}$ will reduce to an integral over space of the field function or function of the field function. This clearly indicates that the change with time of a field function at a point depends on the entire field and again demonstrates the inherently nonlocal nature of a bosonic field.

We may compare the above wave equation with the classical wave equation

$$\nabla^2\mathbf{A} - \frac{1}{c^2}\frac{\partial^2\mathbf{A}}{\partial t^2} = 0.$$

upon which semi-classical radiation theory is also based. It is evident that neither classical nor semi-classical radiation theory can hope to reproduce the results of the quantum theory where the functional derivative of the quantum potential does not approach zero. Classical and semi-classical radiation theories are adequate in many circumstances and, though nobody claims that classical electromagnetic theory is adequate in all circumstances, there are those who claim that semi-classical radiation theory is enough to describe the quantum nature of the electromagnetic field. Apart from our theoretical considerations above, there is an increasing number of experiments which confirm the inadequacy of semi-classical radiation theory (e.g. Mandel *et al.*'s fourth-order coherence experiments: Ghosh *et al.* 1986; Ou and Maudel 1988; Ou 1988).

These considerations complete our ontology and may be summarised by stating that the electromagnetic field (and other bosonic fields) is considered to be an objectively existing field with which we can associate precise values of dynamical field variables such as energy, momentum, angular momentum, spin and their corresponding densities at every instant of time. There are no photon particles. The field, however, differs radically from a classical field in that the field at any given point depends nonlocally on the field at all other points.

Since we have selected a particular gauge, we should comment on the gauge invariance of our model. Specifically, we have chosen the radiation gauge in which only the transverse components of the vector potential are non-zero. In the usual language we may restate this by saying that the field contains only transverse photons, though we do not mean to imply particle properties by use of the language of photons. Gauge invariance is best discussed in terms of the Gupta–Bleuler formalism which admits both longitudinal and scalar photons (in admixtures allowed by the Gupta–Bleuler Lorenz condition). It is shown in this formalism that states containing various allowed admixtures of transverse, longitudinal and scalar photons differ from an equivalent state containing only transverse photons by a gauge transformation. Gauge invariance is then demonstrated by showing that the expectation values of observable field quantities for states containing different admixtures are the same as those obtained from the equivalent state containing only transverse photons. This shows that physical results depend only on the transverse photons and that physical results are independent of the gauge chosen. From this we can conclude that our selection of the radiation gauge is admissible in that it will lead to gauge invariant results, at least as far as the statistical results of experiment are concerned. To demonstrate that the actual values of field observables as defined by our formulae are also gauge invariant requires an extension of our ideas to the Gupta–Bleuler formalism. Though we have not done this explicitly, there is no reason not to expect gauge invariant results.

Gauge freedom in the choice of potentials, both classically and quantum mechanically, has led to a tendency to regard potentials as mathematical tools without ontological significance. In this case, the validity of attempting an ontological interpretation in terms of potentials may be questioned. However, we may point that the Aharanov–Bohm effect strongly indicates (though does not confirm) that potentials have greater ontological significance than has hitherto been realised. Further, the lack of uniqueness of potentials need not force the conclusion that potentials must be relegated to mere mathematical tools. It is enough to suppose that a given physical system is associated with one and only one of a set of gauge related potentials, but that no physical measurement can reveal which one. Another possible position to take is to consider that since only transverse components (photons) contribute to physical results, then only one choice of gauge, the radiation gauge, has ontological significance.

Since we are dealing with a relativistic formalism, we comment briefly on this issue. The first point to note is that since our formalism is mathematically equivalent to the usual formalism (as long the three assumptions we have stated are maintained) and since the usual formalism is known to be consistent with relativity at the statistical level, we conclude that our formalism will also be consistent with relativity at the statistical level and, therefore, with experiment. We have seen, however, that our formalism is based on the field Schrödinger equation (equivalent to the imposition of equal-time commutation relations) which is evidently not covariant. We should expect, therefore, some non-covariant features in the description of individual processes. Although it is possible to cast the field Schrödinger equation into a covariant form by use of the Tomonaga–Schwinger multitime formalism, which leads to identical experimental predictions, this is still not sufficient to guarantee a fully covariant description of individual processes because of the inherent nonlocality of the field. Valantini has presented an interesting view that both quantum mechanics and relativity should be viewed as equilibrium theories valid when the condition $P = |\Phi|^2$ holds (Valentini 1991a; 1991b; 1992). Nevertheless, it must again be emphasised that relativity holds only at the statistical level and not necessarily at the level of the description of individual processes. We may add that, despite the possible breakdown of relativity at the level of the description of individual processes, causal paradoxes cannot be set up. This is because to set up causal paradoxes there must exist a two-way nonlocal connection between two (or more) quantum systems, whereas the nonlocal connection through the quantum potential acts only one way (in the sense that, once a measurement is performed on one member of an EPR correlated pair, the EPR correlation is broken). A further noncovariant element is the existence of a preferred frame defined by the ground state of the field. Nevertheless, it is well known that the selection of a preferred frame, even if not in the spirit of relativity, still leads to results fully consistent with relativity.

To exemplify features of our interpretation, we look at some specific free states of the field. To obtain such states we need to solve the field Schrödinger equation. Mathematically this is done most conveniently by introducing normal mode coordinates. The introduction of normal mode coordinates also provides a different perspective to view the main ideas of our interpretation.

In order to use the simpler and more physically transparent Fourier series rather than the Fourier integral, we imagine the field to be enclosed in a large volume V and impose periodic boundary conditions. Also noting that the divergence condition $\nabla \cdot \mathbf{A} = 0$ allows only two transverse components, the field function can be expanded in terms of normal modes thus:

$$\mathbf{A}(\mathbf{x}, t) = \frac{1}{V^{1/2}} \sum_{k, \mu} \varepsilon_{k, \mu} q_{k, \mu}(t) e^{i\mathbf{k} \cdot \mathbf{x}},$$

$$\Pi(\mathbf{x}, t) = \frac{1}{V^{1/2}} \sum_{k, \mu} \varepsilon_{k, \mu} \pi_{k, \mu}(t) e^{-i\mathbf{k} \cdot \mathbf{x}},$$

where the summation extends from $k = -\infty$ to $k = \infty$, and where $\varepsilon_{k, \mu}$, with $\mu = 1, 2$, are two unit vectors orthogonal to each other and to the wave vector $\mathbf{k}$.

Substituting for $\mathbf{A}$ and Π into the field Schrödinger equation we get the field Schrödinger equation in terms of normal modes

$$\frac{1}{2} \sum_{k, \mu} \left(-\hbar^2 c^2 \frac{\partial^2 \Phi}{\partial q^*_{k, \mu} \partial q_{k, \mu}} + \kappa^2 q^*_{k, \mu} q_{k, \mu} \Phi \right) = i\hbar \frac{\partial \Phi}{\partial t},$$

where $\kappa = |\mathbf{k}|$. Again substituting $\Phi(q_{k, \mu}, q^*_{k, \mu}) = R(q_{k, \mu}, q^*_{k, \mu}) e^{(i/\hbar) S(q_{k, \mu}, q^*_{k, \mu})}$ into the field Schrödinger equation and separating real and imaginary terms gives the continuity equation

$$\frac{\partial R^2}{\partial t} + \sum_{k,\mu}\left[\frac{c^2}{2}\frac{\partial}{\partial q_{k,\mu}}\left(R^2\frac{\partial S}{\partial q^*_{k,\mu}}\right) + \frac{c^2}{2}\frac{\partial}{\partial q^*_{k,\mu}}\left(R^2\frac{\partial S}{\partial q_{k,\mu}}\right)\right] = 0,$$

and the Hamilton–Jacobi equation

$$-\frac{\partial S}{\partial t} = \sum_{k,\mu}\left[\frac{c^2}{2}\frac{\partial S}{\partial q^*_{k,\mu}}\frac{\partial S}{\partial q_{k,\mu}} + \frac{\kappa^2}{2}q_{k,\mu}, q^*_{k,\mu} + \left(-\frac{\hbar^2c^2}{2R}\frac{\partial^2 R}{\partial q^*_{k,\mu}\partial q_{k,\mu}}\right)\right].$$

The last term in the Hamilton–Jacobi equation is the field quantum potential.

Again by analogy with classical Hamilton–Jacobi theory, the total energy and conjugate momenta are given by

$$E = -\frac{\partial S}{\partial t}, \qquad \pi_{k,\mu} = \frac{\partial S}{\partial q_{k,\mu}}, \qquad \pi^*_{k,\mu} = \frac{\partial S}{\partial q^*_{k,\mu}}.$$

With these definitions it is an easy matter to transcribe the formulae for the field dynamical variables (e.g., momentum, angular momentum and spin, in terms of normal mode coordinates).

The continuity equation expresses the conservation of probability in $q_{k,\mu}$-space and we can, as usual, maintain the interpretation of $R^2 = |\Phi|^2$ as a probability density, such that $|\Phi|^2 dq_{k,\mu}$ is the probability at time t that $q_{k',\mu'}$ has a value in the range $q_{k',\mu'}$ to $q_{k',\mu'} + dq_{k,\mu}$. Thus, a probability can be assigned to each $q_{k,\mu}$ and to each value of $q_{k,\mu}$. By insertion into the Fourier expansion of **A**, we can associate a probability with each field configuration and we arrive back at our earlier interpretation.

The free field Schrödinger equation can be easily solved to give the standard Fock states:

$$\Phi_{n_{k,\mu}} = \frac{(a^\dagger_{k,\mu})^{n_{k,\mu}}}{\sqrt{n_{k,\mu}!}}\Phi_0 e^{-n_{k,\mu}\kappa c t},$$

where $a^\dagger_{k,\mu}$ is the creation operator and $n = 0, 1, 2, 3 \ldots$.

The ground state is given when $n = 0$. For this state the momentum, angular momentum and spin are zero so the field in the ground state is static. The total energy is the usual infinite zero-point energy $E = \sum\hbar\kappa c/2$, with the energy density indicating that the field energy is spread uniformly over all space. The energy in the higher Fock states increases in units of $\hbar\kappa c$, and the momentum correspondingly increases in units of $\hbar\mathbf{k}$. The higher Fock states do not look very different from the ground state and, like the ground state, the field energy is distributed uniformly throughout space, indicating a character contrary to the usual textbook statement that the Fock states are particle-like. The ground state is an example of a field for which there is no nonlocality, though the higher Fock states do exhibit nonlocality.

The expectation value of the field operator $\widehat{\mathbf{A}(\mathbf{x})}$ is

$$\langle\Phi_0|\widehat{\mathbf{A}(\mathbf{x})}|\Phi_0\rangle = 0,$$

and that of the operator $\widehat{\mathbf{A}(\mathbf{x})\mathbf{A}}(\mathbf{x}')$

$$\langle\Phi_0|\widehat{\mathbf{A}(\mathbf{x})\mathbf{A}}(\mathbf{x}')|\Phi_0\rangle = \frac{1}{(2\pi)^3}\int \frac{\hbar c}{2\kappa}\, e^{i\mathbf{k}\cdot(\mathbf{x}-\mathbf{x}')}d\kappa,$$

from which we conclude that upon measurement of the field at some point there is an equal probability of finding a positive or negative value for the field at that point and that the value is most likely to be infinite. The expectation value $\widehat{\mathbf{A}(\mathbf{x})\mathbf{A}}(\mathbf{x}')$ is what is usually referred to as vacuum fluctuations, though we see that the fluctuations are not spatial or temporal but are fluctuations in values between different members of an ensemble of fields. For the first excited state the most probable field configuration is that of a plane wave.

By forming a linear superposition of Fock states, with a weighting factor that is a peaked function of the wave vector, we can form a state corresponding to a most probable field configuration which is a wave packet. The energy above zero point forms a pulse which moves through the infinite zero-point energy. For this state the nonlocality is limited to regions within the wave packet.

The coherent state is important in that for high states of excitation it is the closest quantum mechanical approach to a classical electromagnetic wave, and because a single mode laser operated well above threshold generates a coherent state. The coherent states are eigenstates of the annihilation operator, and though not orthogonal they do form a complete set, and arbitrary states can be expanded in terms of them. Like the ground state the coherent states are local and for a coherent state with a single mode excited the most probable field configuration is that of a plane wave (as with the first excited state but without the nonlocality).

The photoelectric and Compton effects are usually regarded as experimental evidence for the particle characteristics of the electromagnetic field. Indeed, quite generally the electromagnetic field in interactions with matter reveals particle-like characteristics. The essential feature that is required to explain such particle-like manifestations on the basis of our model of the electromagnetic field as a continuous field distribution is nonlocal absorption and re-emission of an electromagnetic quantum. This feature emerges from detailed calculations which have been performed for the case of the photoelectric and Compton effects by Kaloyerou (1994). In the case of the photoelectric effect it is the fact that ultraviolet light incident on alkali metals liberates electrons without time delay (within 10^{-9} s of striking the metal surface) that lead to the conclusion that the light quanta must be localised in a particle-like way. Clearly, superluminal absorption is consistent with this experimental fact and a particle concept need not be invoked. In the case of the Compton effect it is that Comptons's formula for the wavelength shift can be derived on the basis of a model of two particles colliding, together with energy and momentum conservation, that leads to the conclusion that electromagnetic quanta are particle-like. Energy and momentum conservation arise in a fully field theoretic calculation and again the model of superluminal absorption and re-emission of a field quantum is consistent with the experimental facts. Again, a particle picture need not be invoked.

Very recently Grangier *et al.* (1986) have introduced a thought-provoking experiment in which they claim to have confirmed the particle nature of the electromagnetic field. The novel feature of the experiment is the experimental realisation of a Fock state containing a single quantum by use of parametric down conversion and a gating system. The quantum is directed at a beam splitter and photo-detectors are placed in the two exit paths. They claimed that the observed perfect anticoincidence confirmed the particle nature of the electromagnetic field. However, our model easily explains this result. The single quantum is split at the beam splitter. For registration to occur an atom in one of the detectors must absorb the entire quantum, which it does superluminally (or nonlocally) from both paths.

Bohm has provided a detailed solution of the measurement problem on the basis of the non-relativistic causal interpretation (Bohm 1952a,b). Solving the Schrödinger equation for the situation in which the apparatus interacts with the quantum system results in a solution which is a superposition of terms. Each term corresponds to an eigenstate of the operator being measured correlated to an apparatus wave packet. The trajectory of the apparatus in the configuration space of the apparatus variables selects one of the terms of the superposition giving rise to a definite measurement result without the need to invoke the mathematically inconsistent collapse of the wave function.

The theory of measurement for the relativistic case follows through in the same way, irrespective of what the ontology for fermions turns out to be (recalling our earlier comments), though the role of configuration space needs to be de-emphasised. For the sake of argument, and as seems likely, suppose that fermions are "fields" as are bosons. In this case, the only new issue for the measurement problem is to explain how definite positions of macroscopic objects arise. In principle the answer is straightforward in that we need only suppose that the field configurations of fermions and bosons which combine to form macroscopic objects are sufficiently localised to give rise to fairly definite macroscopic positions (definite pointer positions). The corresponding packet selection is then in the function space of the apparatus fields. In other words, configuration space is not fundamental to the measurement process; localised field configurations serve just as well.

The description of the two-slit experiment in our model is similar to that of classical wave theory. Consider a state of the field containing one quantum or photon. Since the photon is continuously distributed in space, it passes through both slits. Interference after the slits results in much the same way as in classical theory, except that there is the additional field quantum potential which can be considered as squeezing the field to regions leading to bright fringes. The detection at the screen is, however, entirely non-classical. Registration occurs when one of the atoms in the screen nonlocally absorbs the entire photon leaving the field in its ground state. The process is essentially the same as for the photoelectric effect and the absorption position is most likely to be where the field is most intense. Thus the continuous field distribution appears as a spot on the photographic plate. Notice, however, that the registration of the spot on the photographic plate does not imply the existence of a trajectory prior to detection, and the question of "which slit" becomes irrelevant.

An EPR experiment is one way to observe the nonlocality inherent in the field. Consider an initial state of the field containing two photons,and corresponding to a most probable field configuration which is a wave packet with stationary centre, and hence zero total momentum. In this case, the field is not well localised nor is the momentum of either photon well defined. We follow the original EPR setup and choose to measure position (in the sense of the position at which a photon is absorbed) and momentum. The photoelectric effect may be considered a position measurement in the latter sense. Thus, placing an atom in one arm of the EPR-setup results in one of the photons being absorbed and an electron ejected. The field will be nonlocally changed everywhere in space to a state containing one photon corresponding to a most probable field configuration which is a wave packet with centre correlated to the position of the absorbing atom. The Compton effect results in an electron and a photon each in a definite momentum state and may therefore be used as a momentum measurement. Thus, placing an electron in one of the arms of the EPR-setup results in one of the photons of the EPR-pair being absorbed and re-emitted into a definite momentum state. Again, the field will be nonlocally changed everywhere, but this time the final state will be an eigenstate of momentum containing the remaining photon.

University of Swaziland

NOTES

[1] See, for example, Bohr (1928; 1934, especially p. 96; 1949; 1963) and Jammer (1974, especially pp. 98–99). My views on complementarity are discussed at length in Kaloyerou (1992).

[2] The equivalence of the field Schrödinger equation to the imposition of non-zero commutation relations is not normally stated, but in older field theory texts the field Schrödinger equation (though not called this) is at least written down (Heitler 1954, 61; Schweber *et al.* 1955). The equivalence is pointed to, first, by analogy with the nonrelativistic quantum theory where the Schrödinger equation underlies the operator formalism and, second, by the fact that expectation values calculated from the wave functional obtained by solving the field

Schrödinger equation (in practice, in its normal mode form) are the same as those calculated by the usual operator methods of field theory.

[3] *Notation*: Vectors are written in boldface. The functional derivative with respect to the vector potential is denoted by $\delta/\delta\mathbf{A} = (\delta/\delta A_x)\mathbf{i} + (\delta/\delta A_y)\mathbf{j} + (\delta/\delta A_z)\mathbf{k}$. The volume element for integration is denoted by $d\mathbf{x} = dxdydz$. A field function at the point $\mathbf{x}'$ is denoted by $\mathbf{A}' = \mathbf{A}(\mathbf{x}', t)$.

CHRIS DEWDNEY AND GEORGE HORTON

DE BROGLIE, BOHM AND THE BOSON

1. INTRODUCTION

Until recently most of the discussion concerning the interpretation of quantum theory has been carried out in the context of phenomena that can be adequately treated using the non-relativistic Schrödinger equation. In this context the interpretations of quantum mechanics proposed by de Broglie (1926; 1927) and by Bohm (1952a,b; Bohm and Vigier 1954) formally overlap in the one-particle case in what de Broglie called the "pilot-wave" theory and Bell called the de Broglie–Bohm interpretation.[1] Both Bohm's and de Broglie's approach yield the same guidance condition for the "particle", but they differ in what the particle *is*. For Bohm the particles are point-like whilst their guiding ψ-field exists in configuration space, whereas in de Broglie's theory the particle is considered to be a soliton-like structure in its own ψ-field, consequently the field of each particle must exist in real space (and not the abstract configuration-space of the many-particle non-relativistic Schrödinger equation). The latter idea presented grave difficulties for de Broglie, which were never satisfactorily solved and we do not consider the soliton solutions in this paper.

Of course it is possible to define, given the configuration-space wave function for a system of particles, a conditional wave function for an individual particle which is obtained by specifying the coordinates of all the particles in the system except the one of interest. One could even, using Schmidt decomposition, write down separate particle equations, but these would be highly nonlocal, containing functions of all the particle coordinates. Furthermore the individual wave functions thus defined would not be solutions of the one-particle Schrödinger equation and their evolution would be non-unitary. Consequently the often presented and rather naive version of the pilot-wave theory, which has the wave-particle duality replaced by a unity of waves and particles in real space, has to be treated carefully. The mental image this interpretation implies with both wave and particle travelling in space-time is seductively simple in the one-particle situations that one often meets in discussions of the interpretation of quantum theory, but it should be remembered that such examples are idealisations that mask the subtlety of the theory. For example, the interference observed in the two-slit experiment seems to be accounted for with the image of the de Broglie wave propagating through both slits and thereafter determining a quantum potential which guides the particle (which has passed through one slit only) into the interference fringes. This simple description masks the fact that really there are no single-particle phenomena

J. T. Cushing et al. (eds.), Bohmian Mechanics and Quantum Theory: An Appraisal, 169–189.

of interest, a more complete description would have the particles in the beam interacting with the particles in the slit diaphragm, the system would then be described by a configuration-space wave function and the image of the wave associated with the interfering particle propagating through the slits would be lost. Instead one has a many-particle system with a given configuration and the dynamics of this configuration is dependent on the evolution of the configuration-space wave function. The quantum description cannot be recovered using simple notions of space-time waves and particles. Furthermore, the particle properties, such as mass, charge and so on, cannot be considered to be naively located on the particle. These properties are also parameters that appear in the Schrödinger equation and hence determine the evolution of the configuration space wave and thus are also not localised (Brown, Dewdney and Horton 1995).

In the pilot-wave theory the long-standing interpretative puzzles, for example the measurement problem and wave-particle duality, are resolved simply by completing the quantum formalism through the postulation of individuals (the particles) which maintain their identity through the continuity of their space-time trajectories. Once the particles are reinstated in this way all of a system's dynamical variables are well defined and evolve in a continuous fashion according to continuous equations of motion. Associated with each configuration of particles one has a configuration space amplitude, and this quantity is, despite its configurational nature, a real ontological component of the theory, just as, in classical mechanics, the interaction potential energy of a set of particles is associated with a given configuration and not with the individual particles.

Many details of the approach have been worked out over the last fifteen years since the availability of the computers often necessary to explicitly solve the guidance equations became widespread. It is not our purpose here to review these calculations, as this has been adequately done in review papers and books heretofore (Dewdney *et al.* 1992; Bohm and Hiley 1993; Holland 1993; Dewdney 1997). Rather, we wish to concentrate on the difficulties associated with the extension of the approaches of de Broglie and Bohm beyond the applications covered by the Schrödinger equation, and in particular to the quantum description of boson fields.[2]

Many authors have argued that the most profound problems arise in the non-relativistic context and that, in particular, nothing is to be gained by considering more sophisticated treatments, but more recently it has become clear that certain problems can only be answered satisfactorily in the context of relativistic theories and in the context of quantum field theories (QFT). One obvious problem of this sort is the tension between the non-locality inherent in quantum theory and the requirement of Lorentz invariance. We shall not discuss this particular problem here. Instead, we turn to a set of interpretative problems that arise in the context of quantum field theories which have no counterpart in the non-relativistic Schrödinger context. Many of these problems were discussed in Redhead's "A Philosopher Looks at Quantum Field Theory" (1988) in which he poses eight questions. We shall revisit some of these questions in the context of the differing approaches to QFT proposed by de Broglie and by Bohm. In particular we shall be concerned mainly with Redhead's questions:

1. Can QFT be given a particle interpretation and indeed is there a formal underdetermination between field and particle approaches to the so-called elementary 'particles'? (Or as Redhead puts it later in the paper, Are some elementary entities *really* particles and others *really* field excitations (quanta)?)

2. Does QFT resolve the problem of wave-particle duality in quantum mechanics? and

6. Does QFT allow a distinction between matter and force?

Evidently the above questions can have both a formal and an ontological reference, but they are especially pertinent to those interpretations that attempt to provide an ontology for QFT (as do both de Broglie's and Bohm's). Both de Broglie and Bohm would argue that, even if there is an underdetermination at the formal level between field and particle approaches, this degeneracy is removed for non-formal reasons (as it must be) in an ontological interpretation of QFT. However, taking the ontological reference, de Broglie's answers to these questions are different from Bohm's. In Bohm's approach, question one is answered negatively, bosons must be interpreted as field excitations whereas fermions are genuine particles, there is no underdetermination of the ontology. Hence, question six is answered affirmatively for Bohm. For de Broglie, both bosons and fermions are to be treated on an equal footing through the particle interpretation and question six is then answered negatively. The fact that these two programmes for an ontological interpretation of QFT exist must in itself be a consequence of the formal underdetermination referred to in Redhead's question one.

In this paper we wish to review the theories of Bohm and de Broglie and ask whether or not the possible formal underdetermination of QFT also manifests itself as a possible underdetermination of the ontology. We are also interested in the picture of fundamental quantum-mechanical processes that each approach supplies. One possible motivation for developing the particle ontology for fields is that we seem to have an informal intuitive picture based on the concept of the photon as a particle, and using this picture the quantum mechanical aspects of fields seem to be easily explained. But, as we shall see, the picture that emerges when the particle approach is developed seriously is not so intuitive after all and one has to account for the appearance of continuous fields in the classical limit. Similarly, it is not intuitively clear at first sight how Bohm's continuous field approach can account for the particulate quantum-mechanical nature of the field, although it clearly has an intuitive classical limit. So, part of the motivation for this paper is to see how these alternative approaches work out when they are developed rigourously and applied to simple systems.

We first give a brief resumé of the interpretation of field theories provided by Bohm and by de Broglie and then proceed to compare the descriptions they offer of some concrete but simple phenomena. In particular, we shall be interested in the description afforded by each approach of the transition between energy eigenstates of a fermionic system induced by interaction with a scalar bosonic field and the anti-correlation of two detectors in a single-photon field. Ultimately we will find that the models employed in this paper provide only partial insight and that more sophisticated models need to be used to provide a complete picture of emission and absorption of radiation in a quantum transition. Finally, we discuss the difficulties inherent in each approach and reconsider the possible underdetermination of the ontology.

2. BOHM'S QUANTUM FIELD THEORY FOR BOSONS

Bohm (1952b) first outlined the essential details of his approach to quantum field theory, in the appendix of his paper "A Suggested Interpretation of Quantum Mechanics Using Hidden Variables II". Belinfante (1973) then provided a much more detailed account in *A Survey of Hidden-Variables Theories*. More recently, further details of the approach have been given by Bohm, Hiley and Kaloyerou (1987), Lam and Dewdney (1994a,b; Dewdney and Lam 1991), Dewdney *et al.* (1992), Dewdney (1994), Holland (1993) and Kaloyerou (1994).

As Bohm pointed out, his theory of fields is very similar to his theory of Schrödinger particles when the former is expressed in terms of normal modes. The theory of fields can, after all, be presented as the theory of a set of oscillators (one per mode) which together determine the behaviour of the field. Classically each oscillator coordinate is well defined and obeys a deterministic equation of motion. In Bohm's approach to quantum fields, the oscillator coordinates remain well-defined when the field is quantised and the motion of the oscillators becomes subject to a "super-Schrödinger" equation. Although at the level of the "super-Schrödinger" equation there are no interaction terms for these oscillators, in general their motion is correlated in Bohm's approach since the expression for the velocity of each oscillator will contain the coordinates of all the field oscillators. (One may also say that the correlation is accounted for by the action of the quantum potential, which depends on the wave functional of the field, and this accounts for all the non-classical (particle-like) behaviour of the field.)

In Bohm's quantum field theory (BQFT), the field has a well-defined amplitude and phase at each point in space and it evolves according to a deterministic, but nonlinear and nonlocal, field equation. The exact form of this equation depends on the nature of the configuration-space wave function of the field. Given the initial form of the field, the super-wave function, $\Psi(\phi)$ and the Hamiltonian, BQFT provides a continuous and deterministic description of the development of the field, $\phi(x, t)$ under all circumstances. As in the case of quantum particle systems, the causal interpretation of quantum fields shares the observed results of the orthodox theory, but has the advantage of possessing a well-defined model of individual physical systems in all contexts and, of course, the theory does not depend on the existence of an observer for its interpretation.

In Bohm's description of fields, there is no temptation to consider the system (the field) as being accompanied by the propagation of a guiding wave-field in space-time, as there is in the particle case, yet the logical relation between the system and its wave function (or the system and its properties) is the same for both particles and fields. In particular the question, referred to above in the context of the pilot-wave theory, concerning the delocalisation of the particle properties is reversed in the context of Bohm's field theory. Clearly the field properties are distributed in space; nevertheless, they are also in a sense localised as the field also displays particle-like behaviour. In interaction, both particles and fields share a single infinite-dimensional configuration-space wave function which, through the equations of motion that Bohm's theory determines, yields both the particle-like behaviour of fields and the wave-like behaviour of particles.

In Bohm's approach the familiar informal way of describing the quantum properties of fields through particles (photons) takes a radically different meaning and our mental images of, for example, atoms spitting out or engulfing a photon as they jump from one energy eigenstate to another are completely absent. Since the particle-like aspects of the field only become apparent in this type of interaction with matter, it is to be description of quantum transitions that we now turn.

2.1. *The Normal Mode Description of the Field*

Here we shall briefly review the mathematical framework of the approach of Bohm for the case of a real, scalar field, $\phi(x, t)$, confined to a one-dimensional cavity, extending between $x = 0$ and $x = L$ (we shall put $\hbar = 1$ and $c = 1$). The boundary conditions require that the field is zero at $x = 0$ and $x = L$, so that the normal modes of the field in this cavity are

$$(1) \qquad \phi_k(x) = \left(\frac{2}{L}\right)^{\frac{1}{2}} \sin kx,$$

An arbitrary state of the field may then be written as

$$(2) \qquad \phi(x, t) = \left(\frac{2}{L}\right)^{\frac{1}{2}} \sum_k q_k^{(t)} \sin kx,$$

$$(3) \qquad \left(k = \frac{n\pi}{L}\; n = 1, 2, \ldots\right).$$

As is well known, the classical equation of motion for the coordinates,

$\ldots, q_k \ldots,$ of the field is

$$\frac{\partial^2 q_k}{\partial t^2} + k^2 q_k = 0, \tag{4}$$

where Π_k is the conjugate momentum for the mode k. This is the equation of motion for a harmonic oscillator of coordinate q_k, frequency $\omega = k$ and mass $m = 1$, so the field can be represented by an infinite set of harmonic oscillators each obeying Eq. (4). The set of oscillators is quantised by the introduction of a Schrödinger equation, termed by Bohm *et al.* the super-Schrödinger equation.

$$i\hbar \frac{\partial \Psi}{\partial t}(\ldots, q_k, \ldots, t) = \frac{1}{2}\sum_k \left(-\frac{\partial^2}{\partial q_k^2} + k^2 q_k^2\right)\Psi(\ldots, q_k, \ldots, t). \tag{5}$$

It is this equation which governs the behaviour of all the field oscillators.

The usual interpretation of the wave function, $\Psi(\ldots, q_k, \ldots, t)$, is simply that $|\Psi|^2$ gives the probability density for "finding" a particular set of q_ks. To obtain a causal interpretation for the quantised field we write, just as is done for the case of quantum particle systems,

$$\Psi(\ldots, q_k, \ldots, t) = R(\ldots, q_k, \ldots, t)\exp(iS(\ldots, q_k, \ldots, t)), \tag{6}$$

where R and S are real and Ψ is a function of a configuration space of infinite dimensions, which is spanned by all he oscillator coordinates. One finds, by substituting this form of Ψ into the Schrödinger equation, that

$$\frac{\partial S}{\partial t} + \frac{1}{2}\sum_k \left(\left(\frac{\partial S}{\partial q_k}\right)^2 + k^2 q_k^2\right) - \frac{1}{2}\sum_k \frac{1}{R_k}\frac{\partial^2 R}{\partial q_k^2} = 0, \tag{7}$$

$$\frac{\partial P}{\partial t} + \sum_k \frac{\partial}{\partial q_k}\left[P\frac{\partial S}{\partial q_k}\right] = 0, \tag{8}$$

where $P = |\Psi|^2$. Taking Eq. (7) as the Hamilton–Jacobi equation for the field, $\phi(x, t)$, implies that $\partial S/\partial q_k$ is the rate of change of q_k. However, there is an additional energy term, not present in the classical equation, which is termed the super-quantum potential

$$\frac{1}{2}\sum_k \frac{1}{R_k}\frac{\partial_k^2 R}{\partial q_k^2}. \tag{9}$$

If $\partial S/\partial q_k$ is taken as the rate of change of the mode coordinate q_k, then Eq. (8) expresses the conservation of probability. If it is assumed that, at some point in time, the probability density $P = |\Psi|^2$ gives the actual distribution of the coordinates q_k in an ensemble, then P will continue to give the distribution for all subsequent times.

In general the field itself, $\phi(x, t)$, is the "hidden variable" (or beable), but within this normal mode formulation the mode coordinates, . . . , q_k, . . . , may be thought of as the "hidden variables". Each q_k has a well-defined value and therefore, according to equation (2), the field $\phi(x, t)$ is also well-defined. Each coordinate, q_k, evolves according to

$$(10) \qquad \frac{\partial q_k}{\partial t} = \Pi_k(t) = \frac{\partial S}{\partial q}(\ldots, q_k, \ldots, t),$$

so that its motion is determined in the configuration space spanned by all the cavity modes, . . . , q_k, ... Given the initial values of the mode coordinates (the set of which is referred to collectively as the initial "system point" in configuration space) and the initial wave function for the system, $\Psi(\ldots, q_k, \ldots)$, the trajectory of the system point is determined and it is this trajectory which determines the evolution of the field $\phi(x, t)$ in real space.

Evidently, in this approach there are no photon trajectories and the "photon" itself is simply an excitation of a field mode. Instead there are well-defined quantum fields which evolve continuously and deterministically, according to Eq. (10). The particle-like aspects of the field arise as a result of the influence of the super-quantum potential on the evolution of the system point, as the field interacts with matter. Bohm, Hiley and Kaloyerou (1987) have discussed this in a general way, and Lam and Dewdney (1994a,b) have carried out detailed modelling for cavity fields. These results are discussed below, but first we turn to the alternative many-particle view initiated by de Broglie.

3. DE BROGLIE'S MANY-PARTICLE THEORY FOR BOSONS

De Broglie (1960) gave details of his approach to quantum theory in the context of the Klein-Gordon equation in his book *Non-Linear Wave Mechanics* published after he returned to his original ideas as a result of becoming aware of the work of Bohm. De Broglie gave the standard approach for treating boson fields as particles,[3] he also considered some of the difficulties. In de Broglie's approach, one proceeds with a polar decomposition of the field, e.g. for scalar fields

$$(11) \qquad \phi = R \exp(iS),$$

and the wave equation is then decomposed into a continuity equation

$$(12) \qquad \partial_\mu(R^2\partial_\mu S) = 0,$$

and a Hamilton–Jacobi equation

$$(13) \qquad \partial_\mu S \partial^\mu S - \frac{\Box R}{R} = m_0^2.$$

The problem with the use of $\partial_\mu S$ to define the flow lines is that $\partial_\mu S$ is not always a time-like four vector. This fact was, of course, known to de Broglie, but he did not consider it a serious pathology. After all, no experimental consequences follow from the implied superluminal motion of the particles. The predictions of the theory do not contravene relativistic covariance, but, nonetheless, one remains uneasy with such a formulation and, if an alternative definition can be found for which the particle motion remains sub-luminal, then this would be preferable.

Another approach based on relativistic hydrodynamics has been developed (de Broglie 1926; 1927; Halbwachs 1960; Garuccio and Vigier 1981), but in this work also there is an implied assumption that $\partial_\mu S$ is a timelike four-vector.

It turns out, however, that at least for massive bosons there is a possibility to define time-like flows and hence particle trajectories which follow the flows and so one objection against a particle formulation of boson fields may be overcome. We outline this possibility in the following.

Dirac (1958b,c) has shown that the Hamiltonian density H for any field splits into two pieces, each piece being proportional to a lapse or shift vector (N, N_i). The two pieces have a simple physical significance as shown by Schutz (1971), namely

$$(14) \qquad N\frac{\partial(H(-g)^{1/2})}{\partial N} = (-g)^{1/2}\eta^\mu T_{\mu\nu}\eta^\nu N \equiv \varepsilon N,$$

$$(15) \qquad N_i\frac{\partial(H(-g)^{1/2})}{\partial N_i} = (-g)^{1/2}g^{ij}\eta^\mu T_{\mu j}N_i \equiv P^i N_i,$$

$$(16) \qquad \eta^\mu = -N({}^4g^{0\mu}),$$

where η^μ is a unit normal to a space-like hypersurface, $({}^4g^{\mu\nu})$ is the four dimensional metric and g^{ij} is the three-dimensional metric. ε and P_i are, respectively, the coordinate densities of energy and momentum measured by an observer at rest on the hypersurface.

One can define the flow lines for the bosonic fields through the energy-momentum tensor T^μ_ν. If one defines a time-like vector η^ν then $T^\mu_\nu\eta^\nu$ is a true four vector defining the flow of rest energy. This flow, however, is dependent on an arbitrary choice of η^ν and is not therefore suitable, as it stands, if one wishes to ascribe beables to the field.

If however a unique time-like eigenvector of T^μ_ν exists then a natural vector is provided by the field itself through the equation

$$(17) \qquad T^\mu_\nu W^\nu = \lambda W^\mu,$$

It has been shown (Edelen 1963) that, given such a unique W^μ, an intrinsic energy density (ρ) exists and does not have to be introduced ad-hoc into the Einstein theory. The intrinsic energy density satisfies

$$(18) \qquad (\rho W^\mu)_{;\mu} = 0.$$

Edelen establishes, given the unique W^{μ}, that

(19) $$T^{\mu\nu} = \lambda W^{\mu} W^{\nu} + \phi^{\mu\nu},$$

where $\phi^{\mu\nu}$ is any symmetric tensor field with space-like support. Using also $T^{\mu\nu}_{;\,\nu} = 0$ one gets

(20) $$(\lambda W^{\nu})_{;\,\nu} = \sigma^{\mu\nu}\varepsilon_{\mu\nu},$$

where $\varepsilon_{\mu\nu}$ is the Born rate of strain tensor. One now looks for a path density S such that its Lie derivative along W^{μ} equals $(g)^{1/2}\sigma^{\mu\nu}\varepsilon_{\mu\nu}$. It follows straightforwardly that

(21) $$S = \left(\exp[-\int_a^b \varepsilon d\tau]\right)\left(\int_a^b (\exp[\int_a^b \varepsilon d\tau])\,(-g^{1/2})\sigma^{\mu}\nu\varepsilon_{\mu\nu} d\tau\right),$$

all integrations being performed along the W^{μ} and $\varepsilon = \varepsilon_{\mu\nu} g^{\mu\nu}$. S will, in general, be path dependent and is set to zero on an initial space-like hypersurface.

One now has a unique scalar density

(22) $$\Phi = (-g)^{1/2}\lambda - S,$$

or, if one prefers, a unique ρ given by

(23) $$\Phi = (-g)^{1/2}\rho.$$

One can say that $\lambda(-g)^{1/2}W^{\mu}$ is a flux of rest energy density created in amounts equal to $\sigma^{\mu\nu}\varepsilon_{\mu\nu}$ per unit geometrical volume. $\sigma^{\mu\nu}$ are the generalised stresses giving rise to the energy flow. S then represents a path-dependent flux which combined with the flux of rest energy density gives a conserved flux Φ. Φ can thus be thought of as an intrinsic energy density flux.[4]

The approach used above circumvents a number of objections to the very notion of boson trajectories, either using the de Broglie approach or using the energy-momentum tensor to define flow lines (Bohm, Hiley and Kaloyerou 1987). Firstly, one must form true 4-vectors to define the flow of energy momentum, e.g. $T_{\mu\nu}\eta^{\nu}$ (Holland 1993). One will now clearly obtain a Lorentz covariant description of the particle motion.

The other objection relates to the suggestion that if one has a conserved four vector j^{μ} then the time component j^0 should be taken as proportional to the probability density. In this suggestion use is made of the theorem (Good and Nelson 1971) that if S^{μ} is zero except in a finite region of space (e.g., a box) then $\int d^3x j^0$ is a scalar with respect to Lorentz transformations and $^d/_{dt}\int d^3x j^0 = 0$. However the theorem will not apply inside the box to an elementary region of space since one cannot localise relativistic wave functions. One would therefore get different probability densities in different frames. The probability density thus defined is not, therefore, a true scalar; neither is it positive definite. The approach we propose in this paper does not use the standard connection of j^0 and "probability density". Instead, one defines a true scalar density which is, over some initial hypersurface, the eigenvalue of T^{μ}_{ν} and is positive definite for the time-like W^{μ}.

3.1. *Time-like Eigenvectors of T^{μ}_{ν} for a Complex Scalar Field*

The eigenvectors of T^{μ}_{ν} are defined by Eq. (17). In the case of massive spin 0 and spin 1 fields unique time-like eigenvectors exist (Minkowski space-time). Corresponding to our discussion of Bohm's field theory we discuss here only the scalar case.

Writing a solution ϕ of the Klein–Gordon equation as

$$\phi = \exp[P + iS], \tag{24}$$

and setting

$$\partial_{\mu} P = P_{\mu}, \tag{25}$$

$$\partial_{\mu} S = S_{\mu}, \tag{26}$$

one gets (Masden 1988; Wald 1994)

$$T^{\mu}_{\nu} = m|\phi|^2[1 - (P^{\mu}P_{\mu} + S^{\mu}S_{\mu})]\delta^{\mu}_{\nu} + 2|\phi|^2[(P_{\mu}P^{\nu} + S_{\mu}S^{\nu})]. \tag{27}$$

The second term can be written as $2|\phi|^2$ times

$$\begin{bmatrix} \vdots \\ P^{\mu} \\ \vdots \end{bmatrix} [\ldots P_{\mu} \ldots] + \begin{bmatrix} \vdots \\ S^{\mu} \\ \vdots \end{bmatrix} [\ldots S_{\mu} \ldots], \tag{28}$$

the eigenvectors of T^{μ}_{ν} must be a linear combination of $[P^{\mu}]$ and $[S^{\mu}]$ (or orthogonal to the term above). Some algebra shows that the two solutions are of the form (unnormalised):

$$S^{\mu} + \exp^{\theta} P^{\mu}, \tag{29}$$

$$S^{\mu} - \exp^{-\theta} P^{\mu}, \tag{30}$$

where

$$\sinh \theta = \frac{P^{\mu}P_{\mu} - S^{\mu}S_{\mu}}{2P^{\mu}S_{\mu}}, \tag{30}$$

The two vectors are easily shown to be orthogonal four vectors using the expression for sinh θ. One can conclude that one is time-like and the other space-like (or both null). The overall eigenvalue λ is given by

$$\lambda = m|\phi|^2[1 \pm ([P^{\mu}P_{\mu} - S^{\mu}S_{\mu}]^2 + 4[P^{\mu}S_{\mu}]^2)^{1/2}]. \tag{30}$$

3.2. *Quantisation of the Scalar Field*

The natural approach from de Broglie's starting point is to build up the Fock space sector by sector, as a sequence of Hilbert spaces, each with its appropriate wave functions specified over an n-particle configuration space.

If, at least initially, one considers the non-interacting field case, then in a particular sector the total energy-momentum tensor will be a sum of tensors each having the same structure as the one-particle tensor. However, the field ϕ is now a function of all the particle coordinates and so any entanglement will affect the W^{μ}. The "n-particle" Fock space amplitude of the vector $|\Phi\rangle$ is

$$\Phi^{(n)}(x_1 \ldots x_n) = (n!)^{1/2}\langle 0|\phi^{\dagger}(x_1) \ldots \phi^{\dagger}(x_n)|\Phi\rangle, \tag{33}$$

(we consider here a real scalar field for simplicity). $\Phi^{(n)}$ satisfies the Klein–Gordon equation in each of its coordinates since $\phi^{\dagger}(x_i)$ obeys the Klein–Gordon equation. One can decompose $\Phi^{(n)}$ as

$$\begin{aligned}\Phi^{(n)}(x_1 \ldots x_n) &= \langle x_1 \ldots x_n|\Phi\rangle \\ &= \sum\langle x_1 \ldots x_n|n_1 \ldots n_i\rangle \langle n_1 \ldots n_i|\Phi\rangle,\end{aligned} \tag{34}$$

where n_i is such that $\sum n_i = n$. The transformation functions

$$\langle x_1 \ldots x_n|n_1 \ldots n_i\rangle, \tag{35}$$

are given by

$$\sqrt{\frac{n!}{n_1!n_2! \ldots}}\,\frac{1}{n!}\sum_k P(f_{\alpha_1}(x_1) \ldots f_{\alpha_n}(x_n)), \tag{36}$$

where the permutations P run over all $n!$ permutations of the labels of the x's. $\Phi^{(n)}$ is thus given in the form of a sum of symmetrised products of one-particle positive energy states. $[f_{\alpha}(x)]$ are a complete set of normalisable, positive energy solutions of the Klein–Gordon equation[5]

$$(\Box + \mu^2)f_{\alpha(x)} = 0. \tag{37}$$

One may also ensure that one is dealing with positive energy solutions by solving

$$i\hbar\frac{\partial h}{\partial t} = \sqrt{\mu^2 - \hbar^2\nabla^2}f, \tag{38}$$

(see Fulling 1989). One may proceed to construct a two-particle equation, following the example of the two-particle Dirac case discussed by Bohm and Hiley (1993)

$$\left(i\hbar\frac{\partial}{\partial t_1} - \sqrt{\mu^2 - \hbar^2\nabla^2} + i\hbar\frac{\partial}{\partial t_2} - \sqrt{\mu^2 - \hbar^2\nabla^2}\right) f(1, 2) = 0. \tag{39}$$

As we have seen in the field picture used by Bohm, one deals with a denumerable infinity of harmonic oscillators. The eigenstates of the Hamiltonian are infinite products of Hermite functions, but one only ever deals with a finite sequence of non-zero excitation numbers, n_i. The Fock space associated with a Hilbert space $\mathcal{H}$ is defined to be

$$(40) \qquad \mathcal{F}(\mathcal{H}) = \oplus_{n=0}^{\infty} \otimes^n \mathcal{H},$$

and one defines $\otimes^0\mathcal{H} = \mathcal{D}$, i.e. the domain of the complex numbers (Wald 1994; Schweber 1961; Fulling 1989). The notion of a tensor product for $n = \infty$ has been given by von Neumann (1938).

The correspondence between the field picture and the configuration space picture is complete except for the zero-excitation configuration space wave function which is equal to a complex number by definition (see Wald 1994).

In order to deal with the problems of varying numbers of particles and the apparent lack of correspondence between field and configuration space in the zero excitation sector, it is worthwhile using Dirac's original approach. Dirac uses the idea of a zero state and it is worth quoting him directly on this point:

The second generalisation which we can make is to allow the total number of systems in the assembly to vary. This generalisation is necessary when the theory is applied to photons, since any emission or absorption of a photon by an atomic system results in a change in the total number of photons in existence. We can get the theory for a varying number of systems from the theory for a fixed number by postulating in the latter theory a *zero state* for the systems, in which they are not physically observable in any way. Variations in the total number of observable systems can then be interpreted as arising from systems making transitions into or out of the zero state. We must suppose the number of systems in the zero state to be infinitely great, in order to allow the number of observable systems to increase without limit. (Dirac 1935, 236–237)

The configuration-space wave function corresponding to a particle in the zero state will be a constant function normalised in an appropriate way and written $f_g(x)$. The standard formalism will be modified so that the transformation functions will include a reference to the particles in the zero state. Provided one insists, as above, on only dealing with a finite sequence of non-zero n_i's, no difficulties will occur in the construction of the Hilbert spaces required. Any realistic physical state would have a maximum excitation number, n say, so that only products of n modes occur. The Fock-space wave function will now appear as a superposition of wave functions each with the same number of particles. For example the one-excitation sector will have a configuration space wave function proportional to

$$(41) \qquad \sum_P P(f_\alpha(x^{(1)}) \prod_{i=2}^{n} f_g(x^{(i)}).$$

One may now calculate trajectories in a standard way by putting $\Psi = \exp[P + iS]$ and finding

$$(42) \qquad \frac{\partial S}{\partial x_\mu^{(i)}},$$

and

$$(43) \qquad \frac{\partial P}{\partial x_\mu^{(i)}},$$

It therefore appears that in the non-interacting field case one can calculate trajectories.

3.3. *Interacting Fields*

The interacting field case requires more careful consideration as one cannot, in general, start with bare particles and apply perturbation theory (Wightman and Schweber 1955; von Neumann 1938; van Hove 1952).

We only consider the simplest soluble model of interacting fields; namely, a neutral scalar field and a single nucleon of mass m_0. We follow the analysis of Schweber (1961). The Hamiltonian can be taken to be

$$(44) \qquad H = m_0 + \int d^3k\omega(\vec{k})a^*(\vec{k})a(\vec{k}) + \frac{\lambda}{(2\pi)^{3/2}}\int d^3k \frac{f(\vec{k}^2)}{\sqrt{2\omega(\vec{k})}}(a(\vec{k}) + a^*(-\vec{k})),$$

when $f(\vec{k}^2) = 1$ one is dealing with a point source and $\omega(\vec{k}) = \sqrt{\vec{k}^2 + \mu^2}$. H can be rewritten as

$$(45) \qquad H = m + \int d^3k^*(\vec{k})c(\vec{k})\omega(\vec{k}),$$

where

$$(46) \qquad c(\vec{k}) = a(\vec{k}) + v(\vec{k}),$$

$$(47) \qquad v(\vec{k}) = \frac{\lambda f(\vec{k})}{\sqrt{2(2\pi)^3\omega^3(\vec{k})}}$$

$$(48) \qquad m = m_0 - \int d^3k|v(\vec{k})|^2\omega(\vec{k}).$$

The operators $c(\vec{k})$, $c^*(\vec{k})$ satisfy the same commutation relations as $a(\vec{k})$, $a^*(\vec{k})$. Although the two sets of operators have the same algebra they do not have a common no-particle state unless

$$(49) \qquad \int d^3k|v(\vec{k})|^2 < \infty,$$

which is not the case for a point interaction with $f(\vec{k}^2) = 1$ (see especially Wightman and Schweber (1955) and van Hove (1952)). The representations of the $c(\vec{k})$ and $a(\vec{k})$ are not related by a unitary transformation. It is, however, possible to relate two Fock spaces using the notion of the Bargman–Fock space representation (Itzykson and Zuber 1978). a and $a^\dagger$ are represented in a space of analytic functions

(50) $a \to \frac{\mathrm{d}}{\mathrm{d}\bar{z}}$,

(51) $a^\dagger \to \bar{z}$,

with a scalar product

(52) $$\langle g|f\rangle = \int \frac{\mathrm{d}z\mathrm{d}\bar{z}}{2\pi} \exp[-z\bar{z}]\bar{g}(z)f(\bar{z}).$$

An orthonormal basis in this space is

(53) $$f_n(\bar{z}) = \frac{\bar{z}^n}{\sqrt{n!}}$$

The c and $c^\dagger$ will be similarly represented in terms of $z' = z + \alpha$, α a complex number. One sees that the basis functions in the two spaces are non-unitarily related by a lower triangular matrix

(54) $$f'_n(\bar{z}') = \sum_{i=2}^{n} (\bar{a})^{n-m} \frac{\sqrt{n!}}{\sqrt{m!}} \frac{1}{(n-m)!} f_m(\bar{z}).$$

A unitary transformation maps these spaces on the conventional spaces of square integrable functions of the oscillator configuration variables q. Using a standard Gaussian integration formula

(55) $$\int \prod_{i=1}^{N} \frac{\mathrm{d}_{z_k}\mathrm{d}_{\bar{z}_k}}{2\pi_i} \exp[-\bar{z}Az + \bar{u}z + u\bar{z}] = (\det A)^{-1} \exp[\bar{u}A^{-1}u],$$

where z and u stand for column vectors, one sees that no problems arise in dealing with an infinite number of modes, no restriction on $\int \mathrm{d}^3k|v(\vec{k})|^2$ being necessary.

Although one only deals directly with the physical particles and fields, the field description can be transcribed into the bare particle picture. A configuration-space can then be constructed and trajectories established. One could say that one has an apriori ontology of bare particles and fields and what one is tempted to call the synthetic apriori of the physical particles and fields. A more general result is needed to establish how far the notion of a transcription is valid in dealing with more general interacting field theories.

4. COMPARISON OF THE DESCRIPTIONS PROVIDED BY BOHM AND BY DE BROGLIE IN THE CONTEXT OF SIMPLE MODELS

4.1. *Free States of the Field*

The vacuum state of the field has the wave function

(56) $$\prod_{i}^{\infty} H_0(q_i),$$

where H_i is the ith hermite polynomial. In Bohm's approach, each oscillator is in its ground state and has some well-defined coordinate. The resulting field is spatially chaotic but stationary. In the de Broglie-type particle picture, the configuration space wave function is

$$(57) \qquad \Phi(x_1, \ldots, x_N) \propto \prod_{i=1}^{\infty} f_g(x_i),$$

all the particles are in the zero state and are stationary.

A number state of the field, with a definite number of excitations (N) of a single mode (λ) has a wave functional given in the field picture by

$$(58) \qquad H_N(q_\lambda) \prod_{i \neq \lambda}^{\infty} H_0(q_i).$$

In Bohm's approach, the field is stationary, but the mode containing the excitations tends to be elevated above the random background (due to presence of all the other contributions). The corresponding configuration-space wave-function in the particle trajectory picture will be, apart from normalisation factors

$$(59) \qquad \Phi(x_1, \ldots, x_N) \propto \sum_P p f_\lambda(x_1) f_\lambda(x_2) \ldots f_\lambda(x_N) \prod_{i \neq \lambda}^{\infty} f_g(x_i).$$

Each factor has a time dependent part $\exp[-i\omega t]$ times a real spatially dependent mode function. One therefore has, with $\Phi = \exp[p + iS]$

$$(60) \qquad \frac{\partial S}{\partial t} = -\omega\lambda,$$

$$(61) \qquad \frac{\partial S}{\partial x_i} = 0,$$

$$(62) \qquad \frac{\partial P}{\partial x_i} = \frac{1}{f_\lambda(x_i)} \frac{\partial f_\lambda(x_i)}{\partial x_i},$$

so that S_μ is a time-like vector and P_μ is a space-like vector and the particles are stationary.

An approximation to a single-mode coherent state can be formed by superposing the ground state and first excited state for one of the mode oscillators. One has the wave functional

$$(63) \qquad (H_1(q_\lambda) + H_0(q_\lambda)) \prod_{i \neq \lambda}^{\infty} H_0(q_i),$$

and this gives rise to an oscillatory motion for the q_λ and so the field-mode tends to behave classically. For the corresponding configuration-space particle wave function, one has

$$(64) \qquad \sum_P p[f_\lambda(x_1) + f_g(x_1)] \prod_{i \neq 1}^{\infty} f_g(x_i),$$

which has the particle oscillating back and forth in the cavity.

When the wave functional for the field is a single excitation superposed in two modes, i.e.

$$(65) \qquad [H_1(q_\lambda)H_0(q_\kappa) + H_1(q_\kappa)H_0(q_\lambda)] \prod_{i \neq \lambda, \kappa} H_0(q_i),$$

the cavity field in Bohm's approach has a maximum which tends to oscillate back and forth in the cavity. The corresponding particle configuration space wave function is

$$(66) \qquad \sum_p p[f_\lambda(x_1)f_g(x_2) + f_g(x_1)f_\lambda(x_2) + f_g(x_1)f_\kappa(x_2) + f_\kappa(x_1)f_g(x_2)] \prod_{i \neq 1, 2} f_g(x_i),$$

which has the two particles oscillating in the cavity.

4.2. *Interaction of the Cavity Scalar Field with a Two-level System*

The simplest model of a quantum field interacting with matter is furnished by the Jaynes–Cummings model, in which a tuned cavity-field interacts with a two-level material system. In our version both field and matter are confined to one dimension. The cavity field is tuned so that one of its modes has exactly the right frequency to cause transitions between the two energy levels of the matter system. In fact, for the purposes of illustration, we take the material system to be a particle confined to a box (also referred to hereafter as the atom), with the parameters chosen so that there is a resonant interaction between one particular field mode q and one particular transition of the particle in the box (we chose the ground state to the third excited state). Because of the exact resonance condition, the particle in the box will behave essentially as a two-level system. We assume that initially the field mode is in its first excited state, that is, the resonant mode contains one photon, whilst the particle in the box is initially in its ground state. With this initial state, it is easy to show that the system evolves in such a way that the excitation is transferred periodically between the field mode and the two-level system. In fact, the mathematical description of the situation we have described is just the quantum mechanics of coupled oscillators with one oscillator representing the field.

4.2.1. *The Field Approach*

The configuration space wave function can be written

$$(67) \qquad \chi(q, x, t) = U_1(q, x)\cos(\Omega t) - i\sin(\Omega t)U_2(q, x),$$

where Ω is determined by the coupling between field and matter and

$$(68) \qquad U_1(q, x) = \theta_1(q)\psi_0(x),$$

and

$$(69) \qquad U_2(q, x) = \theta_0(q)\psi_1(x),$$

where $\theta_0(q)$ is the ground state of the resonant field mode and $\theta_1(q)$ is the first excited, or single photon, state of the field mode; similarily $\psi_0(x)$ and $\psi_1(x)$ are the ground state and first excited state of the two-level system. The results of detailed calculations for this system have been given in Dewdney and Lam (1991) and in Lam and Dewdney (1994a,b) and computer animations can also be seen in Dewdney (1997). Here we just present the bare essentials. Each trajectory traces out a unique evolution of the configuration of the system as the system-point moves in the configuration spacetime: no system-point trajectory ever crossing another. The way in which the field develops with time depends not just on the value of its own ode coordinate but on the value of the particle coordinate as well.

At $t = \pi/4\Omega$ the system wave function is in the superposition state

$$\chi\left(q, x, \frac{\pi}{4\Omega}\right) = \frac{1}{\sqrt{2}}\left[U_1(q, x) - iU_2(q, x)\right]. \tag{70}$$

In the usual interpretation it is not possible, with this wave function, to say anything about the state of either the field or the two-level system alone, all that one may say is that if the state of the atom is measured, then the probability of finding it excited, in an ensemble of cases, is one-half. In the usual approach, on measurement, one has no way of selecting one particular branch of the superposition (70) and so there is no definite outcome in an individual case (this, of course is the measurement problem in this context). If, when the state of the system is given by (70), the energy of the particle in the box is measured by removing the confining potential, then the particle either comes out rapidly with the excited-state wave function or more slowly with the ground-state wave function. The whole particle-field system evolves in such a way that the position (and hence energy) of the particle *become* correlated with the state of the field. So that (as always in de Broglie–Bohm-type theories) it is the measurement process which actually brings about an evolution of the variables towards their measured values. If the packets corresponding to the different outcomes (associated with U_1 or U_2) remain separated in the configuration space, then the system point remains within whichever packet it has entered, and the effective state of the system will remain the one associated with that packet alone.

In this simple model one can see that when the particle in the box undergoes a transition between energy states there is a concomitant emission or absorption of the field. But strangely, in this model, for some choices of the initial coordinates of the field-mode and the particle, the amplitude of the field may be greater after it has given up its photon of energy than it was before. On average in an ensemble of such processes the tendency is for the field amplitude to reduce, but in individual cases this need not be so. Of course, for single photon processes, such cases would be hidden in what is usually called the vacuum fluctuations. Since in this model the interaction of the particle is with only a single mode of the field, the model is inherently nonlocal in its formulation. Thus the field changes over the whole cavity simultaneously and one does not see the field changing at first in the vicinity of the atom and then the change propagating outwards. This makes the model rather artificial, a more informative description would ensue if one considered the transient behaviour when the interaction between the cavity and the field first starts. Such behaviour would require a multi-mode description in which one would see the "emission" of the field from the vicinity of the particle. A study of this process using Bohm's approach is currently being carried out.

In spite of its simple nature, the model of atomic transitions described here does serve to demonstrate that a particle ontology for the field, with random quantum jumps, is not necessary if one wants to account for atomic transitions. The particle-like behaviour of the field can be adequately accounted for using continuous well-defined fields.

4.2.2. *The Particle Approach*

The wave function for the field-mode coordinates given in Section 4.2.1 above, can be transformed into the corresponding particle configuration space wave function Ψ defined on the particle coordinates x_s for the scalar field and x for the atom, namely

$$\Psi(x_s, x, t) = f_g(x_s)\psi_1(x)\cos(\Omega t) - i\sin(\Omega t) f_\lambda(x_s)\psi_0(x). \tag{71}$$

One can now calculate $\partial_\mu P$ and $\partial_\mu S$ with respect to x_s and x and hence calculate trajectories for both the scalar field particle and atom. The trajectories are somewhat similar to those for the mode-coordinate and atom given in the field picture and the measurement on the atom, described above, procedes in the same way. If the particle emerges rapidly then as it does so the field particle becomes stationary (and unobservable) in the zero state.

4.3. *Anticorrelation of Detectors in a Single-photon Field*

4.3.1. *The Field Approach*

At first sight it would seem that a theory based on boson trajectories would give a simple and intuitive description of the anticorrelation of two detectors in a single photon field, since a localised photon on a trajectory could only be in the vicinity of one detector (atom). It is fascinating to see that the observed results can also be recovered deterministically and causally using extended, but well-defined fields, whilst the picture provided using the trajectory approach is (as we see below) not so intuitive after all. To describe this case we simply extend the above model, in which a single two-level system interacts with a single mode of a cavity, to include two two-level atoms in the cavity with the same coupling.

$$V_i = q\sin(\kappa x_i + \chi_i), \tag{72}$$

where atoms 1 and 2 are labelled by $i = 1$ and 2, respectively. In Eq. (72), q is the coordinate of the field mode, κ is the wavevector of the mode and χ_1 and χ_2 are phases which depend on the position of each atom in the cavity. Although the atoms are considered to be in different regions of the cavity, it is assumed that the dimensions of the atoms, and their separation, are small in comparison ith the wavelength of the field. We wish to look at the situation in which the sizes of the interaction of each atom with the field are approximately equal, so we put $\chi_1 = \chi_2 = 0$. Without the presence of the interaction, the field-matter system is triply-degenerate, and the effect of the coupling of the atoms to the field can be found using degenerate perturbation theory. If the ground and excited state wave functions of the two atoms are denoted by $\psi_0(x_{1,2})$ and $\psi_1(x_{1,2})$, respectively, and the ground and first-excited states of the mode oscillator are represented by $\theta_0(q)$ and $\theta_1(q)$, then the three degenerate levels are

$$U_1 = \theta_1(q)\psi_0(x_1)\psi_0(x_2)$$

$$U_2 = \theta_0(q)\psi_1(x_1)\psi_0(x_2)$$

$$U_3 = \theta_0(q)\psi_0(x_1)\psi_1(x_2).$$

If the system is in the state $U_1(x_1, x_2, q)$ at $t = 0$, then the subsequent development of the system wave function during the period of interaction is given, from degenerate perturbation theory, by

$$\text{(73)} \qquad \Psi(x_1, x_2, q, t) = U_1(x_1, x_2, q) \cos \Omega t - \frac{i}{\sqrt{2}} \sin \Omega t \, [U_2(x_1, x_2, q) + U_3(x_1, x_2, q)],$$

where Ω depends on the overlap integrals between the unperturbed states, for the interaction $V_{1,2}$. When Ωt is a half-integer multiple of π the wave function of the two atoms is

$$\text{(74)} \qquad \Psi(x_1, x_2, q, t) = \frac{\pm i}{\sqrt{2}} [U_2(x_1, x_2, q) + U_3(x_1, x_2, q)],$$

which is an energy-entangled state, involving the two separated atomic systems. For arbitrary values of t, the system wave function is a superposition and it is not possible, in the usual approach, to give any definite description of the state of the field or of the atoms. The description of Bohm's theory is in terms of the field-matter configuration-space wave function and the precise values for the field coordinate and the positions of the two atomic electrons.

After a period of interaction, the field and the two atoms are decoupled and the final part of the detection process takes place; this is the performance of a momentum measurement on each atom in the same way as described above in Section 4.2.1. The decoupling and the removal of both atomic potentials occurs at time $t = \pi/2\Omega$. After which time, the development of the field is independent from that of the atoms, and the final part of the detection of the field quanta may be modelled using the two-atom entangled state alone, which at $t = \pi/2\Omega$ is

$$(75) \quad \psi\left(x_1, x_2, \frac{\pi}{2\Omega}\right) = \frac{-i}{\sqrt{2}}\left[\psi_1(x_1)\psi_0(x_2) + \psi_0(x_1)\psi_1(x_2)\right].$$

It is interesting to note that if the semiclassical treatment, given in Dewdney and Lam (1991), is extended to the case of two atoms interacting with a single mode cavity field, then the quantum state of the atoms remains a simple product; no anticorrelation can be encompassed by the semiclassical description. Here, in the fully quantised description, the overall state of the two atoms becomes entangled through the interaction with the single photon field. When the energy of the two atoms is measured it is clear that one of the two alternatives given below occur.

1. particle 1 comes out rapidly, whilst particle 1 remains in the vicinity of its potential, or
2. particle 2 comes out rapidly, whilst particle 2 remains in the vicinity of its potential

Which alternative actually occurs in any given case depends on the initial values of the system's coordinates and so which atom will be found to be excited on measurement is already determined at the outset. But note that the observation that, for example, particle 1 has moved away quickly does not allow us to say that therefore particle one had made a jump to the excited state before the measurement took place. The wave function before the measurement was energy-entangled and so according to the usual approach neither atom alone could be ascribed a definite energy. Once more we see how the expected values are brought about by the measurement process.[6] In this section, we have outlined a causal description of the anticorrelation observed between two detectors interacting with a one-proton field. The description has a well-defined motion for the quantum field, but no photon trajectories.

4.3.2. *The Particle Approach*

The wave functional given above for this system translates into the particle configuration space wave function according to

$$(76) \quad \Theta_1(q) \rightarrow f_\lambda(x_s),$$

$$(77) \quad \Theta_0(q) \rightarrow f_g(x_s).$$

As discussed above after a time lapse $\Omega t = \pi/2$ the field is de-excited and the two atoms are energy entangled. No correlation will be observed however between the scalar field particles and the excited atom in this model since $\Theta_0(q)$ is a factor in U_2 and U_3. The field particles have trajectories that, if initially distributed according to $|f_\lambda(x_s)|^2$ will become distributed at $\Omega t = \pi/2$ according to $|f_g(x_s)|^2$, i.e. uniformly in the cavity. So it is not the case that the atom which is observed to be excited is the one near the actual field-particle. The measurements on the atoms goes just as in the field case discussed above.

5. CONCLUSION

We conclude that the formal underdetermination between field and particle approaches to many-boson theories naturally gives rise to an undetermination of the ontology in a realist interpretation of the de Broglie–Bohm type. There do not seem to be any compelling formal reasons for choosing the field approach over the particle approach. Both approaches give a description of a well-defined reality at the quantum level and both give plausible accounts of the simple processes that we have examined. Both approaches also give a reasonable account of the classical limit. This being the case it seems to be a matter of aesthetics which approach one prefers.

University of Portsmouth

NOTES

[1] Bell often presented the pilot-wave theory by simply stating the guidance condition for the particles and this is in contrast to the way in which the theory is often presented in terms of the modified Hamilton–Jacobi equation with the introduction of "quantum potentials" and "quantum forces". Bell's approach has been followed in the work of Dürr *et al.* (1992a), whilst others have continued to developed the full quantum Hamilton–Jacobi formalism. Here we take no side in what seems to us largely a matter of preference in how the theory is presented. If Bell's approach is preferred then obviously one need only be concerned with the velocity conditions in what follows.

[2] The questions discussed here clearly have their counterpart in fermion theories where the problem is more to develop a field picture but we do not discuss these problems here. See Bell (1987a), Holland (1993) and Valentini (1996a,b).

[3] We shall give a different way of defining the particle trajectories and shall use a configuration-space description, but nevertheless shall refer to the thus modified theory as a de Broglie-type theory.

[4] In the following we will not pursue the general case where the field generates a significant gravitational field, but the above results are important in establishing a coherent framework.

[5] See Schweber (1961) for a full account.

[6] Detailed trajectories for this example can be found in Lam and Dewdney (1994a,b).

TREVOR M. SAMOLS

A REALISTIC FORMULATION OF QUANTUM FIELD THEORY

1. INTRODUCTION

Bohm's work has shown that it is possible to give a precise, realistic account of quantum theory. In his formulation, no recourse is made to ill-defined notions of "observation" or "measurement", and the relation to classical mechanics is unproblematic, the latter simply being a good approximation to the "true" (Bohmian) dynamics under certain precisely definable conditions.

However, while the approach is quite compelling as applied to non-relativistic mechanics, the extension to relativistic field theories is less satisfactory. In particular, there is a problem in reconciling it with fundamental relativity. A Bohmian field theory, such as the scalar field theory of Bohm, Hiley and Koloyerou (1987), or the stochastic generalisation due to Bell (1984), may certainly be relativistic at the phenomenological level, where it reproduces the standard quantum predictions, but the formulation of the theory is still dependent on the choice of a preferred frame of reference. This dependence enters through Bohm's guidance condition (or Bell's stochastic generalisation thereof), which ties together the field configurations on consecutive time slices in a manner that depends on how the slicing is done. The situation is thus rather like Lorentz's electrodynamics, with an unobservable aether playing an indispensable role.[1]

The purpose here is to briefly describe an approach to quantum theory similar in spirit to Bohm, but which does not depend on a preferred frame. As in the Bohm theory, the mathematical structure of quantum theory is left intact, but is supplemented by a more detailed dynamics governing the evolution of a particular set of variables. However, unlike Bohm, the procedure may only be applied to a relativistic theory – or more precisely, one with a non-trivial causal structure, and in consequence, a local dynamics; the implications for the non-relativistic regime must be derived afterwards, by taking the appropriate limit. As we shall see, the richer structure of the relativistic world allows for a very simple prescription. In particular, there is no need for the guidance condition; the only inputs required are the local, unitary evolution law of the underlying quantum theory, and the standard Born probability rule applied in a particular way. A fuller treatment may be found in Samols (1995).

2. A SIMPLE QUANTUM FIELD THEORY

The discussion will be based on a simple (1+1)-dimensional quantum field theory defined on a (light-cone) lattice (Destri and de Vega 1987). This

J. T. Cushing et al. (eds.), Bohmian Mechanics and Quantum Theory: An Appraisal, 191–196.

allows us to avoid the technical difficulties associated with working in the continuum, while at the same time retaining the essential non-trivial causal structure we require. Although on the lattice one does not have full Lorentz symmetry, there is still the notion of a spacelike hypersurface, and what we shall mean by "relativistic" in this context is that the dynamics does not depend on any preferred choice of such surfaces. Given an appropriate lattice theory, one expects that full Lorentz symmetry will be restored in the continuum limit.

The light-cone lattice representing spacetime is depicted in Figure 1. It is a mesh of null rays, with the obvious causal structure; a spacelike surface is a cut that intersects each future-directed path on the lattice exactly once. On each link l of the lattice there is a field variable α_l, taking the values 0 or 1. At each vertex, the (local) evolution law of the quantum theory is encoded in a 4-dimensional unitary matrix – an "R-matrix" – whose entries, $R_{\alpha_1\alpha_2}^{\alpha_1'\alpha_2'}$, are the amplitudes connecting the four possible configurations on the ingoing and outgoing pairs of links at that point (see Figure 2).

The quantum state Ψ of the system is specified by the wave function on any spacelike surface σ, i.e., a complex function of the variables on the links cut by the surface, and which we write as $\Psi(\alpha|_\sigma)$. To evolve the wave function to another surface σ', one applies R-matrices in the obvious way. The simplest case is when σ' is obtained from σ by a local deformation through a single vertex, v say – what we shall call an "elementary motion"

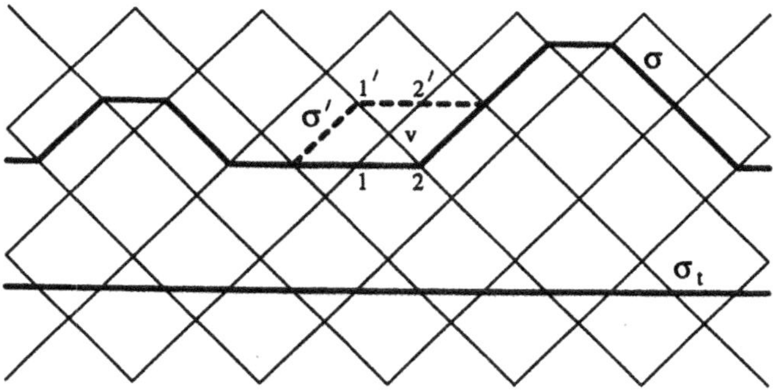

Fig. 1. The light-cone lattice. σ_t is a constant time slice; σ is a general spacelike surface, and σ' one obtained from it by an elementary motion.

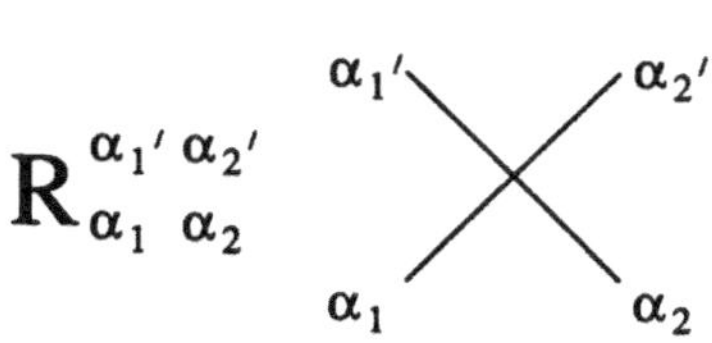

Fig. 2. The R-matrix associated with a vertex.

of the surface. Denoting by $l = 1, 2$ and $1', 2'$ the ingoing and outgoing links respectively at v, as shown in Figure 1, the evolution of the wavefunction associated with the elementary motion is given by

$$(1) \qquad \Psi(\alpha|_{\sigma'}) = \sum_{\alpha_1\alpha_2} R^{\alpha_{1'}\alpha_{2'}}_{\alpha_1\alpha_2} \Psi(\alpha|_{\sigma}).$$

To obtain the evolution between two arbitrary surfaces one may compose a sequence of elementary transformations of this type, though of course the result depends only on the initial and final surfaces and not on the particular sequence employed.

Equation (1) encapsulates the mathematical structure of the quantum theory. Assuming that all measurements can ultimately be reduced to measurements of the α_l, the standard interpretation is summarised by the usual Born rule applied to these variables: in a state Ψ, the probability of finding the configuration $\alpha|_\sigma$ on the surface σ is given by

$$(2) \qquad |\Psi(\alpha|_\sigma)|^2.$$

Of course, to complete the definition of the quantum theory one needs to specify the precise form of the R-matrices. Here we may leave them completely general and need only note that with a suitable choice one can obtain a non-trivial relativistic quantum field theory in the continuum limit (e.g., the massive Thirring model).

3. A SIMPLE REALISTIC FORMULATION

We now turn to the realistic formulation of the quantum theory just described. The first step is to specify the "beables". For a local theory, the natural choice is to take the local field variables, which in the present case are the α_l's. Thus we suppose that in the time evolution of the system each variable α_l realises a definite value $\hat{\alpha}_l$ with some probability, as part of an objectively defined physical process. Given a state Ψ, there will then be an associated joint probability distribution $p_\Psi(\hat{\alpha})$ for the realized values $\hat{\alpha}_l$ over the entire spacetime lattice. Our task is to provide a law that generates $p_\Psi(\hat{\alpha})$ in such a way as to reproduce the predictions of the standard quantum theory, and without recourse to some preferred choice of surfaces.

Most naïvely we might demand that $p_\Psi(\hat{\alpha})$ be chosen so that its marginal distribution on every spacelike surface be given by the Born rule, i.e., for all σ,

$$(3) \qquad \sum_{\hat{\alpha}|_{\sigma^c}} p_\Psi(\hat{\alpha}) = |\Psi(\hat{\alpha}|_\sigma)|^2,$$

where $\hat{\alpha}|_{\sigma^c}$ denotes the variables on all the links of the lattice not cut by σ. In general, however, this requirement is impossible to meet, as follows directly from the Bell inequalities interpreted as the conditions for the existence of a joint distribution.

We are therefore forced to consider a more subtle prescription. In fact, we already know from Bohm that imposing (3) for all surfaces is unnecessarily strong, and that weaker conditions will suffice to ensure agreement with standard quantum theory. In a Bohmian theory, such as Bell's, the guidance condition has two roles: (a) it preserves the validity of the Born probabilities on every constant time slice; and (b) it ensures that any conserved local quantity changes continuously as one moves from slice to slice. Taken together (a) and (b) constrain the theory to be empirically equivalent to the quantum theory interpreted in the standard way. We shall now see that for a relativistic quantum theory, such as the one we are considering, it is possible to achieve the same result, but in a rather simpler manner, and without giving preference to a particular set of surfaces.

To describe the scheme it is convenient to provisionally pick out some initial surface σ_0 and consider the distribution $p_\Psi(\hat{\alpha}|_{\sigma_0^+})$ over the variables to the future of σ_0; at the end we shall remove the dependence on σ_0 by pushing it back into the infinite past. We shall denote by γ any sequence of spacelike surfaces (σ_n: $n = 0, 1, \ldots$) generated from σ_0 by successive elementary motions, and for definiteness suppose that the motion $\sigma_{n-1} \to \sigma_n$ is between the pairs of links $l = 1, 2$ and $l = 1', 2'$ as in Figure 1. If we think of the pair of realisations $(\hat{\alpha}_{1'}, \hat{\alpha}_{2'})$ on the newly cut links as an "event", then γ may be interpreted as a way of building up a spacetime history of $\hat{\alpha}_l$'s as a sequence of such events.

The realistic formulation is based on two simple assumptions: (i) all sequences γ occur with equal probability; and (ii) *given* a particular sequence, $\gamma = (\sigma_n)$, the conditional probability $f_\Psi(\hat{\alpha}|_{\sigma_{n-1}\to\sigma_n})$ of the nth event in the sequence, $(\hat{\alpha}_{1'}, \hat{\alpha}_{2'})$, given all the events up to that point, is independent of the events to the past of σ_n. (ii) allows us to use the Born rule to fix this conditional probability uniquely as

$$(4) \qquad f_\Psi(\hat{\alpha}|_{\sigma_{n-1}\to\sigma_n}) = \left.\frac{|\Psi(\alpha|_{\sigma_n})|^2}{\sum_{\alpha_{1'},\,\alpha_{2'}}|\Psi(\alpha|_{\sigma_n})|^2}\right|_{\alpha|_{\sigma_n} = \hat{\alpha}|_{\sigma_n}},$$

where, by unitarity, the denominator can also be written as $\sum_{\alpha_1,\alpha_2}|\Psi(\alpha|_{\sigma_{n-1}})|^2$. Taking the probability on the initial surface σ_0 to again be given by Born's rule, then, given γ, we obtain the probability distribution

$$(5) \qquad p_\Psi^\gamma(\hat{\alpha}|_{\sigma_0^+}) \equiv |\Psi(\hat{\alpha}|_{\sigma_0})|^2 \prod_{n\geq 1} f_\Psi(\hat{\alpha}|_{\sigma_{n-1}\to\sigma_n}).$$

Since by (i), all possible γ contribute with equal weight, we have

$$(6) \qquad p_\Psi(\hat{\alpha}|_{\sigma_0^+}) \sim \sum_\gamma p_\Psi^\gamma(\hat{\alpha}|_{\sigma_0^+}),$$

or, letting σ_0 recede arbitrarily into the past,

$$(7) \qquad p_\Psi(\hat{\alpha}) \sim \sum_\gamma p_\Psi^\gamma(\hat{\alpha}),$$

where, in each case, $\sim$ indicates the need for a normalization constant.

This then is how we generate the distribution $p_\Psi(\hat{\alpha})$. Comparison with Bohm allows us to understand rather easily why it has the same predictive content as the standard quantum theory. For consider the probability distribution conditional on a particular γ, $p_\Psi^\gamma(\hat{\alpha})$. This conditional distribution shares the two essential properties of the dynamics of a Bohmian theory: (a) its marginal distributions on all the $\sigma_n \in \gamma$ are just given by the Born rule

$$(8) \qquad \sum_{\hat{\alpha}|_{\sigma_n^c}} p_\Psi^\gamma(\hat{\alpha}) = |\Psi(\hat{\alpha}|_{\sigma_n})|^2, \qquad \forall \sigma_n \in \gamma;$$

and (b) if the quantum evolution law (i.e., the R-matrices) conserves a local quantity (e.g., the α-number), then $p_\Psi^\gamma(\hat{\alpha})$ assigns non-zero probability only to spacetime histories $\hat{\alpha}$ for which this quantity is continuous on the lattice. (a) is true by construction, and (b) is an immediate consequence of the fact that the sequence of surfaces $\gamma = (\sigma_n)$ is generated by local (i.e., elementary) motions. Together, just as for Bohm, (a) and (b) mean that were we to employ any $p_\Psi^\gamma(\hat{\alpha})$, we would obtain the same empirical predictions as the standard theory. Since the unconditional distribution $p_\Psi(\hat{\alpha})$ is simply the sum over all the $p_\Psi^\gamma(\hat{\alpha})$, the same is true for it as well. Note that it is the richer structure that one has in a relativistic theory that has enabled us to derive both (a) and (b) essentially as consequences of the Born rule; the guidance condition of the Bohm theory that connects configurations on successive disjoint surfaces is not required.

4. LOCALITY

As one would expect on general grounds, the scheme described here exhibits a certain amount of non-locality. More precisely, a marginal distribution of $p_\Psi(\hat{\alpha})$ associated with the variables in a region of spacetime will generally depend on the R-matrices at points spacelike separated from that region. However, one can identify a weaker locality property which is useful for understanding how the theory does its job.

Consider the marginal distribution of $p_\Psi(\hat{\alpha})$ associated with a general set of mutually spacelike-separated α_l. Generally, this will not be given by the Born rule, since in the sum (7) there will be contributions from γ's for which the variables in the set do not lie on any σ_n in γ. However, in the special case of the *local* marginal distributions – i.e., the two-variable distributions associated with the ingoing or outgoing pair of links at a vertex – the variables lie on some $\sigma_n \in \gamma$ for every γ, and so the Born rule *does* apply. It is then a standard result of quantum field theory (a simple consequence of local unitarity) that these local distributions obey locality in the sense described above – i.e., they are independent of the evolution law (i.e., the R-matrices) at spacelike separated points.

This "locality of local marginal distributions" is the only form of locality that can be tested experimentally and thus all that is required to avoid possible pathologies such as superluminal signalling. For note that any test of the theory will reduce to measuring the marginal distribution associated with some set of α_l. Such a measurement requires that the $\hat{\alpha}_l$ in this set be compared, and this in turn means that records of these outcomes be brought together to the (neighbourhood of the) same point in space. Thus, ultimately, experimental results rest only on the local marginal distributions. Since, as we have seen, the Born rule and so also the locality property do apply to these distributions, the proper working of the scheme is assured.

5. FURTHER REMARKS

The construction described here only makes use of the causal structure and the local unitary evolution law of the underlying lattice field theory, together with a particular application of the Born probability rule. It is therefore rather robust and should be applicable to any local quantum theory that can be regularised on a causal lattice.

Expressing a particular quantum theory on such a lattice should not in itself pose special problems. The more important question is whether there is a sensible continuum limit. In principle, a continuum Lorentz covariant quantum theory can be obtained by coarse graining and rescaling an appropriate lattice system as one approaches a fixed point. One would hope that the same procedure applied to the associated joint distribution $p_\Psi(\hat{\alpha})$ would similarly produce a continuum Lorentz covariant probabilistic theory. Clearly it would be very desirable to have a formulation that employed continuum concepts from the outset, allowing regularisation to be performed in a convenient way at a second stage.

University of Cambridge and King's College

NOTE

[1] To restore fundamental relativity in the Bohmian framework it has been suggested in Dürr *et al.* (1990) and in Berndl *et al.* (1994) that the guidance condition should be made to act with respect to a foliation of spacetime that is itself determined by a relativistic dynamics.

DON N. PAGE

ATTACHING THEORIES OF CONSCIOUSNESS TO BOHMIAN QUANTUM MECHANICS

The de Broglie–Bohm theory of quantum mechanics (here simply called Bohmian Mechanics or BM) (de Broglie 1928; 1953; 1956; 1970; Bohm 1952a,b; 1953b; Bell 1987a; Albert 1992; 1994; Dürr *et al.* 1992a,b; 1993; Bohm and Hiley 1993; Holland 1993; Cushing 1994; Berndl, Daumer, Dürr, Goldstein, and Zanghì 1995) is an augmentation of "bare" quantum mechanics (the bare theory being given by an algebra of operators and a quantum state that sets the expectation values of these operators) that includes a definite history or Bohmian trajectory. This definite trajectory gives BM a somewhat more classical flavor than most other forms of quantum mechanics (QM) (though the trajectory certainly has highly nonlocal and other nonclassical aspects in its evolution), but to see whether or not this makes a difference for observations by conscious beings, one needs to attach theories of conscious perceptions to BM and other forms of QM. Here I shall propose various forms of theories of consciousness for BM, which I shall call *Sensible Bohmian Mechanics* (SBM), and compare them with a proposal I have made for a theory of consciousness attached to bare QM, which I call *Sensible Quantum Mechanics* (SQM) (Page 1995a–d, 1996). I find that only certain special forms of SBM would give essentially similar predictions as SQM, though a wider class might be in practice indistinguishable to any single observer. I also remain sceptical that a viable complete form of SBM will turn out to be as simple a description of the universe as a viable complete form of SQM, but of course it is too early to know yet what the form of the simplest complete theory of our universe is.

I should explain at the outset that my attempt to incorporate consciousness within physical is nonstandard, since most physicists would probably consider this attempt premature, or simply not part of physics. I agree that it *is* probably premature to try to give a complete detailed theory of consciousness, but since we have developed mathematical frameworks like mechanics for describing other aspects of the universe, even though we do not yet know the correct complete detailed form of the mechanics describing our universe, it surely would be helpful to try to develop a mathematical framework for describing consciousness, even though we are a very long way from a correct, complete, detailed theory for it. The objection that consciousness is not part of physics may be historically valid as a sociological analysis of what most physicists actually study (for the reason, I would guess, that physics concentrates on the simplest fundamental descriptions of aspects of our universe, and so far consciousness does not seem to be simple enough to be included in physics). However,

J. T. Cushing et al. (eds.), Bohmian Mechanics and Quantum Theory: An Appraisal, 197–210.

physics has historically continued to be extended to describe a broader and broader range of aspects of our universe, so that in some branches of physics there is even talk today of 'theories of everything.' Thus it seems plausible to me that physics should eventually attempt to describe consciousness itself. Certainly talk of 'theories of everything' in physics seems rather hollow if physics is required to refuse to consider consciousness.

Furthermore, physics, unlike mathematics, is generally seen to be rooted in experiment and observations, and these fundamentally come down to conscious experience. This is particularly true in quantum mechanics, where in many formulations observations are crucial. If one refuses to consider conscious observations, then there seems to be nothing wrong or lacking in bare QM, though then it is just a beautiful mathematical theory for an unconscious world that is totally divorced from observations. However, precisely at the stage at which one wants to explain what is consciously observed, one needs at least some glimpse of a framework for connecting consciousness to observations. Under the assumption that classical mechanics (CM) correctly described our universe, many physicists usually imagined, I suspect, some simple idealized form of psycho-physical parallelism in which the content of one's conscious perception is very similar to certain aspects of the configuration of the classical universe. Then, say, if one looked at a clock whose hands were at the 12 o'clock position, one would have a conscious perception that the clock read 12 o'clock.

One of the problems of ordinary QM is that this simple form of psycho-physical parallelism that seemed adequate in CM does not work in such a naïvely straightforward way. Even though many physicists do not explicitly wish to consider any theories of consciousness, I suspect that much of the trouble they have with QM arises from the fact that they actually implicitly have something like this simple form of psycho-physical parallelism sketched above, and then they find difficulty fitting it with QM. This is indeed one of the main motivations for BM, since it has a definite trajectory, with a definite position in configuration space (e.g., definite particle positions) at each time, to which the simple form of psycho-physical parallelism can readily be attached. Since the sketch of a theory of consciousness is probably implicit in many physicists' assumptions, one does not need to refer to it explicitly when one extols the merits of a definite trajectory in BM, which one can thus do without making other physicists uncomfortable by mentioning the subject of consciousness that physicists do not really understand very well yet.

However, since Sensible Quantum Mechanics (Page 1995a–d, 1996) gives a glimpse of a framework for a psycho-physical parallelism to the quantum world that appears to be just as adequate (though of course just as sketchy in detail) as the simple form implicitly assumed for a classical world, one can see at least this one general possibility for overcoming the problem that many physicists have in relating QM with conscious observations (a problem made worse by a reluctance to consider consciousness explicitly and to acknowledge one's naïve preconceptions about it). SQM seemed extremely obvious when I hit upon it, so initially it was surprising to me that it had not been developed many years ago (though I did later find that Lockwood (1989) had a few years ago expressed highly concordant ideas in less mathematical form). It occurred to me that the novelty of SQM was perhaps simply because of physicists' reluctance to consider consciousness explicitly and because of the strong psychological hold on their assumptions of the simple form of psycho-physical parallelism possible for the classical world.

In view of the apparent success of SQM in sketching how QM may be combined with consciousness with virtually none of the traditional interpretive problems (at least in my own eyes; most physicists I have talked to still seem to think that any theory of consciousness is either unnecessary or premature), it may be of interest to develop also an SBM theory to combine BM in a similar way with consciousness. Then one can have two similar frameworks, containing consciousness explicitly rather than merely implicitly as has unfortunately too often been done, for comparing ordinary QM with BM. Since in my nonstandard view I consider SQM (which is the bare QM of an algebra of operators and a quantum state, with no measurement hypothesis or collapse of the wave function, augmented by a theory of consciousness connected to the bare QM by a nonclassical form of psycho-physical parallelism) to be superior to all other forms of ordinary QM, I shall not bother comparing my SBM extension of BM to other forms of QM, but only to what in my biased opinion I consider to be the best form of ordinary QM with consciousness, SQM.

Since BM is even at the unconscious level an augmentation of bare QM, it would be simplest to describe first SQM and then show how it can be augmented to SBM. SQM is given by the following three basic postulates or axioms (Page 1995c):

QUANTUM WORLD AXIOM: The unconscious "quantum world" Q is completely described by an appropriate algebra of operators and by a suitable state σ (a positive linear functional of the operators) giving the expectation value $\langle O \rangle \equiv \sigma[O]$ of each operator O.

CONSCIOUS WORLD AXIOM: The "conscious world" M, the set of all perceptions p, has a fundamental measure $\mu(S)$ for each subset S of M.

QUANTUM-CONSCIOUSNESS CONNECTION: The measure $\mu(S)$ for each set S of conscious perceptions is given by the expectation value of a corresponding "awareness operator" $A(S)$, a positive-operator-valued (POV) measure (Davies 1976), in the state σ of the quantum world:

(1) $\mu(S) = \langle A(S) \rangle \equiv \sigma[A(S)]$.

Here the Quantum World Axiom is the basic axiom of the mechanics of bare QM, and the Conscious World Axiom is the basic postulate I shall make about the conscious world, even when I go from SQM to SBM. Perhaps I should remind the reader (Page 1995c) that a perception p is in this context taken to mean the entirety of a single conscious experience, all that one is consciously aware of or consciously experiencing at one moment, the total "raw feel" that one has at one time, or (Lockwood 1989) a "phenomenal perspective" or "maximal experience."

Besides a modification of the Quantum World Axiom in BM, in SBM I shall also modify the Quantum-Consciousness Connection, though in both SQM and SBM I shall assume that the measure for each set of conscious perceptions is given by some functional of the corresponding quantum or mechanical world. In SQM the measure is a functional of the operators and the quantum state, and I made the simplest assumption that this functional is linear in the quantum state and is the expectation value of a particular operator (an "awareness operator" $A(S)$) for each set S of perceptions. Of course, one could readily contemplate generalizations in which the measure is a *nonlinear* functional of expectation values (Page 1995c), but I shall not do that here.

Bohmian mechanics is most simply given in the case in which a quantum state is given by a time-dependent wave function over some configuration space (say, for simplicity, of particle positions). One augments this quantum state by a history or time-parametrized trajectory in this configuration space with the velocity vector at each point being given by a suitable functional of the wave function (a normalized gradient of its phase for spinless particles) (de Broglie 1928; 1953; 1956; 1970; Bohm 1952a,b; 1953b; Bell 1987a; Albert 1992; 1994; Dürr *et al.* 1992a,b; 1993; Bohm and Hiley 1993; Holland 1993; Cushing 1994; Berndl, Daumer, Dürr, Goldstein and Zanghì 1995). This gives a first-order differential equation for the trajectory, so it is uniquely determined by this equation (which is itself uniquely determined by the wave function except at its zeros, but those isolated points do not cause any trouble) and by the position of the trajectory at any one (e.g., initial) time.

Because BM thus has a trajectory as well as a quantum state, there are in principle more things on which the measure for conscious perceptions can depend. Again assuming a linear (if nontrivial) dependence on the quantum state, I shall propose that Sensible Bohmian Mechanics (SBM) is given by the following three axioms:

BOHMIAN WORLD AXIOM: The unconscious "Bohmian world" B is described by a normalized time-dependent Hilbert-space wave function $\psi(x, t)$ (obeying an appropriate Schrödinger equation that determines the wave function at all times t once it is given at one time) on a configuration space with coordinates x (indices suppressed) and by a Bohmian trajectory T in this configuration space whose velocity is given by a suitable functional of the wave function.

CONSCIOUS WORLD AXIOM: The "conscious world" M, the set of all perceptions p, has a fundamental measure $\mu(S)$ for each subset S of M.

BOHMIAN-CONSCIOUSNESS CONNECTION: The measure $\mu(S)$ for each set S of conscious perceptions is given by the expectation value of a corresponding "awareness operator" $A(S, T)$, a positive-operator-valued (POV) measure (Davies 1976) that depends on the Bohmian trajectory T, in the wave function of the Bohmian world:

$$(2) \qquad \mu(S) = \int dx\psi^*(x, t_0)A(S, T)\psi(x, t_0).$$

(Here the integral can be taken at any time t_0, since the Schrödinger equation determines the wave function at all times from its form at time t_0, though of course the form of $A(S, T)$ would depend on the particular t_0 chosen. Also, for simplicity, I am suppressing possible spin indices on the wave function as well as the coordinate indices on the x that denotes a point in the multi-dimensional configuration space.)

As in SQM (Page 1995c), so in SBM it is convenient to hypothesize that the set M of all possible conscious perceptions p is a suitable topological space with a prior measure

$$(3) \qquad \mu_0(S) = \int_S d\mu_0(p).$$

Then, just as for SQM, the linearity of positive-valued-operator measures over sets allows one to write the awareness operators for SBM as

$$(4) \qquad A(S, T) = \int_S E(p, T)d\mu_0(p),$$

a generalized sum or integral of SBM "experience operators" or "perception operators" $E(p, T)$ for the individual perceptions p. Similarly, one can write the measure on a set of perceptions S as

$$(5) \qquad \mu(S) = \int_S d\mu(p) = \int_S m(p)d\mu_0(p),$$

in terms of a measure density $m(p)$ that is the quantum expectation value of the experience operator $E(p, T)$ for the same perception p:

$$(6) \qquad m(p) = \int_S dx\psi^*(x, t_0)E(p, T)\psi(x, t_0).$$

It is simplest to consider the two extreme cases in which, for each S or p, $\mu(S)$ or $m(p)$ depends only on the wave function $\psi(x)$ or on T. In the first of these cases, one essentially has a form of SQM with the quantum world consisting of a state that has the particular form of a time-dependent Hilbert-space wave function on some configuration and of operators on such wave functions, and with the Bohmian world consisting of this quantum world augmented by the Bohmian trajectory T. However, in this case the Bohmian trajectory has absolutely no effect on the conscious world with its perceptions and measures. Since our perceptions are the only direct contact we have with the world, if they are completely unaffected by the Bohmian trajectory, there would then seem to be no motivation to add it to one's theory of the quantum world. Although it indeed fits my own prejudice not to augment bare QM for the unconscious aspects of the world and thus not to bother with any Bohmian trajectory in theories of the quantum world that I prefer, for the purpose of the present discussion on BM and its extension to SBM it would seem better to consider a Bohmian-Consciousness Connection in which the Bohmian trajectory really does have an effect on the measure for sets of perceptions. As discussed above, a motivation for the trajectory in BM is that its point in configuration space at each time seems closer to the content of conscious perceptions than does the quantum state or wave function itself.

The second extreme possibility is that for each S or p, $\mu(S)$ or $m(p)$ depends only on the trajectory T. This can be accomplished in Eqs. (2) or (6) by having $A(S, T)$ or $E(p, T)$ not have any nontrivial dependence on x but simply be an S- or p-dependent numerical function purely of T, say $A(S, T) = a(S, T)$ (i.e., a function purely of S and of T times the identity operator in the wave function Hilbert space, so for a normalized wave function, one gets $\mu(S) = a(S, T)$), or, say, $E(p, T) = e(p, T)$ (so then $m(p) = e(p, T)$).

In this case one might say that he could simply dispense with the wave function $\psi(x, t)$, since the measure for one's perceptions would then depend only on the Bohmian trajectory T. However, the time dependence of this trajectory depends in a simple way upon the wave function, so the latter could still be a useful element in giving the simplest description of the trajectory upon which the measure for perceptions depends. (Similarly, one could in principle dispense with the quantum or Bohmian world altogether and consider merely the existence of the conscious world, or actually even only one's own present conscious perception within it, as that is all he has direct experience of, but the description of even just one's own present conscious perception by itself may be simpler if one postulates a whole conscious world with many measured sets of perceptions, and a description of this conscious world by itself may also be simpler if one postulates the existence of a quantum or Bohmian world on which it depends. I personally also believe that a description of the full SQM or SBM world would be simpler if one postulates the existence of an omniscient, omnipotent God as the Creator of this world, but of course this further extrapolation from direct conscious experience takes one beyond what is traditionally called physics to metaphysics.)

If for each S and p, $A(S, T)$ and $E(p, T)$ are the purely numerical functions (or functionals) $a(S, T)$ and $e(p, T)$ of the trajectory T, with no dependence on the wave function, there is still the question of what kind of functions of T they are. In particular they could be nonlocal functionals of the entire trajectory, so that one's perception depended on the entire Bohmian history. The fact that we do not normally perceive much detail about the future might be taken to suggest that there is a time $t(p)$ associated with each perception, and that $e(p, T)$ depends only on the part of the trajectory at times at and earlier than $t(p)$. The dependence on the part of T at times earlier than $t(p)$ might then be postulated to give the memory components of perceptions.

However, the fact that drugs and physical conditions such as Alzheimer's disease seem to have a strong effect on memory suggests to me that it is more plausible to assume that the memory components of present perceptions are directly caused by present physical conditions (perhaps having something to do with neural connections in the brain) rather than being caused directly by past conditions such as the past part of the Bohmian trajectory. Thus even if the measure density $m(p)$ for a perception p were determined entirely by the Bohmian trajectory T rather than by it and by the quantum state (or rather than entirely by the quantum state, as in SQM, which I personally prefer), I would think it to be the simplest plausible assumption that $m(p)$ is determined by the trajectory at the single time $t(p)$, which can be considered to be the time of the perception (something not well defined in SQM except in rather *ad hoc* ways (Page 1995c)). In other words, I am claiming that it seems simplest to suppose that each conscious perception is directly only of the present, though that present can include the records that produce the perception components of memory.

Of course, it may be an extreme idealization to suppose that the dependence is on precisely that single time rather than being spread out over, say, one Planck time. But I see no evidence that each of our perceptions need be directly affected by anything in the unconscious quantum or Bohmian world spread out over a time long in, say, seconds. Although a second is admittedly very long compared with the Planck time, it is certainly much shorter than cosmological times, and there is no evidence that I see that puts a lower limit on the time over which the unconscious quantum or Bohmian world affects a single conscious perception (unless one adopts a theory in which instants and time periods less than some lower limit simply do not exist).

Even if the measure density $m(p)$ for a perception p is determined entirely by the Bohmian trajectory at the instant $t(p)$, there is still the question of what aspect of the trajectory at that instant determines $m(p)$. In principle, it could be determined by the position, velocity, acceleration, and/or higher time derivatives of the position in the configuration space x as a function of time t, evaluated at the time $t(p)$.

If one wants to get the closest agreement between an SQM theory and a version of SBM with $m(p)$ depending purely on the Bohmian trajectory, it seems that one should select the following features in one's choice of the SQM and the SBM:

(1) The SQM and SQM should have the same bare quantum world, a normalized time-dependent Hilbert-space wave function on some configuration space.

(2) The position x_0 of the Bohmian trajectory T at time t_0 should be chosen randomly with the probability distribution given by the quantum "probability" density $\psi^*(x, t_0)\psi(x, t_0)$. (The Bohmian equation for the velocity of the trajectory as the gradient of the phase of the wave function ensures that if one selects a continuum ensemble of trajectories such that at time t_0 their measure density in the configuration space is the quantum "probability" density $\psi^*(x, t_0)\psi(x, t_0)$ at that time, then the measure density of this ensemble of trajectories is carried forward in time such that at time t it is precisely the quantum "probability" density $\psi^*(x, t)\psi(x, t)$ at the new time t. In this sense the "random" probability distribution $\psi^*(x, t)\psi(x, t)$ preserves its functional dependence on the wave function $\psi(x, t)$ at all times.)

(3) The SQM should have each of its experience operators $E(p)$ being a projection operator $P(s(p), t(p))$ onto a subset $s(p)$ of the configuration space and at a time $t(p)$ that both depend on the perception p.

(4) The SBM should have each of its experience operators being of the form $E(p, T) = e(p, T)$ and having a nonzero constant value (say 1) if the trajectory T at the time $t(p)$ (the same function of the perception p as in the corresponding SQM theory) is within the subset $s(p)$ of the configuration space (the same subset for each p as in the corresponding SQM theory), and having the value zero if the trajectory at time $t(p)$ is not within $s(p)$.

Even in this idealized correspondence between SQM and SBM theories, one will not get for each the same measure density $m(p)$ for individual perceptions or the same measure $\mu(S)$ for sets of perceptions. However, if one enlarged the BM and SBM theories to *Continuum Bohmian Mechanics* (CBM) and *Sensible Continuum Bohmian Mechanics* (SCBM) theories respectively that are hereby defined to include a whole continuum ensemble of trajectories, with their measure density given by $\psi^*(x, t)\psi(x, t)$ at any time t (whose preservation at all times is guaranteed by the Bohmian equation of evolution for each trajectory in the ensemble), then one would get agreement if one integrated, over the measured continuum ensemble of all trajectories, the $m(p)$, and hence also the $\mu(S)$, that one gets from each trajectory in the particular SBM theory with its own individual trajectory. In contrast, for a particular SBM theory with its own particular trajectory T, one would get a zero measure for any set S of perceptions p which all give configuration-space subsets $s(p)$ that the particular trajectory T is not in at the corresponding time $t(p)$. (On the other hand, any set of perceptions with nonzero measure in any particular SBM theory would necessarily have a nonzero measure in the corresponding SQM theory when features (1)–(4) hold for the pair of corresponding theories.)

Nevertheless, the fact that each perception p that actually occurs (i.e., that has a nonzero measure density) has no direct awareness of any other perception means that one cannot absolutely test whether or not one is in an SBM universe in which another perception p', with the particular trajectory T not in the configuration-space subspace $s(p')$ at the time $t(p')$ and hence with $m(p') = 0$, does not actually occur, or whether one is in the corresponding SCBM or SQM universe in which $m(p') > 0$ so that this other perception p' does actually occur. It seems that perhaps the best one can do is to compare what the corresponding SBM and SQM theories give for the *typicality* of the perception p (the fraction of the measure of other perceptions p' that have a measure density not greater than that of p, or some modification of this such as the *dual typicality* (Page 1995c)). Although if the features (1)–(4) proposed above hold, the corresponding SCBM and SQM theories would give the same typically for each perception, even the typicality for an existing perception in the SBM theory and that for the same perception in the corresponding SQM theory would differ. However, one might hope that for a reasonably large fraction of individual Bohmian trajectories, chosen randomly out of the continuum ensemble with the $\psi^*(x, t)\psi(x, t)$ measure density, most of the measure of the actually existing perceptions in the resulting particular SBM theory would occur for a set of actually existing perceptions whose typicalities would not be too low in either that particular SBM theory or in the corresponding SQM theory. If this were true (and its truth might well depend on the form of the underlying bare quantum theory, i.e. on the wave function, so that this could well be a question worth further investigation), then the typicality of most existing perceptions would not give a strong test between a particular SBM theory that predicted its existence and the corresponding SQM theory.

The fact that one's perception p cannot absolutely rule out any SBM or SQM theory that gives it a nonzero measure density $m(p)$, and the fact that one can apparently only use something like the typicality of the perception (interpreted as the likelihood or conditional probability for the perception given the theory) to weight the prior probability assigned to the theory in a Bayesian analysis to get the posterior probability for the theory (Page 1995c), means that even if one has a pair of SQM and SBM theories that do not share the features (1)–(4) above, one's actual perception will not necessarily rule out either of these theories. But, crudely speaking, the greater the degree to which the features (1)–(4) are not held by the pair of theories, the greater the possibility apparently is that the typicalities assigned by the two theories to one's perception will differ sufficiently greatly that one can use them to deduce a very low posterior probability to one or the other of the pair of theories.

Of course, I should emphasize that the "one" who is postulated to be doing the comparison of typicalities must be one who can indeed calculate them from the SQM and SBM theories in question. In practice this might be possible only for a being of such intelligence that he, she, or it (or whatever is the correct pronoun for such a being that may not have any sex and yet is more intelligent than we usually ascribe to "things") can exist only outside our universe. Those of us within the universe might be expected to be able to get only a very crude estimate of such typicalities, but if one can ever see roughly that they differ by many orders of magnitude between two theories, that would be sufficient to get a reasonably good idea of which theory to reject (except in the case in which the prior probabilities differed by roughly the same number of orders of magnitude in the opposite direction).

There is also the apparently completely subjective question of what prior probabilities to assign different SQM and SBM theories before weighting them with the typicalities of one's perception in a Bayesian analysis to get the posterior probabilities for the theories. (I say "apparently," since conceivably there is an actual existing measured set of different universes, each described by a different SQM or SBM theory, so that the actual measure of these universe gives an ontologically objective prior frequency-type probability to the different theories describing the different universes, but since we certainly do not have access to all of these conceivable universes or know their measure even if it does exist objectively, for us epistemologically the prior measure must surely be subjective.) Here I should lay my cards on the table and explain the prejudices that I have against assigning BM and SBM theories high prior probabilities.

First, I should say that I would prefer to assign higher prior probabilities to simpler theories. Perhaps most physicists would agree, but they might differ on how to weight or even how to rank the simplicity of different theories (I suspect that this may lie at the core of more metaphysical disagreements as well, such as whether theism or atheism is true.)

I myself think that adding the Bohmian trajectory to bare quantum mechanics reduces the simplicity and thus the prior probability I would assign to the theory. Perhaps for nonrelativistic QM, the extension to CBM with its continuum of trajectories that have the $\psi^*(x, t)\psi(x, t)$ measure density is not too great a loss of simplicity, since this ensemble of trajectories does not take too much additional information to specify and has certain nice properties (such as the measure density's remaining $\psi^*(x, t)\psi(x, t)$ for all time). However, to pick out a single trajectory from this ensemble for a particular BM (as opposed to CBM) theory would seem to require much more *ad hoc* information. I can think of certain choices that do not require too much additional information, such as choosing the trajectory whose time integral of $\psi^*(x, t)\psi(x, t)$ evaluated along the trajectory itself is the maximum out of all possible Bohmian trajectories, but even these relatively simple choices seem rather *ad hoc* and ugly.

Incidentally, I might point out that one could have theories of *Generalized Continuum Bohmian Mechanics* (GCBM), and its augmentation of *Sensible Generalized Continuum Bohmian Mechanics* (SGCBM) to include the Conscious World Axiom and the (suitably generalized, as discussed above for SCBM) Bohmian-Consciousness Connection, in which there is a continuum of Bohmian trajectories with a measure density that is a general normalized nonnegative function over the configuration space at any one time and which is transported to other times by the Bohmian equation for the velocity of the trajectories. One could then say that CBM is the special case (probably the simplest) in which this measure density is $\psi^*(x, t)\psi(x, t)$, and ordinary BM with its single trajectory is the special case in which the measure density at any one time is a delta-function distribution (not so simple, as discussed above, since one must specify the location of this delta-function at one time.) However, even within the wide set of possibilities of GCBM or SGCBM theories, I do not see any so simple as simply leaving out the trajectories altogether, as in bare QM and SQM theories.

Now a supporter of BM might object that although BM is almost certainly more complicated than bare QM in requiring the extra element of the Bohmian trajectory, in another way it is simpler in not requiring any operators on the Hilbert space of wave functions (except presumably for the Hamiltonian that generates the evolution by the Schrödinger equation). However, once the Hilbert space of wave functions is defined, it takes very little additional information to define the set of operators on this space that maps it into itself. Perhaps this additional information is comparable to that for the continuum ensemble of trajectories in CBM, so I would admit that CBM is of comparable simplicity to bare QM for the same Hilbert space (though specifying a precise single trajectory in ordinary BM seems to me in general more complicated). But even in a comparison with CBM, the framework of bare QM allows a much greater range of possibilities for the quantum state than the restriction to a wave function over configuration space (or some slight generalizations to include spin, etc.) that seems to be necessary for present formulations of BM and CBM, and it might turn out that a quantum state that cannot be written as a wave function will be at the heart of the simplest complete description of the universe.

A supporter of BM might also object that even if BM is indeed more complicated than bare QM in having its Bohmian trajectory, when one attaches a theory of consciousness the result is simpler if one can attach it to a theory with a trajectory. For example, in the Sensible theories outlined above, a supporter of SBM over SQM might say that BM allows simpler awareness operators $A(S, T)$ (e.g., of the form $a(S, T)$, with no nontrivial dependence on the wave function) in the Bohmian-Consciousness Connection of SBM than the awareness operators $A(S)$ allowed in the Quantum-Consciousness Connection of SQM, and that the increased simplicity of $A(S, T)$ over $A(S)$ overbalances the extra complication of the trajectory in SBM. I will admit that it is hard to answer this objection when we are as ignorant as we are about the connection of consciousness to the rest of physics (e.g., about what the awareness operators are in Sensible theories), but I should say that at present I for one am rather sceptical that a viable SBM theory will turn out to be as simple as a viable SQM theory.

Another problem I have with BM theories is that when I imagine applying them to relativistic fields instead of to nonrelativistic particles, I find that even if the quantum state is Lorentz invariant (e.g., the vacuum state), almost all the trajectories are not, and neither is the continuum ensemble of CBM. For example, consider for simplicity a free scalar field. Each plane-wave mode of the field can be considered to be an harmonic oscillator, and the vacuum state of the field can be considered to be a product of the ground states of all of these modes. For such a static state of zero energy, the wave function has a constant phase with zero gradient in the configuration space (the space of amplitudes of the modes), so the Bohmian trajectory is static. When one superposes the static amplitudes for the modes, one gets an arbitrary space-dependent static configuration of the scalar field as its Bohmian trajectory. Except for the special homogeneous cases in which this static configuration is precisely the configuration of constant field everywhere (which in the case of zero field is the one proposed above whose time integral of $\psi^*\psi$ evaluated along the trajectory itself is the maximum out of all possible Bohmian trajectories), these Bohmian trajectories are not Lorentz invariant, for the spatial dependence in the frame in which their evolution was calculated by the Bohmian equation of motion becomes a temporal dependence in any other frame.

Furthermore, in any other frame the trajectory does not even satisfy the Bohmian equation of motion, so the situation is worse than the version of spontaneous symmetry breaking in which the quantum state is not invariant under the symmetry group of the equations which it solves. In that case the action of an element of the symmetry group on the nonsymmetric solution yields another nonsymmetric solution, but for the inhomogeneous Bohmian trajectories, the action of a Lorentz transformation on them yields time-dependent trajectories that do not even satisfy the Bohmian equations for trajectories. (I am grateful to Shelly Goldstein for pointing out the importance of this distinction.)

Of course, one could say that perhaps Lorentz invariance is just a useful approximation to certain aspects of the world and need not apply to the Bohmian trajectories. After all, in our perceptions there are objects that seem to have fairly definite velocities (with the bulk of nearby ones, such as the earth, being fairly near zero in our local frame) that appear to break Lorentz invariance, so presumably there is nothing blatantly inconsistent with observations to have Bohmian trajectories breaking Lorentz invariance even when the quantum state is Lorentz invariant. However, this lack of Lorentz invariance for all but the homogeneous Bohmian trajectories for a quantum field in the Lorentz-invariant vacuum state, and the need to define a preferred velocity in order to define the Bohmian equations for the trajectories, seems at least aesthetically rather ugly.

I suspect that there would be an even greater degree of ugliness if one attempted to devise a Bohmian version of quantum gravity, say for the quantum cosmology of closed universes. If one has a solution of the

Wheeler–DeWitt equation for canonically-quantized general relativity, this gives a wave functional of three-geometries that is invariant not only under coordinate transformations of the spatial hypersurface, but also under the equivalent of time translations that are arbitrary at each point of space. Unlike the case of quantum field theory in Minkowski spacetime, in which only one quantum state (the vacuum) has the full Lorentz invariance of the Minkowski spacetime itself, in canonical quantum gravity *every* physical state has the diffeomorphism invariance (including local Lorentz invariance) of general coordinate transformations. It is not clear to me how to construct a Bohmian trajectory for such a theory of quantum cosmology without breaking this diffeomorphism invariance in a complicated and ugly *ad hoc* way.

I would also suspect that if one tries to overcome the severe technical problems of constructing a finite theory of gravity by going from a field theory of general relativity to a superstring theory, it would be even more cumbersome and ugly to try to construct a Bohmian trajectory. Of course, the bare QM of superstrings at the nonperturbative level is hardly understood at the moment, and virtually nothing is known of how to produce consciousness out of it (say by some form of awareness operators), so one cannot pretend to be certain that this task will be easier without constructing a BM version of superstrings, but I personally suspect that the latter would be simply an onerous burden that is unnecessary. It would seem much more plausible that one should go directly from a bare quantum theory of superstrings (or whatever the ultimate quantum theory is, assuming that it *is* a quantum theory) to a theory of consciousness (say given in terms of the unconscious quantum world by expectation values of awareness operators in SQM), without bothering with trying to find a BM theory with a definite trajectory.

In conclusion, I have outlined how a theory of consciousness may be attached to the Bohmian version of quantum mechanics (BM) to give what I call Sensible Bohmian Mechanics (SBM), in a way highly analogous to the way in which I have proposed attached a theory of consciousness to ordinary quantum mechanics (QM) to give Sensible Quantum Mechanics (SQM) (Page 1995a–d, 1996). Because BM has a Bohmian trajectory that bare QM does not have, SBM for a fixed wave function has more possibilities for awareness operators for giving the measure of sets of conscious perceptions. (Conversely, BM seems to require a wave function on a configuration space, or something like it, which QM does not, so in the latter way QM is more flexible.) However, if a SBM is to give mostly typical perceptions that would also be typical in a corresponding SQM theory, it seems that there should be at least four features of the correspondence (listed above), though it is not completely clear that these are entirely sufficient. Although SBM might conceivably allow a simpler set of awareness operators than SQM, this is by no means obvious, and in other respects SBM seems to me to be more complicated and ugly than SQM.

I especially appreciate many helpful discussions with Shelly Goldstein, though he does not share my subjective negative evaluation of Bohmian mechanics. Financial support has been provided by the Natural Sciences and Engineering Research Council of Canada.

University of Alberta

MARA BELLER

BOHM AND THE "INEVITABILITY" OF ACAUSALITY

1. INTRODUCTION

Referring to the orthodox denial of the possibility of quantum ontology, John Bell invoked a vivid image: "It is," he wrote, "as if our friends could not find words to tell us about the very strange places where they went on holiday. We could see for ourselves whether they came back browner or fatter" (Bell 1987a, 171).

Why were quantum physicists mute for so long on the issues of quantum reality? Why did they voluntarily give up an exciting, though admittedly inconclusive, quest to unveil the mystery of the quantum world? Why did they hold fast to the notion of the lack of objectivity and inevitable acausality, despite the existence of Bohm's alternative version?

David Bohm himself explained this seemingly puzzling situation by the inability of the orthodox physicists to conceive of any other concepts, except classical (waves and particles) and by their commitment to the philosophy of positivism:

> . . . There appears to have existed, especially among physicists such as Heisenberg, and others, who first discovered the quantum theory, a rather widespread impression that the human brain is, broadly speaking, able to conceive of only two kinds of things, namely fields and particles". (Bohm 1957, 96)

Another reason for the refusal to consider a "subquantum-mechanical level," according to Bohm, has been the adoption of the positivistic thesis of elimination of unobservables from physical theorizing, "a general philosophical point of view containing various branches such as 'positivism,' 'operationalism,' 'empiricism,' and others, which began to attain a widespread popularity among physicists during the twentieth century" (*ibid.*, 97).

I will reiterate Bohm's insight about the intimate connection between the finality of acausality, indispensability of classical concepts, and positivism, but will provide a different account of their interrelatedness, along more sociological lines. I will also analyze Bohm's own views on the relation between classical and quantum concepts, and I will argue that even before Bohm's conversion took place, an unbridgeable difference of opinion existed between Bohm and Bohr.

From a certain perspective, the stand of the foundational acausality of quantum theory can be a natural and a legitimate one. If our best scientific theory is de facto acausal, what good reasons, except of an unwarranted a priori belief in classical causality, would urge physicists to search for a causal version? After all, many physicists, including the late Bohm himself, did not find determinism either indispensable or appealing. Heisenberg's

J. T. Cushing et al. (eds.), Bohmian Mechanics and Quantum Theory: An Appraisal, 211–230.

argument – that, if physicists are convinced in the correctness of the quantum theory, which is built on probabilistic foundations, then the conclusion of lack of causality follows (Heisenberg 1958, 89) – is a characteristic example of a sound reasoning which Arthur Fine (1986, 87) coined by the term "entheorizing." The metaphysical question of whether a certain property is a property of our world is translated into the question of fruitfulness, or correctness, of a theory, which entails this property.

Yet such sober "entheorizing" reasoning is rare. The Copenhagen writings, Heisenberg included, are permeated by the heavy rhetoric of finality, of inevitability of acausality, of impossibility of a causal alternative. It is this argumentation of inevitability which calls for an explanation.

It is helpful to divide the historical issue of acausality in quantum mechanics into a few distinct, though interrelated issues: how and why did probabilistic arguments become foundational in quantum mechanics? How and why were strong anticausal sentiments introduced and advocated? How did the claims of inevitability evolve and by what means were they argued? And why were they seemingly so effective and persuasive? Though briefly attempting to deal with all these issues, it is the anatomy of inevitability which constitutes the main theme of this paper.

1. ACAUSALITY AND MATRIX MECHANICS

I have treated the first issue, the question of how the probabilistic arguments became central in quantum mechanics, elsewhere, so I will mention here only some relevant highlights.

The first general, confident and unequivocal stand against causality occurred in Heisenberg's uncertainty paper (1927b). In this paper Heisenberg, countering Schrödinger's accusation of unintuitiveness of matrix mechanics, restored the regular space-time and resurrected the visualization of particles in spatio-temporal statistical terms. In the original matrix theory, the meaning of the matrix elements was a solely electromagnetic one, connected with frequencies, intensities, and polarizations of the emitted atomic radiation (Beller 1983b). The original matrix theory was a discrete deterministic one: the problem was fully determined when the Hamiltonian of the system was known.[1] Nobody hurried to draw any categorical indeterministic conclusions from the fact that matrix elements represented transition probabilities. Many theorists, including Born and Jordan, did not exclude the possibility that this might be a temporary weakness of the theory – a stand remarkably close to Einstein's. In November 1925 Pauli declared in a letter to Bohr that a satisfactory physical theory should contain no probability concepts in its fundamental propositions. As late as the beginning of 1927, just before the uncertainty paper, Jordan argued that as long as quantum theory is not reduced to a theory of elementary independent probabilities, there can be no verdict on the issue of determinism (Beller 1985; 1990). No wonder Max Born did not feel at first that his statistical interpretation of the wave function had any far-reaching philosophical implications (Beller 1990). Nor did the quantum physicists, Heisenberg included, initially express any belief in an a priori status of discontinuities, or of a quantum postulate, in Bohr's terms: in fact, Heisenberg initially rejoiced that the matrix theory does not postulate dogmatically the discrete energy levels – their existence follows from the mathematical formalism. Heisenberg, Jordan and Born described this as a "notable advantage" of the new matrix mechanics as compared to the old quantum theory (Beller 1985).

The stand against causality, especially in Heisenberg's able hands, crystallized as a response to the threat of Schrödinger's competitive theory. Schrödinger's declared aspirations were to eliminate the "monstrous" – as he called them – quantum jumps, replacing the Göttingen-Copenhagen research program with a continuous causal alternative. It was in response and in opposition to Schrödinger's threat that Heisenberg resurrected the foundational status of discontinuity, and swiftly developed arguments against causality. It is around the status of "quantum jumps," which Schrödinger aimed to eliminate, that the controversy and the shaping of the Copenhagen philosophy took place – the indeterminism and the wave-particle duality were, to a large extent, derivative issues (Beller 1992).

The probabilistic formalism of quantum mechanics was developed by Dirac and Jordan, as a direct extension of Heisenberg's fluctuation paper (1927a), in which, against Schrödinger and his supporters, Heisenberg claimed that for two atoms in resonance, the exchange of energy must occur via discrete jumps. One cannot obtain the energy of an atom as a function of time, argued Heisenberg, but only the time mean of energy, or probability of energy to have a particular value. Similarly, for any dynamical variable, both Dirac and Jordan postulated, one can only calculate the probability that it lies between two specified values. In this way, the discontinuity and acausality were entrenched into the very axiomatic basis of quantum mechanics to a degree that made them appear as "hard facts" of nature (Beller 1992).

This reading dispels two widespread misconceptions concerning the history of quantum mechanics. The first misconception assumes that an abstract quantum formalism developed prior to any interpretive concerns, simply as a disinterested mathematical activity. According to the second misconception, the Copenhagen-Göttingen interpretation came first, and only later Schrödinger's competing interpretive attempts followed. Both claims, I believe, must be substantially qualified. From the beginning, the orthodox interpretation and the formalism itself developed as a defense project of the past achievements that Schrödinger threatened to undermine, and as a philosophical legitimization of the Copenhagen–Göttingen research program against any challenge, or alternative. The ambitious Heisenberg, who laid the foundations of matrix mechanics, was one of the central players in this enterprise. Strong sentiments against causality quickly arose precisely because causality became intrinsically tied up with a threat of opposition.

Perhaps another few words about why it was Heisenberg who initially led the campaign against causality are in place here. The interpretive attempts of quantum theorists took place against the background of philosophical controversies around the change of status of space and time concepts implied by Einstein's relativity. Young Heinsenberg witnessed the emotional, and politically charged, confrontation between Einstein and the Neo-Kantians (Cassidy 1992, 96–98). The question whether the same fate did not await the a priori status of causality, as happened with the a priori status of Newtonian-Euclidean space and time, was a natural one. An intensely ambitious individual, who aspired, following Einstein's example, to leave his mark on the philosophy of Nature and the public imagination, and not only to solve scientific "puzzles," must have kept an eye on this issue. No less than Heisenberg's own teacher, Arnold Sommerfeld, posed the high-stakes challenge. Discussing "the present state of physics" and referring to the causality issue just a few months before Heisenberg's uncertainty paper, Sommerfeld called for a "new Kant" (Sommerfeld 1927, 235). Heisenberg wanted to be the new Kant – in his initial presentations of the uncertainty principle for academic audiences, he always described the abandonment of the "Kantian category of causality" as a natural continuation of the Einsteinian overthrow of Kantian space and time as forms of intuitions (e.g., Heisenberg 1928; 1931).

How opportune for Heisenberg that the ambition to be the new Kant and the aim to conclusively counter the opposition so neatly coincided!

3. ACAUSALITY AND THE "INDISPENSABILITY" OF CLASSICAL CONCEPTS

The philosophical framework of the orthodox interpretation arose initially as a forced response to the challenges of opposition and developed subsequently into the huge project of legitimization of the Göttingen-Copenhagen program.[2] As the argumentation for "finality," or inevitability, is perhaps the strongest way to advance any sort of claim, the rhetoric of finality and inevitability permeates many philosophical arguments in the Copenhagen framework. Both Bohr and Heisenberg present a search for a causal quantum theory not merely as hopeless, but as downright meaningless. Heisenberg argues (1934, 17) that "there seems to be the strongest evidence" for the "final" renunciation of causality and objectivity. The concept of "causality" after the discovery of quantum mechanics became as devoid of meaning as the concept of the "end of the world" after the discoveries of Columbus and Magellan (*ibid.*). Bohr, claiming that acausality is "unavoidable," ridiculed the "odd wish" for causal description (Bohr 1937a, Lecture 6): "It appears to me to be exactly that, that we would say that we are not satisfied with descriptions in English, and we might like, because we can't get the word 'causality'. . . to translate it into Sanskrit, to get everything different . . ." And Heisenberg, quoting Bohr, asserts that the structure of Bohm's "strange" arguments for hidden variables is identical to the hope that "$2 \times 2 = 5$, for this would be of great advantage for our finances" (1958, 132).

One would expect that the proponents of the Copenhagen interpretation are in possession of some very strong arguments, if not of inevitability, at least of high plausibility. Yet a critical reading reveals that all the far-reaching, or one might say far-fetched, claims of inevitability are built on shaky circular arguments, on intuitively appealing, but incorrect, statements, on metaphorical allusions to quantum "inseparability" and "indivisibility," which have nothing to do with quantum entanglement and non-locality. Acausality, in the writings of Bohr, Heisenberg, Pauli and Born, is not a sharply defined concept.[3] Precisely because acausality is used as a tool of legitimization, as a sword to pin down the opponents, its meaning changes from text to text, from context to context.

Among Heisenberg's many argument for the finality of acausality, one of the more frequent and careless ones is the argument of the mobility of the "cut" between the quantum object and classical measuring device. Heisenberg argues that the statistical aspect enters only on the dividing line, or on "the cut," (Heisenberg 1979, 15, 49; 1958, 48–49), because on both sides of the line there is a deterministic lawfulness (on one side, because of the laws of classical physics, and on the other side, because of the differential equations of quantum mechanics). If we supplement the variables at the dividing line by some "hidden variables" and then "move" the "cut" to a new place, we will come into conflict with the causal laws that prevailed at this place to begin with (the situation becomes overdetermined).

This argument is, of course, unconvincing. If we move the cut into what was the classical domain, it does not necessary imply that the situation becomes overdetermined – the "hidden" variables for the cut can be the existing variables in macrodomain, such as Bohm's supplement of the quantum description by the position variable. It is not too hard to find serious flaws in other arguments advanced by Heisenberg along similar lines. The more interesting question is: when, for what purposes, and in what contexts did Heisenberg use such arguments?

The crucial part of all the "cut" arguments is the necessity of describing the measuring apparatus by classical concepts, which is connected with a more far-reaching Bohrian claim of the denial of "quantum concepts" and "quantum ontology." In many of his writings Heisenberg distances himself from this position. Thus, he often says that the measuring apparatus "can" (as opposed to the Bohrian "has to") be described by classical concepts. Nor does he agree with Bohr that no new concepts and no new language beyond classical can be conceived – in fact, while discussing the lessons of the theory of relativity, he emphasizes that new concepts arise and a new language develops "which adjusts itself after some time to the new situation" (1958, 175). For example, in "the theory of general relativity, the language by which we describe the general laws actually now follows the scientific language of the mathematicians" (1958, 177).

I have argued elsewhere that mathematical physicists, especially those who had a first-hand familiarity with the relativity theory, such as Born, Fock and Heisenberg, did not accept the Bohrian special status of the classical concepts (Beller 1996). Instead, they held that justification of concepts arises from their place in the coherent mathematical formalism of the physical domain – be it classical or quantum. There is no warrant, therefore, for giving a special status to classical notions. David Bohm was of a similar opinion: "Mathematics in general and the differential calculus in particular have played a key role in guiding the development of a clear concept of accelerating motion . . . it is practically impossible to gain such a clear concept on the basis of everyday experience alone, or even on the basis of the laboratory experience alone" (1957, 97–98). Similarly, our concepts of waves come not from watching water waves, but from the mathematical treatment of interference and wave propagation (*ibid.*). Bohr, a "natural philosopher, and not a mathematical physicist" (Heisenberg's expression) was, I believe, quite unique among quantum physicists in his denial of the possibility of the new concepts. One cannot then avoid the question: why and when did Heisenberg support this peculiar, unnatural for a mathematical physicist, doctrine of Bohr?

In the cut argument, analyzed above, Heisenberg emphasizes the mobility of the cut as "particularly important." The next sentence reveals why it is so important: "an appreciation of this fact also helps to dispose of an objection frequently made against the *finality* [italics mine] of quantum mechanics" (1934, 16). Heisenberg immediately proceeds to the "more general question of the finality" of changes promoted by the Copenhagen proponents, and concludes that the renunciation of objectivity and causality will be final (*ibid.*, 17).

Similarly, Heisenberg presented the flawed cut argument in another lecture a year later, again using it as an argument for "the finality of the statistical character of quantum mechanics," and against "any hope" of "completing quantum mechanics on a determinist [*sic*] basis" (1936, 48). Again, in another article, Heisenberg claimed that the "root of the statistical character of quantum theory" is the "tension" between the necessity to use the concepts of the classical physics and "from the knowledge that these concepts do not fit accurately." Heisenberg mentions that "it has been suggested that one should depart from classical concepts," searching for radically new ones that "might possibly lead back to a nonstatistical, completely objective description of nature" (Heisenberg 1958, 56). He immediately denies such a possibility, repeating the Bohrian argument that "the concepts of classical physics are just a refinement of the concepts of daily life" (*ibid.*). Heisenberg's denial is a forceful one: "We cannot and should not replace these [classical] concepts by any others . . . we can not and should not try to improve them" (1958, 44).[4]

I have argued elsewhere (Beller 1996) that the Copenhagen interpretation cannot be reconstructed as a coherent philosophical framework, for it is more adequately described as a collection of local, often contradictory, arguments in changing theoretical and socio-political circumstances. Heisenberg's contradictory statements about the necessity of classical concepts in different contexts is another example for my claim. From our discussion it becomes clear that Heisenberg insists on the indispensability of classical concepts in those contexts, and, as far as I am able to determine, only in those contexts, when he attempts to argue the finality of the indeterministic Copenhagen version.

Returning to Bohm's claim that it is Heisenberg's belief in the indispensability of classical concepts (and Heisenberg's inability to conceive of other categories than waves and particles) which lead him to belief in finality, my claim is rather the opposite one. It is the stand of, or vested interest in, finality that led Heisenberg to insist on the indispensability of classical concepts in a local and opportunistic fashion. There was neither belief nor commitment on Heisenberg's part – only a selective and opportunistic use of Bohrian doctrine in the circumstances when Heisenberg's aim was to argue against the opposition and for the finality of the Copenhagen orthodoxy. This is a characteristic example of powerful social strategy of legitimization, disguised as an abstract theoretical argument.

It is quite instructive to see how Heisenberg argues in those articles in which he discusses the lessons both of relativity theory and of quantum physics. For the main lesson of relativity, as Heisenberg presents it, is the denial of the Kantian a priori status of classical concepts, while the defense of the Copenhagen interpretation forces him to claim their indispensability and finality in an a prioristic Kantian sense. Ingenious Heisenberg has no problem avoiding the impasse: he invents the notions of "practical a priori," and "relative synthetic a priori," which are, in fact, contradictions in terms. The only function of these self-contradictory notions, as they appear in Heisenberg's works, is to argue for an a priori status of classical concepts in the context of the defense of the Copenhagen framework, against Heisenberg's own better judgment when discussing relativity (1958, 90–92).

David Bohm, after his "conversion" had taken place, understood that arguments of inevitability of acausality cannot be compelling. They are, in fact, circular arguments: "the conclusion that there is no deeper level of causally determined motion is just a piece of circular reasoning, since it will follow only if we assume beforehand that no such level exists" (Bohm 1957, 95).

4. OPERATIONALISM: FROM CONSISTENCY TO "INEVITABILITY" ARGUMENTS

Bohr's arguments of the "inevitability" of acausality and complementarity similarly cannot withstand close scrutiny. I have argued elsewhere that, in the best case, Bohr's arguments of inevitability are disguised arguments of consistency (Beller 1993). "The uncontrollable disturbance," or uncontrollable interaction between the measuring and the measured, is the crucial component of all these arguments, and the basic building block on which Bohr's web of arguments for complementarity rests. Bohr employed different strategies for making this "uncontrollable interaction," founded on the "indivisibility of the quantum of action," a fundamental concept in quantum philosophy.

Yet how exactly does the "uncontrollability of the interaction" follow from the "indivisibility of the quantum of action," and what do these terms mean? Their use and their interconnection vary, in fact, significantly, before and after 1935 (the EPR challenge.) Before 1935, Bohr's arguments are incorrect, and are based on the discredited later idea of disturbance. After 1935, the arguments are tautological, or circular: textual analysis reveals a web of terms, where "indivisibility of the quantum of action," "individuality of quantum phenomena," "unsurveyability of interaction," or "uncontrollability of interaction" have the same meaning, and are interchangeable.

The following lines are typical in Bohr's pre-1935 writings:

> Quantum of action means that it is not possible to distinguish between the phenomena and observation . . . we cannot make the interaction between the measuring instrument and the object as small as we like. Now, that means that phenomena are influenced by observation. (Bohr 1930)

Both talks about the "difficulty of distinguishing between the object and measuring instrument . . . when it is necessary to take the atomic constitution of measuring instruments into account" (Bohr 1931).

Before 1935 "indivisibility of the quantum of action" simply meant "finitude" of the quantum of action. Measurement "disturbs" the phenomena precisely because of the atomic constitution of measuring devices. The underlying intuition was that, because in the quantum domain phenomena and measuring interaction are "of the same order of magnitude," such finite interactions, contrary to the classical case, cannot be neglected or accounted for. Such intuitively appealing argumentation was, of course, incorrect. The concept of disturbance is an inconsistent one: it presupposes the existence of objective exact values that are changed by measurement, contrary to the desired conclusion of their indeterminateness. Moreover, the finitude of interaction does not assure uncontrollability – many classical measurements involve finite interactions that cannot be neglected, yet can be calculated.

After 1935, when the EPR challenge undermined the notion of disturbance (Fine 1986; Beller and Fine 1994), Bohr's reasoning undergoes an about-face. The "uncontrollability of the interaction" is now inherently tied to the necessity of describing measuring devices by classical terms only, neglecting "in principle" their atomic constitution:

> The essentially new feature is a fundamental distinction between atomic objects and measuring devices . . . it is this fundamental distinction that excludes any separate account of interaction just because in the description of the measuring apparatus we have to entirely look apart from the quantum. (Bohr 1958a)

In Bohr's later writings the "uncontrollability" of the interaction follows from its "unsurveyability," which in turn follows from the dogmatically postulated principal difference between the quantum object and the classical measuring apparatus. If, in pre-1935 Bohr, the uncontrollability necessitated acausality, in the later writings it is the statistical character of quantum theory that leads to inseparability of the measuring and the measured, and to the "uncontrollability" and "unsurveyability" of the measuring interaction.

This reversal of what is to be explained and what is foundational, of explanans and explanandum, is frequent in Bohr's and Heisenberg's writings. This reversal is an eloquent illustration of the circularity and ad hocness of many of their arguments. One striking example is Heisenberg's reversal of the idea of disturbance – if the original idea of disturbance was based on indivisibility, or finitude, of microscopic particles, now the "unavoidable disturbance" is seen as the reason for their indivisibility:

> Nature thus escapes accurate determination . . . by an unavoidable disturbance which is part of every observation . . . In atomic physics it is impossible to neglect the changes produced on the observed object by observation . . . The supposition that electrons, protons and neutrons, according to modern physics the basic properties of matter, are really the final, indivisible particles of matter, is only justified by this fact. (Heisenberg 1979, 73)[5]

Heisenberg's and Bohr's arguments of a foundational, a prioristic status of "uncontrollable interaction," are illegitimate. The uncontrollability of interaction has one, and only one reason – it follows from the mathematical framework of the quantum theory. It is necessary to ensure the consistency of quantum theory, when measuring possibilities are taken into account. David Bohm, after his conversion took place, understood this clearly. Thus, discussing the "uncontrollability" of the interaction in Heisenberg's γ-ray experiment, Bohm pointed out:

> the disturbance is unpredictable and uncontrollable because *from the existing quantum theory* [italics are his] there is no way of knowing or controlling beforehand what will be the precise angle, with which the light quantum will be scattered into the lens. (Bohm 1957, 82 (footnote)).

Bohm realizes that the indeterminacy relations are "primarily . . . a deduction from the quantum theory in its current form," and that the attempts to prove its "absolute and final validity" are based on nothing else but on an aprioristic and illegitimate insistence on the finality of the quantum theory (*ibid.*).

The legitimate function of "uncontrollability" arguments is to establish the consistency, rather than the inevitability, of uncertainty and complementarity. Bohr does discuss certain canonical thought experiments along these more legitimate lines (Bohr 1935).[6] He argues, for example, for the impossibility of measuring the position of a particle in a set-up measuring its momentum. Using a moveable diaphragm, we can measure a particle's momentum (by using an appropriate test body) before and after the passage of a particle through the slit (*ibid.*). By applying the law of conservation of momentum to a system consisting of a particle and a moving diaphragm, we can calculate the momentum of a particle passing through the slit. However, because we have exact knowledge of the momentum of the diaphragm during, or immediately after, the particle's passage, we block the possibility of knowing also the position of the diaphragm, in accord with uncertainty relations. Consequently, we are denied a knowledge of particle's position, which is identical with that of diaphragm location. Why does the diaphragm obey the uncertainty relations? Either we just state it (it applies to quantum objects), and then we deduce the uncertainty for the particle (this is, of course, an argument of consistency), or we say that while measuring the momentum of the diaphragm with a test body, the diaphragm undergoes uncontrollable interaction. We deduce consequently the uncontrollable interaction between the particle and the diaphragm – again, these are elementary considerations of consistency.

The crucial point of these thought experiments is that momentum and position cannot be *measured* simultaneously. This is however not satisfactory for Bohr. Why is it not satisfactory? The answer is clear from Bohr's own words. He warns against "misunderstanding by expressing the context of uncertainty relations . . . by a statement as 'the position and momenta cannot simultaneously be measured with arbitrary accuracy' . . . such a formulation would not preclude the possibility of a future theory taking both attributes into account" (Bohr 1937b).

What is the strategy of precluding a theory, such as Bohm's where particles have a well-defined positions and momenta, which however cannot be measured simultaneously according to the uncertainty relations? How do we argue that Bohm's theory is impossible in principle? How do we transform consistency arguments into those of inevitability?

With the benefit of hindsight the answer is obvious. Simply postulate that what cannot be measured – does not exist. By defining concepts operationally through a procedure for their measurement, and then applying the quantum formalism to the analysis of the measurement procedure, we will obtain nothing else but deductions from the quantum formalism (such as, for example, the uncertainty relations). In this way an illusion is created that the features of the theory (such as uncertainty) belong to the very definition of the concepts used, and that they inevitably follow from a logical analysis of conditions of experience.

The initial use of thought experiments by Heisenberg and Bohr was indeed to argue for the consistency of the new quantum-mechanical scheme. This was the original thrust and aim of Heisenberg's analysis in γ-ray Gedanken experiment. Bohr, following Heisenberg, argued that uncertainty relations are "the basis of the thorough investigation of the logical consistency of quantum mechanics" (Bohr 1929).

Bohr's considerations of mutually exclusive experimental arrangements were initially aimed at proving the consistency of quantum theory: the wave and particle aspects "can never be brought into direct contradiction . . . since they demand mutually exclusive arrangements" (Bohr 1933, 422). This mutual exclusivity, aided by operationalism, soon developed into the arguments of inevitability of uncertainty and complementarity. This connection between operationalism and inevitability becomes substantially strengthened in Bohr's response to EPR. As Arthur Fine and I have argued (1994), the only way Bohr was able to counter the challenge of EPR was by the operational and relational definition of concepts. After EPR, operationalism became the central tool for the strongest case of legitimization arguments for the inevitability of quantum theory.[7] It is in this context that Bohr often uses the word "logical" – the inevitability of complementarity and uncertainty follows, in his words, from "the simple logical analysis" and leads to "the situation where the very concept of determination loses its possibility of logical application" (Bohr 1957).

With the "dictatorial help," to use Schrödinger's (1935) apt expression, of operationalism, the central Copenhagen dogmas become unassailable and final.

Now we can return to Bohm's claim that it is the positivistic commitments of the founding fathers of the Copenhagen interpretation that led them to believe in the "finality" of the quantum theory, and the impossibility of revising it along more objective or deterministic lines. As in the case of the "inevitability" of classical concepts, the situation is a reversed one: it is the need, or the desire, to argue for finality against the threat of opposition, that led Heisenberg and Bohr to forcefully employ the operationalistic stand. I have argued (Beller 1996) that the proponents of the Copenhagen interpretation do not have, collectively or even individually, a consistent stand on the realism-antirealism issue, and that they employ different, often contradictory arguments, depending on the local context. We now have identified this context in which positivistic operationalistic arguments are selectively employed. Positivism in quantum philosophy has two central strands – that of operationalism, and of instrumentalism. Instrumentalism implies that the quantum formalism is merely a tool, an algorithm for the description and prediction of the measurement results. The instrumentalist stand, denying the possibility of a quantum ontology, while emphasizing the Kantian inevitability of the classical one, is, of course, directly connected with the doctrine of the indispensability of classical concepts. Both varieties of positivism – instrumentalism and operationalism – are aimed at arguing for the finality of quantum mechanics.

Were physicists aware that, unless one takes an extreme and untenable positivistic stand, what the Copenhagen interpretation offered were at best considerations of consistency, and not of inevitability? After his conversion, I believe, David Bohm was aware of this. According to Bohm, Bohr's essential contribution was to demonstrate that one can "*consistently* [italics mine] . . . renounce the notion of unique and precisely defined conceptual models in favor of that of complementary pairs of imprecisely defined models" (1957, 93). What Bohr was able to demonstrate was that such a use of complementary concepts provides " . . . a *possible* [italics mine] way of discussing the behavior of matter in the quantum-mechanical domains" (*ibid.*).

Not only such dissidents as Bohm, but orthodox quantum physicists were also aware that many of the philosophical pronouncements on complementarity and uncertainty are merely consistency arguments. Wolfgang Pauli announced that the central function of the notion of complementarily is "to symbolize the fundamental limitation of our . . . idea of phenomena as existing independently of the means by which they are observed. . . . It is indeed this limitation that makes the theory *logically consistent*" [italics mine] (Pauli 1945, 99). Paul Dirac, who was fond neither of pictures nor of complementarity, conceded that one may "extend the meaning of the word 'picture' to include any way of looking at the fundamental laws which make their *self-consistency* [italics mine] obvious" (quoted in Jammer 1974, 13).

Both Feynman and von Weizsäcker realized that Heisenberg's analysis of the γ-ray experiment is only a consistency argument, despite the fact that it is often presented as an "inevitability" one. Feynman announced in his *Physics Lectures*: "it was suggested by Heisenberg that the new laws of nature could only be *consistent* [italics mine] if there were some basic limitation on our experimental capabilities not previously recognized" (Feynman *et al.* 1965, 1–9). And, more eloquently, von Weizsäcker: "the impossibility of simultaneously observing precise values of position and momentum does not mean the pseudo-positivistic nonsense 'what cannot be observed does not exist,' but the test of consistency: 'what does not exist in the theory cannot be observed in an experiment describable by the same theory'" (von Weizsäcker 1987; 280–281). In this argument von Weizsäcker was following his teacher Heisenberg who revealed that he considered all of Bohr's complementarity elaborations as demonstrating the consistency of the quantum formalism by an analysis of the possibilities of measurement. Heisenberg considered all these arguments "trivial": "Many experiments were discussed, and Bohr again successfully used the two pictures, wave and particle, in the analysis. The results confirmed the validity of the relations of uncertainty, but in a way this outcome could be considered trivial, because if the process of observation is itself subject to the laws of quantum theory, it must be possible to represent its results in the mathematical scheme of this theory" (Heisenberg, 1973). This position undermines, of course, all of Heisenberg's own "inevitability," or "finality," arguments.

The project of arriving at true, final, certain, indisputable knowledge was, of course, not a new one. Classical physics, despite the empiricist revolt against scholasticism, inherited the Aristotelian ideal of science as demonstration, of arriving at certain conclusions from certain premises. The "father of empiricism" himself, Galileo Galilei, used such scholastic demonstrative arguments: two bodies of unequal weight must fall at the same rate, for if they would not (if heavy bodies fall faster), we will arrive at the logical contradiction.[8] Heisenberg's cut arguments are peculiarly reminiscent of such pseudological proofs.

It was Kant, of course, who provided the most grandiose philosophical demonstration of an indubitable knowledge – by his analysis of Newtonian space and time as forms of intuition, and causality as a necessary category. Newtonian mechanics received the status of synthetic a priori knowledge – knowledge that, though empirically arrived at, had the same certainty as the analytic knowledge of Euclidean geometry.

Bohr's attempts to arrive at a true, certain, final knowledge by a mere analysis of the conditions of experience is thoroughly Kantian in spirit. What is striking, however, is that all of these Kantian arguments were advanced after the full realization of the bankruptcy of such a priori knowledge (the theory of relativity being a final stroke), and by those physicists who, except Bohr himself, had a thorough knowledge of relativity theory. Only extra scientific reasons can, I believe, plausibly explain this fact. No wonder that what the orthodox physicists advanced as arguments of inevitability were, in the best case, arguments of consistency. It is enlightening to mention here the views of the Göttingen mathematician Hilbert, with whom mathematical quantum physicists were in close contact. Itamar Pitowsky referred to Hilbert's stand succinctly: ". . . consistency, or rather the avoidance of a paradox, is ultimately the only safeguard in the pursuit of knowledge. We would have liked to do better for sure, but we have very little choice in the matter" (Pitowsky 1994, 119).

The Copenhagen interpretation, and its rhetoric of inevitability, crucially depends on two central pillars – positivism and the doctrine of the necessity of the classical concepts. Both are unnatural for working mathematical physicists (Beller 1996). No wonder, Copenhagen-Göttingen physicists, Heisenberg included, often denied their adherence to positivism. No wonder, contradictory statements abound concerning the doctrine of the necessity of the classical concepts.

5. BOHM ON CLASSICAL VERSUS QUANTUM CONCEPTS AND ON INDETERMINISM

How did the outstanding mathematical physicist David Bohm, fluent in relativity theory, deal with these issues before his conversion took place?

Bohm's *Quantum Theory* is rightly considered as the best Bohrian textbook. Indeterminism, indivisibility and wave-particle duality – all are elaborated with rare clarity and force. The arguments are perfected as far as they possibly could – precisely because Bohm labored hard to make them as convincing as possible, he could not escape seeing their weak points.[9] It is perhaps not incidental that both David Bohm and Alfred Landé, another prominent dissident, defected the orthodoxy after completing a textbook on quantum theory in the Copenhagen spirit.

While the central pillars of the Copenhagen interpretation are elaborated faithfully in Bohm's textbook, a careful reading reveals substantial undercurrent tensions with Bohr – tensions that, when taken to their logical conclusion, should have, and historically could have, been the reason for Bohm's final break with the orthodoxy. Not surprisingly, these discrepancies concern Bohm's discussion of classical and quantum concepts. Bohm treats these issues in Chapter Eight, revealingly titled "An Attempt to Build a Physical Picture of the Quantum Nature of Matter" – wording which contains the hint at, and possibly a seed of, his future quantum ontology.

As opposed to Bohr, Bohm does not argue that classical concepts are either indispensable, or even natural. In fact, a Bohmian analysis undermines Bohr's arguments that classical concepts are an extension of common sense. Nor does Bohm seem to agree with Bohr that there can be in principle "no quantum concepts," only a quantum-mechanical abstract algorithm of prediction and description of experimental results. Bohm calls for the development of a new language, which will depart from the classical intuitions and will be adequate to the new quantum conceptual framework.

Bohm undertook a thorough critical analysis of the classical idea of continuity of motion and of classical determinism. His aim is to demonstrate that "there is no a priori logical reason for their adoption" (Bohm 1951, 144). Quantum concepts of indivisible transitions and incomplete determinism are "much more analogous to certain naive concepts that arise . . . in common experience" (*ibid.*). Not surprisingly, Greek philosophers were unable to grasp the idea of continuous motion, as Zeno paradoxes eloquently remind us (*ibid.*, 147). In fact, our intuitions revolt against thinking of a motion in which the position and velocity are both precisely determined, and are thus in close harmony with the uncertainty principle. By the use of photography, this becomes especially clear. Our simplest intuition of an object with a definite position is that the object is not moving. To obtain a picture of movement, we must allow the position to blur slightly. It is the blurred picture of the speeding car which suggests that some space was covered during some time, that the car is moving. If we use a very fast camera and will get a sharp picture of the moving car, we will not deduce motion. Our intuitive presupposition is that "a continuously moving object has a somewhat indefinite range of positions" – we simply "cannot visualize simultaneously a particle having a definite momentum and position" (Bohm 1951, 146). Bohm concludes that "quantum theory . . .

gives a picture of the process of motion that is considerably closer to our simplest concepts than does classical theory" (*ibid.*).

But does not our concept of an object at rest, with fixed position and zero velocity, contradict the uncertainty principle, according to which an object with a precise position has a completely indefinite momentum? Bohm argues that in fact there is no contradiction. When we think of an object at a definite position, we can imagine that in a short time later it will move to a different position. Which position it will move to we cannot infer from the original position – all positions are equally consistent with it. This means that any velocity, or any value of momentum, is equally consistent with a picture of a particle in a definite position. Put differently, "if we think of an object in a given position, we simply cannot think of its velocity at the same time" (Bohm 1951, 146).

We see that the classical idea of continuity of motion is an unnatural, non-commonsensical idea. On a purely logical basis, there is no reason to choose the concept of a continuous trajectory in preference to that of discontinuous trajectory. The idea of continuous motion developed through experience with planetary orbits, gun trajectories, and by the theory of differential calculus, aimed at dealing with such motions. It is only by studying "such things for a while" and getting used to them that the "succeeding generation began to take the basic ideas for granted" (Bohn 1951, 148). This analysis is of course a far cry from Bohr's position of the necessity of classical concepts because of the alleged proximity of classical to common sense notions, and of a direct accessibility of classical reality to our immediate sense perception.

According to Bohm, there are quantum concepts that are inherently different from the classical ones. Quantum energy and momentum, despite the same *names* as in the classical physics, are such concepts. In classical theory, we regard energy and momentum as fundamental properties of matter, because they obey conservation laws. But from a logical point of view they need not be considered fundamental, because energy and momentum can be expressed as functions of positions and velocities, so they are in certain sense redundant. All the classical laws of motion can be expressed in terms of space-time relationships alone.

This is not the case in quantum theory. In quantum theory, velocity cannot be defined as a limit of distance divided by time. In quantum theory, momentum, defined by the de Broglie relation, is to be considered as an independent, rather than a derived, property of matter – "the statement that an election was observed to have a given momentum . . . stands on the same footing as the statement that it had a given position. Neither statement is subject to further analysis" (Bohm 1951, 154).

Bohm departs from Bohr also in his relation to the issue of classical versus quantum mechanical reality. He uses the word "convenient," rather than the Copenhagen "necessary," to differentiate between classical and quantum processes: ". . . it is convenient to make a distinction between classically describable processes and those essentially quantum-mechanical in nature" (*ibid.*, 165). Such a distinction is convenient when we deal with large objects, whose features of interest "do not critically depend on the transfer of few quanta more or less" (*ibid.*). In a direct, though unacknowledged, contradiction to Bohr, Bohm asserts: "In the last analysis all processes are, of course, quantum-mechanical in nature" (*ibid.*).

It is very enlightening to see how Bohm argues against classical Laplacian determinism. By presenting classical determinism as an historically contingent, rather than a necessary notion, Bohm aims to ease the way for the acceptance of indeterminism. Yet, precisely because of its historical contingency, Bohm implicitly undermines the Copenhagen stand of "inevitability" of indeterminism. Bohm argues for replacing the notion of classical determinism with a more limited notion of causality as determining general tendencies (but not determining the behavior of systems completely). Such a notion is much closer to the way we relate to forces, or causes, in everyday life: "The idea most likely to have been used in connection with common experience is that a particular force or cause produces a tendency toward an effect, but that it does not guarantee the effect" (Bohm 1951, 152). Quantum lawfulness, which determines only statistical trends, is much closer to our common sense than complete determinism (*ibid.*, 152).

Bohm offers a brief critical historical analysis of the idea of classical determinism. The Greek idea of "fate" – the course of events which is beyond the power of humans to change – belongs to one of the earliest ideas of complete determinism. Yet it is only with the invention of machinery, which allowed the control of motion with precision, that the idea of a mechanistic and deterministic world view developed in the sixteenth through the nineteenth centuries. When work was done mostly by hand or by aid of animals, one could not control motions exactly – one could push in the right general direction, and push backward if one had gone too far. "It is very likely that the modern form of the idea of complete determinism was suggested . . . by its resemblance to complex and precisely constructed machines, such as clocks" (Bohm 1951, 150).

Bohm's analysis of machines – material means of production – as determining the basic conceptual framework of Newtonian mechanics, resembles the analysis of the Soviet historian of science Boris Hessen, who in 1931 at the Congress in London, launched the Marxist historiography of science by an analysis of Newtonian mechanics along similar lines (Hessen 1931). It would be rewarding to know whether Bohm was acquainted with the works of Marxist historians of science in general, and with Hessen's work in particular. Yet the crucial point here is the following: Bohm's analysis of historical rootedness, of socio-historical contingency of our most cherished, most uncritically accepted concepts, does not agree with the rhetoric of finality of the Copenhagen interpretation. What Bohm's analysis rather discloses is that nothing more than continuous habit endows concepts with the aura of naturalness, necessity, inevitability. It is only a small step from here to question whether the same kind of critical analysis cannot be extended, or will not be extended in the future, to the Copenhagen quantum framework itself. Even the idea of "hidden variables" surfaces favorably in this discussion (Bohm 1951, 171). Let us also mention that Bohm, in his later quantum ontology, is open on the issue determinism-indeterminism. Bohm's non-local ontology can be either causal or stochastic. No wonder Bohm was soon led to his idea of understanding nature in terms of an inexhaustible diversity and multiplicity of things, and of understanding of scientific theory as approximate, conditional and relative truth (1957, 164–165).

How strikingly different from Bohr's ideas this trend of thought was! Bohr, who considered classical concepts as natural and necessary, and who considered quantum theory "a rational generalization" of the classical one, could honestly believe (as opposed, perhaps, to the more opportunistic Heisenberg), that his analysis of quantum theory is really the only one possible. He was impatient with those who questioned him: "How could philosophers not understand, that his was an objective description, and the only possible objective description" (Interview with Kuhn and letter to Born, 1953, at AHOP). As an official spokesman of the orthodox interpretation, who had to fight against the threat of opposition, such as Bohm', Bohr preached closure, finality, inevitability, the impossibility of an alternative. How fitting for Bohm, a victim of such an approach, to develop a diametrically opposed view of science – the one pleading for tolerance, for creative plurality, for peaceful theoretical coexistence, for a free play of imagination, for friendly, open-minded and joyful scientific cooperation and communication (Bohm and Peat 1987).

6. COMMENTS ON BOHR'S AUTHORITY AND CONCLUSION

Why were Bohm's valuable ideas ignored? Why were Niels Bohr's ideas considered indisputable, and why, with rare exception, did those, who might have had reservations about them, not challenge Bohr openly?[10] Why did those scientists, who realized that certain arguments for uncertainty and acausality are those of consistency, and not of inevitability, not extend the critical analysis to the whole Copenhagen framework?

The answer lies neither in physics nor in philosophy, but, as Alfred Landé perceptively put it, in "social psychology" (Interview with Landé, AHQP). The nature of Bohr's authority is a crucial ingredient of such a social analysis. I cannot attempt this analysis here, providing instead a few characteristic examples of the grip of Bohr's authority – examples that are eloquent enough to speak for themselves.

Hendrik Casimir, a physicist from Bohr's circle, wrote a comical poem for Bohr's 50th birthday. He described Bohr's infamous theory about the advantages of defensive, as opposed to aggressive shooting, because supposedly making a voluntary decision takes more time than reacting in a purely mechanical way. The poem concludes with the attempt to give Bohr a gun so that he will prove his theory "experimentally" by defending himself:

Eine [Pistol] haben wir dem Bohr gegeben,
Der sollte schiessen für sein Leben,
Der Bohr blieb aber unverdrossen
Und hat uns resloss abgeschossen.
Und die Moral von der Geschict:
Bezweifle Bohrsche Weisheit nicht. (Casimir 1867, 113)

The moral of the story is clear. Nothing positive awaits those who will dare to challenge Bohr's authority. But is not such a literary interpretation inconclusive and tendentious, misrepresenting the joyful and humorous atmosphere in Copenhagen? Clearly not, for let us quote Casimir's own sober interpretation of poem: ". . . the moral of the story that one should not doubt the wisdom of Bohr applies to more important things than shooting gunmen in Westerns" (*ibid.*)

Bohr's unpublished correspondence discloses an overbearing guilt experienced by those physicists who dared to challenge Bohr. Thus, Bopp argued in a letter to Bohr (Feb. 4, 1962, AHQP) that one cannot obtain a contradictory-free interpretation of quantum theory by relying on thought experiments which explicitly use the concepts of the classical particulate mechanics – exactly the concepts that the new physics gave up. Bopp's position strikes, of course, are the very heart of Bohr's life-long enterprise. Bopp prefaces his argument with the following words:

Ein junger japanischer Kollege hat einmal gesagt: wir müssen etwas zu tun, was unsere Eltern nicht verstehen, und das schmerzt uns sehr. (*ibid.*)

Such was the parental authority of Bohr, that many physicists, among them Pauli and Jordan, referred to Bohr by the word "Vater" (father).

Many could not trespass the psychological barrier of even beginning to criticize Bohr. Thus, Du Mond wrote to Bohr of his numerous but unsuccessful attempts to grasp the meaning of Bohr's writings:

This is of course not the first time I have read and tried to grasp your point of view . . . regarding your viewpoint I have never been able to get a clear answer from others. I hasten to say that I do not mean to imply any shortcomings in your own written exposition and am very ready to admit that the difficulty is entirely the fault of my own slowness and dullness. (Du Mond to Bohr, March 7, 1961, AHQP)

Yet von Weizsäcker's reminiscences about his first encounter with Bohr constitute perhaps the best evidence for the overpowering, almost disabling, impact of Bohr's authority. After the meeting with Bohr, von Weizsäcker, in his own words, asked himself:

What had Bohr meant? What must I understand to be able to tell what he meant and *why was he right?* [italics mine] I tortured myself on endless solitary walks. (von Weizsäcker 1985, 185).

The question was not: "was Bohr right?" or, "to what an extent was Bohr right?" or, "on what issues was Bohr right?," but, quite incredibly, "what one must assume and in what way one must argue in order to render Bohr right.[11]

The times when it was unimaginable, or at least "politically incorrect," to challenge Bohr's wisdom, have passed. The present volume on Bohmian mechanics is a telling testimony to this fact.

The Hebrew University of Jerusalem

NOTES

[1] By a "principal axis transformation," which diagonalized the Hamiltonian matrix, one obtained in principle the matrices for position and momentum. Though astonishingly simple in principle, the procedure was useless, due to the technical difficulties in practice (Beller 1983b).

[2] About the "Copenhagen hegemony" see Cushing (1994), and about the "diffusion of the Copenhagen spirit" Heilbron (1988).

[3] Despite Bohr's life-long preoccupation with causality, his use of this concept is unsystematic and sometimes contradictory. Bohr had a very "thick" notion of causality: sometimes it is a cause-effect relationship, sometimes it is "determinism," sometimes it is an epistemological – other times ontological – definition, sometimes causality is equated only with applicability of conservation laws of energy and momenta, other times with simultaneous applicability of both space-time and energy-momenta concepts. Sometimes Bohr's understanding of causality is a probabilistic one, applying to an individual system, and sometimes, it is a statistical ensemble interpretation. Bohr often conflates determinism and predictability, which are substantially different notions. I will follow in this paper Bohr's, Heisenberg's and Bohm's use of "determinism" and "causality" as interchangeable notions, despite the differences between them. These differences are not important for the issues discussed here.

[4] David Finkelstein's pointed comment on these lines strikes to the heart of the matter: "This seems like too much protest; if we cannot then the should-not is surely otiose, and if we should not then we probably can" (1987, 291).

[5] In Bohr's and Heisenberg's writings, even when the original idea of disturbance is explicitly rejected, its imagery is used implicitly in metaphorical, intuitively appealing allusions.

[6] This example is analyzed in depth in Beller and Fine (1994).

[7] It is instructive to mention here that Bridgman's original aim of introducing operationalism was to prevent the "embarrassing" need to basically revise physical theories, ensuring their certainty by strict operationalistic definition of concepts (Beller 1988).

[8] If we combine the two bodies of unequal weight, the composite body should fall at a slower rate than the heavy one between the two, because the lighter of the two bodies impedes the fall of the heavier one. It also should fall faster than the heavy body, because its weight is greater than that of each of the two separate bodies. Hence – all bodies, regardless of their weight, must have the same rate of fall.

[9] Many physicists who did not trouble to go into the Bohrian argument in depth had an easier time to follow them uncritically. Heisenberg complained as early as 1930 that physicists have more faith in the Copenhagen interpretation than a clear understanding of it (Heisenberg 1930, preface).

According to Bohm himself, "the whole development [of a search for a hidden variables alternative] started in Princeton around 1950, when I had just finished my book *Quantum Theory*. I had written it from what I regarded Niels Bohr's point of view. . . . I had taught a course on quantum theory . . . and had written the book primarily in order to try to obtain a better understanding of the whole subject. However, after the work was finished, I looked back over what i had done and still felt somewhat dissatisfied" (Bohm 1987, 33).

[10] Einstein and Schrödinger were, of course, the most prominent exceptions.

[11] Not surprisingly, von Weizsäcker did not attempt to generalize his conclusion that certain Copenhagen arguments were merely those of consistency, and not of inevitability, to the whole orthodox framework.

ARTHUR FINE

ON THE INTERPRETATION OF BOHMIAN MECHANICS

Now the concept of "physical reality" presents itself as problematic and the question arises as to what, then, it really is and what theoretical physics (through quantum mechanics) tries to describe and to what its established laws refer. Certainly this question could be answered in different ways. (Einstein 1953, 35. My translation)

1. INTRODUCTION

Terms like "complementarity," "potentia," the "collapse" of the wave packet, "phenomena" identified with the whole experimental arrangement, and so forth, mark the standard interpretation of quantum mechanics. Despite different public faces there is a core unifying theme. The theory is probabilistic and, in contrast with statistical mechanics, the standard interpretation regards the probabilities as objective, in the sense that it does not ground them in human limitations concerning knowledge of the finer details of things. The objectivity of the probabilities makes for indeterminism and, more fundamentally, for some sort of irrealism; since, according to the standard view, in significant situations there just *are* no finer details of things. The irreducibility of the probabilities might be thought to constitute a realism of a higher order (with respect to the probabilities themselves) except that in the standard interpretation the probabilities are entirely instrumental. They express a relation between a physical system and acts of measurement; they are probabilities for measurement outcomes. Thus, in a curious turn about, despite the objectivity of the probabilities the observer enters quantum theory in a fundamental way. On Bohr's view we are required to divide each experimental situation into an observer part, that is treated classically and to which we do not apply the quantum formalism, and a quantum part, to which we do. On Heisenberg's view the probabilities in the wave function somehow objectively represent both real "potentialities" and also subjective knowledge. Standing outside the causal order, an act of measurement "actualizes" a potentiality and, when we take account of this actualization, our changing knowledge is again objectively represented by a new "collapsed" wave function. Either view makes a mystery of how any object ever comes to possess any property; that is, of how anything at all actually happens. Standardly, we are cautioned not to inquire further. Physics stops here.

There have always been dissenters from this quantum metaphysics. Some would re-interpret the formalism, others would amend it. Both de Broglie and Einstein were early amenders, each proposing a hidden variables extension of the quantum formalism and each subsequently retracting his effort.

J. T. Cushing et al. (eds.), Bohmian Mechanics and Quantum Theory: An Appraisal, 231–250.

In Einstein's case the retraction was swift and permanent. With his field-theoretic orientation, a glimpse of quantum entanglement in his 1927 attempt to extend the quantum formalism convinced Einstein that a theory with quite different concepts would have to supersede quantum mechanics entirely and that fundamental physics would not advance by supplementing the existing quantum theory from within, in the manner of hidden variables (Fine 1986; Belousek 1996). Unlike de Broglie, after 1927 Einstein's interest in hidden variables was never born again, not even by David Bohm's re-introduction of them in 1952.

Although Bohm's ideas failed to get Einstein's endorsement, they did interest John Bell whose indoctrination into quantum metaphysics was never complete. Bell's own investigations of nonlocality were sparked by the nonlocal character of the Bohm theory and when Bell came to believe that nonlocality was in the phenomena not just in our theories, Bohm's way appeared more palatable and interesting to him than it had before. I believe that the impetus for the current exploration of Bohm's program is due to Bell's taking it seriously. Whatever the sociological currents, clearly we have reached a stage where we can consider Bohmian mechanics on its own, and even distinguish it from Bohm's original ideas just as we distinguish Newton's mechanics from Newtonian mechanics.

Let me recall some salient differences. The kinematics of Newton requires absolute space and time. Dynamically, Newton requires forces: external forces to account for change in the state of motion but also internal forces (*vis insita* and *vis inertia*) to account for resistance to change of state, and sometimes even maintenance of state. The whole edifice was intended to serve theological functions which constrain its interpretation (e.g., absolute space as God's sensorium, or the distinction between relative and absolute motion which is necessary to preserve the truth of Scripture). Supposedly, it is built to fixed methodological prescriptions (the famous rules of reasoning) intended to be exemplary for all of natural philosophy. Newtonian mechanics is a looser and more pragmatic framework used to treat bodies in motion by an analysis involving the fiction of point particles obeying Newton's laws. As above, Newton's mechanics is committed both to a specific kinematical ontology and a dynamical one; Newtonian mechanics has no such commitments. One can dispense with absolute space in favor of 3+1 "laboratory frames" that are near enough to inertial. One can treat motion in a flat neo-Newtonian 4-space, or absorb Newtonian gravitation in a curved 4-space. One can dispense with forces entirely in favor of energy, as the nineteenth century energetics movement insisted. One can take gravity as an action-at-a-distance force, or one can introduce a potential field and dispense with action-at-a-distance. One can use a minimal-action formulation that employs a teleological rather then a causal mode of explanation, and even make this serve theological or metaphysical ends – thinking of cosmic design or final causes. One can take formulations that highlight symmetry and conservation principles, and pursue the deeper significance of that. In short, there is a wide range of formulations and interpretations of Newtonian mechanics involving quite different answers to the question of what "really" exists, what concepts are "really" central, and what does it "really" mean (Jones 1991). If scientists or philosophers seek answers to such questions, it is clear that they will have to attach additional premises (from metaphysics or semantics or whatever) that go beyond the physics. Newtonian mechanics does not pick out a unique interpretation; it is not self interpreting.

Thanks to David Bohm's decisive counterexample, we now know that quantum mechanics is not self-interpreting either, despite preachings and "proofs" to the contrary. In this essay I want to develop the theme that theories do not pick out a unique interpretation by treating Bohmian mechanics seriously and examining some of the interpretive latitude to be found there. This is as important a task for Bohmian mechanics as disentangling theological commitments was for the development of Newtonian mechanics. Certainly Bohm himself had an ideological commitment to a radically holistic world view that his physics was intended to serve.

> [T]he ambiguities which arise in the measurement process are only a special case of a universal ambiguity in the very mode of existence of everything. . . . An essential aspect of what each thing *is* therefore lies in all other things, with which it is thus indivisibly related. (Bohm 1961, 199–200)

Moreover, in the eyes of some, the way Bohmian mechanics undoes traditional quantum metaphysics is by substituting a different metaphysics of its own. By disentangling Bohmian mechanics from Bohm and exploring the latitude for different interpretations of Bohmian mechanics itself I hope to insure that Bohmian mechanics will continue to be a vehicle of genuine liberation from old quantum myths – rather than a source of new ones.

2. HIDDEN VARIABLES

Bohmian mechanics is a hidden variables theory, so it may be useful to say some things about hidden variables in general in order to see what is distinctive about the Bohmian version. The usual interpretation of the physical quantities, the "observables" of quantum mechanics, embodies a rule known as the "eigenvalue-eigenstate link" (Fine 1973, 20): only in an eigenstate (or a proper mixture of eigenstates) does an observable have a value. In Bohr's interpretation this principle regulates quantum discourse, functioning as a Rule of Silence: don't ask about – indeed don't even talk about – the values of observables in non-eigenstates. (By course graining, or introducing the notion of an approximate eigenvalue, we can adjust the rule for observables with a continuous part in their spectrum.) Much of the mystery of quantum mechanics is contained in these rules. In the first instance they make superposed states mysterious. What, for example, do we say about the spatial location of a system if experimentally we have narrowed its momentum to a nearly sharp value? It seems to be practically everywhere and also nowhere. Wise indeed, not to ask. The insolubility theorem for measurements shows that after a measurement interaction the pointer observable is similarly stricken (Fine 1970). This is the mystery of actualization referred to above, and hidden variables are supposed to be its solution.

> If one wishes to deal with this problem in a more adequate way it would seem almost inevitable that one will have somehow to introduce further physical conceptions and associated mathematical functions which could represent the state of the "physical" actuality as existing outside the mind of the observer who perceives it. These new "actuality functions" would supplement the wave function, which latter would then constitute only one side to the theory. (Bohm 1961–62, 203)

Thus hidden variables retain the wave function and supplement the theory so as to produce objectively existing values for observables in (some, at least) non-eigenstates. There is a canonical way to implement such a program (anachronistically called a "noncontextual" hidden variables theory). Relative to a given state function ψ of the system, introduce a probability space $\mathbf{H}^{\psi}$ of "hidden states", with a probability measure μ defined on a suitable algebra of subsets. Then associate with each quantum observable Q a random variable $\mathrm{val}_{\psi}Q$ on $\mathbf{H}^{\psi}$ (that is, a μ-measurable function from $\mathbf{H}^{\psi}$ to real numbers). These are the "actuality functions." For $s \in \mathbf{H}^{\psi}$, $\mathrm{val}_{\psi}Q(s)$ is the "actual value" at s of Q in state ψ. Now we have broken the eigenvalue-eigenstate link and broached the rule of silence. To get the statistics of quantum mechanics back it is sufficient to require for every observable Q that the probability distribution of the random variable $\mathrm{val}_{\psi}Q$ on $\mathbf{H}^{\psi}$ match the quantum mechanical distribution (call this $|\psi(Q)|^2$) for Q in state ψ. It follows, inter alia, that only eigenvalues will turn up in eigenstates. This program reinterprets all the quantum probabilities as probabilities for actual values, and thus removes the observer and any reference to measurement. Exactly as in statistical mechanics, the quantum probabilities can be taken to represent our ignorance of the fine grained states. Put succinctly, then, a hidden-variables interpretation represents the observables of quantum mechanics as random variables over a common space whose (single) distributions for actual values match those of quantum mechanics for measured values. One can see that this program is trivial for any one observable Q, as follows. Quantum mechanics gives us a probability distribution $|\psi(Q)|^2$ for Q in state ψ, which is a measure defined on the Borel subsets of the reals. The support of this distribution (roughly, all the real numbers whose quantum mechanical probability is non-zero) are the possible values that could turn up in a measurement of Q in state ψ. Let these be the "hidden states", with measure μ being the quantum mechanical $|\psi(Q)|^2$ restricted to these possible Q-values. If we let the random variable associated with Q be the identity function (that is, set $\mathrm{val}_{\psi}Q(s) = s$), the quantum-measurement statistics for Q become the distribution function for $val_\psi Q$. A random-variables representation is not quite so trivial if we deal with more than one observable, but it is not difficult to carry through for any number – indeed for all the observables at once. (One simple way to do so is to make the preceding construction for each observable separately, and then form the product space endowed with the product measure. That will do it for all observables simultaneously, interpreting them as independent random variables over a common space but nevertheless getting all the quantum statistics exactly right.)

If hidden-variables interpretations are easy to come by as random-variables representations, what are we to make of the "necessity" of the rule of silence, and the other elements of quantum metaphysics? More pointedly, what do the no-go theorems show not to go? The answer about the no-go theorems is straightforward, although I think not well known. These theorems are usually divided into Kochen–Specker-type theorems, which impose algebraic constrains on the hidden-variables representations of observables (sum or product rules, or the equivalent), and Bell-type theorems that impose locality constraints (including the recent generation of Greenberger–Horne–Zeilinger theorems). To be sure, there are significant distinctions that can be drawn between these two types of theorems. But there is also an important similarity; indeed a mathematical equivalence between them (Fine 1982). Namely, what all the theorems do is to impose one further constraint on the random-variables representation. Both the algebraic and the locality constraints are equivalent to requiring that for a suitable (finite) family of observables in some state the joint probabilities of quantum mechanics, where defined, be identical to the joint probabilities of the corresponding random variables over the common space. It is easy to see that an additional constraint of this sort may run into trouble. For example, the construction above makes all the random variables independent, but certainly not all compatible quantum observables are stochastically independent. On the other hand, it is not clear, from a general point of view, why this additional constraint is reasonable. After all, any attribution of a quantum joint probability is identical to the quantum probability for a single observable (namely, the product of appropriate special projections of the original observables). Since the distributions of all single observables come out exactly right on a random-variables representation, so do all attributions of joint probability. Why require further that the joint distribution structure of the random-variables space match that of quantum mechanics? This question is especially appropriate since we know in advance that there can not be a complete match. Incompatible observables have no quantum joint distributions, but their random variables counterparts always do. Thus, inevitably, hidden variables introduce excess joint probability structure. Indeed, it is this very fact that makes the structural demand of the no-go theorems impossible to satisfy. The joint probability structure for a family of random variables functions as a whole in that the distributions for random variables representing incompatible observables cast their shadows on the joint distributions for the compatible ones. On these grounds I have long been suspicious of the no-go demands, no matter how dressed up they are by physical sounding "plausibility arguments" (Fine 1974).

What the preceding random-variables representation leaves out is how the "actual values" change with time. There is no canonical way to do this. But, trivially, we can let the spaces $\mathbf{H}^{\psi}$ change with the Schrödinger evolution of ψ, and just impose a random-variables interpretation at every moment. When, according to the orthodox theory, the wave function collapses, a hidden-variables interpretation could allow the wave function to collapse as well, since on this interpretation the collapse does in fact represent an actual shift in our knowledge. The wave function guides the possible distribution of actual values but does not determine the values themselves. So when we learn something new, say by measurement, we must adjust the guiding wave accordingly. This picture is close to that of Heisenberg's but with "actuality" already there waiting to be revealed (not created) by an observation. Alternatively, we could adopt a no-collapse version drawing on the results of investigations into decoherence (e.g., Zeh 1993) to insure that the continuing branches of the wave function, only one of which contains an actual value, do not interfere and so may be effectively reduced. Whatever the choice, the evolution of the hidden states may give us wildly shifting values, constrained by the quantum probabilities but not by contiguity, or other such conditions. One could rationalize this by suggesting that below the level of the quantum probabilities there is no further simple lawfulness. In fact this was the rationale that Einstein sometimes suggested for why one might regard a merely probabilistic theory as fundamental, although Einstein could not bring himself to believe it. Or, one could try to do better and impose some lawfulness. The Bohm theory tries to do better, and it succeeds (replacing wild anarchy with true chaos) – but only by departing from the preceding random variables framework in significant ways.

3. CONCEPTUAL FEATURES OF THE BOHM MODEL

Considered abstractly, Bohmian mechanics is a theory that treats the behavior of objects with respect to a single, special variable. Given a system of N objects ("particles"), at each moment of time the theory associates (1) a determinate value of the variable with each object in the system and (2) a function (the "wave function"), defined on the N-space of associated values, that determines how these values evolve. Concretely, Bohmian mechanics is a theory of motion; that is, of change of position. Thus, it selects spatial location (position in 3-space) as the special variable of interest. As originally presented, the motion of the particles (i.e., the evolution of the values of the position variable) involved a quantum "potential" whose gradient exerted "forces" on the particles, in analogy with Newtonian forces. As in the Newtonian case, it turns out that potentials and associated forces can be eliminated. In the compact version that I will consider here, the motion of the particles is made lawful by a system of first-order differential equations that give the velocities of the particles directly as normalized gradients of the phase of the wave function ψ (in coordinate representation).[1] This determines unique trajectories for the particles in configuration space, fixed by an initial configuration and ψ. A second first-order equation, namely, Schrödinger's equation, determines how ψ itself evolves. The usual dynamical variables can be defined explicitly from position and momentum. Other quantities (e.g., spin) can be introduced quasi-operationally using position as a base; that is, by specifying how the quantities can be measured by measuring positions. This quasi-operational procedure seems appropriate if you regard these quantities not as possessed properties of the system but as dispositions for how the system will behave under appropriate circumstances.[2] This would be all there is to Bohmian mechanics if we did not also require it to be empirically adequate. In order to generate the right statistics, however, the localization of particles cannot be more precise than $|\psi|^2$, which then emerges as a probability density for the ensemble of particles and represents our ignorance of their precise locations. With the particles distributed according to $|\psi|^2$ Bohmian mechanics predicts the same experimental outcomes as does quantum mechanics in experiments for which the quantum mechanical predictions are definite.

How do we know that? How do we know that there can never be a predictive discrepancy between quantum mechanics and Bohmian mechanics? The answer is contained in the preceding section where we discussed the trivial random-variables representation for a single observable Q. With respect to every quantum state function ψ, Bohmian mechanics imposes precisely that trivial representation on the quantum position observable. That representation treats position as a random variable whose "hidden states" are its various possible measured values, which are distributed exactly as dictated by the ψ function of the system (i.e., according to $|\psi(Q)|^2$). So every particle has as its actual value some possible measured value. This representation guarantees that position measurements in Bohmian mechanics will follow the probabilities of quantum mechanics in every state, since the quantum probabilities that we have matched are the probabilities for position measurements. So much for position. Now, and here's the clever argument, every experimental prediction can be formulated as a prediction about the disposition of equipment in the laboratory, and then reformulated (if necessary) as a prediction about how things will be arranged spatially at the conclusion of the experiment (e.g., on the computer printout). Since our trivial representation says of these exactly what quantum mechanics says at the level of measurement statistics, no predictive discrepancy can arise. That is, there can be no conflict just in case we are not able to get below the level of the quantum statistics and know more about the actual positions of things than $|\psi|^2$ allows.

Note that in the foregoing argument all reference to observables other than position has dropped out. Once we agree on how to reflect the measurement of a physical quantity in the positions of things we need not give any special theoretical account of the quantity itself. Consider momentum. Bohmian mechanics makes it a determinate function of ψ and position (and of mass, of course). The actual values of momentum could be anything at all, perhaps not even in the spectrum of the momentum operator (as indeed they are sometimes not; see Section 6 below). All that we require from Bohmian mechanics is that the *measured* momenta fall within the range allowed by quantum mechanics and with the frequency that quantum mechanics predicts. If this were not true, then, by the above argument, it would be reflected in the deviation of some position measurements from the quantum-mechanical distribution. In so far as that never happens, we need not worry about momentum measurements coming out wrong. Thus, from the point of view of experimental practice, Bohmian mechanics might treat all the quantum "observables" other than position quasi-operationally. If, as in the case of momentum, it does not and instead defines the dynamical variables in terms of ψ and position then the preceding argument – clever though it is – has a gap. For the argument to work one needs to show that the way a measurement of a quantum observable (corresponding to a dynamical variable) is reflected in the positions of things is the same as the way a measurement of its Bohmian counterpart is tied to position. One might describe this task as the measurement problem for this version of Bohmian mechanics. Unlike the quantum problem this one is solvable (Holland 1993, Chapter 3).

In order to insure that shifting position values, as the wave function evolves, will yield continuous trajectories, the Bohmian momentum, for example, always has a definite value but not necessarily a value that could turn up in a measurement of the quantum-mechanical momentum operator. This construction does not interpret the quantum momentum probabilities as probabilities for the actual values of Bohmian momentum. To get the quantum-mechanical momentum distribution requires measurement. Thus Bohmian mechanics departs from the random-variables program. Relative to a given state, that program makes all the quantum observables into random variables over a common space in such a way that the quantum probabilities for measured values can be derived from the underlying space as probabilities for actual values. Bohmian mechanics holds onto this ideal of truly exorcising measurement only with respect to position.

We are now ready to contrast Bohmian mechanics with quantum mechanics and explore some of the interpretive latitude that exists with respect to dualism, determinism and realism.

4. DUALISM

In the old quantum theory, wave-particle duality was the primary conceptual puzzle. The new quantum mechanics, despite its experimental success, did not advance our understanding of this. In Einstein's words it "hardly brings us closer to the secret of the Old One" (1926). Einstein and others were looking for a new conceptual framework that would synthesize both wave and particle concepts. In typical instrumentalist fashion, however, quantum mechanics avoided commitment to any such ontological enterprise. Instead, it claimed that its task was simply to predict and coordinate the results of experimental practice, where those results are interpreted in the old framework of classical physics. In addition, Bohr's doctrine of complementarity was designed as philosophical therapy to diagnose any lingering need that might be felt for radically new quantum concepts, and to cure it. Reversing the usual meaning of the word, in Bohr's hands, "complementarity" prescribed segregation not synthesis. Like those of the quantum theory, the fundamental concepts employed by Bohm also derive from classical point mechanics. However, where quantum mechanics merely flirts with dualism but avoids commitment, Bohmian mechanics embraces it. Bohmian mechanics requires both wave (= ψ function) and particle (= position coordinates) in order to specify the state of a system. This development makes Einstein, writing in 1931, seem prescient, "In a particular department of physics the continuous field thus appeared side by side with the material point as the representative of physical reality. This dualism remains even today, disturbing as it must be to every orderly mind" (1954, 268).

Beyond the wave-particle duality, in Bohmian mechanics the wave function itself acquires a multiple personality. In one guise it behaves as what Einstein dubbed a *Führungsfeld* (guiding field), directing the particles along trajectories determined by solutions to the velocity equations. Paraphrasing Einstein, this is the role to which Born called attention in his second fundamental paper on collision processes ("the waves are present only to show the corpuscular light quanta the way" (1926; Ludwig 1968, 207).) In satisfying Schrödinger's equation, however, the wave function also assumes the role of a true *Gespensterfeld* (spook field), acting as a "spook" or secret agent that reflects and transmits salient information about the environment, including measurement interactions.[3] The "spooky" action-at-a-distance of ordinary quantum mechanics derives from this latter role. In Bohmian mechanics that nonlocality acquires the velocity field as an extra guide. Finally, as indicated above, ψ has a third role to play; namely, that of a probability measure limiting the localization of particles. Echoing Bohr's style and approach, Bohm gave a philosophical rational for this limitation, connecting it with "unpredictable and uncontrollable disturbances introduced by the measurement apparatus" (1952a, 171). He also grounded it more concretely with thermodynamic considerations to show, in illustrative cases, that the spatial distribution of a system is driven to $|\psi|^2$ as an equilibrium condition. (Dürr *et al.* (1992a) and Valentini (1991a) develop this idea differently. See the discussion in Cushing (1994, Chapter 9).)

While in some sense Bohmian mechanics certainly requires both particle and wave is it really so clear that this is a dualism of a sort to disturb an orderly mind? For the wave is defined on configuration space and that means it is essentially dependent on all the possible particle positions. Measuring particle positions actually determines it. But the particles, as above, are informed and guided by the wave which makes their possible positions likewise dependent on the wave. The line of actual influence is one way, since the wave affects the actual motion of the particles, and not conversely. Still, this interdependence suggests that we might think more holistically not of two distinct kinds of entities, but of one new kind. If we think globally then we have a single entity everywhere at once whose nature makes "what each thing *is*" depend on "all other things, with which it is thus indivisibly related." Thus it is a Bohmian X without separable parts. It changes over time but it does not move since, being everywhere at all times, it has no place to move to. Certainly this is a new synthetic conception and, despite its obvious tensions with relativistic spacetime, not something for which "dualism" seems an appropriate label.[4]

Here, then, is nice interpretive latitude. We can think of Bohmian mechanics dualistically as involving the same old things, a one-way dualism of waves guiding particles, or a we can think of it monistically as involving a new kind of unitary world-stuff. Let us see if there is a similar latitude with respect to determinism and realism. Conventional wisdom has it that quantum mechanics is both irrealist and indeterministic. What does this mean and in what sense does Bohmian mechanics overcome it?

5. DETERMINISM

Consider determinism first. Quantum mechanics is deterministic with respect to the evolution of the state function. Its indeterminism, if any, shows up only if we look at the values of the quantum observables. The problem here, however, is not that the initial values fail to determine the final values, or that they only do so stochastically. It is, rather, that we cannot specify the initial "state" of the system in the classical way, by giving a complete set of values for all the variables of state. In particular, there is no specification of simultaneous values for position and linear momentum. A standard formulation of determinism holds that for a closed system initial values for all the state variables determine final values. This formulation presupposes *determinateness*: it is possible to assign initial values to all the relevant variables at the same time. In the first instance, the failure of determinism in the quantum theory is the failure of determinateness. Bohmian mechanics overcomes the problem in the simplest possible way; namely, by placing primary emphasis on exactly one variable and letting the others trail in its footsteps. As we saw in preceding sections (2 and 3) if we look at a single variable in quantum mechanics, say Q, we can always insist that Q has a value at all times. The only possible conflict with this assumption would be in states of a single particle where a complementary variable, say P, is sharp. Then, according to the uncertainty formulas, ΔQ should be very big (indeed, the whole infinite line). This would be problematic if ΔQ represents an actual spread in Q values (which, for determinate Q, would be zero). However, if we interpret ΔQ as the spread in measured values of Q when a measurement of P is sharp, then the possibility arises of explaining the spread as a result of "uncontrollable" disturbances in Q created by the P measurement and codified in the $|\psi|^2$ distribution. Heisenberg's well known treatment of measurement disturbance in terms of the γ-ray microscope does exactly this. For that analysis, like any honest account of a *disturbance*, presupposes some determinate pre-measurement value to be disturbed. Thus in making Q determinate Bohmian mechanics nicely exploits possibilities and techniques inherent in regular quantum mechanics. But there is more to quantum indeterminism than the failure of determinateness.

Consider a free electron prepared in a z-spin eigenstate. Measurements of x-spin would yield spin-up half of the time. This is usually understood as a manifestation of genuine randomness, or chance. There is no property of the electron, one would say, that determines the result of an x-spin measurement. In this respect a spinning electron is thought to be different from a fair coin. In the case of the coin the 50:50 odds for heads or tails result from sensitivity to a small range of initial conditions of the toss. If one could control those initial conditions (which a well-built mechanical coin flipper can do) then one could control the outcome – say, arrange for heads-up every time. There is no little "$p = 1/2$" in the coin. But there is in the electron, at least on the conventional view. According to that view, identical experimental arrangements will generally produce different outcomes that, taken together, exhibit a definite probabilistic pattern. The indeterminism here consists in the fact that identical experimental arrangements produce different outcomes. (That these "undetermined" outcomes follow a probabilistic law is a nice bonus, showing, perhaps, that God is not malicious.) Here again Bohmian mechanics overcomes the difficulty by restricting attention to exactly one variable. As we saw above (Section 2), for a single variable any probability distribution can be looked at as the distribution function of a well defined random variable over some domain. We could think of the fair coin, for example, as having a set of "hidden states" (in fact, just its coordinates!) in half of which (for a typical toss) it would definitely land heads-up and in the other half definitely tails. Similarly for the spinning electron. Once the measurement problem for Bohmian mechanics is solved, and we have reduced all measurements to measurements of position, the clever argument of Section 3 shows how all the probabilities for measurement outcomes are just like this deterministic account of the fair coin. That "identical" experimental arrangements produce different outcomes is because we can only individuate the initial particle positions up to $|\psi|^2$. There is no indeterminism at the level of the individual particle positions themselves. They are just inaccessible.

Certainly that is the conventional view of Bohmian mechanics, but not the only one possible. To find another we might go back to a conception of determinism as a regulative principle, the venerable Laplacian idea that determinism is about seeking the possibility of accurate prediction. Recent philosophy typically scorns this conception (e.g., Earman 1986, Chapter 1), which it regards as conflating ontology (the structure of the world) with epistemology (how we can know it). The philosophical generation closest to the new developments in physics, however, recognized that physics does not thrive on fine distinctions. They held that there was no conflation here, but rather good scientific judgment concerning the significance of determinism in theory development. They held that maintaining philosophical categories was not an end in itself, and that anyway the quest for an "ontology of the world" often appeared more quixotic than sensible. In an essay for *Die Naturwissenschaften* in 1931, Moritz Schlick – the leader of the Vienna Circle and a student of Planck's – summed up the impact of quantum mechanics on causality (which he does not distinguish from determinism) this way.

> The principle of causality ["All events are in principle predictable."] . . . constitutes an imperative to seek regularity, to describe events by laws. . . . And what quantum physics teaches us is just this: that the principle is *bad*, useless, impractical within the limits precisely laid down by the principle of indeterminacy (Schlick 1931, 107).

Rudolf Carnap drew much the same conclusion some thirty-five years later.

> Quantum theory is fundamentally indeterministic in that it does not supply definite predictions for the results of measurements. . . . Classical physicists were convinced that, with the progress of research, laws would become more and more exact, and that there is no limit to the precision that can be obtained in predicting observable events. . . . Determinism in [this] classical sense has been abandoned (Carnap 1966, 286, 288).

The guiding idea here is to regard determinism (or causality) as standing or falling with the possibility of exact prediction. Determinism can be accepted to the extent to which practical prediction is possible, and rejected to the extent that quantum uncertainty, for example, places limits on predictability.

If we follow the conception of determinism that regards predictability in principle as the central ingredient, then what is the verdict with respect to Bohmian mechanics? Is exact predictability there possible in principle or not? The answer depends critically on how one understands the limitation imposed by $|\psi|^2$ on particle localization. If one regards this as an absolute uncertainty, as one must if the assertion of observational equivalence with quantum mechanics is true, then determinism is no different for Bohm than for Bohr. But even if we take the limitations as one for all practical purposes, then does determinism come out any better off for Bohm? The issue here is very much like the issue raised in nonlinear dynamics where, for systems sensitive to initial conditions, the evolution equations are formally deterministic, but useless for prediction. Shall we accept chaotic systems as deterministic or not? One could find some justification for going either way. The situation is similar in Bohmian mechanics (Dürr *et al.* 1992a,c.) One could say that Bohmian mechanics is deterministic in principle but that for all practical purposes we find the same limitations on predictability there as in quantum mechanics, and hence the same degree of indeterminism. Since determinism no longer functions to insure ever higher degrees of predictability, is this a distinction without a difference? (See related discussions in Cushing (1994, Chapter 11) and in Kellert (1993).)

There is a clear sense in which Bohmian mechanics and quantum mechanics are in the same boat with respect to determinism and a clear sense in which they differ. If one takes determinism to be about ontology and evolution equations, one could view Bohmian mechanics as a theory of a deterministic world that, ironically, we may only grasp as though it were indeterministic. Alternatively if determinism is epistemological and regulative, one could view Bohmian mechanics as utilizing the resources of deterministic evolution in a *reductio*; viz., to argue for the abandonment of determinism by showing that it is fruitless to seek after exact predictability. In trying to balance between ontology and epistemology, we are broaching questions of realism, our next topic.

6. REALISM

One aspect of Bohr's complementarity is marked by his dictum that descriptions of the "phenomena" in quantum mechanics must always include the whole experimental arrangement. We have seen that this is precisely what Bohmian mechanics requires in order to get spreads of values compatible with the uncertainty formulas. More fundamentally the measured values of the dynamical variables depend on their being linked to position variables during a measurement interaction. The Schrödinger evolution of the wave function during the measurement establishes the linkage. Thus ψ, acting as spook, scouts out the whole experimental arrangement. In Bohr's version this holistic conception of a "phenomenon" is tied to his rather mystical idea of the "feature of wholeness" or the irreducible "individuality" of the quantum phenomena. We saw in Section 4 that Bohm's writings show a similar mystical tendency that can be exploited to motivate a monistic interpretation of the theory as an account of a truly undivided universe. But none of this has to intrude into Bohmian mechanics, except in so far as we must calculate the effects of the experimental conditions in order to determine the experimental outcome.

Although there is nothing necessarily mystical in Bohmian mechanics about the need to look at the whole experimental set-up, there is something of importance about it from an epistemological point of view. In standard quantum mechanics the "problem of observation" is often assimilated to the technical measurement problem. This is the problem of reconciling the linear dynamics with the registration of specific measurement outcomes by the measuring apparatus. Bohmian mechanics has no technical problem of this sort since the assumed determinateness makes existing values of position every-ready to record and be recorded. The observational problem, however, is actually much broader than this. A thoroughly classical model of an observation pictures the observing subject as an outside spectator viewing an objective reality, an onlooker who mirrors determinate aspects of the world that are of interest. The act of observation may, of course, make some difference to the observed system. But one can take that disturbance into account so as to eliminate observer effects or, at any rate, to discount for them. Thus in the classical picture what observation does is to reveal an observer-independent realm. Moreover, an important ingredient of this classical picture is the assumption that the observed-world is not very different from the unobserved world. How things are does not depend very much on whether things are observed. Observation is more passive than active. Reality is recorded not made. This picture of observation as disclosure is part of the "independence" that nature has from observation and it lies behind the idea that knowledge is objective. According to quantum mechanics, however, the observer does make a difference, indeed a difference so big that one cannot really discount the effect of observations so as to frame an observer-independent picture of the world. This is a standard epistemological route to quantum nonrealism. Although Bohmian mechanics looks more realist, insofar as it postulates an underlying reality, the role played by observation there is even more prominent than in standard quantum mechanics.

Einstein first called our attention to this aspect of Bohmian mechanics (Einstein 1953). He considered a wave function for the one dimensional motion of a "particle" moving back and forth between two perfectly rigid and reflecting walls. Applied to a macro-particle we know that, to a good approximation, it has a definite position and a (non-zero) velocity at (almost) each moment. Quantum mechanics does not yield that conclusion, and indeed no quantum state function will yield simultaneous approximate positions and momenta that are stable in the classical way. What quantum mechanics does provide, via Born's rule, are correct predictions for P or for Q upon their respective measurement. Bohm's treatment of this example provides those same correct predictions, but also a picture of the situation before measurement. The picture it provides, however, is not the classical one of a particle moving to and fro. Since the phase of the wave function confined between the walls does not depend on the position coordinate of the particle its gradient vanishes. According to the Bohmian prescription for velocity, the unmeasured particle is actually standing still! Measurement disturbs the situation, freeing the wave function, which guides the particle into motion. Einstein thought the Bohmian picture sufficiently bizarre to reject this prescription for introducing determinate P and Q.

From a purely logical standpoint there is, in my opinion, no principled objection to be made against Bohm's completion of the quantum theory. Looked at from a physical standpoint, however, Bohm's way out does not seem to me acceptable. In all cases where the ψ function is not approximated in the neighborhood of each point by a *travelling* wave, one obtains values for the momenta that violate the requirement that the quantum theory should go over to classical mechanics *in the limit*. It is connected with this, that the Bohmian rule determines the momentum values not through a Fourier transform but rather through a *local* regularity in coordinate space. (Letter to Aron Kupperman, November 14, 1953. My translation from the German draft.)

Bohm countered that this theory did not contradict any known experimental facts and he declared that only contradiction by experiment could be sufficient grounds for rejecting a consistent theory. Moreover Bohm urged that it was not reasonable to require exact correspondence with classical predictions in a domain where those predictions cannot be experimentally verified. (Bohm 1953c)

Ironically, Bohm's response to Einstein's criticism employs the same positivist lines of defense that Bohr and others had used to turn back criticism of standard quantum mechanics, and to marginalize deviant "interpretations" – including Bohm's! Characteristically, Einstein paid no more attention to Bohm's methodological refrain than he had to the same tune when broadcast from Copenhagen. Note the difference, however, between Bohm and Copenhagen. On the latter view a rule of silence takes over and we should not frame any hypotheses about the state of motion of the unobserved "bouncing" ball. According to Bohm we need not be silent since there is a determinate reality, regardless of observation. That reality, however, is significantly different when observed than when unobserved. Moreover, observation does not reveal the unobserved situation, whose status is that of a theoretical postulate. For the pre-measurement state is highly unstable and observation necessarily triggers velocity effects that cannot be discounted without knowing the exact position of the unmeasured particle. That knowledge, however, is bounded by $|\psi|^2$. Thus, in order to overcome the quantum rule of silence, Bohmian mechanics pays a price. It departs from the classical picture of observation as disclosure, thereby modifying the conceptions of observer-independence and objectivity which that picture supports. It is important to understand this as clearly as we can in order to understand how the "realism" of Bohmian mechanics compares with the irrealism of quantum mechanics.

Classical scientific realism is a blend of metaphysics and epistemology. It makes two claims: metaphysically, that there is an observer independent world; epistemologically, that in our theories and observations we gain access to that very world. To paraphrase Kant: reality without disclosure is blind; disclosure without reality is empty. Metaphysically, with respect to certain quantities Bohmian mechanics is realist; certainly with respect to position, but also with respect to momentum and the classical dynamical variables that can be defined from these two. Moreover, observations of position really disclose position. As we have seen, this is not the case for momentum. Rather, as in Einstein's non-bouncing ball, typically momentum is created by the act of measurement, and pre-existing momentum values (while "real") are not disclosed by the act of momentum measurement. Thus for momentum (and any similarly defined variables), "The value of the measurement only arises in the act of measurement itself in accord with the unique probability given it by the ψ function" (Einstein 1949, 82, 84. My translation). This is the position of the physicist called "B" (no doubt "B" was for Bohr but here too for Bohm) in Einstein's "Autobiographical Notes". Nonlocality is supposed to drive B to admit that quantum mechanics is incomplete. Leaving aside Einstein's concerns with nonlocality and incompleteness, his characterization brings out exactly what is peculiar about Bohmian realism. From a realist we do not expect measured values to arise "in the act of measurement itself". That is, beyond the existence of pre-measurement values (determinateness), we expect that those are the values disclosed by measurement. As Einstein's "A" physicist puts it, "The individual system (has before measurement) a definite value . . . , and more specifically *that* value which is determined by a measurement of this variable" (1949, 83). Thus what Einstein demands is that realism blend metaphysics with epistemology, as above. This is a demand for the right sort of concepts. They must be both determinate and accessible. In Bohmian mechanics only "position" is like that. That is to say that with respect to all the other variables Bohmian mechanics is not realist in the expected sense.

There is a familiar way of defusing this departure from realism; namely, to label it contextualism and to assert that it is required by the no-go theorems and therefore by the very structure of quantum mechanics itself. One difficulty with this move is that we do not know exactly what contextualism is. Like Bohr's "complementarity" it is a balm in search of an itch. For some, contextualism is a formal feature: one operator can represent many physical quantities. For others, it marks the fact that specification of a value for a variable is always relative to a context. In this latter sense probably every physical variable (and certainly every dispositional property) is contextual. I believe that contextualism first arose in Bell's development of Gleason's theorem into a Kochen–Specker type no-go result (Bell 1966). There Bell identified an assumption of the proofs, which was that the measured values of variables were functions only of the hidden parameters and of the quantum state. One way of contravening this, and so possibly of avoiding the no-go results, was to make those values also depend on what variables (if any) were co-measured. Subsequently, this sort of co-measurement contextualism came to be seen as a cost of doing hidden variable business, as though quantum mechanics plus hidden variables somehow implies contextualism. If so, the criticism raised by Einstein above, and the desire for a classically realist version of quantum mechanics would be in vain. But, of course, one can get around the no-go theorems in other ways. For example, we can simply deny the functional connections (e.g., the sum and product rules) which those proofs posit (Fine 1973; 1974), and we can allow variables to have values outside their spectrum. In Bohmian mechanics pre-measurement values are not confined to the spectrum, but also in Bohmian mechanics measured values depend sensitively on what variables (if any) are co-measured (Bell 1987a, 164–65). So Bohmian mechanics gets around the Kochen–Specker type no-go theorems both by allowing values other than eigenvalues and by exhibiting co-measurement contextualism for the dynamical variables other than position.

Co-measurement contextualism is one way in which measured values can arise in the act of measurement. If the result of measuring A depends critically on the co-measurement of a some B that is spatially separate from A (in particular if the separation of the measurement events is space-like), then this sort of contextualism amounts to nonlocality, which is how Bohmian mechanics gets around the Bell type theorems. These resources, by virtue of which Bohmian mechanics by-passes the Kochen–Specker and the Bell type no-go theorems, do not by themselves entail a conflict with realism. In particular, if the only determine variable were position, and all the other dynamical variables were treated as pure dispositions (recall the discussion in Section 3), then there would be no problem in Bohmian mechanics over the disclosure of pre-measurement values. That problem is generated by the particular Bohmian mixture of determinateness and co-measurement contextualism.

The culprit here is the guidance condition that associates velocities with the wave function. That condition insures that the classical dynamical variables have pre-measurement values outside their spectrum and therefore values that could not possibly be disclosed by measurement (at least if observational equivalence with quantum mechanics is to be maintained). Moreover, by making the velocities a function of ψ in coordinate representation, initial data on particle positions is required in order to fix the velocities. Unlike dispositions, the velocities are already there in pre-measurement situations, but hidden. The question naturally arises as to whether they are (or ought to be considered) "real", in the usual sense. If reality without disclosure is truly blind, perhaps it would be better to treat Bohmian velocity, as what Vaihinger (1925) called a "scientific fiction"; that is, an expedient means to certain ends; namely, a means for calculating particle trajectories and positions. Born's original picture comes to mind here.

[T]he guiding field, represented by a scalar function ψ of the coordinates of all the particles involved and the time, propagates in accordance with Schrödinger's differential equation. Momentum and energy, however, are transferred *as if* the corpuscles . . . actually moved. (Born 1926; Ludwig 1968, 207. Emphasis mine.)

Taking Born's "as if" seriously would interpret the dynamical variables as calculational tools. The measured values of these fictional variables in a spatio-temporal region would depend on the particle positions observed there and on co-measurements, some distant. So the variables would be subject both to co-measurement contextualism and to nonlocality. For fictions, however, neither feature seems worrisome. Unmeasured values would be treated as conceptual devices for calculating the results of future measurements of position and not as the possession of some property, not even a "relational" property.

The fictionalist interpretation suggested above, and by Born's seminal remarks on the significance of the ψ function, differs from approaches to Bohmian mechanics that regard the dynamical variables as what one might call quasi-real (e.g., Holland, 1993). Here the dynamical variables are assigned pre-measurement values signifying some intrinsic-looking property of the system. But the assigned values are not the ones disclosed by measurement. Once one appreciates that classical observer-independence and objectivity are weakened in a quasi-real interpretation it seems fair enough to highlight the best candidates for what may be really real, and to fictionalize the rest. This difference over whether to treat the dynamical observables other than position as quasi-real or as fictions marks a nice interpretive fork for Bohmian mechanics. The question that now emerges from our attempt to understand realism with respect to Bohmian mechanics is just this: how shall we regard position and the particle trajectories?

So far I have been emphasizing a specific two-part criterion of reality; namely, (observer-independence) if something is real then it exists whether or not it is observed and (accessibility) measurement reveals it's pre-existing situation. I urge this not as a sufficient but as a necessary condition for judging that something is real. Is there something else to require, in addition to this criterion, which would be sufficient and do we find that something else in Bohmian mechanics? I do not think that there are universal, subject neutral requirements. Historically, issues over what is real tend to fade away rather than to be settled by principled demonstrations. On pragmatic grounds on comes to accept something, and the question as to its reality goes away when it works well and fruitfully over time in a variety of experimental and theoretical contexts in which it seems to play an essential role. For that to happen the concept must mesh appropriately with others whose reality-status is not in question. A problem with the position variable and the trajectories in Bohmian mechanics is that they do not have anything else "real" with which to mesh, except the wave function itself. Since the wave function is only defined by reference to position coordinates, this generates an uncomfortably small circularity. Moreover, trajectories conjure up the picture of moving particles. But the reality-status of the ordinary dynamical concepts associated with that picture, the momentum along the path or the energy carried from point to point, as we have seen, is open to question and can not be used to ground the reality of the particle positions. The reverse is true. The velocity field, it seems, is there just so that we may interpolate the particle trajectories. Moreover, for experimental purposes, that is for obtaining experimental predictions, there is little value in actually integrating the velocity equations to obtain the exact trajectories. Once we accept that particle localization follows the $|\psi|^2$ rule, ordinary quantum mechanical reasoning about the effective part of the wave function for a given experimental situation gets us all the predictions that are possible there. Thus it seems reasonable to ask how seriously the particle trajectories themselves ought to be taken. Do the particles really move, or do the calculated trajectories function as intervening variables that could be eliminated? As Born seems to suggest, do the trajectories too fall under Vaihinger's *as if*?

We know the answer. Certainly the particle trajectories are eliminable, for that is how one gets ordinary quantum mechanics! Cushing tells a wonderful and illuminating tale of how the Bohm theory, rather than quantum mechanics, might have won the allegiance of the scientific community and how physics in this century might have developed on that basis (Cushing 1994, Chapter 10). Given the entanglement of physics with positivistic philosophy in the first half of this century, a process initiated by Einstein and relativity, we might embellish Cushing's counterfactual history a bit. We might conjecture that if Bohmian mechanics had indeed been the first theory of the quantum domain, subsequent epistemological discussions might well have inspired at least some participants to wield Occam's Razor (along lines suggested above) to pare away everything that is not simply quantum mechanics! Well, perhaps not quite everything. For, although the continuous trajectories provided by the velocity field might be eliminated as "real", if the association of position coordinates with the wave function also goes we are left with the problem of actuality. Surely one would be reluctant to move to a theory in which it is a mystery that anything happens at all, especially if there is an alternative. What emerges then is a random variables representation of the position variable involving particles that are *as if* they move according to the velocity field. (See Section 3.) On this interpretation what is real is the wave function and that there are determinate particle positions, and nothing else. For the particular trajectories are a convenience. What we actually need is that in a superposition in coordinate representation the particles are spread across some of the branches. Spatial separation of the branches will then suffice to define an effective part of the wave function with which to work. At the heart of Bohmian mechanics is the wave function and determinate particle positions, and perhaps we need be realist about nothing else.

In this interpretation the wave function does not actually guide the moving particles. It just assigns particle positions and shifting distributions. Thus one might regard ψ as a kind of global hidden hand with just these functions. One version of this might be the monistic picture in Section

4. Maybe reality consists of a unitary global X that organizes particle aspects everywhere and at every moment in accord with a changing wave aspect. Of course one might go the other way and treat ψ simply as a tool for how we are to take account of the environment in setting up a model of an experimental situation. This is an instrumental alternative not likely to be attractive to those whose interest in Bohmian mechanics is realist-driven. It is a genuine alternative nevertheless, and may well correspond with Born's original picture of the ψ function.

7. CONCLUDING REMARKS

In Newtonian mechanics we are comfortable with a wide range of interpretive options. My hope was to display some similar options for Bohmian mechanics. In Section 4 we saw that Bohmian mechanics is a dualistic theory (wave + particle) that could also be monistic (an undivided X). In Section 5 we saw that the evolution equations of Bohmian mechanics certainly make it deterministic, except in so far as determinism really has to do with predictability, in which case it is indeterministic. In Section 6 we looked at how momentum and perhaps other dynamical variables of "realist" Bohmian mechanics depart from the realist ideal of an accessible, observer independent realm. That marks these variables at best as quasi-real or, more radically, as fictional. The particle motions, the Bohmian trajectories in configuration space, are subject to similar interpretations. This leaves the wave function, which could be read as a holistic Bohmian hidden hand, or as an instrumental Bornian probability guide. Moreover Bohmian mechanics can be done with quantum potentials and forces, or (as here) without. There certainly seem to be choices sufficient for different philosophical creeds and ideologies. More importantly, we know from the Newtonian case that interpretive latitude contributes to heuristic power. Thus Bohmian mechanics emerges as a real theory with good prospects.

Northwestern University

NOTES

[1] The velocity field associated with ψ falls nicely out of symmetry considerations; namely, Galilean and time reversal invariance (Dürr *et al.* 1992a).

[2] Here and below we follow Bell (1987a, Chapter 17) in treating spin and other discrete quantities quasi-operationally. Treatments that regard them as more intrinsic are also possible. See Cushing (1994, Chapter 5) for a discussion. These possibilities would make spin, for example, come out more like momentum. The discussion in Sections 3 and 6 below would then apply.

[3] See Pais (1982, 442–443) for Einstein's terminology of *Führungsfeld* and *Gespensterfeld*.

[4] One version of this holistic conception might correspond to Bell's "Everett(?)" version of many words (Bohm without trajectories), hopefully without the radical solipsism that concerned Bell, "But if such a theory were taken seriously, it would hardly be possible to take anything else seriously." (Bell 1987a, 136).

MILLARD BAUBLITZ AND ABNER SHIMONY

TENSION IN BOHM'S INTERPRETATION OF QUANTUM MECHANICS*

1. INTRODUCTION

David Bohm and his predecessor Louis de Broglie (1927) were exploratory men, each of whom proposed and examined a number of theories. Their various theories shared some central ideas but also differed in significant respects, and an assessment of all of these variant theories is beyond the scope of this paper. We shall mainly restrict our attention to what we shall call "Bohm's initial theory," which was presented in his pioneering papers (Bohm 1952a,b), and in two letters (Bohm 1952c; 1953a) and a paper (Bohm 1953b) which clarified and defended his initial presentation. Our main thesis is that two different physical views coexist uneasily in Bohm's initial theory, which we shall call "the causal view" and "the guidance view." We shall cite passages indicating that he was aware of this fact and that he pondered ways of relieving the tension between the two views. Because of the great historical importance of Bohm's initial theory, there is obviously value in a careful *explication de texte*. In addition, however, we have a conceptual thesis: that what makes Bohm's initial theory attractive philosophically and scientifically is a conjunction of some features of the causal view and some of the guidance view. We shall present some reasons for skepticism about the prospects for a coherent combination of the two views. We shall also note that some of the most influential advocates of Bohmian mechanics – including Dürr *et al.* (1992a), Valentini (1991a), and Cushing (1994) – treat the tension between the two views in the initial theory of Bohm by surgery, for they retain the guidance view and discard whatever of the causal view is incompatible with it. We shall not undertake to assess these recent expositions of Bohmian mechanics, because of mathematical difficulties which we find very formidable, but we shall express some skeptical doubts in hope of evoking clarifications and amplifications from the authors.

2. BOHM'S INITIAL THEORY

Section 4 of Bohm's paper (Bohm 1952a, 169), entitled "New Physical Interpretation of Schrödinger's Equation," restricts its attention to a single particle, but there is a straightforward generalization to many-particle systems in a later section. Since the presentation in the initial paper (Bohm 1952a) is somewhat diffuse, we shall quote from a compact summary, which for simplicity concerns only a single particle, given in a later paper (Bohm 1953b).

J. T. Cushing et al. (eds.), Bohmian Mechanics and Quantum Theory: An Appraisal, 251–264.

Passage I:

. . . the author has proposed a causal reinterpretation of the quantum theory, based on the following hypotheses:
(a) a quantum-mechanical system, such as an electron, consists basically of a particle having a precisely defined position, which varies continuously as a function of the time.
(b) This particle is acted on not only by the classical potential $V(\mathbf{x}, t)$ but also by an additional quantum-potential $U(\mathbf{x}, t)$, which is important at the atomic level but negligible at the macroscopic level.
(c) If we write $\psi = R\, e^{iS/\hbar}$, where ψ is the wavefunction and R and S are real, then the quantum potential is given by

$$(1) \qquad U(\mathbf{x}, t) = -\left(\frac{\hbar^2}{2m}\right)\frac{\nabla^2 R(\mathbf{x}, t)}{R(\mathbf{x}, t)}.$$

The equation of motion of the particle then takes the form

$$(2) \qquad \frac{m d^2\mathbf{x}}{dt^2} = -\nabla\{U(\mathbf{x}, t) + V(\mathbf{x}, t)\}.$$

To obtain the same predictions for all experimental results as are obtained from the usual interpretation of the quantum theory it is necessary, however, to make the following additional special assumptions . . . : (1) The ψ field satisfies Schrödinger's equation. (2) The particle velocity is restricted to $\mathbf{v} = \nabla S(\mathbf{x})/m$. (3) We do not predict or control the precise location of a particle, but have a statistical ensemble of particles with a probability density, $P(\mathbf{x}) = |\psi(\mathbf{x})|^2$. (Bohm 1953b, 458–459)

A striking feature of this passage is its hierarchical structure. Bohm asserts that the causal interpretation is "based upon" the hypotheses (a), (b), (c). The propositions (1), (2), (3) are important, of course, since they are needed to "obtain the same predictions for all experimental results as are obtained from the usual interpretation of the quantum theory," but their secondary status relative to (a), (b), (c) is signalized by calling them "special assumptions." In the earlier paper (Bohm 1952a) propositions (1), (2), (3) were given almost verbatim (but with $\mathbf{p} = \nabla S$ instead of $\mathbf{v} = \nabla S/m$), and also there they were called "special assumptions." Furthermore, a comment was added to (3): "The use of statistics is, however, not inherent in the conceptual structure, but merely a consequence of our ignorance of the precise initial conditions of the particle."

There are other passages showing that the hierarchical structure which we have noticed is deliberate. In the earlier paper (Bohm 1952a) he characterizes the equation $\mathbf{p} = \nabla S$ as a "restriction" on the equation of motion.

Passage II:

. . . this restriction is consistent, in the sense that if it holds initially, it will hold for all time. Our suggested new interpretation of the quantum theory implies, however, that this restriction is not inherent in the conceptual structure. We shall see in Sec. 9, for example, that it is quite consistent in our interpretation to contemplate modifications in the theory, which permit an arbitrary relation between $\mathbf{p}$ and $\nabla S(\mathbf{x})$. The law of force on the particle can, however, be so chosen that in the atomic domain, $\mathbf{p}$ turns out to be very nearly equal to $\nabla S(\mathbf{x})/m$ [*sic*], while in processes involving very small distances, these two quantities may be very different. In this way, we can improve the analogy between the ψ-field and the electromagnetic field (as well as between quantum mechanics and classical mechanics). (Bohm 1952a, 170–171)

Furthermore, we cite a passage (Bohm 1952c), which is a reply to a criticism (Halpern 1952, 389) of Bohm's characterization of the function $S(\mathbf{x}, t)$ as "the Hamilton–Jacobi function" (Bohm 1952a, 170). Halpern objects that if a mechanical problem is to be solved via the Hamilton–Jacobi equation, with f being the number of degrees of freedom of the mechanical system, then "it is necessary to have a solution S which depends on the coordinates of the system, the physical parameters entering into the expression for kinetic and potential energy and f nonadditive integration constants. Bohm has failed to show that such a function can be found" (Halpern 1952, 389). We shall return in Section 5 to the matter of these nonadditive constants. What we want to stress here is the hierarchical ordering of principles expressed by Bohm in his answer to Halpern. It is noteworthy that he does not object to Halpern's attribution to him of the "opinion that there exists the possibility of looking on nonrelativistic quantum mechanics as a form of classical mechanics involving special quantum forces" (Halpern 1952, 389). Indeed, Bohm seems to accept the characterization of his theory "as a form of classical mechanics" in the following passage (especially when he says "the theory itself was to be based directly on the equations of motion").

Passage III:

It was stated in paper I, Sec. 4, however, that the Hamilton–Jacobi theory was being used only for the purpose of indicating in a simple way how one might arrive at a causal interpretation of the quantum theory; while the theory itself was to be based directly on the equations of motion . . .

$$\frac{m d^2\mathbf{x}}{dt^2} = -\nabla\{U(\mathbf{x}) + V(\mathbf{x})\},$$

where $V(\mathbf{x})$ is the classical potential, and $U(\mathbf{x})$ is the "quantum-potential." Now, the solution of these equations by the Hamilton–Jacobi technique would certainly require the f nonadditive integration constants referred to by Halpern, but the author intended to employ another method of solution in his papers; namely, that of guessing a function and verifying it by direct substitution in the differential equations. Since the treatment given in the papers is perhaps not completely explicit, it may be useful to amplify it here. Guided by the Hamilton–Jacobi theory, we guess tentatively that if the momentum were equal to $\mathbf{p} = \nabla S(\mathbf{x})$, this function would satisfy the equations of motion. Now,

$$\frac{d\mathbf{p}}{dt} = \frac{\partial \mathbf{p}}{\partial t} + (\mathbf{v} \cdot \nabla)\mathbf{p} = \frac{\partial(\nabla S)}{\partial t} + \frac{\nabla(\nabla S)^2}{2m}.$$

But since S is defined as the phase of the wave function, we have (paper I, Eq. (6))

$$\frac{dS}{dt} = -\frac{(\nabla S)^2}{2m} - V - U, \qquad \frac{d\mathbf{p}}{dt} = -\nabla\{V - U\}.$$

Thus, we verify that a particle travelling with velocity $\mathbf{v} = \nabla S/m$ will satisfy our equations of motion. This means that the condition $\mathbf{p} = \nabla S(\mathbf{x})$ is a consistent subsidiary condition. It was then pointed out in paper I, Secs. 4 and 9 that if we assumed this subsidiary condition to be satisfied initially, we would be able to obtain a mechanical explanation of the results of the quantum theory; if not, we obtained a wider theory. Since it was shown to be possible to choose this wider theory in such a way that **p** approached $\nabla S(\mathbf{x})$ at the atomic level, but differed significantly from it at the level of 10^{-13} cm, it was concluded that the wider theory might be needed in the domain of very small distances, where present theories seem to be inadequate. (Bohm 1952c, 389–390)

The one important point that we have omitted so far in our summary of Bohm's initial theory is his insistence upon the physical reality of the wave function.

Passage IV:

Since the force on a particle now depends on a function of the absolute value, $R(\mathbf{x})$, of the wave function, $\psi(\mathbf{x})$, evaluated at the actual location of the particle, we have effectively been led to regard the wave function of an individual electron as a mathematical representation of an objectively real field. This field exerts a force on the particle in a way that is analogous to, but not identical with, the way in which an electromagnetic field exerts a force on a charge, and a meson field exerts a force on a nucleon. In the last analysis, there is, of course, no reason why a particle should not be acted on by a ψ-field, as well as by an electromagnetic field, a gravitational field, a set of meson fields, and perhaps by still other fields that have not yet been discovered. (Bohm 1952a, 170)

3. TENSION BETWEEN THE CAUSAL AND GUIDANCE VIEWS

The passages quoted in Section 2 will permit us to document our thesis that Bohm's initial theory contains both a "causal" and a "guidance" view and that there is a tension between them.

What we call "the causal view" consists of a Newtonian interpretation of Bohm's equation of motion in hypothesis (c),

$$\frac{m\mathrm{d}^2\mathbf{x}}{\mathrm{d}t^2} = -\nabla\{U(\mathbf{x}, t) + V(\mathbf{x}, t)\}, \tag{3.1}$$

the only non-Newtonian element being the quantum-mechanical potential $U(\mathbf{x}, t)$. The central feature of Newtonian mechanics is that the acceleration of a particle is proportional to the force on it (the second law of motion). The velocity of the particle is underdetermined by the force. There are, of course, infinitely many sets of data sufficient to fix the trajectory of the particle, but one of these sets consists of the position $\mathbf{x}_0$ and velocity $\mathbf{v}_0$ at an arbitrary initial time t_0, together with the force on the particle at all times. (We abstain from the mathematical complications arising from the fact that a differential equation only determines the trajectory locally (Thirring 1992, 91–95)). Other sets of sufficient data are mathematically equivalent to the "standard" set of fixing $\mathbf{x}_0$ and $\mathbf{v}_0$, and therefore we restrict our attention to the standard set. Thus in the causal view, the initial velocity is a contingency in the same sense as the initial position of the particle – neither is implicit in the equation of motion or in the law that gives the force entering the equation of motion. Observation, or some other source of information over and above the equation of motion and the force law, is required to fix $\mathbf{x}_0$ and $\mathbf{v}_0$.

We now inquire whether the special assumption (2) in Bohm's summary of his theory, i.e.,

$$\mathbf{v} = \frac{1}{m} \nabla S(\mathbf{x}), \qquad \text{(\textit{guidance condition})}$$

is compatible with the causal view. There is no doubt that this equation for $\mathbf{v}$ is *consistent* with the equation of motion stated in hypothesis (c). Indeed, Bohm is justified in making a much stronger assertion: it is consistent to assume $\mathbf{v} = \nabla S/m$ at t_0, and once that assumption is made, the equation of motion together with the time-dependent Schrödinger equation (Bohm's special assumption (1)) ensures that $\mathbf{v} = \nabla S/m$ at all times. Bohm's argument to this effect (Bohm 1952c, 389–390) is given in Passage III. The crucial point, however, is that according to the causal view $\mathbf{v}_0$ is a *contingency*, something contained neither in the equation of motion nor in the force law. Is this the case for the special assumption that $\mathbf{v}_0$ equals $\nabla S/m$ at t_0? A positive answer to this question could be given by a semantical trivialization, which we wish to avoid. It might be said that $\mathbf{v}_0$ is determined by the function S that occurs in the exponent when the wave function $\psi(\mathbf{x}, t)$ is expressed in the polar form

$$(3.2) \qquad \psi(\mathbf{x}, \mathrm{t}) = R(\mathbf{x}, \mathrm{t})e^{iS(\mathbf{x}, t)/\hbar},$$

whereas the quantum-mechanical potential $U(\mathbf{x}, t)$ depends only upon the modulus $R(\mathbf{x}, t)$; consequently, the argument goes, $\mathbf{v}_0$ is determined by something over and above the equation of motion and the force law. But the quantum-mechanical potential U implicitly depends upon the entire complex wave function ψ, because the time evolutions of the functions R and S are coupled, their evolutions being contained in the time-dependent Schrödinger equation governing $\psi(\mathbf{x}, t)$. As Bohm says in Passage IV, it is the ψ-field which is objectively real and determines the novel force on the particle (Bohm 1952a, 170). Therefore, when we ask whether $\mathbf{v}_0$ is a contingency, we may properly inquire whether information over and above the wave function $\psi(\mathbf{x}, t)$ is needed to fix it. The answer to this question is negative if

$$(3.3) \qquad \mathbf{v} = \frac{1}{m} \nabla S(\mathbf{x}) \equiv \frac{\hbar}{im} \nabla[\log \psi].$$

The non-contingency of $\mathbf{v}_0$ is dramatized by considering an ensemble of particles all associated with the same wave function $\psi(\mathbf{x}, t)$. Presumably, the histories of these electrons varied greatly, and it should make sense to speak of the histories of individual electrons in Bohm's initial theory, since all particles have definite trajectories. In a subensemble with all electrons at t_0 located within a small region $d\mathbf{x}$ (the origin of the coordinate system for each electron being the nucleus with which it is associated), there is no reason to deny a variety of histories. And yet, in this subensemble,

$\mathbf{v}_0 = \nabla S/m$ is almost exactly the same for all electrons. Thus the "special assumption" (2) overrides the dependence of $\mathbf{v}_0$ upon previous history, and this dependence can hardly be separated from the causal view.

Before turning to what we call "the guidance view," we wish to comment on the clause in Passage II that "it is quite consistent in our interpretation to contemplate modifications in the theory, which permit an arbitrary relation between $\mathbf{p}$ and $\nabla S(\mathbf{x})$. The law of force on the particle can, however, be so chosen that in the atomic domain, $\mathbf{p}$ turns out to be very nearly equal to $\nabla S(\mathbf{x})/m$" [*sic*] (Bohm 1952a, 170–171). Bohm amplifies this remark by writing the equation of motion as

$$(3.4)\qquad m\frac{d^2\mathbf{x}}{dt^2} = -\nabla\left\{ V(\mathbf{x}) - \frac{\hbar^2}{2m}\frac{\nabla^2 R}{R}\right\} + \mathbf{f}(\mathbf{p} - \nabla S(\mathbf{x})),$$

and the force $\mathbf{f}$ is described in the following quotation.

Passage V:

> . . . $\mathbf{f}(\mathbf{p} - \nabla S(\mathbf{x}))$ is assumed to be a function which vanishes when $\mathbf{p} = \nabla S(\mathbf{x})$ and more generally takes such a form that it implies a force tending to make $\mathbf{p} - \nabla S(\mathbf{x})$ decrease rapidly with the passage of time. (Bohm 1952a, 179)

Obviously, Bohm is contemplating an important modification of his theory because $\mathbf{f}$ is a radically different force both from the classical force ∇V and the force ∇U derived from the quantum-mechanical potential. Nevertheless, this modification remains within the causal view since $\mathbf{p}_0$ (and hence $\mathbf{v}_0$) is clearly a contingency; the virtue of the new force is that it ensures a rapid effective forgetting of the contingent $\mathbf{v}_0$.

By "the guidance view" we mean the view that makes special assumption (2), $\mathbf{v} = \nabla S/m$, fundamental and exact (not derivative and approximate, as in the variant theory contemplated by Bohm in Passage V). Of course, the guidance condition does not by itself constitute a complete theory, but when it is accompanied by hypothesis (a) in Passage I (that every particle has a definite position at each time) and by special assumption (1) (that the wave function satisfies the time-dependent Schrödinger equation), and when the requisite generalization is made to a many-particle system, one does have a well formulated dynamical theory. We do not include in the guidance view special assumption (3), Born's rule equating probability density $P(\mathbf{x}) = |\psi(\mathbf{x})|^2$, since that rule concerns statistics rather than dynamics, and it will be considered separately. What is deliberately omitted from the guidance view is hypotheses (b) and (c), the expression for the quantum-mechanical potential U and the statement of an equation of motion employing U. The equation of motion in hypothesis (c) in Passage I is, in fact, redundant in the guidance view; a derivation of it was given in Passage III (Bohm 1952c, 389–390). We emphasize, however, that even though the equation of motion is an integral part of what we have called "the causal view," it by no means exhausts the conceptual content of the causal view, for essential to that view – as discussed in the preceding paragraphs – is the contingency of $\mathbf{v}_0$. The tension that we find in Bohm's initial theory is the contingency of the initial velocity in the causal view and its non-contingency in the guidance view.

Before we proceed to Section 4 to point out the strengths and weaknesses of the causal and guidance views, it may be useful to the reader to answer the question: who among the avowed followers of Bohm are the advocates of the causal view and who of the guidance view? Unequivocally there are advocates of the guidance view, notably Bell (1987a), Dürr, Goldstein and Zanghì (DGZ) (1992a), Valentini (1991a) and Cushing (1994). They extract from Bohm's initial theory only the time-dependent Schrödinger equation, the attribution of a definite position to a particle at every time, and the guidance equation. The conjunction of these three principles is called "Bohmian mechanics" by DGZ, and their terminology is widely followed. They do not include the Born rule $P(\mathbf{x}) = |\psi(\mathbf{x})|^2$ as one of the assumptions of Bohmian mechanics because they claim to be able to derive it mathematically from their austere set of principles. They dispense with Bohm's hypotheses (b) and (c), because "For them, the quantum potential appears as an artificial means to cast this essentially non-classical mechanics into a superficially Newtonian form" (Cushing 1994, 166 – said of DGZ, but with apparent intention of application to Valentini and himself). To our knowledge, there is no follower of Bohm who advocates the causal view in its purity. This fact does not imply, however, that the causal view is uninteresting, because it is an ingredient in the overall theory of Bohm and Hiley (1993), Dewdney *et al.* (1987), Holland (1993), Kyprianidis (1988), and perhaps others. The endurance of the causal view in the thought of these Bohmians might be explained by the historical contingency of Bohm's heuristics in his initial theory. But we feel that there is another reason, going beyond historical contingency – viz., the conceptual strengths and attractions of the causal view.

4. STRENGTHS AND WEAKNESSES OF THE CAUSAL AND GUIDANCE VIEWS

The greatest strength of the causal view is that it subsumes quantum mechanics under classical mechanics, once the innovation of the quantum potential is accepted. Since classical mechanics is an "intuitive" and "intelligible" framework – in part because of its inherent qualities and in part because of its internalization by post-Newtonian physicists – much of the mystery of quantum mechanics is thereby exorcised. The indefiniteness of physical quantities, which the orthodox interpretation of quantum mechanics requires us to accept, is removed by this subsumption under classical mechanics, since a definite position and a definite velocity are attributed to each particle at every time. The causal view clarifies the status of the wave function by making it objective and physically real, an entity which acts upon particles by a force law formulated in the quantum mechanical potential, analogous to the way the electromagnetic field acts upon a particle by the Lorentz force law. The "guidance" of a particle by the wave function is, in principle, no more mysterious than the guidance of a charged particle by the electric and magnetic fields. In both cases there is no direct determination of the velocity of the particle by the force, but rather a determination of an acceleration.

The greatest weakness of the causal view is its difficulty in recovering the Born rule

$$(4.1) \qquad P(\mathbf{x}, t) = [R(\mathbf{x}, t)]^2 \equiv |\psi(\mathbf{x}, t)|^2.$$

Suppose that the Born rule holds at an initial time t_0, by special preparation, accident, equilibration process, or whatever. One would like to be able to infer that it holds for all time. The time-dependent Schrödinger equation provides one step along the way towards achieving this desideratum, since it implies that

$$(4.2) \qquad \frac{\partial R^2}{\partial t} + \nabla \cdot \left(\frac{R^2 \nabla S}{m} \right) = 0$$

(and the generalization of this equation to a system of n particles). This equation for the time evolution of R^2 looks very much like an equation of continuity, and indeed it becomes the equation of continuity

$$(4.3) \qquad \frac{\partial R^2}{\partial t} + \nabla \cdot (\mathbf{v}R^2) = 0,$$

provided that the guidance condition $\mathbf{v} = \nabla S/m$ holds. But in the causal view $\mathbf{v}_0$ is a contingency and therefore it may or may not be the case that the guidance equation holds at t_0. If it fails to hold, then the derivation of the equation of continuity breaks down and with it fails the justification of the Born rule for all time.

The unavailability of the guidance condition to the causal view makes it difficult to see how the standard quantum-mechanical predictions for entangled systems can be generally recovered. Consider, for example, the measurement of the orbital angular momenta of a pair of spin 0 particles, with $\mathbf{L}_1 + \mathbf{L}_2 = 0$, but with quantum numbers $L_1 = L_2 = 1$. The standard quantum-mechanical prediction is that if particle 1 goes into the beam corresponding to $M_1 = 1$, then particle 2 goes into the beam corresponding to $M_2 = -1$ and conversely, and if particle 1 goes into the $M_1 = 0$ beam, then particle 2 goes into the $M_2 = 0$ beam. If the guidance equation holds, then it is relatively straightforward (Holland 1993, 465–471) to show that the momenta of the two particles after interaction with the appropriate inhomogeneous magnetic fields are equal in magnitude and opposite in direction, thus yielding the standard correlations of the output beams. But if the guidance equation is not assumed and one relies instead on the equation of motion, one can only infer that the quantum mechanical potential will impart *impulses* that are equal in magnitude. Whether this correlation will guarantee the proper correlated behavior of a pair of particles into output beams will depend, of course, upon the initial velocities $\mathbf{v}_1(t_0)$ and $\mathbf{v}_2(t_0)$.

A corollary of the difficulty of the causal view to recover quantum-mechanical correlations is a lacuna in its treatment of measurement. If an observable of the system represented by the quantum-mechanical operator Q (with eigenvalues denoted by the variable q) is measured, Bohm (1952b, 181) proposes a suitable apparatus and suitable interaction Hamiltonian such that the final quantum state of the particle plus apparatus is

$$\text{(4.4)} \qquad \psi(\mathbf{x}, y, t) = \sum_q c_q \psi_q(\mathbf{x}) g_0\left(y - \frac{aqt}{\hbar^2}\right),$$

where $\psi_q(\mathbf{x})$ is an eigenfunction of Q with eigenvalue q and g_0 is a sharply peaked function of the position variable y of the apparatus – the sharpness being important in order to guarantee that for different values of q the shifted functions $g_0(y - aqt/\hbar^2)$ are orthogonal. The $\psi_q(\mathbf{x})$ are orthonormal, and for simplicity we shall assume that they actually have non-overlapping supports so that q can be inferred from $\mathbf{x}$. According to standard quantum mechanics a correlation is established by Eq. (4.4) between q and the position variable y of the apparatus, even though both are indefinite in the state given by Eq. (4.4): if, somehow, a definite value of y is picked out by the mysterious process of "reduction of the wave packet," then a definite value of q is picked out in tandem. Bohm's theory is supposed to remove the mystery from the reduction of the wave packet by ascribing definite values to the position variable $\mathbf{x}$ of the particle and the position variable y of the apparatus. But in the causal view what guarantees that the correlation postulated by standard quantum mechanics is recovered? The quantum-mechanical potential does indeed impart correlated impulses to the particle and to the pointer needle of the apparatus, but if a range of initial velocities is allowed by both particle and apparatus, the correlation between $\mathbf{x}$ and y needed to obtain agreement with quantum mechanics may not be obtained. It is relevant to remark at this juncture that in both the causal and guidance views, the positions of the particles do not affect the temporal evolution of the wave function. There is a causal asymmetry, in that the wave affects the particles, but the particles are causally irrelevant to the wave. This asymmetry is troublesome (Anandan and Brown 1995, 349), but we shall not base our skepticism upon it. Rather, we shall tentatively accept the fact that correlations established by the quantum-mechanical potential can be a surrogate for the causal efficacy of the particles. The difficulty is that even after this surrogate is accepted, the casual view is insufficient to recover the quantum-mechanical correlation of q and y required by the standard interpretation of Eq. (4.4), because the initial velocities of the particle and the apparatus can act as "noise" that disrupts the correlation required by a reliable measurement process. In sum, the quantum-mechanical account of the measuring process is mysterious but robust, while the causal view offers a theory of measurement that is not mysterious but also not robust.

Another weakness of the causal view is its treatment of the *s*-states of a hydrogen atom, which are real spherically symmetrical eigenfunctions. These states were troublesome since de Broglie's expositions in 1927 and Bohm's in 1952, because the guidance condition implies that $\mathbf{v} = \nabla S/m$ is 0 – i.e., the electron is at rest. But since we are now studying the causal view, we are not bound by this equation. Instead, we have to see what force is implied by the right side of the equation of motion (Bohm 1953b, 458) in Bohm's hypothesis (c), which contains a contribution from the ordinary Coulomb potential and another contribution from the quantum mechanical potential. The answer is that these two contributions produce compensating forces (Bohm 1952a, 173), and hence the acceleration of the particle – in this case the electron of the hydrogen atom – is zero. But this result is even more paradoxical than the inference from the guidance condition that the electron is at rest, because if the ensemble is governed by a distribution of initial velocities, presumably with only a set of measure zero having velocity exactly zero, then when the wave function is a quantum-mechanical *s*-state almost all the members of the ensemble will drift away from their nuclear binding centers. The stability of the stationary state, which after all was one of the triumphs of quantum theory, is objectively lost: stability is maintained only in the wave, but not in the entire state $(\mathbf{x}(t), \psi(t))$.

We turn now to the guidance view, and we find that its strengths and weaknesses are complementary to the strengths and weaknesses of the causal version. The greatest strength of the guidance view is agreement with the predictions of quantum mechanics, once the Born relation is given at an initial time. The reason is that the guidance condition guarantees the equation of continuity (4.3), and hence the validity of the Born relation at one time ensures its validity at all times. Once the guidance condition is granted, the correlation between particles governed by a quantum-mechanical entangled state is no longer disrupted by the contingencies of initial velocities of those particles – because the guidance condition eliminates these contingencies and fixes the velocities for all time. The guidance view indeed requires that we accept the counterintuitive conclusion that the electron is at rest when the wave function is a quantum mechanical *s*-state, but at least this counterintuitive proposition guarantees stability of atoms and prevents the automatic ionization that the causal view entailed.

The weakness of the guidance view is that it abandons the intuitive framework of Newtonian mechanics. The wave function is indeed a kind of force field in the guidance view, but it exerts a force on the particle not by providing a quantum mechanical potential U to supplement the ordinary potential V, but rather by directly influencing the velocity $\mathbf{v}$ through the guidance condition. It is an Aristotelian doctrine, rejected by the classical mechanics of the seventeenth century, that there is a direct connection between force and velocity. The guidance view, in postulating a direct relation between the guiding field and velocity, is thus philosophically closer to pre-Newtonian than to post-Newtonian mechanics.

We have deferred the discussion of one important weakness. In listing the strengths of the guidance view we tentatively accepted the validity of Born's relation between the probability density in configuration space and the absolute square of the wave function at an initial time. But is there a promising justification of Born's relation within the guidance view? A concrete example provides ground for skepticism. Consider a particle confined to a circle with wave function

$$(4.5) \qquad \psi(\theta, t) = (2\pi)^{-1/2} e^{i(n\theta - Et/h)},$$

where n is an integer to guarantee single-valuedness. The guidance condition together with Eq. (4.5) implies that every particle in the ensemble moves with the same velocity regardless of initial position. Hence, an arbitrary localized distribution will be translated at a uniform rate about the circle, and there will be no convergence towards the quantum mechanical distribution, which of course is uniform about the circle. In this example the causal view, which permits an initial velocity distribution not constrained by the guidance condition, is more successful than the guidance view in recovering the Born relation. Indeed, the causal view predicts an approach to a uniform distribution in configuration space for "most" initial velocity distributions, and the reasoning is analogous to a classical explanation of the uniform distribution of the asteroids.

The objection may be raised that the example just presented is too simple to be relevant, and that the justification of Born's relation can only be obtained for "a sufficiently complicated system" (Cushing 1994, 164). Both the H-theorem argument of Valentini (1991a, 5) and the "typicality" argument of Dürr *et al.* (1992a, 843) rely on complexity. We are not in a position to analyze these arguments in detail, but we wish to point out a fact that troubles us. Even though very simple examples are excluded by directing one's attention to complex systems of interacting particles, and therefore simple counterexamples to Born's relation cannot be adduced, it must be recognized that the many-particle correlation asserted by

$$(4.6) \qquad P(\mathbf{x}_1, \ldots, \mathbf{x}_n) = |\psi(\mathbf{x}_1, \ldots, \mathbf{x}_n)|^2,$$

is much stronger than the relation for a single particle, since implicit in Eq. (4.6) are n single-particle distributions, $n(n-1)/2$ two-particle distributions, etc. In other words, even if the resources of the guidance view of Bohmian mechanics are greater when one is dealing with a complex system rather than a simple system, the proposition which one is committed to justifying is commensurately stronger, and it is by no means clear that the promissory note can be redeemed.

5. CAN THE CAUSAL AND GUIDANCE VIEWS BE COMBINED?

In Passages II and V (quoted in Section 2) Bohm recognizes the tension between the causal and the guidance views of his novel mechanics, gives primacy to the former, but suggests a way of recovering the latter in the atomic domain by proposing a force of the form $\mathbf{f}(\mathbf{p} - \nabla S(\mathbf{x}))$. One may wonder how the law of force can have the remarkable property (desirable from the standpoint of achieving agreement with standard quantum mechanics is usual experimentation) of implying that in the atomic domain $\mathbf{p}$ is very nearly equal to $\nabla S(\mathbf{x})$, since the deviation between these two quantities in the causal view is due to the contingency of initial momenta regardless of the force law. The force $\mathbf{f}(\mathbf{p} - \nabla S(\mathbf{x}))$ considered in Passage V is velocity dependent, and seems to be dissipative and not invariant under time reversal. It is not easy to see how this new force can be incorporated into fundamental theory, because of its tension with Hamiltonian mechanics. Speculations about the new force led Bohm to further speculations (Bohm 1952a, 179) about a level of very short distances where the linear Schrödinger equation breaks down and must be replaced by a nonlinear equation. These speculations are beyond the scope of the present paper. We feel justified, however, in asserting that Bohm's proposal in his 1952 papers for combining the causal and guidance views of his mechanics led him beyond the simple principles of his initial theory which constitute much of its attraction.

Kyprianidis (1988) has a mathematically ingenious proposal for reconciling the causal view and the guidance view of Bohmian mechanics. (Note: in this discussion we shall adopt Kyprianidis's notation.) At first sight his proposal appears more conservative than that of Bohm discussed in the preceding paragraph, but this appearance will be seen to be misleading. Kyprianidis formulates the causal version as a Hamilton–Jacobi theory, in which Hamilton's principal function $W(q, P, t)$ depends upon the initial positions and momenta of the system. The "new" momenta P are constants in time and depend upon the initial values of the "old" momenta p. Since initial conditions are contingencies, the force law leaves a range of possible values for the "new" momenta P. Kyprianidis assumes that there is an appropriate probability distribution of these contingencies, and hence an appropriate probability distribution $\rho(P)$. What it is physically that makes the distribution appropriate is obscure, but the mathematical condition that he imposes is clear:

$$(5.1)\qquad \int dP\rho(P) \exp\left[\frac{i}{\hbar} W(q, P, t)\right] = R \exp\left[\frac{i}{\hbar} S(q, t)\right] = \psi(q, t).$$

Thus, the quantum mechanical exponent $S(q, t)$ is a function on configuration space obtained by averaging the phase space function $W(q, P, t)$ with the appropriate momentum distribution. The guidance condition thus determines a "mean trajectory" with mean momentum

$$(5.2) \qquad p = \frac{\partial \langle W \rangle}{\partial q}(q, t),$$

rather than the true trajectory of an individual system. What is attractive about Kyprianidis's proposal is that it begins with the causal point of view, recognizing that a force law underdetermines the initial momenta, and it concludes formally with the guidance condition. But there are obscurities in Kyprianidis's exposition. (1) How does nature provide an "appropriate" momentum distribution that satisfies Eq. (5.1)? (2) Is the satisfaction of the "mean" equation (5.2) sufficient to guarantee the equation of continuity (4.3), without which Born's relation is not preserved in time? (3) Since Kyprianidis's calculations are carried out only for free particles, do we know that there exists mathematically a distribution $\rho(P)$ that satisfies Eq. (5.1) when there are interactions? (Note that in posing this question we are not asking for an explicit calculation of $\rho(P)$.) Kyprianidis promised a sequel to his paper of 1988 to deal with interactions, but we have found no indication that it has been published.

Although Kyprianidis does not give a detailed answer to question (1), he makes an important statement about the conceptual framework within which he anticipates a solution:

A far more interesting aspect of the approach is tied to the assumption of P being a random variable and having a characteristic distribution $\rho(P)$. This fact cannot be assigned to a lack of knowledge of the exact initial (and actual) momenta of the particle, but implies a constant perturbation of the latter throughout the motion. This can only happen if one assumes a perturbing underlying medium where $\rho(P)$ indirectly manifests its characteristics via the randomness of P: it reflects the degree of organization of the medium, as it becomes explicit in the perturbation of the motion of the system imbedded in it. (Kyprianidis 1988, 416).

It is evident from this passage that Kyprianidis does not expect to achieve a combination of the causal and guidance views within the range of Bohm's simple initial formulation. Instead, he looks ahead to the ideas of sub-quantum fluctuations, as proposed by Bohm and Vigier (1954) and later works of Vigier and his school. This program is beyond the scope of the present paper.

6. CONCLUSION

According to Bell, "the de Broglie–Bohm version of non-relativistic quantum mechanics is . . . experimentally equivalent to the usual version insofar as the latter is unambiguous. But it does not require, in its very formulation, a vague division of the world into 'system' and 'apparatus' . . . Indeed the de Broglie-Bohm theory is sharp where the usual one is fuzzy, and general where the usual one is special" (Bell 1987a, 111). Bell's endorsement is very strong. Other writers (e.g., Cushing 1994, 191–193) have maintained that only historical circumstances have made Bohmian mechanics heterodox, since it has all the virtues of the orthodox interpretation of quantum mechanics and some other virtues besides.

Our main conclusion is a dissent from these endorsements. We question the claim that resistance to Bohmian mechanics is irrational, due primarily to authority and indoctrination. What we have found is that there are two distinct views implicit in Bohm's initial theory, the causal and the guidance, and each has some virtues and some weaknesses. We are dubious that a simple choice of one version will suffice, as maintained by Dürr *et al.* (1992a), who reject the causal version in favor of the guidance view. If the two versions could be consistently combined so as to preserve the strengths and eliminate the weaknesses of each singly, then the strong praise accorded by Bell, Cushing, and others would be well deserved. But we have no idea how this combination can be achieved without drastic elaborations of Bohm's initial theory. It is the intention of one of us (Baublitz 1996) to study the subquantum fluctuation theory of Bohm and Vigier (1954); judgment on the prospects of the elaborated formulations of Bohmian mechanics will be postponed at least until this study.

We wish to add that we are adherents neither of the orthodox Copenhagen interpretation of quantum mechanics nor of the view that the quantum formalism needs no modification. In our opinion, the quantum-mechanical measurement problem cannot be solved without a modification of the formalism itself, perhaps by a stochastic modification of the time-dependent Schrödinger equation, perhaps by an even more radical modification. We doubt, however, that the solution to the measurement problem and to other conceptual difficulties of quantum mechanics has already been achieved by Bohm and his followers.

ACKNOWLEDGMENT

We wish to thank Profs. James Cushing, Sheldon Goldstein, and Howard Stein for valuable suggestions. The research for this paper was supported in part by the National Science Foundation, grant no. PHY93-21991.

Boston University

NOTE

* This paper is dedicated with deep appreciation to Professor Hans Primas on the occasion of his retirement from the ETH.

ROBIN COLLINS

AN EPISTEMOLOGICAL CRITIQUE OF BOHMIAN MECHANICS

I. INTRODUCTION

In this paper, I argue that, as currently developed, Bohm's theory does not provide a better explanation of observable phenomena than a *phenomenalist account* of quantum mechanics, that is, an account that merely takes the phenomenological laws of quantum mechanics as its unexplained given. Thus, by a version of Ockam's razor, I conclude that we do not have sufficient reason to believe that Bohm's theory is (approximately) true – that is, that it provides a basically correct account of the way the world is. Although not everyone interested in Bohm's theory is concerned with its truth, it is an issue that should be of concern to those trying to understand what quantum mechanics tells us about the nature of the world.

To avoid misunderstanding, however, I should stress several things about the aims and scope of my argument. First, my argument only applies to the deterministic, non-relativistic version of Bohm's theory. Second, it does not address the issue of whether Bohm's theory might have other virtues besides truth, such as that of providing a predictively and conceptually useful (though perhaps false) model of an underlying physical reality, as for instance, Bohr's "planetary" model of the atom does. And finally, it does not address the issue of whether Bohm's theory provides the best realist (as opposed to instrumentalist) "interpretation" of quantum mechanics.

II. SOME EPISTEMOLOGICAL PRINCIPLES

Every theory has a set of fundamental postulates that it leaves unexplained and by which it tries to account for the observed phenomena (though what exactly these fundamental postulates are might vary depending on how one formulates the theory). For example, when realistically interpreted, classical atomic theory's most fundamental postulates can be thought of as being that atoms exist which obey the laws of classical mechanics, along with either the equiprobability or ergodic assumption of classical statistical mechanics. Now, even from a realist perspective, to be justified in believing a theory, there must be some truth-indicating theoretical advantage in accepting the fundamental postulates of the theory as unexplained *givens* instead of simply accepting its phenomenalist counterpart – that is, accepting the data and the phenomenological laws that the theory tries to explain as unexplained (though not necessarily unexplainable) givens. Accepting the

J. T. Cushing et al. (eds.), Bohmian Mechanics and Quantum Theory: An Appraisal, 265–276.

fundamental postulates of atomic theory, for instance, supposedly had the theoretical virtue of simplifying our understanding of the physical world: instead of having to postulate as separate unexplained givens the gas laws, the laws governing chemical combinations, the laws governing Brownian motion, and the like, we now only had to postulate two givens, the existence of atoms that obey the laws of classical mechanics and the equiprobability (or ergodic) assumption underlying statistical mechanics, and from these derive everything else. If instead, for each of these observational laws, we had to hypothesize a different set of laws at the atomic level, the atomic explanation would not have simplified or unified our understanding of the physical world, and thus we would not have had sufficient reason to believe it.

The above example illustrates an epistemological rule that can be stated as follows: *in order to have sufficient reason to believe that a theory is (approximately) true, the theory must at least offer some truth-indicating theoretical advantage – that is, make some explanatory progress – over its phenomenalist counterpart.* Since the phenomenalist account is the account that makes the minimal possible ontological commitments, this rule is actually a specific case of a more general epistemological principle which seems to be widely accepted by those who philosophers call "scientific realists" – that is, roughly those who want to claim that our fundamental physical theories are not merely useful predictive devices, but really tell us the way the world is. This principle, a version of Ockam's razor, can be precisely stated as follows. Suppose we have two hypotheses, H_1 and H, such that (i) H_1 both entails H and makes more ontological commitments than H does, and (ii) H accounts for all the observable phenomena that H_1 does. Then, we can only have sufficient reason to believe H_1 is approximately true (instead of, for instance, merely believing that H is true) if H_1 makes some explanatory progress over H: that is, if the unexplained givens of H_1 have some truth-indicating theoretical virtue – such as simplicity – that the unexplained givens of H do not have.[1]

I shall argue below that, when compared with the phenomenalist account of quantum mechanics, Bohm's deterministic, non-relativistic hidden-variable theory fails to meet the above epistemological rules, and thus that we do not have sufficient reason to believe it is (approximately) true. Before doing this, however, we first need to explicate the theoretical givens of Bohmian mechanics.

III. BASIC ONTOLOGY OF DETERMINISTIC BOHMIAN MECHANICS

I shall consider Bohm's theory to be a global theory applying to the whole universe, for as Dürr *et al.* (1992a, 858) point out, this is the only way of guaranteeing that Bohm's theory is consistent.

The following two postulates summarize the ontology of Bohmian mechanics:

(1) An ontology of particles each having a definite position and velocity $V(q, t)$, where $V(q, t) = (\mathbf{v}_1, \mathbf{v}_2 \ldots \mathbf{v}_n)$ and $\mathbf{v}_k(q, t) = \nabla_k S/m$. (Here, $q \equiv (\mathbf{q}_1, \mathbf{q}_2, \ldots, \mathbf{q}_n)$ represents the joint configuration space coordinate for all the n particles in the universe; $\mathbf{v}_k$ and ∇_k represent the velocity vector and gradient for the kth particle, respectively; and S is given by $\Psi = \mathrm{R}e^{iS/\hbar}$, where Ψ is the wave function of the universe and is assumed to obey the formalism and rules of quantum mechanics, such as the Schrödinger equation.)

(2) A probability density, $P(q, t)$, exists such that at some initial time t_0, $P(q, t_0) = |\Psi|^2(q, t_0)$, where $P(q, t)\, dq$ denotes the probability that, at time t, the system consisting of all the particles in the universe is in the region $(q, q + dq)$ of its configuration space.[2]

Because the particles are moved in accordance with the velocity field $V(q, t)$, the probability function $P(q, t)$ must satisfy the following conservation equation:

$$\frac{\partial P}{\partial t} + \nabla \cdot (PV) = 0, \tag{1}$$

where ∇ is the generalized version of the gradient for the $3n$ dimensions of configuration space. Equation (1) is a key equation of Bohmian mechanics, since once the velocity field, $V(q, t)$, and the initial value of the Bohmian probability, $P(q, t_0)$, are specified, this equation determines the value of $P(q, t)$ for all time t.

Now, Bohm's choice for $V(q, t)$'s dependence on Ψ is such that if for some time t_0, $P(q, t_0) = |\Psi(q, t_0)|^2$, then by Equation (1) it follows that $P(q, t) = |\Psi(q, t)|^2$ for all time t. Thus, postulates (1) and (2) imply that $P(q, t) = |\Psi(q, t)|^2$. Hence, since the probability $P(q_1, t)$ for measurements of position for each subsystem in the universe can be derived from $P(q, t)$,[3] it follows that Bohmian mechanics makes the same statistical predictions as quantum mechanics for all measurements of position. But, since quantum mechanics makes statistical predictions for more observables than position, if it is going to reproduce the predictions of quantum mechanics, Bohmian mechanics must introduce one additional assumption:

(3) All measurements, and hence observables, are ultimately measurements of position, namely the positions of those aspects of measurement apparatuses that register the measurement result in terms of position. (I shall call these aspects the "pointers" of the measuring apparatuses.)

With this assumption, it follows that Bohmian mechanics is empirically indistinguishable from quantum mechanics. For the purposes of this paper I will grant (3) and thus *shall assume that postulates (1) and (2) comprise the basic ontological assumptions of Bohmian mechanics.*

IV. BOHMIAN MECHANICS VERSUS A PHENOMENALIST ACCOUNT

The basic phenomenological law of quantum mechanics is simply the well-known formula that expresses the probability that an observable R will yield a certain measurement result r_i when the system has a density operator ρ. For convenience, however, we shall follow Bohmian mechanics by first only considering pure estates, and second by assuming that all measurements ultimately reduce to measurements of position. Given these assumptions, the basic phenomenological law of quantum mechanics reduces to the following claim:

> (P*) A probability density $P(M, t)$ exists such that $P(M, t) = |\Psi(M, t)|^2$, where $P(M, t)\, dM$ denotes the probability that, at time t, the pointers of a system of measurement apparatuses are in the region $(M, M + dM)$ of their joint configuration space, and $\Psi(M, t)$ is the wave function for the pointers.

Claim (P) constitutes the fundamental, and sole, postulate of our phenomenalist account of quantum mechanics, with the stipulation that the probability P(M, t) is to be interpreted in the same way as the probability in Bohm's theory,*[4] and Ψ is assumed to obey the rules and formalism of quantum mechanics. Given this stipulation, we will be able easily to compare postulate (P*) of our phenomenalist account with postulate (2) of Bohmian mechanics.

Now, the only difference between postulate (P*) of the phenomenalist account and postulate (2) of Bohmian mechanics is the range of their respective probability distributions: (2) makes a claim about the probability distribution for the positions of all particles at some initial time t_0 whereas (P*) makes a claim about the probability distribution for the positions of the pointers of measuring apparatuses for all times t. Based on this commensurability, I shall argue that postulate (2) of Bohm's theory has no truth-indicating theoretical advantage over postulate (P*) of the phenomenalist account. Consequently, since Bohmian mechanics makes more ontological commitments than the phenomenalist account, it follows that Bohm's theory ends up increasing our ontology without offering us any truth-indicating theoretical advantage in return. Thus I conclude, by the epistemological principles elaborated in Section II, we do not have sufficient reason to believe that Bohm's theory is (approximately) true.

The crucial premise in the above argument is that postulate (2) has no truth-indicating theoretical advantage over postulate (P*). To make good on this claim, I shall consider five ways in which one could claim that (2) has such an advantage over (P*), and argue that each of these ways fail. Of course, this will not prove that (2) has no such advantage, since there could be truth-indicating virtues that I have overlooked. But, I believe, it does put the burden to find such an advantage on those who want to claim that Bohm's theory is (approximately) true.

Way 1

One might argue that postulate (2) is preferable to postulate (P*) because the initial probability distribution in (2) can be justified in a way that the probability distribution in (P*) cannot. I shall now argue that this is not the case, at least for the justifications of (2) that have so far been offered.

Several arguments for postulate (2) have been given in the literature. One type of argument begins by assuming that P differs from $|\Psi|^2$ at the beginning of the universe, and then attempts to show that P eventually will approach $|\Psi|^2$ for purely statistical reasons, thus justifying the assumption that $P(q, t_0) = |\Psi(q, t_0)|^2$ for some initial time t_0 not too far in the past. Since for any arbitrary choice of P at the beginning of the universe, Equation (1) determines the value of $P(q, t)$ for all time, *this approach must show that for typical choices of P at the beginning of the universe, Equation (1) yields solutions for $P(q, t)$ that eventually will approach* $|\Psi|^2$.

In this regard, one might initially think of Bohm's (1953b) paper purporting to show that the hidden-variable probability density, ρ, will eventually approach $|\Psi|^2$. For our purposes, the problem with Bohm's approach is that it is not generalizable to the universe as a whole (which our probability density $P(q, t)$ applies to), for Bohm's approach crucially assumes that the ensemble of quantum systems each undergo a different time evolution due to chaotic interactions with an *external* environment (Bohm 1953b, 462), something the universe obviously does not have. Thus, I will not consider his approach further but instead turn to Valentini's (1991a) recent attempt along these lines.

In trying to show that $P = |\Psi|^2$, Valentini first notes that, since both $|\Psi|^2$ and P obey Equation (1), the ratio $P/|\Psi|^2$ remains constant in time along any trajectory in configuration space. This implies that if initially P differed from $|\Psi|^2$, it will always differ from $|\Psi|^2$ for all time along every trajectory. To get around this problem, Valentini considers a coarse-grained probability distribution for P and $|\Psi|^2$: that is, a probability distribution obtained by dividing configuration space into cells and then defining $P_c(q, t)$, the coarse-grained analogue of P, to be a constant for each cell that is equal to the average value of $P(q, t)$ over the cell, and analogously for $|\Psi|_c^2$, the coarse-grained analogue of $|\Psi|^2$. Then, Valentini defines an entropy $S(P_c, |\Psi|_c^2)$ and shows that if initially $P = P_c$ and $|\Psi|^2 = |\Psi|_c^2$, then S (which is always ≤ 0) monotonically increases with time: that is, if time $t_2 >$ time t_1, then $S_{t2} \geq S_{t1}$. Valentini then says that he will take the monotonic increase of S as "proof" that S converges to zero and hence that P_c converges to $|\Psi|_c^2$ (since $P_c = |\Psi|_c^2$ when $S = 0$), even though he recognizes that strictly speaking it is not a proof.

Now, the key premise in Valentini's argument is that S monotonically increases for those highly atypical cases in which $P = P_c$ and $|\Psi|^2 = |\Psi|_c^2$, at some initial time t_0. Presumably, however, the fact that S monotonically increases in these atypical cases is supposed to give us good reason to believe that S monotonically increases for all cases, for otherwise his argument would give us no reason to believe that P_c converges to $|\Psi|_c^2$ for those typical cases for which $P \neq P_c$ and $|\Psi|^2 \neq |\Psi|_c^2$ at some initial time t_0. I will now prove, however, that, based on the time-reversal invariance of Bohmian mechanics (see Dürr *et al.* 1992a, 853), S cannot monotonically increase with time for all cases. *Proof*: Suppose that S monotonically increases for all P that obey Equation (1), all Hamiltonians, and all Ψ. Consider an arbitrary P and Ψ such that $S(P_c, |\Psi|_c^2)$ monotonically increases with time. Now, when applied to closed systems such as the universe as a whole, the Schrödinger equation is usually assumed to be time-reversal invariant (except for k-meson decay): that is, if $\Psi(q, t)$ is a solution of the Schrödinger equation, then the time-reversal of $\Psi(q, t)$, $\Psi_{rev}(q, t) \equiv \Psi^*(q, -t)$, is also assumed to be a solution, where Ψ^* represents the complex conjugate of Ψ. Hence, it follows that the velocity field corresponding to Ψ_{rev}, V_{rev}, is simply $-V(q, -t)$, as can be explicitly verified from the equation linking V with Ψ (see postulate (1) of Bohm's theory). Thus, $P(q, -t)$ will be a solution to Equation (1) corresponding to $\Psi^*(q, -t)$. But, $S(P_c(q, -t), |\Psi(q, -t)^*|_c^2)$ must monotonically decrease since, by assumption, $S(P_c(q, t), |\Psi^*(q, -t)|_c^2)$ monotonically increases. Unless S remains constant, however, it cannot both monotonically increase and decrease. Consequently, since in general S will not remain constant, it follows that S does not monotonically increase for all P and Ψ that obey Equation (1).[5]

Second, Valentini presents a heuristic understanding of his argument in which he compares the process of P_c approaching $|\Psi|_c^2$ to the mixing of two fluids of density P and $|\Psi|^2$ respectively. According to this understanding, at the coarse-grained level the two fluids will become so thoroughly mixed that they will be indistinguishable from each other: that is, P_c will be approximately $|\Psi|_c^2$. Although Valentini does not present this heuristic understanding as a separate argument, one might think that it nonetheless gives us good heuristic reasons to believe Valentini's overall claim that P_c approaches $|\Psi|_c^2$, even though his mathematical argument fails as I showed above. A major problem with this line of reasoning, however, is that, by the same sort of "mixing" argument, at equilibrium the fluid with density $|\Psi|^2$ will also become so thoroughly mixed with itself that the value of $|\Psi|_c^2$ would be essentially uniform. But, for the systems we are concerned with, $|\Psi|_c^2$ is not uniform. Thus it follows that these systems cannot yet be in their equilibrium state, and consequently such a heuristic mixing argument cannot give us any reason to believe that $P_c \approx |\Psi|_c^2$ at our present time.

Finally, it can be easily shown that any time t^* there will be an infinite set of functions, $P(q, t)$, that are solutions to Equation (1) such that $P_c(q, t)$ significantly differs from $|\Psi|_c^2$ at time t^*. (I will leave this notion of significance vague, but it is easy to see how to apply the following argument to a precise notion of significance.) *Argument:* Consider some arbitrary time t^*. For any $\Psi(q, t)$ and any non-trivial definition of "significantly differ," there exists an infinite number of probability distributions, $P^*(q)$, over configuration space such that their corresponding coarse-grained analogues, $P^*_c(q)$, differ significantly from $|\Psi(q, t^*)|^2_c$. Now, since for every $P^*(q)$, there exists a unique solution $P(q, t)$ to Equation (1) such that $P(q, t^*) = P^*(q)$, it follows that at any arbitrarily chosen time t^* there will exist an infinite number of solutions $P(q, t)$ such that $P_c(q, t^*)$ significantly differs from $|\Psi(q, t^*)|^2_c$.

To circumvent this last result (and also the time-reversal objection above), Valenti would have to explicitly introduce a measure over the space of probability distributions over configuration space – for example, one in which the above solutions for $P(q, t)$ have zero measure. This is effectively what Dürr *et al.* do in their exposition of Bohmian mechanics (1992a, 859, 873–4). Thus, I will now consider their arguments.

Dürr *et al.* do not try to prove that P approaches $|\Psi|^2$ but instead they simply assume that the most natural probability measure over configuration space – which they call a measure of typicality – is $P = |\Psi|^2$. In defence of this assumption, they claim that, because P obeys Equation (1), this measure is the simplest, and perhaps only, measure that is equivariant – that is, remains the same function of Ψ, namely $P = |\Psi|^2$, for all time. But why should the Bohmian probability have to be equivariant? That is, Why must there be a temporally invariant connection between Ψ and P? It is certainly difficult to understand what there could be in Bohmian ontology that could explain this connection. To elaborate on this point, suppose we take the ontological significance of Ψ to be that of determining the velocity field for the particles, as is typically done. Then the question immediately arises, Why would such a velocity field be connected by a temporally invariant set of equations with the probability P? Certainly this connection, whatever it is, cannot be merely the result of the nature of the velocity field, for one can construct scenarios within the framework of Bohmian ontology in which this connection does not hold. For example, suppose that the universe began in some arbitrary state Ψ and that the initial positions of the particles were determined by some "particle generator" that randomly produced particles in an equiprobability distribution with respect to spacial volume. In such a scenario, the subsequent probability for P typically would not be $|\Psi|^2$. Thus, Bohmian mechanics must make some additional hypothesis, such as about the origin of the initial probability distribution or about our ignorance of the initial condition, to account for why $P(q, t_0) = |\Psi(q, t_0)|^2$. But, it is difficult to imagine what that postulate could be. One could not, for example, merely apply Shannon's well-known "negative entropy" formula for measuring information, and then attempt to derive $P(q, t_0) = |\Psi(q, t_0)|^2$ by stipulating that we are in a state of ignorance, and hence of minimal information, regarding the initial distribution of particles. For, when applied to a continuum of alternatives, the formula requires that one specify an *a priori* probability measure M (van Fraassen 1989, 344–346), which for Bohmian mechanics must be $M(q, t_0)$

$= |\Psi(q, t_0)|^2$ in order to obtain $P(q, t_0) = |\Psi(q, t_0)|^2$. Consequently, unless one could further justify this *a priori* measure, this approach would beg the question. Yet no other approach to obtaining a probability from a condition of ignorance appears to do any better, and no non-question begging method of justifying the above *a priori* measure seems available. (To simply require, for example, that the measure be a temporally invariant function of Ψ would beg the question since this sort of requirement is the point under question here.) Thus, until some additional postulates are made and justified, the postulate that $P(q, t_0) = |\Psi(q, t_0|^2$ will be left unexplained, just as the corresponding postulate that $P(M, t) = |\Psi(M, t)|^2$ is left unexplained in the phenomenalist account.

Of course, Dürr *et al.* could respond by claiming that the initial probability distribution $(P(q, t_0) = |\Psi(q, t_0)|^2$ is to be preferred over other possible distributions on the grounds that it is the most mathematically elegant or natural distribution, since given this distribution P's dependence on Ψ turns out to be temporally invariant. To this argument, however, the phenomenalist could respond by pointing out that the quantum-mechanical formalism, along with the claim that $P(M, t) = |\Psi(M, t)|^2$, is the most elegant or natural mathematical model that adequately meshes with the observational data. Thus, considerations of mathematical elegance will not give us reason to prefer the probability distribution of Bohmian mechanics over that of the phenomenalist account.

Perhaps at this point, however, they could claim that the particles in the universe simply happen to be in a configuration that makes the distribution $P = |\Psi|^2$ the appropriate choice for experimental purposes. But, then, of course, the phenomenalist could respond in kind by saying that the results of measurement just happen to be distributed in such a way that $P(M, t) = |\Psi|^2$ turns out to be the experimentally appropriate distribution. Thus, I conclude, the justification that Dürr *et al.* offer for $P(q, t_0) = |\Psi(q, t_0)|^2$ does not appear to give Bohmian mechanics any theoretical advantage over the phenomenalist account.

Way 2

Another way one might try to argue that postulate (2) of the Bohmian account is preferable to postulate (P*) of the phenomenalist account is as follows: assuming that there are fundamental particles, atoms, and the like, it is simpler to suppose that the probability distribution $P = |\Psi|^2$ ranges over all particles, not just those composing the pointers of measurement apparatuses.

A major problem with this argument is that it treats the existence of fundamental particles and atoms as established. But, I shall argue, if Bohmian mechanics is our best account of the behavior of fundamental particles and atoms, then we cannot take their existence as given. I shall focus my argument on atoms, though it applies more generally. As is well known, from about the middle of the nineteenth century until the early part of this century, a heated debate occurred between those who proposed a phenomenalist approach to physics and those who advocated the atomic theory, a debate that was eventually decided in favor of atomic theory around the turn of the century (see Nye 1972). What decided the debate was that by using atomic theory, along with the equiprobability assumption of statistical mechanics, one could derive a whole host of divergent phenomenological laws such as gas laws, the laws governing Brownian motion, the laws governing chemical combinations, the laws of thermodynamics, and the like. Thus, atomic theory allowed one to account for a large body of phenomena by accepting fewer unexplained givens (such as the existence of atoms and the equiprobability or ergodic assumption) than the phenomenalist account, which required one to accept a host of phenomenological laws as givens. Put simply, atomic theory had the virtue of hypothesizing a simpler set of givens than the phenomenalist account. Moreover, atomic theory proved more fruitful than the phenomenalist program.

Thus, based on considerations of its theoretical virtues, a realist had good grounds for accepting classical atomic theory and its accompanying statistical mechanics, even though classical statistical mechanics has been plagued with foundational difficulties. If, however, we conjoin Bohm's theory with atomic theory, we can no longer simply assume that we have such grounds since Bohm's theory introduces new (and problematic) postulates. Rather, we must compare the theoretical advantage of Bohmian mechanics, *with* its attendant assumption that microscopic particles exist, to the phenomenalist account, as we are doing. To merely assume the existence of microscopic particles would beg the question.

Way 3

One might argue that, unlike Bohm's theory, the phenomenalist account has the classic problem confronting instrumentalists of drawing a distinction between observable and unobservable entities. This is not the case, however. First, the phenomenalist account does not presuppose this distinction because it does not presuppose the existence of unobservable entities. But, second, neither does it presuppose, as classical instrumentalists seem to, that theoretical terms can be classified into those that, if they referred, would refer to unobservable entities and those that, if they referred, would refer to observable entities. For, unlike classical instrumentalism, the above phenomenalist account does not say anything about what terms in a theory refer, or are likely to refer, simply because it is not an account of the range of reference of scientific theories. Of course, the phenomenalist account must presuppose that we have some idea of what counts as a measurement result, but this idea need not admit of a precise definition any more than we need a precise definition of a chair in order to make intelligible statements about, for instance, the probability of a randomly selected chair holding a 500 pound human being.

Way 4

The literature on Bohmian mechanics leaves one with the impression that there is something particularly special about quantum mechanics, such as its dynamics, that allows one to construct a hidden-variable model in which each particle has a definite position and momentum. Thus, based on this impression, one might conclude that Bohmian mechanics is likely to be (approximately) true. Here, I shall counteract this line of reasoning by showing how to construct such a model for any non-relativistic statistical theory in which some everywhere differentiable probability density, $P_s(q, t)$, can be defined such that $P_s(q, t)\,\mathrm{d}q$ represents the probability that upon performing a simultaneous measurement of position for all the particles in the universe, they would be found in the region $(q, q + \mathrm{d}q)$ of their joint configuration space at time t. (In the case of quantum mechanics, $P_s(q, t) \equiv |\Psi(q, t)|^2$.) Of course, any statistical theory that applies to the whole universe and that covers all observables will have such a function (since position is an observable), barring the requirement that it be everywhere differentiable.

The construction of such a model goes as follows. Consider a statistical theory, S, in which a probability function $P_s(q, t)$ can be defined that meets the above conditions. Now consider a hidden-variable model for theory S in which $V(q, t)$ represents the velocity field for the particles and $P(q, t)$ represents the probability density for the differential region around point q *being occupied* at time t. This hidden-variable velocity field $V(q, t)$ and probability $P(q, t)$ will obey Equation (1). Now, it can easily be shown that there exists a function $V^*(q, t)$ such that $P(q, t) = P_s(q, t)$ is a solution of Equation (1) when $V(q, t) = V^*(q, t)$.[6] Moreover, for any specification of the velocity field $V(q, t)$ and specification of $P(q, t_0)$ for some time t_0, Equation (1) has a unique solution for $P(q, t)$.[7] It follows, therefore, that if we let our velocity field $V(q, t)$ equal one of these fields $V^*(q, t)$, then if $P(q, t_0) = P_s(q, t_0)$ for some time t_0, $P(q, t) = P_s(q, t)$ for all t. Consequently, in analogy to Bohmian mechanics, we can construct an empirically indistinguishable hidden-variable model for our statistical theory S by postulating that $V(q, t) = V^*(q, t)$ and that $P(q, t_0) = P_s(q, t_0)$ for some time t_0.

Way 5

One could attempt to argue that Bohmian mechanics makes explanatory progress over the phenomenalist account because it *explains* the relation $P(M, t) = |\Psi(M, t)|^2$ whereas the phenomenalist account merely assumes that $P(M, t) = |\Psi(M, t)|^2$. The problem with this response is that although

Bohm's theory might explain why $P(M, t) = |\Psi(M, t)|^2$, it does so at the cost of hypothesizing an unexplained given (namely postulate (2)) which is, as we have been arguing, at least as much in need of explanation as the postulate that $P(M, t) = |\Psi(M, t)|^2$. Hence, in the end Bohm's theory does not make any real explanatory progress in this regard.

V. CONCLUSION

So far I have argued that there are no reasons to prefer postulate (2) of Bohm's theory to postulate (P*) of the phenomenalist account. Succinctly put, the argument essentially is that, as things currently stand in the literature, insofar as one finds the probability distribution in (P*) puzzling or in need of explanation, one should find the distribution in (2) as puzzling and as much in need of explanation. Thus, Bohm's theory does not make any explanatory progress over the phenomenalist account. Moreover, since Bohmian mechanics entails $P(M, t) = |\Psi(M, t)|^2$, it entails (P*) and thus must make at least the same ontological commitments as the phenomenalist account. Consequently, since it also hypothesizes an ontology of microscopic particles and a somewhat mysterious velocity field, its ontological commitments turn out to be greater than that of the phenomenalist account. It follows from the epistemological rules of Section (II), therefore, that we do not have sufficient reason to believe Bohm's theory (instead of, for instance, merely believing the phenomenalist account), since it seems that accepting Bohm's theory over the phenomenalist account increases our ontology without offering us any explanatory or truth-indicating theoretical advantage in return.

Of course, those advocates of Bohm's theory who believe in its (approximate) truth could nonetheless insist that (2) does have a truth-indicating theoretical advantage over (P*). But, in this case I think that the burden is on them to show us what this advantage is. For the two postulates only differ concerning the range of times and objects over which they postulate a probability distribution, and as we saw above, it is difficult to see how these differences give (2) an advantage.

Finally, I should emphasize that I have neither attempted to show that Bohmian mechanics is false, nor that it is not worth pursuing as a research program, nor that its fundamental postulates will not one day receive further justification. Neither have I tried to argue that we should believe the phenomenalist account of quantum mechanics. Rather, I have only intended to argue that at present, we do not have sufficient reason to believe that Bohmian mechanics is (approximately) true.

Messiah College

NOTES

[1] Of course, even among scientific realists it will be a controversial matter in many cases whether a virtue, such as simplicity or some aesthetic virtue, is really a truth-indicating theoretical virtue.

[2] For the purpose of this paper, the word "probability" is taken to denote any measure over a space of outcomes that obeys what is known as Miller's rule, a rule according to which one's degree of expectation (that is, one's subjective probability) that a sequence of events E has occurred, conditioned on the supposition that the objective probability of E is x, is x. (See van Fraassen 1989, 82.) I shall leave open the question as to whether the probability in Bohmian mechanics is to be further interpreted as an epistemic, logical, theoretical, or statistical probability, or as a measure of typicality as Dürr *et al.* do, or as something else. (For various other interpretations of probability, and how they apply to probability as used in statistical mechanics, see Sklar (1993, Chapter 3).)

[3] $P(q_1, t) = \int P([q_1, q_2], t)\, dq_2$, where $q = [q_1, q_2]$ and q_1 represents the position coordinates of the particles in the subsystem and the integral is taken over all values of q_2.

[4] For example, if one interprets the probability in Bohmian mechanics as an epistemic probability, then one should interpret the probability in this account as an epistemic probability (see footnote (2)).

[5] This reversibility objection is the same sort of objection raised against Bolztmann's original H theorem by Loschmidt and Zermelo, an objection that was finally answered only by assuming an equiprobability measure over phase space. (Davies 1974, 56–61.) I shall show below that in order to justify $P(q, t) \approx \Psi^2$, one cannot avoid introducing an explicit probability measure.

[6] See Courant and Hilbert (1989, 28–32).

[7] See *ibid.*

DAVID Z ALBERT

ELEMENTARY QUANTUM METAPHYSICS

Once upon a time, the twentieth-century investigations of the behaviors of sub-atomic particles were thought to have established that *there can be no such thing* as an objective, observer-independent, scientifically realist, empirically adequate picture of the physical world.

And it was part and parcel of thinking things like that, it was (you might even say) the *essence* of thinking things like that, that one looked at quantum-mechanical wave functions *not* as representing physical objects *directly*, but (say) as representing *what observers know* of such objects, or as representing imaginary *ensembles* of such objects, or as representing *the probabilities of the outcomes of measurements* on such objects, or something like that.

And it has consequently been essential to the project of digging one's way *out* of those sorts of confusions, it has been essential (that is) to the project of quantum-mechanical *realism* (in *whatever* particular form it takes – Bohm's theory, or modal theories, or Everettish theories, or theories of spontaneous localization), to learn to think of wave functions as physical objects *in and of themselves*.[1]

And of course the space those sorts of objects *live* in, and (therefore) the space *we* live in, the space in which any realistic understanding of quantum mechanics is necessarily going to depict the history of the world as *playing itself out* (if space is the right name for it – of which more later) is *configuration*-space. And whatever impression we have to the contrary (whatever impression we have, say, of living in a three-dimensional space, or in a four-dimensional space-time) is somehow flatly illusory.

I learned all this (insofar as I can reconstruct it now) from a few casual remarks scattered here and there in various papers and private communications of John Bell. And it has seemed so straightforward and so ineluctable to me since then as not to merit any further discussion.

But it turns out not to have seemed that way to everybody. It turns out (as a matter of fact) that this sort of talk still frequently manages to *surprise* people, even to *appal* them.

Maybe there's a point, then, in writing it all down in some detail – just to see how it looks when we're done, and to give its critics something convenient to shoot at.

J. T. Cushing et al. (eds.), Bohmian Mechanics and Quantum Theory: An Appraisal, 277–284.

1. REALITIES

The sorts of physical objects that wave functions *are*, on this way of thinking, are (plainly) *fields* – which is to say that they are the sorts of objects whose states one specifies by specifying the values of some set of numbers at every point in the space where they live, the sorts of objects whose states one specifies (in *this* case) by specifying the values of *two* numbers (one of which is usually referred to as an *amplitude*, and the other as a *phase*) at every point in the universe's so-called *configuration* space.[2]

The values of the amplitude and the phase are thought of (as with all fields) as intrinsic properties of the points in the configuration space with which they are associated. And so (for example) the fact that the integral over the entirety of the configuration of the square of the amplitude of the universe's wave function is invariably equal to one is going to have to be thought of not as following analytically from the sorts of physical objects wave function *are* (which it certainly can not), but as a *physical law*, or perhaps as an *initial condition.*

What physical role this object *plays* in the world, precisely, will depend on precisely how the measurement problem gets solved.

On Bohm's theory, for example, the world will consist of exactly two physical objects. One of those is the universal wave function and the other is the universal *particle.*

And the story of the world consists, in its entirety, of a continuous succession of changes of the *shape* of the former and a continuous succession of changes in the *position* of the latter.

And the dynamical laws that *govern* all those changes – that is: the Schrödinger equation and the Bohmian guidance condition – are completely deterministic, and (in the high-dimensional space in which these objects live) completely *local.*

The correspondence with our ordinary, three-dimensional, multi-particle language is trivial. Every particle invariably has a perfectly determinate position, and (more particularly) the x, y, and z co-ordinates of particle m are, respectively, the $(3m - 2)$th, the $(3m - 1)$th, and the $(3m)$th co-ordinates of the world-particle.

On the GRW theory (or for that matter on *any* theory of collapse), the world will consist of exactly *one* physical object – the universal wave function. What happens, *all* that happens, is that that function changes its shape in accord with the theory's dynamical laws. And those changes are *not* entirely continuous, and the laws which govern them are not entirely deterministic,[3] and (even in the high-dimensional configuration space) not entirely local.[4]

The correspondence with the three-dimensional multi-particle language is a bit more complicated. Talking about particles, in the context of *these* sorts of theories, is a way of talking about the wave function's shape. In particular, locutions like "particle *m* is located in the (finite) region bounded by x_1 and x_2 and y_1 and y_2 and z_1 and z_2" mean nothing more or less than that the universal wave function is (at the moment in question) bunched up in such a way that almost the entirety of its squared amplitude is confined to the (infinite) region between the x_1 and x_2 values of the $(3m - 2)$th co-ordinate, and between the y_1 and y_2 values of the $(3m - 1)$th co-ordinate, and between the z_1 and z_2 values of the $(3m)$th co-ordinate.

Note, by the way, that the above understanding of locutions like "particle *m* is located in the region bounded by x_1 and x_2 and y_1 and y_2 and z_1 and z_2" counts such locutions as ineluctably *vague*. And there is no point in denying that this comes as something of a surprise. And it is surely worth recalling, at this juncture, that the meanings of locutions like that are perfectly precise on *Bohm's* theory. But there's nothing enormously *mysterious* here either, and (more particularly) none of this represents any fundamental impediment to the project of describing the world *exactly*. The world invariably has a perfectly exact description, on these theories, in the language of the *wave function*; and particle-talk is inexact purely and simply because it amounts to an inexact description of *that*.[5]

One more example will suffice, I think.

Consider, then, what you might call a 'maximally atomistic' version of a modal interpretation of quantum mechanics, in which a determinate property is assigned, at every instant, to every separate co-ordinate-space degree of freedom, by means of a bi-orthonormal decomposition.[6]

On *that* sort of a theory, the physical world will consist, in its entirety, of *two* wave functions. One of those evolves in accord with the Schrödinger equation, and the other (which will have its own, separate, laws of evolution) is at every moment a simultaneous eigenfunction of all of the (commuting) Hermetian operators picked out by the above full set of decompositions.

Particle-location talk will supervene, here, on the *second* of these wave functions; and the *rules* of that supervenience will be precisely the ones we have just now discussed, precisely the same (that is) as the rules whereby particle-location talk supervenes on the *unique* universal wave function in the *GRW* theory.[7]

2. APPEARANCES

Those, then, are the rules whereby the three-dimensional multiple-particle language of our everyday lives supervenes on the exact and complete and fundamental language of the world, which is the language of *wave functions* (and whatever else) in *configuration* space.

But what is it about the world that can have suggested that everyday language to us – that *false* language, that *mirage* – in the *first* place?

What it is (I think) is the world's *Hamiltonian*; what it is is that the world has the sort of potential-energy operator whose expectation-value gets big exactly when (on that *three*-dimensional language; but not on any *two*-dimensional language, and not on any *one*-dimensional language, and not on any *five*-dimensional language) the expectation values of the "distances between particles" gets *small.*

Let's think that through in some detail.

Let's think first (because it's a little simpler) about the *classical* case.

Let's think about how it happens that a Newtonian-mechanical universe acquires the look of having some particular number of ordinary spatial dimensions.

Consider, then, an N-dimensional classical-mechanical configuration space, in which a single world-particle is floating around.

And suppose, to begin with, that that world-particle is floating around *freely*; suppose (that is) that its Hamiltonian is:

$$(1) \qquad H = \sum_{i=1}^{N} (p_i)^2.$$

And note that the trajectory of a world-particle like this one can patently contain *no suggestion whatever* as to whether we are dealing here with a single material particle moving freely in an N-dimensional physical space, or (say) $N/3$ distinct material particles moving freely in a *three*-dimensional physical space, or $N/2$ distinct material particles moving freely in a *two*-dimensional physical space, or N distinct material particles moving freely in a *one*-dimensional physical space. *Nothing* about a trajectory like that (to put it slightly differently) can make it natural or make it plausible or make it reasonable or make it simple or make it elegant or make it any other desirable thing to suppose that any particular *one* of those possibilities, as opposed to any one of the *others* I mentioned, or any one of the others I *didn't* mention, actually *obtains.*

The very question (in this context) seems pointless.

But try adding something to the dynamics – something along the lines of an *interaction.*

Imagine, for example, an N-dimensional classical configuration-space in which a world-particle is floating around not under the influence of the free Hamiltonian in (1), but under the influence of a more complicated Hamiltonian of the form:

$$(2) \qquad H = \sum_{i=1}^{N} (p_i)^2 + \sum_{\substack{k,j=1 \\ k \neq j}}^{N/3} V_{jk}([(x_{3j-2} - x_{3k-2})^2 + (x_{3j-1} - x_{3k-1})^2 + (x_{3j} - x_{3k})^2]^{1/2}).$$

Now, all of a sudden, *one particular* hypothesis about the number of physical dimensions this space actually has, and about the number of distinct material particles which are actually floating around *in* it, can be distinguished as uniquely natural and reasonable and elegant and whatever else you might like: the hypothesis that we are looking at a *three*-dimensional space, in which $N/3$ distinct material particles are floating around; the hypothesis (that is) is that the potential terms in this Hamiltonian represent an interparticle force whose intensity depends on the *distance* between the particles in question.

Now (to put all this slightly differently) there is suddenly an available conception of inter-particle distance – the *three-dimensional* one – which is capable of playing a *meaningful physical role*; now there is an available conception of inter-particle distance which (other things being equal) reliably measures the degree to which the particles in question can *dynamically affect* one another, can *alter* one another's *trajectories*.

Now we're dealing with a world that might reasonably be expected to look to its inhabitants (if there are any) as if it's *three dimensional*; now (that is) we are dealing with a world whose laws might well entail, over some reasonably broad range of initial conditions, that the world-particle will eventually make its way into regions of the configuration-space which correspond – in the *three-dimensional* language – to the existence of physics textbooks, written by reputable experts, which contain inscriptions like "the dimensionality of our world appears to be three".

Moreover, different sorts of modifications of the Hamiltonian can patently generate *different* such appearances. If, for example, we had replaced (1) not by (2) but by:

$$(3) \qquad H = \sum_{i=1} (p_i)^2 + \sum_{\substack{j,k=1 \\ j \neq k}}^{N/2} V_{jk}([(x_{2j-1} - x_{2k-1})^2 + (x_{2j} - x_{2k})^2]^{1/2})$$

we would have made a world which appears to its inhabitants not to have *three* dimensions, but *two*.

Good. Let's get back (with all this in mind) to quantum mechanics.

Consider, to begin with, a quantum-mechanical world (*any* quantum-mechanical world you like: a GRW world, a Bohm world, a modal world, a many-minds world, whatever) whose configuration space is N-dimensional, and whose Hamiltonian (considered now, of course, as an *operator*) is the one in Equation (1).

Questions about the dimension of this world's classical counterpart (remember) turned out to be silly; it turned out that nothing in any of the possible trajectories of a world like that could imaginably tip the scales in any particular direction. But the *quantum mechanical* case – as we discussed a few pages back – is going to be different: the set of all possible trajectories of a quantum-mechanical world with the Hamiltonian in (1) is simply not going to be *representable* on a space whose dimension is smaller than N.[8] And precisely the same thing will be true of a quantum-mechanical world with the kind of Hamiltonian in Eq. (2) (notwithstanding the fact that a *classical* world with a Hamiltonian like that turns out to have *three* dimensions); and precisely the same thing will be true of a quantum-mechanical world with the kind of Hamiltonian in Equation (3) (notwithstanding the fact that a classical world with a Hamiltonian like *that* turns out to have *two* dimensions).

But appearances are patently going to be another matter. How any particular quantum-mechanical world *appears to its inhabitants*, or at any rate how it appears to them *if they don't look too closely*, is going to be part and parcel of all of the usual arguments that insofar as so-called 'familiar macroscopic objects' under so-called 'familiar macroscopic circumstances' are concerned, the predictions of classical mechanics and the predictions of quantum mechanics are going to *correspond* with one another. Part of what has always followed from arguments like that (although we have perhaps not been in the habit of *thinking* about them in these terms) is that quantum-mechanical worlds are going to appear (falsely!) to their inhabitants, if they don't look too closely, to have the same number of *spatial dimensions* as their *classical counterparts* do. Part of what has always followed from them is (for example) that a quantum-mechanical world with a Hamiltonian like the one in (3) will appear to its not-too-closely-looking inhabitants to have *two* dimensions, and that a quantum-mechanical world with a Hamiltonian like the one in (2) will appear to *its* not-too-closely-looking inhabitants (just as *our own* does) to have *three*.

That's what (I think) I think. But now that it's all on the table, maybe there's a point in trying to put it a bit more diplomatically.

There are (you might say) *two* ideas we're accustomed to having in mind when we think of 'physical space'.

There is, to begin with, the space of possible *interactive distances*, the space (if there is one) that one reads off of the formal Pythagorean relations among the individual terms in the world's Hamiltonian – irrespective of whether those terms are considered as classical variables or quantum-mechanical operators. The space of interactive distances of any classical or quantum-mechanical world with a Hamiltonian like the one in (2), then, is three-dimensional; and the space of interactive distances of any classical or quantum-mechanical world with a Hamiltonian like the one in (3) is two-dimensional, and the space of interactive distances of any classical or quantum-mechanical world with a Hamiltonian like the one in (1) has *no* particular dimension.

And then there's an altogether different idea (and an altogether more *fundamental* one, it seems to me; but let's not squabble about that for the moment) of an *arena* within which the dynamics does its work, a *stage* on which whatever theory we happen to be entertaining at the moment depicts the world as *unfolding*: a space (that is) in which a specification of the local conditions at every address at some particular time (but not at any proper *subset* of them) amounts to a complete specification of the physical situation of the *world*, on that theory, at that time.

And it just so happens that in the context of classical physics those two spaces invariably *coincide*;[9] and it just so happens that in the context of *quantum* physics they *don't*.[10] And so the habit of thinking of the two of those spaces together, and of calling them by a single name, turns out to be one more of the things that we are apparently going to have to learn to give up.

Columbia University

NOTES

[1] That wave functions have the ontological status of physical objects is often claimed as a recent scientific discovery, as something that follows (more particularly) from the work of Aharonov and Viadman (1993) on so-called "weak" measurements.

But that can't be right. Aharonov and Viadman's very beautiful discoveries can (after all) perfectly well be accommodated within any *anti*-realist interpretation of quantum mechanics; and insofar as we are committed to *realism*, there was simply never anything other than physical objects that wave functions *could* have been.

[2] Or at any rate that's how things would work for a universe of elementary particles whose spins are all zero. In more realistic cases, we shall of course have to specify the values of some *larger* set of quantities at every point, but you get the idea.

[3] Note, however, that to say that the laws of the evolutions of wave functions are probabilistic (which is perfectly true, on theories like GRW), is *not at all* to say that those wave functions are somehow probabilities *themselves*, or that quantum mechanics somehow confronts us with a new and utterly mysterious modality of 'potentia' or 'possibilia' (which is gibberish). On any realistic understanding of quantum theory, wave functions are never anything more or less than perfectly actual, perfectly low-brow, *field*-configurations.

[4] The probability that some particular GRW collapse will multiply the value of the world's *wave function*, evaluated at some particular point in the world's *configuration space*, by some particular *number*, will depend (after all) on the wave function of the world throughout the *entirety* of that configuration space (and *not* just at the single point in question) at the instant just before the collapse occurs.

[5] Barry Loewer and I have gone into these matters in some detail (Albert and Loewer 1996).

[6] This is a theory which perhaps nobody will be tempted to take very seriously (on this theory, for example, the laws governing which Hermitian operators are assigned determinate values, at any particular moment, will *not* be invariant under rotations in ordinary three-dimensional space) but it will do perfectly well for our purposes here – it will leave the reader in no doubt whatever, I think, about how this sort of talk is to be extended to any modal theory she likes.

[7] Note that all of the pictures of the world I have been discussing here are what Paul Teller (1989) calls 'particularist' ones: ones in which (as Teller puts it) the world is composed of individuals, and the individuals have non-relational properties, and all relations between individuals supervene on the non-relational properties of the relata. The individuals of which the world consists on the GRW theory are the wave-function fields at all points in configuration space, and the individuals of which it consists on Bohm's theory are those fields together with the world-particle, and the individuals of which it consists on *Modal* theories are *two* fields like that; and pinning the conditions of all of those individuals down, on *any* of those theories, pins down *everything*.

What Teller thinks, of course, is that *quantum-mechanical* worlds *can't* be particularist ones; but (on the picture of the world we've been trying out here) that's because he misunderstands what the *individuals* of worlds like that *are*, that's because he supposes that the individuals of worlds like that have got to be the same sorts of things that they are in *classical* physics, things like elementary particles, or fields at points in 3-space, or something like that.

[8] That is: for any co-ordinatization whatever of the configuration space in question, that set will include trajectories which pass through states which are *completely non-separable* in those co-ordinates, states which cannot be expressed as products of functions of any proper *subset* of those co-ordinates.

[9] It just so happens (that is) that insofar as classical-mechanical theories of the world are concerned (and this is perhaps a way of putting one's finger on what it is that's *distinctive* about such theories) whatever amounts to a space of possible interactive distances amounts to a *stage*-space too.

[10] The geometries of the interactive-distance-spaces of quantum theories depend (by definition – and precisely as they do in the classical case) on those theories' *Hamiltonians*, but the geometries of their *stage*-spaces *don't*. The *stage*-space appropriate to a quantum theory can (alas) never be anything other than that theory's *configuration*-space.

That things can appear otherwise, if one doesn't look too closely, is a peculiarity of the wave functions of macroscopic objects. What happens is that the wave function of the world (on theories like GRW) or the *effective* wave function of the world (on theories like Bohm's) tends to be approximately *separable* in the centers-of-masses of objects like that, that the wave function of the world, or its effective wave function, tends to be approximately of the sort that can in fact be represented in the theory's *interactive-distance* space. And what *makes* that happen, in the GRW case, is the *collapse* part of the dynamics; and what makes it happen in the *Bohm* case is the part that governs the interactions of such objects with their *environments*.

TIM MAUDLIN

SPACE-TIME IN THE QUANTUM WORLD

1. NON-LOCALITY

Consider pairs of electrons produced in the singlet state and allowed to separate to a very great distance. We know from the work of Bell that no theory can predict violations of Bell's inequality for spin measurements on those pairs if it satisfied two conditions. The first (*condition a*) is that the theory ascribe either a single initial state or a convex sum of states to the ensemble of pairs such that the initial states are statistically independent of the spin measurements later carried out on the electrons. The second (*condition b*) is most easily understood if expressed differently for deterministic and for stochastic theories. For a deterministic theory we require (*condition b′*) that the theory determine the results of each measurement solely on the basis of the initial state and the details of the measurement carried out on that particle. For a stochastic theory we demand (*condition b**) that the theory assign probabilities for measurement results based solely on the initial state and the measurement carried out on a single electron, which probabilities are unchanged when one conditionalizes on the measurements and results obtained on the other particle in the pair. (Conditions *b′* and *b** are really the same condition, expressed in the one case appropriately to a deterministic theory, in the other for an indeterministic one. A theory which violates *b′* also violates *b** since conditionalizing on information about the measurement carried out on the second electron can render the result of the first measurement certain.) Einstein *et al.* (1935) had already pointed out the impossibility of a non-deterministic theory which obeys *a* and *b** to recover the predictions of quantum theory, and so argued in favor of a deterministic theory. Bell showed that no deterministic theory obeying *a* and *b′* could recover all of the predictions of quantum mechanics. The later work of Greenberger *et al.* (1989) showed that the reference to ensembles is otiose: there can be individual triples of particles such that no initial state of type *a* can recover all of the predictions of quantum mechanics if the theory is of type *b′* or *b**.

These results take on a slightly different cast when embedded into space-time theory. Since the settings of the measurement devices can be performed in the absolute future of the creation of the electrons, and since the settings can be determined independently of the process which creates the electrons (e.g., the setting could be determined by a computer program running a complex algorithm), any theory which denies condition *a* must posit *backwards causation*: the ineliminable asymmetric dependency of absolutely earlier states on absolutely later events. And since the measurement events

J. T. Cushing et al. (eds.), Bohmian Mechanics and Quantum Theory: An Appraisal, 285–307.

can be arranged to take place at spacelike separation, a deterministic theory which denies b' or a stochastic theory which denies b^* must posit *superluminal influence*: the ineliminable dependency of certain outcomes on events at spacelike separation. In short, any theory that predicts violations of Bell's inequality for measurement outcomes at spacelike separation must be *non-local* in the following sense: if deterministic, the theory cannot locate all of the determining causes of every event in the past light-cone of the event; if stochastic, cannot assign probabilities to events conditional on everything in their past light-cone such that those probabilities are unchanged by conditionalizing on events at spacelike separation. (Of course, there is nothing wrong with the probabilities in a local stochastic theory being changed by conditionalization on events in the future light-cone: those events could be *effects* which indicate how the measurement came out. There would, though, be some sort of backward causation if one changed the probabilities by conditionalizing on a future event which could not be interpreted as an effect, such as a future measurement setting controlled by a computer.)

So any theory which predicts violations of Bell's inequality for events at spacelike separation must, in this sense, be non-local. And if there *are* reliable violations of Bell's inequality for events at spacelike separation, then any physically adequate theory must be non-local. This is the moral of Bell's work, though it has taken a remarkably long time for that moral to be widely recognized. Bell's work is *not* particularly about "hidden variables" theories, nor particularly about deterministic theories, nor particularly about "realistic" theories (whatever that might mean). The theorems apply to hidden variables theories and to deterministic theories and to realistic theories because they apply to *all* theories. But the bottom line is that *locality* (not hidden variables or determinism or realism) cannot be reconciled with reliable violations of the inequalities at spacelike separation.

So far, we have only a conditional result: if the inequalities are violated for measurements at spacelike separation, then no local theory can be predictively adequate. And it seems a small step to discharge the condition. Do not the experiments of Aspect *et al.* demonstrate violations of the inequalities at spacelike separation? Not every interpretation of the quantum formalism entails this result. The many-minds theory of Albert and Loewer (1988), for example, holds that measurements only have determinate outcomes in individual minds (not in the purely physical apparatus), and that the inequalities are only violated for sets of measurements that are experienced by a single mind.[1] Since the relevant statistics govern only experiences in particular minds there is no violation of the inequalities for events at spacelike separation. Of course, all of this requires a peculiarly radical reinterpretation of Aspect's laboratory reports: they do not record particular outcomes on apparatus separated by 12 meters, with the outcomes being recorded within 5 nanoseconds of one another, and so on. Rather, they "record" only the experience of some mind, which experience does not reflect any unique course of physical events in the laboratory.[2]

Let us call any interpretation of the quantum formalism according to which there was a unique set of outcomes in Aspect's laboratory, which outcomes occurred at spacelike separation, a "standard" interpretation. In this sense, any collapse theory, such as the GRW theory, is standard. Even a theory which ties collapses to consciousness can be standard, so long as a conscious observer is stationed at each apparatus to cause the collapse and ensure a definite outcome. The Copenhagen interpretation is also surely standard, as it insists on the necessity of describing the macroscopic laboratory equipment in classical terms, and the classical terms include phrases like "Apparatus A, set at 30°, registered spin up etc. . . .". That is, the classical description of Aspect's experiment entails that Bell's inequalities were violated by events at spacelike separation. Bohm's theory is obviously also, in this sense, standard.[3]

Any non-standard interpretation must do considerable violence to our basic conception of the physical world. That is, if it contains space-time at all, it cannot contain physical events or objects located in the space-time that even vaguely correspond to the world as it appears to us. If it seems to us that a needle on a piece of apparatus moves one way rather than another at a particular time and place, or a particle detector clicks at a particular time and place, a non-standard interpretation must insist that nothing at all which reflects those supposed events takes place in the corresponding regions of space-time.

We shall, for the time being, leave aside consideration of non-standard interpretations of the quantum formalism. The reason is fairly simple. Our basic worry concerns the compatibility of quantum theory with the relativistic account of space-time structure. If we accept a non-standard interpretation, then physical reality contains nothing like a relativistic space-time containing localized events and objects which even approximately correspond to the events and objects we think we see. In such a circumstance, it is hard to see why we would continue to hold the relativistic account of space-time structure seriously, since that account is based on observations which were taken to report objects and events in space-time. In short, it is hard to see why we would seriously believe that we had gotten the deep structure of space-time right if we had gotten questions about whether, for example, a needle on an instrument actually moved to the right or the left wrong.

According to the taxonomy just proposed, both the usual collapse theories and Bohm's theory count as standard interpretations. And we have already argued that any standard interpretation must abandon either *a* or *b*. A glance at the non-relativistic versions of these theories confirms this analysis.

In non-relativistic Bohmian mechanics, the results of one of the spin measurements on our electrons can sometimes depend on the exact nature of the measurement carried out on the other electron. Further, it is only the result of the later measurement which so depends; each measurement outcome is determined solely by events which occur earlier than it. On account of this, the time order of distant measurements with respect to the absolute simultaneity of Galilean space-time plays an ineliminable dynamical role in the theory. For some initial states and spin measurements, the measurements will yield one result if the right-hand measurement precedes the left, and a different result if the left-hand measurement precedes the right. We have not only the dependence of some outcomes on measurements made at arbitrary separation, but dependence on absolute time order (see Albert (1992, 156 ff.) for a clear account of Bohm's theory in this context).

Non-relativistic collapse theories do not display exactly this structure. One cannot, in a collapse theory, say that given the initial state the measurement outcome would have been different had the time order of the measurements been reversed. This is primarily due to the introduction of chance elements (the collapses) into the dynamics: chance elements make evaluation of counterfactuals somewhat tricky. But there still are features of collapse theories which reveal the non-locality inherent in the theory, and also the dependency of non-relativistic theories on absolute time order.

Taking non-locality first, consider the EPR set-up with spin measurements taken in the same direction on pairs of particles in the singlet state. The results are sure to be anticorrelated. Furthermore, the exact outcomes are not determined by the initial state: Give the same initial state and measurement set-up, one could get either up on the left and down on the right or down on the left and up on the right. As Einstein *et al.* (1935) pointed out, one can guarantee such anticorrelation in a stochastic theory only by having some sort of influence of one measurement on the other. The argument is painfully obvious, but given the rough reception that the EPR argument has received, it is worth repeating.

The initial complete physical state of the pair (the singlet state, and assuming no additional "hidden" variables) does not determine the measurement outcome on either side. Now consider the state after the first measurement has taken place, and a result been obtained. The new, collapsed state does determine what the result on the unmeasured electron will be: it will be the opposite of the result obtained on the measured electron. Hence the physical state of the unmeasured electron has *changed*, and it has changed *due to the collapse*, and more particularly it has changed *due to the collapse that was caused by the measurement carried out on the other electron.* Here a definite counterfactual still holds: had the first measurement not taken place, the collapse would not have occurred, the state of the pair would still be the singlet state, the state of the unmeasured electron would be different from what it is. Since collapse theories take collapses as fundamental dynamical events, this collapse represents a real physical change in one electron consequent upon a measurement carried out on the other.

So non-relativistic collapse theories endorse two definite counterfactuals which hold only because of the physical effects of collapse. *Outcome Counterfactual:* Given exactly the same initial state and measurement set-up, the measurement results could have been different, and had one been different (i.e., down instead of up), so would the other have been. *State Counterfactual*: Given exactly the same initial state, had the measurements taken place in different time order, the state of the each electron (and the state of the pair of electrons) when it reached its apparatus would have been different. The first counterfactual is perhaps the more immediately compelling since it adverts to *observable* events (the outcomes), while the second involves the dynamics of the wave function. But collapse theories must take the wave function seriously, and must take the collapses seriously, so that changes in the wave function count as real physical changes as much as differences in measurement outcomes. If someone believes that the collapses need not be taken seriously, that they can be understood as merely epistemic, then they should take them out of the dynamics: produce a non-collapse theory (such as Bohm's).

It is interesting to note that what we may call the *Bohm Counterfactual* contains something like the content of both the Outcome Counterfactual and the State Counterfactual: according to non-relativistic Bohmian mechanics it is sometimes true that given the same initial state, the measurement outcomes could have been different and would have been different had the measurements taken place in a different time order. The Bohm Counterfactual points up especially forcefully the dynamical role of absolute simultaneity in Bohm's theory, even when systems are widely separated. And I suspect that the belief that *only* Bohm's theory will have a hard time with Relativity stems from a misguided sort of divide-and-conquer strategy. Collapse theories don't endorse the Bohm counterfactual, they endorse the Outcome and State counterfactuals separately. Then one argues: the Outcome Counterfactual isn't really serious since one can get perfect correlations at arbitrary distances without any non-local effects, as the correlations which are produced upon opening the envelopes if we tear a dollar bill in half and send one half in a sealed envelope to a distant correspondent.[4] And one further argues: The State Counterfactual isn't really serious since after all the wave function is just a mathematical artifact that represents our knowledge, it doesn't represent anything real, and it changes at distant locations when we learn something new (e.g., when I open my envelope I suddenly come to know what is in the distant envelope). But neither of these arguments is compatible with a collapse theory. Distant correlations of the dollar bill variety can only be produced by a local *deterministic* theory (EPR again!), and merely epistemic collapses are dynamically otiose collapses, demanding the production of a non-collapse theory. (This rather shameless appeal to *deterministic* models when arguing for the locality of *stochastic* theories is perhaps most evident in Murray Gell-Mann's recent book (1994, 172), in which he mangles Bell's beautiful paper (1981b) by likening the non-locality of quantum theory to the case of Bertlmann's socks!)

In sum: both non-relativistic Bohmian mechanics and non-relativistic collapse theories (e.g., GRW) incorporate ineliminable physical influences between arbitrarily distant events, and both require absolute simultaneity to formulate their dynamics. The question we must now face is how, or whether, one could produce versions of these theories which are compatible with Relativity.

2. THE RELATIVISTIC CONSTRAINT

It is not uncommon to interpret the Theory of Relativity as forbidding something from going faster than light. Elsewhere (1994), I have examined four distinct possibilities: no superluminal matter or energy transport; no superluminal signals; no superluminal causal influences; no superluminal informational exchange. I have also argued that violations of Bell's inequality for events at spacelike separation do not demand violations of the first two constraints and do demand violations of the second two. But none of these constraints really seems to express the fundamentals of Relativity. First, it is obvious that one could construct a manifestly non-Relativistic theory (an ether theory of electromagnetism, say, with a preferred rest frame) which still forbade anything of interest from travelling faster than light (e.g., Maxwell's electrodynamics). Further, one can construct explicitly Relativistic theories (i.e., explicitly Lorentz invariant theories) which violate every one of the four proposed constraints. Relativity is a theory of space-time structure, not a theory of the exact nature or dynamics of the fields or particles which inhabit space-time. So we ought to frame the Relativistic constraint directly in terms of space-time structure, rather than in terms of anything going faster or slower than anything else.

In saying that we ought to frame the constraint in terms of space-time structure, I also mean to rule out formulations based on coordinate transformations. This point is often overlooked because consistency with *Special Relativistic* space-time structure (i.e., Minkowski space-time) turns out to be equivalent to invariance under the Lorentz transformations. Let us pause for a moment to trace the connection between coordinate transformations and space-time structure in Special Relativity.

Minkowski space-time is homogeneous and, with respect to all time-like (or spacelike) directions, isotropic. Like Euclidean space, Minkowski space-time admits of global rectilinear orthogonal coordinate systems. Since the space-time is homogeneous, the space-time structure itself looks the same when expressed *in a coordinate dependent form* relative to rectilinear orthogonal coordinate systems whose origins are shifted with respect to one another. And since it is isotropic (in the time-like directions), the space-time structure also is the same when expressed *in a coordinate dependent form* relative to rectilinear orthogonal coordinate systems whose time axes are rotated with respect to one another. (And similarly for rotations of the spatial axes.) This means that the metric, expressed in terms of the coordinates, takes the same functional form for all Lorentz frames, and hence is invariant under the Lorentz transformations. This invariance is a consequence of the global symmetries of the metrical structure of Minkowski space-time. Any theory which is invariant under those same transformations displays the same symmetries, and so *does not postulate any new space-time structure*. So Lorentz invariance turns out to be a roundabout route to a more fundamental property: the essential fact about Lorentz invariant theories is that their dynamics depend only on the Special Relativistic metrical structure. And once put this way, all reference to coordinate systems and coordinate transformations may be dropped. Given, for example, a coordinate-free formulation of a theory, we may ask whether it postulates only the relativistic space-time structure or whether it posits more.

The advantage of a coordinate-free formulation of compatibility with Relativity is twofold. First, it is immediately extendible to General Relativity. One can ask whether the dynamics of a theory postulate more than the metrical structure of a General Relativistic space-time just as one can ask for Minkowski space-time. Formulations in terms of invariance of coordinate based theories under transformations between rectilinear orthogonal reference systems cannot be extended to the General Relativistic context where no such reference frames exist. Lacking the global symmetries of the Special Relativistic metric, no invariance can guarantee compatibility with relativistic space-time.

But the deeper advantage of the elimination of criteria based on coordinate transformations is simply that coordinate systems do not, in any deep sense, exist. My room may be full of air molecules and electromagnetic fields, but it is not criss-crossed by coordinate curves. Insofar as coordinate systems can be given any physical significance, it is either directly by appeal to the space-time structure (as one can define rectilinearity and orthogonality, and hence rectilinear orthogonal systems by appeal to the metrical structure) or else by appeal to the (possible or actual) results of assigning numbers to space-time points by using, for example, clocks and rods. But clocks and rods are not magical, unanalysable devices that simply reveal underlying space-time structure. There is a measurement problem in Relativity much as there is one in quantum theory: the operations of measuring devices such as clocks must ultimately be explained by the physical theory, and in doing so the theory cannot postulate physical structures which are themselves defined by reference to the measurement results.

Thus, it is not sufficient to render a theory Special Relativistic that it entail that coordinate systems associated with relatively moving clocks and rods are connected by the Lorentz transformations. As Bell shows (1976), an ether theory with a preferred frame can have that consequence:

one simply derives the Lorentz–Fitzgerald contraction and time dilation as real dynamical effects on the measuring instruments. Consistency with Relativity has not to do merely with results of laboratory operations but with the deep ontology of the theory. To make it official,

> *Relativistic Constraint*: a theory is compatible with Relativity if it can be formulated without ascribing to space-time any more or different intrinsic structure than the (special or general) relativistic metric.

Two comments are in order. First, we can immediately make distinctions between theories which are *not*, according to this criterion, relativistic. Some theories may fail to be relativistic because they deny that space-time has the relativistic metrical structure at all. A classical ether theory set in Newtonian space-time would exemplify this approach. As already mentioned, such a theory could still return the Lorentz transformations as the relevant coordinate transformations between moving frames, but the relativistic metric would not correspond to any fundamental physical quantity. Other theories may break the relativistic constraint not by denying the physical significance of the metric but by adding something else to it. For example, we can easily imagine a theory whose dynamics require, along with the metrical structure of space-time, a foliation of the space-time. According to our criterion, such a theory fails to be relativistic, since Relativity itself postulates no such foliation. Of course, retaining the relativistic metric in addition to the new structure may go some way towards explaining the evident success that has attended the demand for Lorentz invariant theories.

The second comment concerns the clarity and applicability of the criterion, specifically of the term "intrinsic". Suppose one postulates, in addition to the metric, a foliation of space-time. Under what conditions should that be considered more intrinsic space-time structure as opposed to being just another field *on* space-time? That is, our basic approach relies on the idea that Relativity is supposed to give the complete account of space-time structure, but by what criterion do we denominate structure *part of* space-time itself?

I doubt that any clear general criterion could be found here, but examples may help. In Minkowski space-time, the postulation of a preferred foliation into flat spacelike hyperplanes, unaffected by all matter fields, would clearly count as more intrinsic structure, indeed as the postulation of absolute simultaneity. It would help make such a foliation look less "intrinsic" if it were coupled to some matter fields somehow, so that the exact form of the foliation depended on the distribution of matter.[5] Attempts to find a preferred reference frame defined by, for example, the cosmological background radiation is a nod in this direction, but as yet falls short of a theory. If the foliation is not intrinsic, then one expects that different initial conditions could have produced space-times with the same metric but different foliations, and one would like to know just what the relevant initial conditions are. But it is doubtful that much of use could be said without concrete theories to hand.

Ultimately, of course, the debate over "intrinsic" or "extrinsic", and hence "relativistic" or "non-relativistic" is scientifically irrelevant. Any theory which can produce violations of Bell's inequality at spacelike separation should be judged in its merits: empirical adequacy, clarity, simplicity, and so on. If it postulates a foliation of space-time which cannot be directly empirically verified, then it will probably be frowned on, irrespective of how "intrinsic" or "extrinsic" the foliation seems. If the theory manages to be accepted nonetheless, then we will add a foliation of space-time to our fundamental picture of the world. Whether the new picture should be accorded the honorific "relativistic", or the (perhaps even greater) title of "post-relativistic" is of no scientific interest.

Now that we have a passably clear criterion for a relativistic theory, we can return to the problem of producing a relativistic quantum theory.

3. HINTS FROM NON-RELATIVISTIC VERSIONS

In Section 1 we saw that any theory which predicts violations of Bell's inequality for widely separated measurements must either deny *a*, that the initial state for the pair of electrons is statistically independent of the types of measurements later carried out or (if a deterministic theory) *b′* that the result of each measurement is determined solely by the initial state of the pair and the type of measurement carried out (i.e., independent of the type of measurement carried out on the other electron), or (if a stochastic theory) *b**, that each measurement is assigned, on the basis of the initial state and the type of measurement carried out, an irreducible objective chance of producing certain results, which probabilities are unaffected by conditionalizing on information about the type of measurement carried out on the other electron or about the outcome of that measurement. This trilemma holds equally for relativistic and for classical domains, so that we can look to the non-relativistic versions of the theories for guidance as to which claim to abandon in the relativistic version.

Interestingly, neither Bohmian mechanics nor collapse theories such as GRW abandon *a*. One can, in the non-relativistic domain, produce adequate interpretations of the quantum formalism which predict the right correlations for distant pairs of measurements about invoking any backwards causation. This suggests (though hardly proves) that any attempt to fit similar theories into relativistic space-time ought also to forego appeal to backward causation (i.e., to statistical dependency of the initial state of the particle pair on the types of measurements to be carried out in their absolute future).

Of course, abandoning *a* cuts the Gordian knot of explaining violations of Bell's inequality for widely separated systems. If the initial state of the particles can be influenced by the measurements to be carried out on them, any statistics, violating Bell's or any other inequality, can in principle be produced. But the gap between "in principle" and a clear theory is extremely wide. Some attempts in this direction, such as the "transactional" interpretation of Cramers which uses both advanced and retarded wave functions, are demonstrably incoherent (Maudlin 1994, 197–201). In any case, since the clear non-relativistic theories we have to hand do not lend themselves to this sort of generalization, we will focus instead on the consequences of doing away with either b' or b^*.

In the non-relativistic collapse theories and non-relativistic Bohmian mechanics, violations of Bell's inequality are secured by instantaneous action (either through the collapse or due to the non-local dependencies in the guidance equation) which allows the results on one side to be influenced by the measurement or outcome on the other. Since relativistic space-time contains no absolute simultaneity, the question is how to model this instantaneous action in a setting where "instantaneous" has no clear meaning. Classical action-at-a-distance laws, such as Coulomb's law or the law of universal gravitation surmounted this difficulty by replacing "instantaneous" with "at the speed of light", taking advantage of the intrinsic light-cone structure of the space-time. Explorations in this direction have been conducted by E. J. Squires (1993), using a retarded version of Bohmian mechanics, but the consequence of this is to render violations of Bell's inequality for measurements *completed* at spacelike separation impossible. Squires argues that Aspect's experiments do not demonstrate the inadequacy of this model since the measurements, although *initiated* at spacelike separation, are not *completed* at spacelike separation (i.e., before one measurement is over, there has been time for a light signal to reach it from the other side). This sort of constraint will also clearly emerge in any collapse theory that uses the forward light-cone for collapses.

Squires' model does not invalidate any of our reasoning, which is predicated in the existence of violations of Bell's inequality for measurements at spacelike separation. Like objections based on detector inefficiencies (e.g., Fine 1989), these are questions which could be resolved by further experimentation. So let us leave this as another open experimental loophole, and press on assuming the worst, viz. that further experiments will find the same behavior as the distance between sides of the experiment are increased.

If the effects of one measurement on the other do not propagate along the forward light-cone, and if we do not wish to countenance backwards causation (i.e., causes in the absolute future of their effects), then we must find some way to secure direct influences at spacelike separation. The most obvious way to do this is to foliate the space-time with spacelike hyperplanes which may be used, in essence, to define a preferred synchronization between widely separated systems. Or, one could just add a synchronization parameter directly, without associating it with a spacelike hyperplane. This latter route is pursued by Berndl, Dürr, Goldstein and Zanghì (1996) for Bohm's theory. The question which faces such attempts is the ontological interpretation of the hyperplanes or the synchronization parameter. If these are entirely new physical quantities, unconnected to other physical fields, then they look suspiciously like new intrinsic space-time structure. If our account of the relativistic constraint is correct, then theories such as these are more accurately described as modifying the relativistic space-time structure to make room for quantum theory than as fitting out a non-relativistic theory to accord with Relativity.

4. WHO'S AFRAID OF ABSOLUTE TIME?

Although the foregoing considerations are filled with gaps, elisions and caveats, they do suggest that the most straightforward route to putting quantum theory into a relativistic space-time is to add something to the space-time, and hence, at least implicitly, to reject Relativity as the complete story. Indeed, given that this is the most obvious way to frame a Bohmian or collapse theory in a non-classical space-time, we ought to take it as a benchmark. What Bohm's theory and the collapse theories seem to need is something like the classical notion of absolute simultaneity: a fundamental physical relation between events at space-like separation. In effect, such a relation would induce a *foliation* of the space-time, a division of the space-time manifold into a stack of space-like hyperplanes. Putting a measure over those hyperplanes yields an absolute time-function in terms of which the Bohmian dynamics or the collapse dynamics can be framed. If there is something objectionable about adding a foliation to space-time, we should consider carefully just how objectionable it is, since there is no point in doing something even *more* objectionable just to retain the relativistic account of space-time. If we are really willing to seriously consider the postulation of backwards causation, for example, then we ought to be convinced that the dislocations attending such an innovation are less severe than would be the addition of a foliation to space-time.

What, then, would be so terrible about slicing up space-time to provide dynamics for Bohmian mechanics or for wave collapse? The most obvious objection is that the true foliation could not be determined empirically. Were the Logical Positivists still with us, they might even object that the proposed structure is *metaphysical* or even *meaningless*. Fortunately, we have outgrown the most restrictive of the empiricist scruples, and so can get beyond mere name calling.

The first thing to note is that although the foliation could not be empirically determined, this is *not* because the foliation would have no effect on other physical quantities (the mark of the truly superfluous). Indeed, *if* one could observe the exact values of all the *other* physical quantities postulated by the theory, one could also fix the next space-time structure. If one could determine the exact trajectories of particles in Bohmian mechanics, one could determine the foliation. As we have seen, the trajectories in non-relativistic Bohmian mechanics sometimes depend on which of two distant measurements occur first. In the relativistic space-time *cum* foliation we are considering, this would mean that the trajectories would depend on which measurement came first with respect to the foliation.[6] So the empirical inaccessibility of the foliations is solely a consequence of the empirical inaccessibility of the exact particle trajectories, and this in turn, as Dürr *et al.* (1992a) have show, is a consequence of the physical dynamics of measurement situations. If the existence of empirically inaccessible physical facts is fatal, then Bohmian mechanics is a non-starter even before Relativity comes into play. Conversely, if one takes Bohm's theory seriously as a possibility (as one should!), then the fact that the foliation is hidden from view cannot by itself imply that theory is untenable. There is, of course, this difference: whereas in Bohmian mechanics the *exact* trajectories cannot be empirically determined, *rough* trajectories can; in the relativistic theory, the structure of the foliation cannot be even approximately measured.

(Of course, the most appropriate analog for the preferred foliation in our hypothetical "relativistic" Bohmian theory is absolute simultaneity in the non-relativistic theory, and here matters are exactly on par. Absolute simultaneity is dynamically important but observationally inaccessible in the non-relativistic theory: even though the results of a particular experiment may depend on which of two measurements happened first, no observable results can inform the experimenters which happened first. Absolute simultaneity in the non-relativistic theory is also hidden from view just as the preferred foliation would be, and could not be operationally determined.)

Similarly, in a collapse theory such as GRW, the foliation would have quite obvious effects on the dynamics of the wave function, and so could easily be observed if the wave function could be. Consider, for example, the original EPR state in which the pair of particles are in an eignestate of particle separation although not in an eigenstate for position of either particle. A position measurement carried out on one of the pair causes the collapse of the other member into a position eigenstate (or, in GRW, nearly a position eignestate). If only we could empirically determine where that sudden change in the partner occurs, we could tell how the foliation cuts the space-time. Here, it is the unobservability of the exact wave function, which we have had to accept from the beginning, that secures the unobservability of the foliation. There is really no point in raising positivist scruples at this point if we have pacified them in the non-relativistic domain.

Another sort of objection to tinkering with space-time structure runs as follows: Adopting relativistic principles, and especially demanding Lorentz invariance in Special relativistic domains, has been an extremely successful strategy. Why should this be so unless Relativity gives an accurate account of space-time structure? Won't any theory according to which

Relativity is *false* involve a very odd sort of conspiracy of physical law so that a non-relativistic space-time appears to be relativistic?

I have myself voiced this sort of concern (Maudlin 1994, 242), but now think that I failed to mark an important distinction. As noted above, there are two quite different ways to depart from Relativity. One is to eliminate the relativistic metrical structure entirely at the fundamental level, and to account for effects of relative motion (time dilation, Fitzgerald contraction) by a close analysis of physical effects on measuring devices. This is essentially the route followed by Bell in his Special Relativity paper (1976): postulate Newtonian space-time and Maxwell's equations, and show how electromagnetic effects in clocks and rods will hide the Newtonian structure from us. The other route is to retain the relativistic metric at the fundamental level, but to add something to it (e.g., a foliation). In this case, relativistic behavior can be explained in the same way it always was: some physical quantities are governed by equations framed in terms of the relativistic metric. It is just that the evolution of some of the physical quantities also depends on the foliation.

If some of the evolution depends on the foliation, though, how have we been able to get so far without it? Why was the rejection of absolute simultaneity by Einstein so successful? Ironically, the answer to this is the very same empirical unobservability that inspired the first objection! *Observable* space-time structure is solely the *relativistic* structure, so it is hardly surprising that Relativity should serve us so well.

It is amusing and instructive to see how this picture could play out in a Bohmian milieu. Since there are no real collapses in Bohm's theory, the wave function has no need of the foliations. So the wave function could evolve in accord with a covariant equation, such as the Dirac equation. The guidance equation for the particles (or other local beables) would be framed with reference to the foliation. The ontological duality already at the base of the theory would map onto a similar duality of space-time structure, with the accuracy and effectiveness of Relativity being explained by the completely relativistic nature of the wave function. The beables would be sensitive to absolute simultaneity, but only in a completely unobservable way.

Nothing quite so symmetrical seems likely for a collapse theory. Since collapse theories tend to be ontologically monist, the best one could do would be for the wave function to evolve sometimes relativistically and sometimes (during collapse) non-relativistically. The ontological duality of Bohm becomes dynamical duality in GRW, with the form of the evolution of the wave function changing during collapses. In the non-relativistic version, it is only the form of evolution that changes, while in the "relativistic" version here envisaged, even the quantities in terms of which the evolution is described would shift.

How high is the price of abandoning Relativity in just this way (i.e., by adding a foliation to space-time)? It is hard to argue that the cost is automatically prohibitive. We would have to postulate some new space-time structure, but it would be exactly the sort of structure one would anticipate given both the experiments and the non-relativistic versions of the theory. There seem to be influences between events at spacelike separation, instantaneous influences in the classical milieu. A spacelike structure such as a hyperplane fairly directly recommends itself. As Bell remarks,

> [I]t may be that we have to admit that causal influences *do* go faster than light. The role of Lorentz invariance in the completed theory would then be very problematic. An 'aether' would be the cheapest solution. But the unobservability of the aether would be disturbing. (1987a, 154–155)

One would think that the cheapest solution would recommend itself to the methodologically frugal. We should try some comparison shopping to see if anything less disturbing can be unearthed.

5. HYPERPLANE DEPENDENCE

Let us briefly review the methodological decisions we have made so far. We have decided, at least provisionally, to set aside non-standard interpretations of the quantum formalism, such as the many-minds formalism, in which classical descriptions of the results of experiments such as Aspect's are radically misleading. Without extremely compelling grounds, we should not deny that descriptions such as "Measurements A and B, which occurred within 5 nanoseconds in the laboratory frame at a distance of 12 meters both had spin up outcomes" reflect corresponding physical facts so located in space-time. We have also set aside doubts raised by Squires which suggest that versions of Aspect's experiment would yield different results if only the apparatus were more widely separated. In short, we are accepting, for the moment, that there are reliable violations of Bell's inequality for measurements both initiated and completed at spacelike separation.

Further, working from the non-relativistic versions of Bohm's theory and GRW, we have set aside interpretations which invoke backwards causation, i.e., causes which lie in the absolute futures of their effects. This mandates some sort of direct casual connection between events at spacelike separation. And in the last section, we suggested that this coordination among spacelike separated events could be secured by the addition to space-time of a foliation by spacelike hypersurfaces. This preferred foliation would constitute a departure from the relativistic account of space-time, but a departure that would entail no conflicts with relativistic predictions. Can a more appealing solution, which is completely relativistic, be formulated?

One route that has been suggested retains the dynamical dependence on spacelike hyperplanes but rejects the postulation of any single preferred foliation. This theory of *hyperplane dependence*, which has been vigorously pursued by Gordon Fleming (1989), claims to retain relativistic space-time structure without emendation. The basic idea is this: standard relativistic space-times contain spacelike hyperplanes. Of course, in standard relativistic theories (e.g., relativistic electromagnetics or in the General Theory) those hyperplanes do no dynamical *work*, in particular, no foliation of space-time plays anything like the role of absolute time in the classical counterparts of those theories. In a sense, the spacelike hyperplanes in these theories have been *disenfranchised*. But the hyperplanes are still *there*, and the objection to a theory that postulates a preferred foliation is that the relativistic metric recognizes no such preferred set. So let us instead enfranchise *all* the foliations, embracing a radical democracy, and thereby demanding no new spatio-temporal structure.

There are really two versions of hyperplane dependence. The Special Relativistic version starts from the facts that Minkowski space-time contains various foliations into *flat* spacelike hyperplanes, and postulates a theory with wave-functions defined with respect to all flat hyperplanes. But the space-times of the General Theory need not contain flat hyperplanes at all, and so seemingly would demand wave-functions defined with respect to all spacelike hyperplanes. Since no General Relativistic version of the theory exists, we will confine our attention to Special Relativity.

Fleming conjoins hyperplane dependence with a (not further specified) collapse theory for the dynamics of he wave function. Collapses are supposed to be "real objective processes" (1992, 109), and presumably, the probabilities one derives from Born's rule are interpreted as probabilities for collapses occurring one way rather than another. As we have seen, it is wave collapse which explains violations of Bell's inequality in standard collapse theories, wave collapse which transmits information between the two wings of the experiment, wave collapse which make the synchronization between distant apparatus essential for the specification of the dynamics of the wave function. Fleming must therefore accept a dynamics for *each* family of hyperplanes which uses that family to play the role of absolute time in the non-relativistic dynamics.

It should be said that Fleming himself does not much focus on *this* role for his hyperplanes. He tends to appeal rather to the hyperplane dependence of quantum *states* or quantum *operators* that one already finds in standard relativistic field theory in order to downplay the radical nature of his innovation. But standard relativistic quantum theory contains no account at all of collapse dynamics (and so is not really an acceptable theory at all), so the analogy is not helpful. If Fleming thinks that collapses are objectively real, then he must take the collapse dynamics seriously, and see how the hyperplanes play a role there. And here there is, I think, a deep and difficult puzzle.

Consider the EPR-Bell experiment starting with electrons in the singlet state, but with the polarizers slightly misaligned, so that the chance of anticorrelated results is, say, 99%. Since Fleming postulates no additional variables, the burden of explaining the anticorrelation of the results falls squarely on the wave collapse. If one electron has been measured then the wave function for the pair has collapsed one way or the other, and the probability for the results of the second measurement have been changed from 50–50 (singlet state) to 99–1 in favor of one result. Let us look at this experiment in space-time, with six hyperplanes indicated (Figure 1). *L* is the spin measurement made on one electron, *R* the spin measurement made on the other. The hyperplanes *a, b* and *c* lie in one foliation, α, β, and γ lie in another.

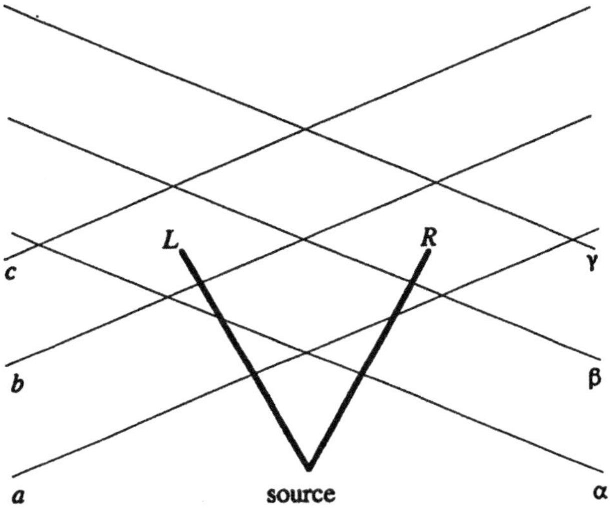

Fig. 1. Pair of electrons in space-time.

If the *L* measurement comes out spin up and the *R* measurement spin down, here is what we can say about the relevant spin wave functions. The wave function of the pair relative to both hyperplane *a* and hyperplane α is the singlet state. The wave function relative to hyperplane *b* is a collapsed state such that the *R* electron has been measured and found spin down and the *L* electron is in an eignestate of spin up in the *R* direction (recall that the *L* and *R* apparatus are slightly misaligned). The wave function relative to hyperplane *c* is collapsed such that both of the electrons have been measured, the *L* electron is spin up in the *L* direction and the *R* electron is spin down in the *R* direction. The wave function relative to hyperplane β is a collapsed state such that the *L* electron has been measured and found spin up and the *R* electron is in an eignestate of spin down in the *L* direction. The wave function relative to hyperplane γ is, in its spin aspect, just like that relative to hyperplane *c*.

Much of Fleming's technical work is devoted to studying how these various wave functions, *defined with respect to given hyperplanes* would be expressed in different frames of reference. But what interests us is how to understand the collapses in this story. There seem to be four distinct wave collapses postulated so far: the collapse that transforms the *a* singlet state into the *b* *R*-spin eigenstate (for both electrons); the collapse that transforms this latter wave function into the *L*-spin eigenstate for the *L* electron and the *R*-spin eigenstate for the *R* electron; the collapse that transforms the α singlet state into an *L*-spin eigenstate; and the collapse that transforms this latter into the γ spin state. Using obvious notation, let us call these the $a - b$, $b - c$, $\alpha - \beta$ and $\beta - \gamma$ collapses.

Some of these collapses are obviously dynamically relevant to others. The singlet state on *a* assigns the probability of 50% to the $a - b$ collapse yielding *R*-spin down. *Once the state has collapsed*, the *new* state on *b* assigns a 99% probability for the $b - c$ collapse to yield *L*-spin up. Thus the nearly perfect anticorrelation between the $a - b$ collapse and the $b - c$ collapse is explained, on this story, by the global, spacelike separated effect of the $a - b$ collapse on the state of the unmeasured *L* electron. Similarly, the distant (from *L*) effects of the $\alpha - \beta$ collapse explains the nearly perfect anticorrelation between the $\alpha - \beta$ and $\beta - \gamma$ collapses.

But there is still an unexplained correlation. The $a - b$ collapse is correlated not only with the $b - c$ collapse, but also with the $\alpha - \beta$ collapse. That is, when the $a - b$ collapse yields a down result on the *R* electron, 99% of the time the $\alpha - \beta$ collapse yields an up result on the *L* electron. What could possibly explain *this*?

Analogies to standard non-relativistic collapse theories give us no help here. In a non-relativistic theory, any pair of collapses is ordered with respect to a single absolute time. Two collapses are correlated because the first collapse determines the wave function which is the initial state for the second collapse, and from which the probabilities of the second collapse are calculated. This mechanism works fine for pairs of collapses which are drawn from the *same* foliation of a relativistic space-time. But pairs associated with *different* foliations pose an entirely new, relativistic problem.

The immediate temptation is to solve the problem this way: if one admits that there is a good explanation for the correlation between the $a - b$ collapse and the $b - c$ collapse, then there must similarly be a good explanation for the correlation between the $a - b$ collapse and the $\alpha - \beta$ collapse since the $b - c$ collapse and the $\alpha - \beta$ collapse *must have the same outcome*. If the $b - c$ collapse yields spin up for the *L* electron so must the $\alpha - \beta$ collapse since both collapses represent one and the same measurement event. But the problem is to understand what, given the ontology of the theory, this last remark could possibly mean.

The natural ontology of Fleming's theory is monistic: the wave function is a complete representation of the physical reality. If Fleming intends to add additional variables to the theory, then we must know just what they are and, especially, how their dynamics are defined in order to be able to assess its compatibility with relativity. (Recall that Bohm's theory, since it contains no collapses, can have a covariant dynamics for the wave function, but not for the particle trajectories.) So there isn't some *extra* item, the measurement, which is *represented* by the various wave functions; rather one must somehow *find* the measurement operation in the wave function and its dynamics. As we have seen, the measurement causes collapse of the wave function. These collapses are governed by a dynamics defined on a family of hyperplanes, and each family of hyperplanes, it seems, operates essentially as an independent universal time for the wave functions associated with them. So how are the collapses in different families coordinated?

One cannot argue that if there is a collapse to, say, a spin-up eigenstate in one family as it passes through region L then there must be a similar collapse in *every* family to maintain *consistency* among the various wave functions which pass through L, because no such consistency is demanded elsewhere. In the region immediately below region L, for example, relative to the $\alpha - \beta - \gamma$ family, the L electron is still in the singlet state, while according to the $a - b - c$ family it is in an up eignestate for spin in the R direction. So why can't it be that *within* the L region while the wave functions on the $a - b - c$ family collapse to a spin down result, the family on the $\alpha - \beta - \gamma$ family collapses to a spin up result? The wave functions on each family of hyperplanes would then completely decouple, yielding an independent world for each foliation.

Consistency conditions relating states defined on different families of hyperplanes can be written down for theories with *non-hyperplane-dependent* local beables, when the global state on a hyperplane is determined by the local beables on the points it passes through. Thus, if I slice up a log using families of cuts a different angles, the cross-sections in each family will differ from those in the other, but the stacks of slices, when reassembled, must yield the same three-dimensional whole. But Fleming's theory doesn't seem to have any non-hyperplane-dependent local beables at all. The wave function on a hyperplane is, as it were, ontologically atomic. Wave functions defined on hyperplanes in the same family can be related by a dynamics which uses the family as a substitute for absolute time, but relations among wave functions from different families are obscure at best. And this remains as true when we are discussing macroscopic events (such as measurement results) as it is when we are concerned with microstates. (*In the absence of collapses*, wave functions on differently oriented hyperplanes may be related by well-defined boost operations, but no such relation connects the wave function on, say, b with that on β, since the rotation from one hyperplane to the other must pass backwards over a collapse.)

From our brief examination of hyperplane dependency, two morals may be drawn. The first is that the non-relativistic mechanisms which secure violations of Bell's inequality offer a clue about where to look for problems in a relativistic theory. If one has no collapses in the non-relativistic theory, then the work must be done in the dynamics of the additional variables, and it is there that problems with Relativity will arise. If one uses a collapse theory, then the dynamics of the collapses must be the key focus: with respect to what dynamical variable are the collapses defined?

The second moral is that fundamental problems of interpretation are always lurking in the background. If a theory has no local beables, then the whole problem of placing the theory into a relativistic space-time is bound to become rather obscure. An instructive example of the careful choice of local beables is provided by Bell in his version of the original GRW theory (Chapter 22 of Bell 1987a). Unlike Fleming's theory, in which every physically real property is hyperplane dependent, Bell proposes a fundamental ontology of events in space-time, localized at the center of the GRW jumps. Bell's theory can be understood (although it is unclear if Bell himself would so understand it) as allowing the demotion of the wave function to a device for fixing probability measures over sets of the fundamental events. The classical analogue of hyperplane dependence (i.e., dependence on the synchronization of separated systems) then washes out of the theory: one gets the same probability measure no matter how distant systems are synchronized (*ibid.*, 208). Bell considers the existence of this *relative time translation invariance* in the classical theory an indication of the possibility of a fully Lorentz invariant relativistic theory. Unfortunately, we do not have the space to take up Bell's interesting suggestion here.

As an illustration of the dangers of launching into investigations of Relativity before having settled basic interpretational problems, let me also mention the decoherence/consistent histories approaches. Robert Griffiths has explicitly claimed that his theory is local (1993, 1601), and Gell-Mann casts aspersions on the very idea that violations of Bell's inequality signal any non-locality at all (1994, 172). But the reason it is hard to get a grip on the relativistic versions of these theories is that the basic ontologies of the theories are not clearly presented.

In both approaches, the authors intend that there be no real collapse of the wave function, and they offer no explicit additional variables beyond the wave function. (Thus: "The consistent history approach is an interpretation of 'standard' quantum mechanics, and not an alternative constructed by changing the laws of motion or introducing additional variables" (Griffiths 1993, 1601).) The principle question that motivates both approaches is the discovery of sets of operators such that, given a wave function, one can assign a probability measure to sets of eigenvalues of the operators. The authors seem to believe that finding such measures solves the basic interpretational problem of quantum theory. That is, they seems to believe that if they can discover such a measure which satisfies the axioms of the probability calculus, they are entitled to interpret it as a probability, viz. the probability that a certain history or set of eigenvalues *occurs*.

But simply obeying the axioms of the probability calculus alone is not sufficient for interpreting a measure as a probability. There are lots of measures defined (e.g., over the surface of my desk) which satisfy the probability calculus under unions and intersections of regions on the desk. But these measures are not thereby *probabilities* since there is nothing (so far as I have said) for them to be probabilities *of*. Similarly, even when the measures found by the decoherentists obey the axioms of the probability calculus, there is nothing yet for them to be probabilities *of*. They cannot, as in GRW, be probabilities of the collapses happening one way rather than another, since there are no collapses. They cannot, as in Bohm's theory, be probabilities of the additional variables taking on one set of values rather than another, since there are no such additional variables. Without further elaboration, there is simply no way for these theories to admit that one "history" rather than another *actually occurs*, and so no way to interpret the nicely behaved numbers as probabilities for such an occurrence.

If one adds additional variables, or adds collapses, then all of the action will be found in their dynamics. But any theory without either one or the other is not a coherent theory, and so further discussion of Relativity is pointless.

6. ADMONITION AND QUERIES

I would like to end with a plea and a series of questions. The plea is motivated by a worrying terminological trend. When Bell's theorem first appeared, it was often characterized as proving the impossibility of local *deterministic* theories or local *hidden variables* theories, implying that locality could be retained if only one abandoned hidden variables or determinism. Bell had to fight long and hard to show that neither determinism nor hidden variables were the issue: in my terminology, Bell's inequality cannot be violated by any local standard theory for events at spacelike separation, that is for any theory that interprets normal laboratory operations as producing measurement outcomes located where we think they are in space and time.

Recently, a new bogeyman seems to have been found: realism. Thus Hardy states: "In 1965 Bell demonstrated that quantum mechanics is not a local realistic theory. He did this by deriving a set of inequalities and then showing that these inequalities are violated by quantum mechanics" (1992, p. 2981). The conversational implication is that Bell's theorem only applies to local *realistic* theories, so that locality (and hence perhaps also consistency with Relativity) can be recovered if one only jettisons realism. But no standard theory, even a non-realistic one (whatever that means) can be local, so we must reject the conversational implication.

Realism in philosophy of science is generally contrasted with instrumentalism or empiricism, which views assert that one can have no grounds to believe that the unobservable ontology of a theory is accurate. In this sense, *theories* are neither realistic nor non-realistic, only interpretations of (or better: attitudes towards) theories. And the strongest argument in favor of instrumentalism, from Osiander onwards, is underdetermination: the existence of many incompatible theories all capable of "saving the phenomena". The beauty of Bell's theorem, of course, is that it is insensitive to the details of the theory suggested: *any* theory which can save the phenomena (if the phenomena include claims about the behavior of macroscopic devices located in space and time) must be non-local. Even a classic instrumentalist would be forced to accept non-locality.

But even supposing that somehow abandoning realism in quantum theory could preserve locality, we would have to wonder about the point of making such a bargain. Physicists have been tremendously resistant to any claims of non-locality, mostly on the assumption (which is not a theorem) that non-locality is inconsistent with Relativity. The calculus seems to be that one ought to be willing to pay *any* price – even the renunciation of pretensions to accurately describe the world – in order to preserve the theory of Relativity. But the only possible view that would make sense of this obsessive attachment to Relativity is a thoroughly realistic one! These physicists seem to be so certain that Relativity is the last word in space-time structure that they are willing even to forego any coherent account of the entities that inhabit space-time.

An instrumentalist would hardly be so sanguine about Relativity. If one doubts our ability to get deep level ontology right, if one insists that the only proper role of theory is to grind out numbers, then it should hardly matter if we add a foliation to space-time in framing a theory, or even reject relativistic space-time structure entirely in favor of Newtonian absolute space and time. Oddly, Bernard d'Espagnat seems to have taken both of these steps: rejecting "ontological" interpretations of quantum theory on account of their non-locality, but then also insisting that we regard Relativity as "weakly objective" (i.e., not as an account of mind independent physical reality) (1995, 320–321). But if the relativistic space-time structure is not objective, why in the world be so concerned with non-locality? It is simply bizarre to man the ramparts in defense of Relativity at the expense of an intelligible account of the material world.

Further, one need not reject the relativistic account of space-time as *accurate*, just as *complete*. If one is certain that the relativistic account of the space-time metric reflects an objective feature of the world, that belief can be retained. As noted above, one can construct a Bohmian theory in which the wave function evolves relativistically, thus accounting for all relativistic effects. The additional foliation, needed for the dynamics of the local beables, is something that the theory of Relativity simply overlooked, as irrelevant to the phenomena it studied.

In conclusion, I would like to pose a series of question which anyone attempting to grapple with these issues must address. My hope is that this might serve to focus the debate.

Question 1: How concerned should we be about closing experimental loopholes, either those based on detector inefficiencies (Fine) or those based on not having the measurements completed at spacelike separation (Squires)? Comment: If the loophole cannot be closed, then we can envision a local standard theory. My own guts say this is unlikely.

Question 2: If we agree to a standard interpretation, and that there will be violations of Bell's inequality at spacelike separation, should we suspect backwards causation or superluminal influence?

Comment: Based on analogy to the non-relativistic theories, I would suspect superluminal influence.

Question 3: Is there a better chance of producing a stochastic theory with superluminal influence (or backwards causation) which is consistent with Relativity than a deterministic theory?

Comment: Although it somehow seems plausible that a stochastic theory will be easier, I haven't found any concrete reason to believe this.

Question 4: Would it ever be rational to abandon a standard interpretation in favor of a non-standard one simply to preserve locality or Relativity?

Comment: Despite much sentiment to the contrary, I do not think this could be rational.

Question 5: If we have to abandon Relativity as a complete account of space-time structure, should we have a preference for a theory which *adds* more structure (e.g., a foliation) as opposed to one that completely replaces relativity (e.g., uses Newtonian Absolute Space and Time)?

Comment: On account of he relative ease of accounting for relativistic phenomena, the strategy of adding structure seems more promising.

Question 6: The wave function is not a classical field defined on space-time, but is rather defined on the much higher dimensional (and more abstract) configuration space. How should that affect our attempts to provide a dynamics for it?

Comment: In a dualistic theory such as Bohm's, the possibility of separating the complete dynamics of the wave function from those of the beables makes this an especially interesting question. Oddly, as mentioned above, it is the dynamics of the wave function, not of the beables, that seems to be relativistic, and hence suited to describing objects in relativistic space-time.

Question 7: Is it premature, and counterproductive, even to begin discussing how to Relativize quantum theory before exploring how to quantize Relativity?

Comment: Good question.

Rutgers University

NOTES

[1] See also Lockwood (1989) and Squires (1990b) for similar views.

[2] The story about lab reports on the many-minds view is really more complicated than this: the ink on pages of journals ends up in a superposition of different states which report different results: a particular "result", though, is experienced in the mind(s) of each reader.

[3] But see the question raised by Squires, discussed below.

[4] An additional flaw in this argument is that it does not provide an instance of the Outcome Counterfactual even for the dollar bill case, since to get different outcomes there one must have a different initial state (i.e., the halves have to have been put in different envelopes).

[5] For a proposal to derive a foliation from a four-vector field on the space-time, see Dürr *et al.* (1990).

[6] For comparison, see how the value of the synchronization parameter in the two-time Bohmian mechanics of Berndl, Dürr, Goldstein and Zanghì (1996) effects the particle trajectories (see especially their Figure 4). The setting of the synchronization parameter corresponds to the foliation of space-time we are considering.

HARVEY R. BROWN, ANDREW ELBY AND ROBERT WEINGARD

CAUSE AND EFFECT IN THE PILOT-WAVE INTERPRETATION OF QUANTUM MECHANICS

1. INTRODUCTION

In the de Broglie–Bohm pilot-wave interpretation of quantum mechanics (QM), the two degrees of freedom associated with a system are its wave function and configuration (i.e., point in configuration space). From a purely dynamical point of view, neither has priority over the other, and it is tempting to treat the interpretation as having a genuinely dualistic ontology – involving both 'ψ-fields' and corpuscles. (We shall use the term 'corpuscle' to designate the hypothetical de Broglie–Bohm point particle, the words 'system' and 'particle' meaning what they do in standard QM.) In so far as the pilot-wave interpretation provides a 'causal' picture, it is in terms of the action of the ψ-field on its associated corpuscle(s). How is this action to be understood? In his admirable 1993 survey of the pilot-wave interpretation, Holland writes that

> . . . the manner in which the ψ-field couples to the particle [i.e., corpuscle] via a quantum force does not on the face of it involve any concepts of cause and effect beyond those of classical physics. It is not essentially different from the action of an electromagnetic field on a charged particle, or a gravitational field on a massive particle (although there is no obvious 'quantum charge' associated with the particle which responds to the quantum potential . . .). (Holland 1993, 83)

It is the main purpose of this brief paper to question Holland's version of the causal story, and to examine what non-classical notions of force are available in its stead. These issues are examined in Sections 5 and 6; several preliminary topics need to be dealt with first.

2. PRELIMINARY REMARKS ON PILOT-WAVE MECHANICS

(i) Before we proceed to discuss the nature of the causal action of the de Broglie–Bohm 'field', it is perhaps worth briefly addressing the question: What entity in the formalism best represents this 'field' structure?

Most commentators regard the 'ψ-field' as the entity in question. Neither the R- nor S-field is of course fundamental (where $\psi = R\exp(iS/\hbar)$) since they are coupled *via* the Schrödinger time-dependent equation. Furthermore, it has been stressed particularly by Holland (1993, §3.2.2) that even with the assumption of the single-valuedness of the ψ-field, the S-field is many-valued. But it is worth recalling in this connection that the single-valuedness of ψ is itself (at least in non-relativistic quantum mechanics) relative to the choice of a spatio-temporal coordinate system and then, when appropriate, of an electromagnetic gauge – and the same goes for the reduced

J. T. Cushing et al. (eds.), Bohmian Mechanics and Quantum Theory: An Appraisal, 309–319.

density operator for the system in question when the system is more realistically treated as a component of a larger system. The supposed objective reality behind the ψ-field must then be considered as represented by an equivalence class of wave functions, elements of which being related by combinations of gauge and coordinate transformations. We are already familiar with a somewhat analogous situation in modern physics: spacetime structure in general relativity is represented by an equivalence class of metric fields, as a result of the general covariance of Einstein's field equations. Indeed the connection between the diffeomorphism 'gauge' freedom in general relativity and its analogue for the projective Hilbert space in quantum mechanics has recently been discussed by Anandan (1993, 514).

(ii) As it is currently presented, the deterministic *corpuscle* mechanics in the pilot-wave theory either emerges out of a Hamilton–Jacobi-type equation that itself follows from the Schrödinger time-dependent equation (Bohm 1952a,b; Bohm and Hiley 1993; Holland 1993) or is given directly by a first-order 'guidance' equation involving the velocities of the hypothetical point corpuscles. (For a clear exposition of the latter approach, see the 'Bohmian mechanics' developed by Dürr *et al.* (1992a; 1992b.) This guidance equation takes the well-known form (in the case of a single body)

$$\text{(1)} \qquad \mathbf{v} = (\hbar/m)\,\mathrm{Im}\,\frac{\nabla\psi}{\psi} = (1/m)\nabla S$$

$\mathbf{v}$ being the velocity of the corpuscle, S the phase of the wave function and m the mass of the system. It is noteworthy that for Bohm in 1952, the precise form of equation (1) was 'not inherent' in the conceptual structure of the new interpretation (Bohm 1952a, §4). Nonetheless, attempts have been made to arrive at (1) from considerations related to spacetime symmetries (see, e.g., Dürr *et al.* (1992a)). Such attempts strike us as having doubtful validity.

In the first place, it is questionable whether spacetime symmetries such as spatial isotropy and Galilean covariance should be granted anything more than phenomenological status (see, e.g., Brown and Sypel (1995)). It is by no means clear therefore that such symmetries have any bearing on an equation such as (1), or its many-body counterpart, which govern the behaviour of 'hidden' processes. Equally, its is far from obvious whether, in the relativistic version of the theory, violation of Lorentz covariance by the relevant guidance equations for the general, many-body case should be regarded as inconsistent with the relativity principle. Some skeptical readers will no doubt recall at this point Einstein's 1905 criticism of the Maxwell–Lorentz treatment of electromagnetic induction, which despite proceeding from Lorentz-covariant field equations, led to explanatory 'asymmetries which do not appear to be inherent in the phenomena' (Einstein 1905). But Einstein, in the electrodynamic part of his 1905 paper, showed how to re-interpret the field quantities so as to remove the asymmetries, as part of his demonstration that the Maxwell–Lorentz ether is 'redundant'. And the analogy simply fails for the relativistic pilot wave theory: the 'hidden' de Broglie–Bohm 'ether' is not redundant. (We shall return to this issue in Section 6 below.) The relativity principle is, nonetheless, not threatened – so long as the standard probability distribution $P = |\psi|^2$ holds.[1] Returning to the non-relativistic case, we cannot agree, in conclusion, with the claim that "covariance determines the form of [the v-field]" (Dürr *et al.* 1992b, 8).

In the second place, it is worth recalling that in the presence of an electromagnetic field (a case not discussed in Dürr *et al.* (1992a,b)) the non-relativistic guidance condition for a charged particle takes the form (see, e.g., Holland 1993, §3.11.2)

(2) $$\mathbf{v} = (1/m)\nabla S - (q/mc)\mathbf{A},$$

where $\mathbf{A}$ is the electromagnetic vector potential, and q is the charge of the system. The $\mathbf{A}$-field may be regarded as a 'compensating field' which renders $\mathbf{v}$ invariant under local gauge transformations. (One sees clearly here that the gauge freedom requires a many-valuedness in S, quite apart from that stressed by Holland (1993, §3.2.2).) But if this method of 'deriving' (2) from (1) is compelling – and note that the status of the gauge freedom is quite different from that of spacetime symmetries – it is only so in the case of the single body. In the many-body case, complications arise owing to the fact that ψ and hence the phase S are functions on configuration space: the compensating field is now not the usual vector field defined on three-dimensional space but a higher-dimensional object.[2]

One putative advantage of the Hamilton–Jacobi approach on the other hand is to motivate (1) and (2) *via* the standard classical connection between momentum, the gradient of the Hamilton–Jacobi S-function and the electromagnetic vector potential. But this requires that the inertial mass of the system is a property of (at least) the corpuscle, a claim we discuss in the next two sections.

3. THE PRINCIPLE OF GENEROSITY

It has recently been argued that the dynamical state-independent parameters associated with mass, charge and magnetic moment cannot be considered as properties which are instantaneously entirely localised at the position of the de Broglie–Bohm corpuscle.

Holland (1993, §3.3.1) has made the simple point that the ψ-field itself depends, as a solution of the time-dependent Schrödinger equation, on these parameters – his conclusion being (in part) that "ψ may be said to be 'massive' (and 'charged' etc.)". This argument has been given an arguably more graphic form by way of an examination of certain specific quantum phenomena (Brown, Dewdney and Horton 1995). Here, the non-localisability argument concerning inertial mass, magnetic moment and the *passive* versions of both gravitational mass and charge – including colour charge in any future pilot wave theory of the strong interactions – is based on certain types of interferometry experiment. The case specifically concerning passive charge involves interference phenomena related to the Aharonov–Bohm (AB) effect, which is mentioned again later. (Note that it is not crucial in the present argument that the classical Lorentz force vanishes on the possible paths of the corpuscle.) The other cases involve a single particle prepared in a coherent superposition of two essentially non-overlapping wavepackets, with a spatially-dependent potential of some kind acting differently on the two wavepackets (one of which of course is 'empty') before they are made to interfere. (This latter situation is also mentioned in Holland (1993, §3.3.9).) Finally, an analogous argument dealing with the *active* versions of both charge and gravitational mass has also appeared, this time having to do with certain types of idealised 'protective measurement' (Anandan and Brown 1995).

All these arguments are essentially intuitive and cannot be considered definitive. We nonetheless take them to be persuasive, and the non-localisability thesis to be true – thus ruling out the possibility that such state-independent parameters are associated with the corpuscle alone. The arguments do not however strictly entail the *parsimonious* view that these parameters are to be associated solely with the ψ-field. (We take it this view is implied by Bell's suggestion that the de Broglie–Bohm corpuscles 'are *not* attributed angular momenta, energies, etc., but *only* positions as functions of time' (Bell 1990, 30). The parsimonious view is at any rate implicit in Brown (1996).) It is possible also to adopt the *principle of generosity* whereby the mentioned parameters are properties attributed to *both* the ψ-field and the corpuscle. This is the stance explicitly adopted by Holland (1993, §3.3.1). What grounds are there for adopting this stance?

It is surely a noteworthy feature of the 'Bohmian mechanics' espoused by Dürr *et al.* (1992a,b) that it can reproduce all the testable predictions of QM – and hence any other versions of the pilot-wave interpretation – without commitment to the principle of generosity. Holland has argued on the other hand that without assigning "energy, angular momentum etc., in addition to a location" to the corpuscles, serious problems arise in the classical limit of quantum mechanics (Holland 1993, §3.2.5). How otherwise are 'classical' bodies to possess such properties? Assuming that they do, it would seem to follow that at least inertial mass is a property of the corpuscle. (Presumably an analogous argument holds for charge.)

It is interesting that Dürr *et al.* (1992b, 8) also accept that the Newtonian concepts of force and energy etc. 'are fundamental to the theory to which Bohmian mechanics converges in the "classical limit", namely Newtonian mechanics', and describe these concepts as 'emergent notions'. But this view seems either to lead to obscurity (at what point in the approach to the classical limit do force and energy become fundamental?) or to essentially concede Holland's point. In our view, the correct response to Holland's argument is to deny the necessity of recovering all the familiar classical ontology when taking the classical limit of QM. Indeed, it would appear to be the case that all the *testable consequences* of the pilot-wave theory in the classical limit are indifferent to the validity or otherwise of the principle of generosity. Is it really necessary in this case to secure anything more than the classical phenomenology in the limit?

Note that the relevance of this question depends on precisely what the technical condition is taken to be in the pilot-wave interpretation for the classical limit. Holland (1993, Chapter 6) accepts as this condition that the 'quantum potential' Q (see Section 5 below) tends to zero. Now it is possible to achieve this condition in the case of a single particle, say, without the ψ-field becoming spatially localised, and without the mass of the system becoming large. It is questionable whether this is consistent with true classical behaviour (see, e.g., Valentini 1992, 21–22), but space permits no further discussion of this issue. Suffice it to say that if stable, sharp localisability of wave packets is also taken to be a necessary condition for classical behaviour, Holland's argument for the principle of generosity again loses force.

What then are the reasons for adopting the principle of generosity? We recognise essentially two, neither of which can be considered definitive.

Consider first the case of the many-body system involving a collection of N distinguishable particles. In this case, there will be attributed to the ψ-field of the composite system, which will generally be entangled (non-factorisable), a *set* of distinct masses, charges (if any), etc. Yet the guidance equation for the kth particle ($k = 1, \ldots, N$)

$$(3) \qquad \mathbf{v}_k = (\hbar/m_k)\,\mathrm{Im}\,\frac{\nabla_k\psi}{\psi} - (q_k/m_k c)\boldsymbol{A},$$

refers specifically to the inertial mass m_k and charge q_k of that particle. Is it not then natural to attribute at least these definite values to the corresponding corpuscle? (Note that this argument is not available in the relativistic field-theoretic version of the pilot-wave theory, where neither mass nor charge appear in the guidance equation of the field; see, e.g., Valentini (1996a).)

The second reason is essentially negative: one avoids the problems associated with the principle of parsimony, to which we now turn.

4. THE PRINCIPLE OF PARSIMONY

According to this principle, as we have seen, the state-independent parameters such as mass are attributes not of the corpuscle but of the ψ-field alone. It would seem to follow that the corpuscle is nothing more than a trajectory painted in space.

Consider the case of an isolated pari of particles consisting of a proton and a neutron, say. Suppose that the composite ψ is factorisable at some instant, and that the factor wave functions, each of which is defined on three-dimensional space, have significant spatial overlap. Suppose further that the proton and neutron corpuscles are located at that instant within the region of overlap. In the light of the principle of parsimony, how is one to explain the ability of the neutron ψ-field, say, *at that instant* to know which of the two corpuscles – these differing only in their locations – it is supposed to act on? We shall call this the *problem of recognition.* To solve this problem, one must suppose that at the instant in question *either* each corpuscle carried a distinct label which can be read by the associated ψ-field, or the corpuscles are recognised through (a sufficiently large part of) their history. This first option is tantamount to accepting that the pilot wave interpretation is incomplete, since no such labels are specified in the theory for the corpuscles if the principle of parsimony is valid. (The fact that the wave function of the joint system is defined on configuration space does nothing, in our view, to remove this incompleteness.) The second option amounts to an admission that the interpretation is *non-local in time as well as space.*

Consider further the case where the spatial paths of corpuscles associated with different (not necessarily distinguishable) systems cross at a certain instant. It would seem that the corpuscle number associated with these systems instaneously collapses to unity. If corpuscle number (in the non-relativistic theory) is to mimic particle number conservation, then the corpuscles must be again identified with trajectories. However, ideally, there will be special situations where for an isolated pair of identical particles even the trajectories may coincide. The pair of particles is now associated with a single corpuscle – since that is now what 'two' corpuscles that possess the same trajectory are. Note that this can even happen for pairs of identical *charged* particles, although in this case the 'coincident' corpuscular behaviour is singular. Nonetheless, the fact that it can happen in principle despite Coulomb repulsion is perhaps less surprising, given the principle of parsimony, than it would be otherwise.

5. NEWTONIAN FORCES

An important element of the Hamilton–Jacobi approach to the pilot-wave interpretation is the appearance of Newton's second law of motion for the corpuscle

$$(4) \qquad m\,\frac{d\mathbf{v}}{dt} = -\nabla(Q + V)$$

where Q is the 'quantum potential' $(-(\hbar^2\nabla^2 R)/2mR)$ and V is the 'classical potential', which is normally taken to be a scalar function of the coordinates. (It is noteworthy again that Bohm appears to have originally viewed equation (3) as more fundamental than the Hamilton–Jacobi equation, the solution of the latter being 'only a matter of convenience'; Bohm 1952a,

§4.) In so far as the *Newtonian notion* of force is involved in (4) – even if part of its provenance is non-classical – it is clear that appeal must be made to the principle of generosity, at least in relation to inertial mass.

Recalling the comments of Holland cited in the introductory section, it is useful to see just how close conceptually the causal role of the quantum potential really is to that of the electromagnetic field. Although Holland is considering this field in the context of classical mechanics, we shall start our discussion, for the sake of completeness, with the case of the *unquantised* field in quantum mechanics. Recall first that the Lorentz force is not actually catered for in (4) even in the classical limit. Although there exists in the 'analytical approach' to classical mechanics a scalar 'work function' from which the Lorentz force on a charged particle can be derived (Lanczos 1986, 30–31), and which is of course a function also of velocity, it is generally not a potential energy and most importantly the Lorentz force cannot be obtained by taking its gradient. At any rate, one simply adds to the r.h.s. of (4) the 'classical Lorentz force' **F** (see for instance Holland 1993, §3.11.2)

(5) $$\mathbf{F} = q[\mathbf{E} + (1/c)\mathbf{v} \times \mathbf{B}],$$

where **E** is the electric field strength and **B** the magnetic field strength $\mathbf{B} = \nabla \times \mathbf{A}$.

Several prominent authors (Bohm and Hiley 1993, 52; Holland 1993, §3.11.2) treat $-\nabla Q$ and **F** on an equal footing – neither force is more nor less real than the other. Furthermore, the Lorentz force is apparently given its full classical interpretation, in which case the charge of the system, as well as its inertial mass, must be attributed to (at least) the corpuscle. The mentioned authors strengthen this impression of treating $-\nabla Q$ as a real force in their similar analyses of the (magnetic, vector) Aharonov–Bohm effect (Bohm and Hiley 1993, §3.8; Holland 1993, §§3.3.4, 5.2). This striking phenomenon involves an electromagnetic interference effect despite the fact that **F** vanishes on all the possible paths of the corpuscle; the shift in the interference fringes is now *explained* in the pilot wave interpretation in terms of the remaining action of the 'quantum force'.

Note that we are dealing here, to repeat, with an *unquantised* electromagnetic field, but even when the so-called 'classical' Lorentz force **F** vanishes, the field (or rather the vector potential **A**) still affects the corpuscle *via* its effect on the evolution of the ψ-field, and hence the 'quantum' potential Q. This is of course a feature of the 'classical' scalar potential V too, even when it has no connection with electromagnetism. The 'classical' force $-\nabla V$ is, again, not its sole manifestation – the presence of V is also felt *via* Q and hence the 'quantum' force (Holland 1993, §3.3.4). Indeed, hypothetically one can also have a case, somewhat akin to the AB effect, of the classical force $-\nabla V$ vanishing, but V still having a non-trivial effect on Q (as when V is time-dependent but spatially constant at every instant).

In his seminal 1952 paper, Bohm had already stressed two significant differences between the quantum potential and the electromagnetic field. The first of these (which he surprisingly called the 'only fundamental difference'; Bohm 1952a, §I.4) is the restriction of the *initial* corpuscle momentum to $\mathbf{p} = \nabla S$. Note here that Bohm appears to be comparing the quantum case where there is no electromagnetic field with the classical case involving electromagnetic interaction. The second difference is a result of the homogeneity of the Schrödinger equation: the ψ-field is neither radiated nor absorbed. It is interesting that Bohm raised the possibility of adding to the Schrödinger equation an inhomogeneous term which has significant effects only at very small distances (Bohm 1952a, §I.9). In this domain, the ψ-field would depend significantly on the position of the corpuscle, in analogy with the electromagnetic field's dependence on the distribution of charge – and the action-reaction principle would be restored in the theory. The violation of this principle in the standard forms of the pilot-wave theory is regarded by some as a serious defect (see for instance Anandan and Brown (1995) and further references therein). It certainly represents an important difference with respect to the case of the electromagnetic field.

A related difference is the property of the quantum potential Q, that Holland himself highlights and which he calls its failure to remain 'aloof' (Holland 1993, §3.4.2) from the processes it influences. He notes that unlike the classical potential V which is a pre-assigned function of the coordinates (and presumably also unlike the electromagnetic field which is well-defined in the absence of the probe particle even in the quantum context) the quantum potential Q depends on the 'total quantum state' of the system, presumably insofar as it depends on the ψ-field. We believe Holland's intuition is sound here, although we have doubts about the argument. After all, one surely expects Q to depend on the ψ-field; what is unexpected rather is *the nature of the dependence of* Q *on the system's characteristic parameters such as mass.*

To see this, consider again the case we mentioned in the previous section of a free pair of particles comprising a proton and a neutron, say, which is assigned a pure ψ-state at some instant. The quantum potential responsible for the quantum force acting on the proton, say, is a function of the inertial mass of the proton; the mass of the neutron is irrelevant here. Suppose again that the composite ψ-field is factorisable at some instant. Then even when the two corpuscles are situated in the region of overlap of the two factor ψ-fields at that instant, each corpuscle is oblivious to the other corpuscle's field. It makes no sense to talk of a quantum potential per unit mass generated by a ψ-field (as it does classically in the case, say, of the gravitational 'field') which acts on any corpuscle whatever its mass, and furthermore exists wherever it does exist whether the corpuscle is there or not. (A similar argument can be made in relation to charge in the case of the electromagnetic interaction when the Lorentz force vanishes.)

However much it would appear within the dualist, 'causal' version of the pilot-wave interpretation that the corpuscle and ψ-field are distinct entities, they are dynamically and ontologically inseparable: one does not exist without the other. This is reflected in the fact that there are two degrees of freedom in the interpretation, which are precisely ψ and the configuration.

6. NON-CLASSICAL FORCES

But the fact that the velocity of the corpuscle is not a degree of freedom in the pilot wave interpretation strongly suggests that acceleration (and hence Newtonian force) should not feature in the fundamental form of the equations of motion. And this is so if these equations are taken to be the time-dependent Schrödinger equation and equation (1)/(2). As Dürr *et al.* (1992b, 8) and Valentini (1992; 4, 34) have correctly argued, in the light of the first-order nature of these equations (in time), the second-order concepts of acceleration and force, work and energy are no longer fundamental. Accordingly, "Bohm's 'quantum potential' is, like the 'luminiferous ether', a concept best abandoned" (Valentini 1996a).

Valentini has called the forces, or causes, involved in equation (1)/(2) 'Aristotelian' (Valentini 1992, 12–17; 1996a), as they give rise to a velocity. (In their 1992a, Dürr *et al.* had also referred to Bohmian mechanics as a 'first-order (Aristotelian)' theory.) What then is the state of *natural* or *unforced* motion? If such a concept is still apposite (and to argue that *no* state of natural motion exists is to diverge greatly from Aristotelian thinking, and not just Newtonian – indeed it is to question the very notion of forces), it surely corresponds to the state of rest. Quite apart from the rather pedantic point that this picture is still not entirely in the spirit of Aristotelian terrestrial dynamics (in which some bodies are naturally fated to fall), it's reconciliation with the non-relativistic from of the pilot wave theory is doubtful. For it is not obvious how the existence of such 'Aristotelian' forces can be a frame-independent matter, in the light of equation (1)/(2). Recall that this equation is Galilean covariant (even in its many-body version) and fails to pick out a privileged inertial frame with respect to which one can define a 'true' state of rest. It is noteworthy that this problem vanishes in the case of the relativistic version of the pilot-wave interpretation, where the (standard) many-body guidance equations of the hidden fields or corpuscles – as opposed to the equation governing the motion of the ψ-field – fail to be Lorentz-covariant. We suggested above in Section 2 that insofar as corpuscle trajectories in the pilot-wave interpretation are *hidden* variables, such non-covariance is perfectly consistent with the relativity principle. Our point now is that such non-covariance, and the consequent existence of a privileged reference frame, is arguably also necessary for making sense of the kind of objective 'Aristotelian' forces that Valentini attributes to the pilot-wave interpretation.

Two final remarks follow regarding Valentini's interesting construal of cause and effect in the quantum world. First, it (like the more questionable Hamilton–Jacobi approach to the pilot wave interpretation) appears to presuppose the validity of the principle of generosity, at least as regards inertial mass and charge. Secondly, it also shares the striking feature mentioned above in connection with the lack of 'aloofness' of the ψ-field. It is clearly irrelevant to the issue concerning the inability of the ψ-field to exist and act independently of its associated corpuscle(s), whether the ψ-field produces forces of a 'Newtonian' or an 'Aristotelian' type. This point has been essentially recognised by Valentini, who reminds us that the result of introducing a test particle to probe the ψ-field is to change the dimensionality of the configuration space on which the ψ-field is defined (Valentini 1996a).

7. CONCLUDING REMARKS

Advocates of the pilot-wave interpretation of QM who appeal to the existence of forces, whether the forces be 'Newtonian' or 'Aristotelian' – and the latter seem more natural – appear to be committed to the *principle of generosity* in regard to the possession of dynamical properties like mass and charge. A fuller discussion than can be given here is needed to establish the full implications and degree of validity of this principle within the pilot wave interpretation. Some of the difficulties inherent in the rival *principle of parsimony* have however been indicated above. It is already clear, at any rate, that the notion of cause and effect, and in particular the role of the ψ-field in generating such 'forces', is in certain *essential* respects *unlike* anything in pre-quantum physics.

ACKNOWLEDGEMENTS

We wish to thank J. Anandan, J. Cushing, O. Johns, T. Maudlin, and S. Weinstein for useful discussions. We are particularly grateful to S. Goldstein and A. Valentini for valuable, detailed comments on an earlier draft of this paper; any remaining errors are of course ours alone. Finally, we wish to thank the editors of this volume for the opportunity to contribute to it.

University of Oxford

NOTES

[1] Valentini (1991b) has given a new and detailed version of the idea in the pilot-wave interpretation that this distribution corresponds to an equilibrium state. In Valentini's scheme, past regimes of non-equilibrium allowed for superluminal signalling, together with true

violation of the relativity principle. According to this picture, the standard structure of Minkowski spacetime emerges only in the near-equilibrium regime. The attractive feature of this picture is that the conspiratorial (equilibrium) aspects of the pilot-wave interpretation – the suppression of superluminal signalling and of any direct evidence of corpuscle trajectories generally – now cease to be fundamental.

[2] We owe this point regarding the many-body case to A. Valentini; further details are found in Valentini (1996b).

MICHAEL DICKSON

IS THE BOHM THEORY LOCAL?

1. THE LAPLACIAN IDEAL

Technicalities aside, Newtonian mechanics is a fully deterministic theory. This fact prompted Laplace's vision of a world whose evolution could be predicted with complete accuracy by a sufficiently powerful demon using only initial conditions and Newton's laws. Forgoing the demon: in a Newtonian world, initial conditions fix *everything*.[1]

Laplace was right to be struck by this determinism, for it is not independent of many familiar physical concepts. The notion of force, or causality, or potential energy, for example, might take a new meaning – or lose meaning altogether – in a fully deterministic world.

The Bohm theory too is fully deterministic. In a Bohmian world, initial conditions fix all matters of physical fact. Again, such determinism may force us to rethink some physical concepts. In particular, familiar questions about locality begin to look unfamiliar in a completely deterministic context. Indeed, it is not even clear whether completely deterministic theories admit an interesting notion of locality at all.

This paper is an attempt to indicate some of the difficult questions that face any evaluation of locality in a deterministic theory, and therefore in the Bohm theory, and to suggest that it is at least not obvious that there is room in a Bohmian world for any interesting notion of locality. The next section gives a formulation of a locality condition in terms of probabilities. Section 3 is a discussion of this condition in the Bohm theory. Section 4 is a discussion of non-local causation, and of whether it happens in a Bohmian world. Section 5 raises the question implicit throughout this paper, namely, whether any interesting locality condition will make sense in the context of the Bohm theory, or indeed in the context of any fully deterministic theory.

2. LOCALITY

Let us consider locality conditions in the EPR-Bohm experiment, in which two particles, α and β, leave a source at a time t_0; α's spin is measured in the i-direction at time t_α, and β's spin is measured in the j-direction at time t_β. (These measurements are assumed to be instantaneous: the measurement-event and the outcome-event are the same event.) If the result of the measurement on α is $+\hbar/2$, then we say that A (an event in a probability space) has occurred, and likewise for B. This paper assumes familiarity with the quantum mechanics of the EPR-Bohm experiment.

J. T. Cushing et al. (eds.), Bohmian Mechanics and Quantum Theory: An Appraisal, 321–330.

A *model* of the EPR-Bohm experiment specifies for each time, t ($t_0 \leq t \leq t_\alpha, t_\beta$) a set of possible complete states, $\{\lambda(t)\}$, of the pair of particles, a probability density, $\rho[\lambda(t)]$, defined over (the σ-field generated by) this set, and a set (indexed by i, j, and $\lambda(t)$) of probability measures, both marginal (denoted '$p_{ij}^{\lambda(t)}[\cdot]$') and joint (denoted '$p_{ij}^{\lambda(t)}[\cdot,\cdot]$'), over the (smallest Boolean algebra containing the) events A, $\bar{A}$, B, and $\bar{B}$, where $\bar{A}$ occurs if and only if A does not occur (given that α's spin is measured in the i-direction at time t_α) and likewise for $\bar{B}$ and B. To say that $p_{ij}^{\lambda(t)}[A] = c$, for example, is to say that *if* α's spin is measured in the i-direction at time t_α and β's spin is measured in the j-direction at time t_β and the complete state of the pair at time t is $\lambda(t)$, *then* the probability that the result will be $+\hbar/2$ is c.

There are at least three ways for a model to be deterministic. First, $\lambda(t)$ could be a deterministic function of time (rather than a stochastic process)[2] Second, both '$p_{ij}^{\lambda(t)}[A] = 0$ or 1' and '$p_{ij}^{\lambda(t)}[B] = 0$ or 1' could be true for all $\lambda(t)$, t, i, j, t_α, and t_β. Third, both '$p_{ij}^{\lambda(t_\alpha)}[A] = 0$ or 1' and '$p_{ij}^{\lambda(t_\beta)}[B] = 0$ or 1' could be true for all $\lambda(t_\alpha)$, $\lambda(t_\beta)$, i, j, t_α, and t_β. (This third way is really a condition on the complete states; it is entailed by the condition that the complete state of any particle include its spin in the direction measured, once that measurement is actually made. This condition holds in standard quantum mechanics with the projection postulate, and indeed in most interpretations of quantum mechanics.) All three of these conditions hold in the Bohm theory.

Now consider locality. Assuming the measurement-events to be space-like separated, one might naïvely suppose (on the basis of special relativity) that they are probabilistically independent, given the initial state, $\lambda(t_0)$. That is, one might naïvely suppose that a local model will contain probability measures $p_i^{\lambda(t)}[\cdot]$ and $p_j^{\lambda(t)}[\cdot]$ such that

$$(1) \qquad p_{ij}^{\lambda(t_0)}[A|B] = p_i^{\lambda(t_0)}[A] \text{ and } p_{ij}^{\lambda(t_0)}[B|A] = p_j^{\lambda(t_0)}[B].$$

However, (1) is clearly a poor locality condition. For example, the occurrence of B might bear on how $\lambda(t)$ evolved between t_0 and t_α, and thereby on the probability for A. Hence, unless the particles left the source at the speed of light, there need not be any non-locality in the violation of (1).

The problem with (1) arises from the use of $\lambda(t_0)$ to calculate probabilities for events at the later times t_α and t_β. One way to fix (1) would be to use $\lambda(t_\alpha)$ to calculate probabilities for α and likewise for $\lambda(t_\beta)$ and β, but this solution ignores the fully dynamical character of the models being considered here. Better is to follow Butterfield (1989) and others (e.g., Bell 1975) by supposing that the complete state of the pair of particles is the union of the backwards light cones of the measurement-events (not including the measurement-events themselves.) In a local theory, one would expect that this state is divisible into a state on each of the backwards light cones. Call these states λ_α and λ_β. Then a fully local theory should contain probability measures $p_i^{\lambda_\alpha}[\cdot]$ and $p_j^{\lambda_\beta}[\cdot]$ obeying the following condition

(2) $p_{ij}^{\lambda}[A|B] = p_i^{\lambda_\alpha}[A]$ and $p_{ij}^{\lambda}[B|A] = p_j^{\lambda_\beta}[B]$.

Equation (2) is a 'screening-off' condition: events in λ_β are screened off from the (space-like separated) event A by events in λ_α.[3]

3. LOCALITY AND THE BOHM THEORY

For the EPR-Bohm experiment, Bell (1981a) has derived an explicit expression for the trajectories of two particles in an entangled state.[4] It is clear from Bell's expression that even when the particles are space-like separated, the instantaneous velocity of one depends on the position of the other. It seems to follow that (2) fails. Because the velocity of α depends on the position of β, λ_β cannot be neglected in calculating the probability of A.

For example, suppose that j is determined by whether a certain atom has decayed. Suppose further that this atom is in the backwards light cone of the measurement for β, but not for α. It is easy to show that in the Bohm theory, the motion of β is partly determined by whether the atom has decayed or not (through its dependence on j). But the motion of α is partly determined by the position of β, and therefore by whether the atom has decayed or not. In this case, although in the Bohm theory the time of decay for the atom is *determined*, that information is not contained in λ_α, and therefore λ_α is not sufficient to fix the evolution of α, which depends on whether the atom decays.

But does the Bohm theory *really* entail the failure of (2)? If the universe were created at t_0 with just the particles, apparatuses and decaying atom, then there is no denying that (2) fails. But let us begin to make the scenario more realistic by supposing that, rather than having been created at t_0, the apparatuses and particles, including the unstable atom, have been around for a long time. In that case, it may be that λ_α *does* contain the state of the atom at some time prior to t_0, and therefore *is* sufficient to determine whether the atom decays, and therefore *is* sufficient to determine the motion of α. If so, then (2) holds.

But suppose now that the polarization of some photon determines what measurement will be made on β. In that case, no matter how far back in time one goes, λ_α might not contain enough information to determine the polarization of the photon; hence λ_α will not contain enough information to determine what the measurement on β will be. (This case is essentially the one envisaged by Maudlin (1994, 134–135).) Is this case a decisive proof of non-locality in the Bohm theory?

On the one hand, there is no denying that the Bohm theory itself allows such a universe. But is our universe one of them? The prevailing opinion in favor of big bang theories seems to say 'no'. In big bang theories, the initial state of the universe is in the backwards light cone of every point of space-time. *A fortiori*, λ_α contains enough information to determine the trajectory of β. Hence in our universe, it looks as though (2) holds.

What conclusion should we draw? Does the Bohm theory allow both local and non-local models? Perhaps, but is it not strange that a condition (namely, locality) that is apparently about *theories* turns out to depend on which model of a theory one chooses? Does the fact that (2) depends on the details of a given universe not hint at the inadequacy of (2) to capture what we mean by 'locality'?

4. DETERMINISM, CAUSATION, AND COUNTERFACTUALS

I think it does. Indeed, considerations such as those in Section 3 lead me to dislike all locality conditions based solely on probabilistic independence, and to prefer a condition that is couched in more metaphysically robust terms, such as in terms of causality. From this point of view, it is unimpressive that (2) holds in some Bohmian universes, and equally unimpressive that it fails in others. The real question is not whether events in the backwards light cone of *A* are sufficient to predict the occurrence of *A*, but rather whether events outside the backwards light cone of *A* were *causally responsible* for its occurrence.[5]

In these terms, it may be that although there is in λ_α enough information to specify what *will* happen, and in particular to specify what measurement will be made on β and what the result will be, the measurement on β does not actually happen inside the backwards light cone of *A*, and yet it is causally responsible (at least in part) for that outcome. The complete information in λ_α is used merely to calculate the future non-local *effect* that the measurement on β will have on the motion of α.

The evidence for a causal influence from β to α in the Bohm theory is presumably the functional relation between the position of β and the velocity of α. The issue therefore comes down to the status of these functional relations. One's initial feeling is likely to be that they indicate some form of non-local causation: the position of one particle causes (in part) the velocity of its space-like separated partner. But to go beyond initial feelings, we need to inquire about the status of the quantities appearing in the functional relations.

Bell (1977), for example, was aware that unbridled determinism makes the discussion of these relations difficult, and therefore preferred to assume that the relevant variables (in this case, *i* and *j*) are 'effectively free'. This assumption is natural, but ultimately requires a *suspension* of the Bohm theory in certain regimes (for example, by allowing that experimenters can make free choices, i.e., choices that cannot be predicted from the past states of the universe). Such a suspension might be warranted, but the hypothesis considered here is that the Bohm theory is universally true, and that the world is therefore completely deterministic.

Nonetheless, even allowing this hypothesis, we are far from having established either the locality or non-locality of the Bohm theory. Two correlated events may be screened off from each other by their common past, but it does not follow that there is no causation between them, even in a deterministic scenario. Consider the following example. You give two people, *A* and *B*, the following instructions: if *A* says 'up', then *B* replies 'down'. Further, you instruct *A* to utter the sequence 'up-up-down'. Then let them converse. The result of the conversation is completely determined ahead of time, and in particular the strict anti-correlation between *A* and *B* is already present in the initial state (your instructions to them). However, it is at least plausible that there is some kind of direct causal influence from *A* to *B*. That is, *B* said 'down-down-up' in part *because* *A* said 'up-up-down'. Might the functional relations be the same way? Might it not be that we can use the initial state of the universe to predict all outcomes because we can use it to predict how each of the particles will *cause* the other to move?

Perhaps, but there are also reasons to doubt whether the relationship between the two particles is causal. Note first that in the example above there is an asymmetry between the two people, *A* and *B*. It is exactly this asymmetry that makes it natural for us to speak about *A*'s causing *B* to say such-and-such: the asymmetry picks out a *direction* in which the causal influence may naturally be said to act. But the relation between Bohmian particles is completely symmetric, and this symmetry makes it more difficult to imagine a causal influence propagating between the particles. In terms of the example, the situation in the Bohm theory is analogous to your instructing *A* to say 'up-up-down' and *B* to say 'down-down-up'. If you then instruct each to say the opposite of the other, you have added nothing. Moreover, even if you *were* to add such an instruction, it is far from clear that in the resulting conversation, either *A* or *B* is *causing* the other to say anything.

This argument should be pushed further. For example, one might believe that a similar argument could be made for the (Newtonian) gravitational effect of one body on another. There is also a kind of symmetry there, but at least some people are nonetheless inclined to attribute some causal relation between the bodies. I do not think that this analogy holds, but it clearly calls for further clarification of the argument in the previous paragraph. Nonetheless, my purpose being only to raise questions and not to answer them, I shall move on to consider the issue from a slightly different point of view.

Why, from the philosopher's point of view, do we want to say that there is causation between *A* and *B* in the first example above? The reason, I think, is that the scenario appears to support counterfactuals. That is, even though *A* was *instructed* to say 'up-up-down', if *A* had disregarded those instructions and had said instead 'down-up-down', then *B* would have said 'up-down-up'. Some popular analyses of causality are exactly along these lines. Lewis (1986a) gives roughly the following counterfactual account: C causes E if and only if: If it had been that not-C, then it would have been that not-E. There are plenty of riders attached to this basic statement, but these need not concern us.

There is no chance here of establishing beyond doubt whether the functional relations in the Bohm theory involve causation. What I shall do is to proceed along broadly Lewisian lines and suggest that whether the functional relations involve space-like causation is contentious. This contentiousness arises from the problem of evaluating counterfactuals in a completely deterministic context. (One may, of course, take this result as evidence against Lewis.)

Consider a two-particle universe, where the particles are in an entangled state at some time, t. Their coordinates are q_α and q_β. We have the following functional relation

$$(3) \qquad v_\alpha(t) = \frac{1}{m}\,\mathrm{Im}\left(\frac{\nabla_\alpha \psi(q_\alpha, q_\beta; t)}{\psi(q_\alpha, q_\beta; t)}\right).$$

One could put in any number of values of q_β, thus, apparently, supporting counterfactuals of the form: If β had been at q_β, then the velocity of α would have been v_α.

But greater care is needed, for what are we to make of the antecedent of such counterfactuals? Recall that the initial state of the universe determines all matters of physical fact, including the location of β at time t. Therefore, the full determinism of the Bohm theory leaves us two ways of evaluating counterfactuals, neither of which is very attractive. The first is to maintain that the initial conditions are the true ones, but to suppose counterfactually that the position of β is other than it is. Call this method of evaluation the 'non-backtracking method'. Using the non-backtracking method, the antecedent of the counterfactual is necessarily false within the Bohm theory – it requires us to imagine, in Lewis' (1986a) words, a 'minor miracle', a temporary suspension of the laws of the Bohm theory. But then even if the counterfactual turns out to be true, what have we learned about the Bohm theory? Can one show that there is causality of some sort in a theory by evaluating a counterfactual whose antecedent entails the falsity of that theory? I return to this question below, and argue that even if one *can* learn something about the Bohm theory from the non-backtracking method, it is not clear that it points to space-like causation.

The second way to evaluate the counterfactual, the backtracking method, is to admit into the antecedent of the counterfactual not only the position q_β, but also the initial conditions that are entailed by it, and in particular a new initial position for α. This new initial position for α will of course in general entail a new velocity for α. Here, in general, one will find that a change in the position of β *does* result in a change in the velocity of α, but again we are faced with the question whether this fact indicates space-like causation. The backtracking method certainly allows the answer 'no'.

That method seems to make it clear that what is responsible for the change in the velocity of α is the change in its initial position.

To illustrate further, consider the following example. Suppose I cut an orange in half – label them 'O_α' and 'O_β' – and I give O_α to you. Now I ask: which half *would* you have had if I had had O_α? Evaluate this counterfactual using both methods.

According to the backtracking method, the answer is 'O_β'. But is there any causation between us (or, our halves) in this case? Strong advocates of the counterfactual analysis will say 'yes', but even they will admit that the causation is indirect. That is, although the truth of the counterfactual 'If I had had O_α, you would have had O_β' indicates some causal relation or other between O_α and O_β, the relation in this case is completely accounted for by an event in the common history of my having O_α and your having O_β. In the Bohm theory, too, the backtracking method yields a true counterfactual: 'if β had been elsewhere, then the velocity of α would have been otherwise'. But the truth of this counterfactual is a consequence of the truth of the counterfactual 'if the initial position of α had been otherwise, then its velocity would have been otherwise', and this counterfactual hardly seems indicative of space-like causation. The *reason* that the velocity of α would have been otherwise is that the new position of β entails a different initial condition, which in turn entails a different velocity for α.

What about the non-backtracking method? I have already indicated that it seems to me less capable of teaching us anything about the Bohm theory, but let us consider it again nonetheless. In the example, the non-backtracking method can be applied in two ways. To see what they are, it is easiest to use Lewis' (1986a) method for evaluation of counterfactuals. Denoting the counterfactual 'If it were that φ it would have been that χ' by '$\varphi \mathrel{\Box\!\!\rightarrow} \chi$', Lewis (1986a) proposes:

> '$\varphi \mathrel{\Box\!\!\rightarrow} \chi$' is true at a world w iff: (1) Three are no φ-worlds (i.e., possible worlds where φ is true) or (ii) *every* φ-world among those worlds closest to w is also a χ-world.

Condition (ii) is the important one here. Which worlds in which I keep O_α are closest to (most similar to) this world (where I give O_α to you)? Probably the intuitive answer is 'the worlds where you get O_β', but remember that we have postulated a minor miracle, and have not changed the initial conditions. In that case, it is less clear. Is the nearest world the one in which the strict anti-correlation between your half and mine is violated (and thus you have O_α also, as the initial conditions imply), or is the nearest world the one in which the correlation is preserved, and a second minor miracle occurs, in which case you have O_β despite the initial conditions?

The answer to this question is not at all obvious. Nor are the consequences for the Bohm theory obvious. If you choose to say that the functional relations may be violated in order not to postulate a second minor miracle (the change in the velocity of α), then the counterfactual analysis yields no causation. If you choose to say that the functional relations must not be violated in moving to the counterfactually nearest world, then you will find that a second minor miracle occurs, and along with the miraculous change in the position of β, there is a miraculous change in the velocity of α. The strict adherent to the counterfactual analysis will say that there is some form of causation here, but the preponderance of miracles makes it difficult to know what to conclude. Does the occurrence of two space-like separated but correlated miracles indicate the presence of space-like causation?

There is plenty more to ask about the Bohm theory. For example, how does the Bohm theory fare according to other analyses of causality? It is a central feature of Salmon's (1984) analysis, for example, that causal processes propagate continuously in space and time. In that case, the functional relations will not point to causation, though other difficult questions arise, some of them about locality. (Salmon is sensitive to the difficulties that his view faces in quantum mechanics.) For example, if the Bohm theory is local – if all (Salmon-) causes of an outcome are in its backwards light cone, and the functional relations are short ways to tell the longer causal story – then is it non-Markovian? The reliance on initial conditions in principle to calculate everything suggests that it might be, in the sense that the initial state inside a light cone (the state of a time-slice of the light cone at some initial time in some reference frame) might determine the state inside the light cone at later times, even though the information contained in the initial state is not contained in the complete state inside the light cone at intermediate times. That is, if we say that events outside the backwards light cone of an outcome do not influence the outcome, then we might be forced to say that the outcome somehow 'remembers' the entire initial state, even though not all of the information in the initial state propagates continuously to the outcome. These questions and many more need further examination in the Bohm theory. My aim in this section was mainly to raise some of them.

5. LOCALITY AND CAUSALITY IN A DETERMINISTIC WORLD?

This paper has raised more questions than it has tried to answer. What is the proper locality condition for fully dynamical models of the EPR-Bohm experiment? (Note that most locality conditions are for models that consider just two times – some 'suitable' initial time, and the time of measurement.) The initially plausible condition (1) is obviously inadequate, and the slightly better condition (2) was also called into question: if locality is a metaphysical condition, then perhaps screening off is too weak a relation to guarantee locality, concerning as it does probabilities rather than, for example, causal responsibility. And here arises the difficult question how to make the notion of causal responsibility precise, though

I ducked this question and adopted Lewis' (1986a) analysis in terms of counterfactuals.

But as it did for locality conditions given in terms of probabilities, the determinism of the Bohm theory raised problems also for locality conditions given in terms of causality. How should one evaluate counterfactuals in a completely deterministic context? Neither the backtracking nor the non-backtracking method led unambiguously to the conclusion that there is space-like causation in the Bohm theory. Indeed, neither method looked very attractive within the Bohm theory, or indeed any deterministic theory.

Do we conclude that the Bohm theory might be local? No. The proper lesson is that the very notion of locality – perhaps even the notion of causality – does not have application in a completely deterministic world. In Sections 3 and 4, the problems that arose for the application of these concepts did so because the Bohm theory is fully deterministic. In a stochastic theory, there will not in general be some initial state of the universe that fixes whether the event A will occur. Nor will the problems with the evaluation of counterfactuals arise. Note also that the discussion of Section 4 was independent of initial conditions. If a Bohmian universe lacks causation (because it is fully deterministic), then it does so regardless of initial conditions.

Much more work needs doing. It may be that even in a universe in which every event is fixed from the start, some interesting notion of 'causality' or 'action-at-a-distance' can be formulated, but what does seem clear is that these will not be just our familiar notions. In the end, this conclusion is perhaps not so surprising. After all, the Laplacian ideal is metaphysically foreign to most of us; that apparently familiar concepts such as locality and causality should not be suited to this ideal is only to be expected.

ACKNOWLEDGMENT

Thanks to Joseph Berkowitz, Jeremy Butterfield, James Cushing, Shelly Goldstein, and Michael Redhead for valuable comments.

Indiana University

NOTES

[1] I am glossing over distinctions between different types of determinism.

[2] Here $\lambda(t)$ is a real number, but it could be a vector or a matrix without ruining the arguments below.

[3] Fans of holism will balk at the assumption that λ is divisible into the separate states λ_α and λ_β. If they do so purely out of a taste for holistic properties, then I have no particular quarrel, but, if they also claim that holistic theories are thereby *local* (because condition (2) simply does not apply to such theories), then I think they are making a mistake. I shall not argue the point here, except to claim that at least as far as the effect on the probabili-

ties for A and B are concerned, the attribution of any supposed holistic complete state can be adequately restated as the attribution of a non-holistic state (one that is separable into λ_α and λ_β), plus the attribution of non-trivial joint probabilities. In that way, even holistic theories may violate (2).

4 When the particles are in the single state, the expression (assuming infinite mass for the particle) is

$$v_\alpha = g_\alpha(t)\,\frac{|\varphi_\alpha^-|^2|\varphi_\beta^+|^2 - |\varphi_\alpha^+|^2|\varphi_\beta^-|^2}{|\varphi_\alpha^-|^2|\varphi_\beta^+|^2 + |\varphi_\alpha^+|^2|\varphi_\beta^-|^2}$$

where $g_\alpha(t)$ characterizes the coupling between the spin of α and its apparatus, and $\varphi_\alpha^\pm$ ($\varphi_\beta^\pm$) is a narrow wave packet moving towards the $\pm$ detector for α (β).

5 At first sight, Bell makes a similar claim: given two spacelike separated regions, 1 and 2, Bell says that 'my intuitive notion of local causality is that events in 2 should not be "causes" of events in 1, and vice versa.' (Bell 1987a, 54) However, Bell implements this condition as a condition of probabilistic independence, thereby apparently assuming that causal independence is the same as, or representable as, probabilistic independence, which is the assumption that I am questioning here.

JEFFREY BUB

MODAL INTERPRETATIONS AND BOHMIAN MECHANICS

The idea behind a 'modal' interpretation of quantum mechanics is that quantum states, unlike classical states, constrain possibilities rather than actualities – which leaves open the question of whether one can introduce 'value states' (van Fraassen 1991, 275) that assign values to the dynamical variables or 'observables' of the theory, or equivalently, truth values to the corresponding propositions. As van Fraassen (1991, 279) puts it:

In other words, the [quantum] state delimits what can and cannot occur, and how likely it is – it delimits possibility, impossibility, and probability of occurrence – but it does not say what actually occurs. The transition from the possible to the actual is not a transition *of* state, but a transition *described by* the state.

Apart from van Fraassen's original (1981) version of the modal interpretation, there are now a variety of other 'no collapse' interpretations of quantum mechanics that can be seen as modal in this sense, for example, the interpretations of Krips (1987), Kochen (1985), Healey (1988), Dieks (1989; 1994) and Bub (1992; 1994b). The interpretations of Kochen, Healey and Dieks exploit the polar decomposition theorem to define value states. All these modal interpretations share with van Fraassen's interpretation the feature that an observable can have a determinate value even if the quantum state is not an eigenstate of the observable, so they preserve the linear, unitary dynamics for quantum states without requiring the projection postulate or 'collapse' of the wave function to validate the determinateness of pointer readings and measured observable values in quantum measurement processes. The connection with hypothetical 'hidden variables' underlying the quantum statistics is clear: hidden variables parametrize value states in the sense of the modal interpretation.

Following my earlier formulation (Bub 1992; 1994b) of a version of the modal interpretation and Rob Clifton's (1995) uniqueness theorem for modal interpretations that exploit the polar decomposition theorem, Clifton and I (Bub and Clifton 1996) have shown that, subject to certain natural constraints, there is a unique class of extensions of quantum mechanics by value states.[1] The following discussion is an exposition of this joint work and its significance for modal interpretations and Bohmian mechanics.[2]

To motivate the Bub–Clifton theorem, recall that there are restrictions on value states, if certain constraints are imposed on the attribution of determinate values to observables. The 'no-go' theorems for hidden variables provide a series of such results.

The constraint imposed by the Kochen and Specker (1967) theorem

J. T. Cushing et al. (eds.), Bohmian Mechanics and Quantum Theory: An Appraisal, 331–341.

requires that the values assigned to a set of mutually compatible observables, represented by pairwise commuting self-adjoint operators, should preserve the functional relations satisfied by these observables. With sums and products defined for mutually compatible observables only, the observables of a quantum mechanical system form a partial algebra, and the idempotent observables form a partial Boolean algebra. Kochen and Specker show that a necessary condition for the assignment of values to all the observables of a quantum mechanical system, in such a way as to satisfy the functional relationship constraint, is the existence of an embedding of the partial algebra of observables into a commutative algebra; equivalently, the embedding of the partial Boolean algebra of idempotent observables into a Boolean algebra. A necessary and sufficient condition for the existence of an embedding of a partial Boolean algebra into a Boolean algebra is that, for every pair of distinct elements p, q in the partial Boolean algebra, there exists a homomorphism h onto the 2-element Boolean algebra $\{0,1\}$ such that $h(p) \neq h(q)$.

As Kochen and Specker show, there are no 2-valued homomorphisms on the partial Boolean algebra of projection operators on a Hilbert space of three or more dimensions. The homomorphism condition implies that for every orthogonal n-tuple of 1-dimensional projection operators or corresponding rays in H_n, one projection operator or ray is mapped onto 1 ('true') and the remaining $n - 1$ projection operators or rays are mapped onto 0 ('false'). This is shown to be impossible for the finite set of orthogonal triples of rays that can be constructed form 117 appropriately chosen rays in H_3: any assignment of 1's and 0's to this set of orthogonal triples satisfying the homomorphism condition involves a contradiction.

Now, an observable A can be compatible with an observable B and also with an observable C, while B and C are incompatible (represented by noncommuting operators). A and B will be representable as functions of an observable X, while A and C will be representable as functions of an observable Y, incompatible with X. If A denotes an observable of a system S, and B and C denote incompatible observables of a system S', space-like separated from S, then the functional relationship constraint, that the value of A as a function of X should be the same as the value of A as a function of Y, becomes a locality condition. In his seminal critical review of 'no-go' theorems, Bell (1966) rejected the general functional relationship constraint as physically unmotivated. Bell's (1964) own 'no-go' theorem shows that the weaker locality condition cannot be satisfied in general in a Hilbert space of four or more dimensions: there are sets of observables for which value assignments satisfying the locality condition are inconsistent with the quantum statistics. A later version of Bell's theorem by Greenberger *et al.* (1989) establishes a similar result without statistical arguments.

Several authors have considered the problem of constructing the smallest set of observables that cannot be assigned values in such a way as to satisfy the functional relationship or locality constraints.[3] An unpublished proof by Kochen and Conway reduces the number of rays in H_3 required to generate a contradiction for value assignments satisfying the Kochen–Specker homomorphism condition from 117 to 31,[4] and Mermin (1993) proves a version of the Kochen and Specker theorem for nine observables in H_4, and a version of Bell's theorem for ten observables in H_8.[5]

The problem of how small we can make the set of observables and still generate a Kochen–Specker contradiction is important in revealing structural features of Hilbert space that impose restrictions on value states, but of no immediate significance for the problem of characterizing all possible modal interpretations or 'no-collapse' hidden variable theories in some well-motivated and well-defined sense. The Bub–Clifton proposal is that, for such interpretations, we want to know how *large* we can take the set of determinate observables *without* generating a Kochen–Specker contradiction, i.e., the relevant question concerns the maximal sets of observables, for a given quantum state, that can be assigned determinate (but perhaps unknown) values by value states, subject to the functional relationship constraint, or the maximal sets of propositions that can be assigned determinate truth values by value states, where a truth-value assignment defined by a value state is a 2-valued homomorphism on the propositions.

More precisely, the projection operators (or corresponding Hilbert space subspaces) of a quantum mechanical system form a lattice L that can be taken as representing the lattice of yes-no experiments or propositions pertaining to the system. We know that we cannot assign truth values to all the propositions in L in such a way as to satisfy the functional relationship constraint, or even the weaker locality condition, for all observables generated as spectral measures over these propositions. That is, we cannot take all the propositions in L as determinately true or false if truth values are assigned subject to these constraints. So the probabilities defined by a quantum state cannot be interpreted epistemically and represented as measures over the different possible value states on all the propositions in L. But we also know that any single observable can be taken as determinate for any quantum state (since the propositions associated with an observable generate a Boolean algebra), so we may suppose that fixing a quantum state represented by a ray $e(\psi)$, spanned by a unit vector ψ in the Hilbert space H, and an arbitrary preferred observable R as determinate, places restrictions on what propositions can be taken as determinate for e in addition to R-propositions.

The natural question for a modal interpretation of quantum mechanics would then appear to be: What are the *maximal* sublattices $D(e, R)$ of L to which truth values can be assigned, where each assignment of truth values is specified by a value state defining a 2-valued homomorphism on $D(e, R)$, and the probabilities defined by e for mutually compatible sets of propositions in $D(e, R)$ can be represented as measures over the different possible value states on $D(e, R)$? As a further constraint, it seems appropriate to require that $D(e, R)$ is invariant under lattice automorphisms that preserve e and R (so that $D(e, R)$ is genuinely selected by e and R, and the lattice structure of H), and that $D(e, R)$ is unaffected if the quantum system S is regarded as a subsystem of a composite system $S + S'$, where S is not 'entangled' with S' by any interaction. Call these the possible 'determinate' sublattices of L.

If the aim is to prove a 'no-go' theorem, then the constraints on value states should be as weak as possible. For the problem considered here, however, the idea is to characterize the maximal sublattices of L, for a given quantum state, that can be assigned determinate truth values by value states, in such a way that the probabilities defined by e on a sublattice can be interpreted expistemically in the usual classical sense. The constraints imposed reflect, roughly, the strongest 'classicality' conditions one can get away with. So $D(e, R)$ is required to be a sublattice of L (rather than a partial Boolean algebra, in which operations corresponding to the conjunction and disjunction of propositions are defined only for compatible propositions represented by commuting projection operators), and the possible truth value assignments characterized by value states are required to be defined by 2-valued lattice homomorphisms (rather than 2-valued partial Boolean algebra homomorphisms, i.e., maps onto $\{0,1\}$ that reduce to 2-valued lattice homomorphisms only on each Boolean subalgebra of $D(e, R)$).

The Bub–Clifton theorem establishes that the sublattices $D(e, R)$ are uniquely characterized as follows: In an n-dimensional Hilbert space H_n, suppose R has $m \leq n$ distinct eigenspaces r_i and the rays $e_{r_i} = (e \vee r_i^{\perp}) \wedge r_i$, $i = 1, \ldots, k \leq m$, are the nonzero projections of the state e onto these eigenspaces.[6] The determinate sublattice $D(e, R)$ is then the sublattice $L_{e_{r_1} e_{r_2} \ldots e_{r_k}}$ generated by the k rays e_{r_i} and all the rays in the subspace $(e_{r_1} \vee e_{r_2} \vee \ldots e_{r_k})^{\perp}$ orthogonal to the subspace spanned by the e_{r_i}. $L_{e_{r_1} e_{r_2} \ldots e_{r_k}}$ can also be characterized as $\{e_{r_i}, i = 1, \ldots, k\}'$, the commutant in L of $\{e_{r_i}, i = 1, \ldots, k\}$, or as $\{p\colon e_{r_i} \leq p \text{ or } e_{r_i} \leq p^{\perp}, i = 1, \ldots, k\}$.[7]

Physically, $L_{e_{r_1} e_{r_2} \ldots e_{r_k}}$ contains all those projections with values strictly correlated to the values of R in the state e. The complete set of (not necessarily idempotent) observables associated with $L_{e_{r_1} e_{r_2} \ldots e_{r_k}}$ includes any observable whose eigenspaces are spanned by rays in $L_{e_{r_1} e_{r_2} \ldots e_{r_k}}$. The set of maximal observables includes any maximal observable with k eigenvectors in the directions e_{r_i}, $i = 1, \ldots, k$.

The existence of determinate sublattices $L_{e_{r_1} e_{r_2} \ldots e_{r_k}}$ calls into question the Kochen and Specker argument that a hidden variable reconstruction of the quantum statistics requires a Boolean embedding of the full partial Boolean algebra of idempotent observables of a quantum mechanical system. If the lattice operations in $L_{e_{r_1} e_{r_2} \ldots e_{r_k}}$ are confined to compatible elements (corresponding to commuting projection operators), then $L_{e_{r_1} e_{r_2} \ldots e_{r_k}}$ can he shown to be a partial Boolean algebra. There exist 2-valued partial Boolean homomorphisms on $L_{e_{r_1} e_{r_2} \ldots e_{r_k}}$, in fact sufficiently many to generate the probabilities defined by the quantum state for the propositions in $L_{e_{r_1}e_{r_2}\ldots e_{r_k}}$, but insufficiently many to provide an embedding of $L_{e_{r_1}e_{r_2}\ldots e_{r_k}}$ into a Boolean algebra. The proposal here is that the determinate sublattices $L_{e_{r_1}e_{r_2}\ldots e_{r_k}}$, for all R, provide a class of modal or 'no collapse' hidden variable interpretations of quantum mechanics, in which the functional relationship constraint (and hence Bell's locality condition) is satisfied without a Boolean embedding.

For such interpretations there is no measurement problem, if R plays the role of a pointer in all measurement interactions (or an observable correlated with the pointer observable). To see this, consider a model quantum mechanical universe consisting of a system S and a system M, which plays the role of a measuring instrument for observables of S. Let R represent the indicator of 'pointer' observable of M. A measurement interaction between S and M, say a dynamical evolution of the quantum state of the composite system $S + M$ described by a unitary transformation that correlates eigenstates α_i of an observable A of S with eigenstates ρ_i of R, will result in a state represented by a unit vector of the form $\psi = \sum c_i\alpha_i \otimes \rho_i$ (assuming initial pure states for S and M). The projections of the ray $e(\psi)$ spanned by ψ onto the eigenspaces ρ_i of R are the rays $e_{r_i}(\psi)$ spanned by the unit vectors $\alpha_i \otimes \rho_i$. So for this state, the determinate sublattice contains propositions represented by the projection operators $P_{a_i} \otimes I_M$, i.e., propositions corresponding to the eigenspaces a_i of A (and also, of course, propositions corresponding to pointer positions, represented by the projection operators $I_S \otimes P_{r_i}$).

It follows that, *without introducing any measurement constraints on the determinate sublattices*, we can *derive* that the propositions corresponding to the observable correlated with the pointer observable in the 'entangled' state arising from a unitary transformation representing a quantum mechanical measurement interaction are determinately true or false. So we can *derive* the interpretation of the probabilities defined by a quantum state for the eigenvalues of an observable A as 'the probabilities of finding the different possible eigenvalues of A in a measurement of A,' where a measurement is represented as a dynamical process that yields determinate values for A.

Different choices for the preferred determinate observable R correspond to different modal interpretations of quantum mechanics. I shall illustrate this with the orthodox (Dirac–von Neumann) interpretation, Bohm's theory, and modal interpretations that exploit the polar decomposition theorem to define value states.[8]

On the orthodox (Dirac–von Neumann) interpretation of quantum mechanics, an observable has a determinate value if and only if the state of the system is an eigenstate of the observable.[9] Equivalently, the propositions that are determinately true or false of a system are the propositions represented by subspaces that either contain the state of the system, or are orthogonal to the state, i.e., the propositions assigned probability 1 or

0 by the state. The orthodox interpretation can therefore be formulated as the proposal that the preferred determinate observable is the unit observable I, and that $D(e, I) = L_e$ is the determinate sublattice of a system in the state e, where $L_e = \{p: e \leq p \text{ or } e \leq p^{\perp}\}$.

Of course, as is well known, the orthodox interpretation leads to the measurement problem. In a measurement of an observable A of a system S by a measuring instrument M with pointer observable R, resulting in a final state of the form $\psi = \sum c_i \alpha_i \otimes \rho_i$, neither R-propositions nor A-propositions belong to the sublattice L_e when $e = e(\psi)$. In order to avoid the problem, Dirac (1958a, 36) and von Neumann (1955, 351, 418) assume the projection postulate, that unitary evolution is suspended in the case of a measurement interaction, and the state of $S + M$ is projected onto the ray spanned by one of the unit vectors $\alpha_i \otimes \rho_i$ with probability $|c_i|^2$, because only in such a state, according to the orthodox interpretation, do the observables A and R have determinate values. Without the projection postulate, the orthodox interpretation can be understood as a modal interpretation with an insoluble measurement problem.

Now, there is nothing in the mathematical structure of quantum mechanics that forces the choice of the preferred determinate observable R as the unit observable I. Indeed, there is every reason to avoid this choice because it leads to the measurement problem. Thus, Bohm's (1952a, b) hidden variable theory or 'causal' interpretation can be understood as a proposal for implementing a modal interpretation in which the preferred determinate observable R is fixed as position in configuration space.

With a fixed preferred determinate observable $R \neq I$, the question arises as to the dynamics of value states on the determinate sublattice $L_{e_{r_1} e_{r_2} \ldots e_{r_k}}$ as the quantum state e evolves in time. Since value states defined by 2-valued homomorphisms (with non-zero measure) are in 1-1 correspondence with the rays e_{r_i}, hence with the eigenspaces r_i of R, and assign the same value to e_{r_i} and r_i, for all i, the evolution of such states is completely determined by the evolution of e and of R. So we want an equation of motion for the determinate values of R that will preserve the distribution of R-values specified by e, as e evolves unitarily in time according to Schrödinger's time-dependent equation of motion.

Vink (1993) has shown how to formulate a stochastic dynamics for any always determinate discrete observable R, consistent with the unitary Schrödinger evolution of the quantum state e, by generalizing a proposal by Bell (1987a, 176–177) for constructing stochastic Bohm-type trajectories for fermion number density regarded as an always determinate observable or 'beable' for quantum field theory.[10] The Vink–Bell dynamics turns out to reduce to Bohmian dynamics for R as position in discrete lattice configuration space in the continuum limit.

Vink considers an arbitrary complete set of commuting observables R^i ($i = 1, 2, \ldots, I$), with simultaneous eigenvectors $|r^1_{n^1}, r^2_{n^2}, \ldots, r^I_{n^I}|$, where the $n^i = 1, 2, \ldots, N^i$ label the finite and discrete eigenvalues of R^i.

Suppressing the index i, these are written as $|r_n\rangle$. The unitary time evolution of the state vector is given by Schrödinger's equation of motion:

$$i\hbar d|\psi(t)\rangle/\mathrm{d}t = H|\psi(t)\rangle$$

or

$$i\hbar d\langle r_n|\psi\rangle/\mathrm{d}t = \langle r_n|H\psi\rangle = \sum_m \langle r_n|H|r_m\rangle\langle r_m|\psi\rangle$$

in the R-representation.

The imaginary part of this equation yields the continuity equation:

$$\mathrm{d}\rho_n/\mathrm{d}t = 1/\hbar \sum_m J_{nm}$$

where the probability density ρ_n and the current matrix J_{nm} are defined by:

$$\rho_n(t) = |\langle r_n|\psi(t)\rangle|^2$$
$$J_{nm}(t) = 2Im(\langle\psi(t)|r_n\rangle\langle r_n|H|r_m\rangle\langle r_m|\psi(t)\rangle).$$

For the non-maximal (degenerate) observables R^i, the probability densities and current matrices are defined by summing over the remaining indices, e.g.,

$$\rho^i_n = \sum_{j\neq i} |\langle r^i_{n^i}, r^j_{n^j}|\psi\rangle|^2,$$

and similarly for J^i_{nm}.

We want a stochastic dynamics for the discrete observable R consistent with the continuity equation. Suppose the jumps in R-values are governed by transition probabilities $T_{mn}\,\mathrm{d}t$, where $T_{mn}\,\mathrm{d}t$ denotes the probability of a jump from value r_n to value r_m in time $\mathrm{d}t$. The transition matrix gives rise to time-dependent probability distributions of R-values. It follows that the rate of change of the probability density $\rho_n(t)$ for r_n must satisfy the master equation:

$$\mathrm{d}\rho_n(t)/\mathrm{d}t = \sum_m (T_{nm}\rho_m - T_{mn}\rho_n).$$

But, from the Schrödinger equation, $\mathrm{d}\rho_n(t)/\mathrm{d}t$ also satisfies the continuity equation

$$\mathrm{d}\rho_n(t)/\mathrm{d}t = 1/\hbar \sum_m J_{nm}$$

so we require

$$J_{nm}/\hbar = T_{nm}\rho_m - T_{mn}\rho_n.$$

We want solutions for the transition matrix T, given ρ and J, with $T_{mn} \geq 0$ Since $J_{mn} = -J_{nm}$ (hence $J_{nn} = 0$), the above equation yields $n(n-1)/2$

equations for the n^2 elements of T. So there are many solutions. Bell's (1987a, 176–177) choice was:

$$T_{nm} = J_{nm}/\hbar\rho_m \text{ if } J_{nm} \geq 0$$

$$T_{nm} = 0 \text{ if } J_{nm} \leq 0.$$

For $n = m$, T_{nn} is fixed by the normalization $\sum_m T_{mn}\, dt = 1$.

Vink shows that Bell's solution for the transition matrix leads to Bohm's theory in the continuum limit, when R is position in configuration space. For example, consider a single particle of mass M on a 1-dimensional lattice. Let $x = nd$, where d is the distance between neighbouring points of the lattice. Taking the derivative, F', of a function F on the lattice as defined by $F'(x) = (F(x + d) - F(x))/d$, Vink shows that to first order in d:

$$J_{mn} = (\hbar/Md)[S'(nd)P_n\delta_{n,m-1} - S'(nd)P_n\delta_{n,m+1}]$$

where S is the phase of the wave function. Thus Bell's solution yields:

$$T_{mn} = (S'(nd)/Md)\delta_{n,m-1},\ S'(nd) \geq 0,$$

$$T_{mn} = (S'(nd)/Md)\delta_{n,m+1},\ S'(nd) \leq 0.$$

For positive $S'(nd)$ the particle can jump from site n to site $n + 1$ with probability $S'(nd)\, dt/Md$, and for negative $S'(nd)$ the particle can jump from site n to $n - 1$ with the same probability. So the nearest neighbour interactions in the Hamiltonian induce jumps only between neighbouring lattice sites for this transition matrix. Since each jump is over a distance d, the average displacement in a time interval dt is:

$$dx = S'(x)\, dt/M$$

i.e., $dx/dt = S'(x)/M$.

As $d \to 0$, $S' \to \partial_x S$, and so in the continuum limit:

$$dx/dt = \partial_x S/M$$

as for the continuous trajectories in Bohm's theory. Vink shows that the dispersion vanishes in the limit as $d \to 0$, and the trajectories become smooth and identical to the trajectories in Bohm's theory as $d \to 0$.

Other solutions for the transition matrix induce jumps between distant sites, so smooth deterministic trajectories are not recovered even in the continuum limit. Nelson's (1966) stochastic dynamics is characterized by one such solution. Bohmian dynamics, as the deterministic continuum limit of a stochastic dynamics, is therefore the unique deterministic dynamics for a modal interpretation of quantum mechanics based on position in configuration space as a fixed preferred determinate observable.

The modal interpretations of Kochen, Dieks and Healey all exploit the polar decomposition theorem to define value states, although Healey's interpretation is somewhat idiosyncratic with respect to part-whole relationships for the properties of composite systems. By the polar decomposition theorem, any pure quantum state ψ of a system $S + S'$ can be expressed in the form:

$$\psi = \sum c_i \varphi_i \otimes \varphi_i'$$

for some orthonormal set of vectors $\{\varphi_i\}$ in $H(S)$ and some orthonormal set $\{\varphi_i'\}$ in $H(S')$. The decomposition is unique if and only if $|c_i|^2 \neq |c_j|^2$ for any i, j. In the non-degenerate case, the basic idea is to take the propositions that are determinately true or false for S in the quantum state ψ as the propositions represented by the Boolean algebra of projection operators generated by the set $\{P_{\varphi_i}\}$. (Similar remarks apply to S', of course.) There are alternative proposals for the degenerate case, but if we include in the set of determinate propositions of S all S-propositions that are assigned probability 0 or 1 by the state then, as Clifton (1993) has shown, the set of determinate propositions of S can be characterized as the propositions represented by projection operators of the form $P_1 + P_2$, where P_1 belongs to the spectral measure of the operator W representing the reduced state of ψ for S, and P_2 belongs to the null space of W (i.e., $P_2W = 0$).

Leaving aside Healey's interpretation, then, the interpretations of Kochen and Dieks can be understood as the proposal that for any quantum state W of a system S, pure or mixed (where W might arise from partial tracing over $H(S')$ if S is a subsystem of a system $S + S'$), the determinate propositions of S are the propositions represented by the projection operators in the set:

$$\{P\colon P = P_1 + P_2,\ P_1 \in \text{spectral measure of } W,\ P_2 \in \text{null space of } W\}.$$

From the perspective of the Bub–Clifton theorem, it is now easy to see that these are modal interpretations based on the preferred determinate observable $W \otimes I'$, i.e., the determinate sublattices of S, as a subsystem of the system $S + S'$, are the sublattices $D(e, W \otimes I')$, where e is the ray representing the quantum state of $S - S'$.[11]

While Bohmian mechanics opts for a fixed preferred determinate observable to provide a modal interpretation of quantum mechanics, the interpretations of Kochen and Dieks derive the preferred determinate observable at time t from the quantum state. So these interpretations involve a preferred determinate observable that changes with time. Several authors (Dieks 1994; Vermaas 1996; Bacciagaluppi and Dickson 1995) have shown how to introduce a dynamics for value states on the sublattices $D(e, W \otimes I')$ that is consistent with the unitary Schrödinger evolution of the quantum state e.

It is usual to think of Bohmian mechanics as an interpretation of quantum mechanics in which particles move along well-defined trajectories according to the influence of a guiding wave. Since the wave influences particle motion as a field in configuration space rather than physical space, Bohmian particles behave rather bizarrely in some circumstances, as both detractors and supporters of this interpretation have pointed out.[12] If we see Bohmian mechanics as a modal interpretation based on position in configuration space as a fixed preferred determinate observable, then what is essential to this interpretation is simply that a determinate sublattice of quantum propositions is selected by the state and the preferred determinate observable at each time, and that the temporal evolution of value states on this sublattice can be described by a dynamics that is consistent with the unitary evolution of the quantum state. The propositions in the determinate sublattice at a particular time represent the propositions that can be taken as determinately true or false, for the quantum state at that time, consistently with position in configuration space having a determinate value, and hence support an interpretation of the probabilities defined by the quantum state in the usual classical sense, in particular an interpretation of the probabilities referring to the alternatives of any interaction we regard as a measurement of an observable other than position.

To be sure, we can describe any physical process in terms of where the Bohmian particles go, but any such description will involve the assignment of determinate values to other observables (such as spin) at particular times, even if we choose to regard the values of these observables as supervening on particle positions and the form of the wave function, and hence as not genuinely attributable to the particles. What Bohmian mechanics achieves, as a modal interpretation of quantum mechanics, is to make sense of quantum probabilities in terms of allowable quantum actualities. As such, it belongs to a class of 'completions' of quantum mechanics satisfying certain constraints, and we should not be too surprised if Bohmian particles fail to conform to commonsense metaphysical intuitions.

University of Maryland

NOTES

[1] This result supercedes the uniqueness theorem in Bub (1994a).

[2] Needless to say, I bear sole responsibility for any foolishness in the following discussion.

[3] See, for example, Clifton (1993), Zimba and Penrose (1993), Peres (1993, Chapter 7), Mermin (1993) and Kernaghan (1994). For an interesting discussion of 'no-go' theorems, especially the Mermin proofs, see Brown (1992).

[4] The proof is discussed in Peres (1993, 197).

[5] Note that the Mermin proofs, which are conceptually very simple, are not nearly so economical as the Kochen and Conway proof in terms of the number of rays.

[6] I use the same lower case letter, with subscripts, to indicate the eigenvalue of an observable and the corresponding eigenspace. So, for example, r_i denotes the i'th eigenvalue of R, or the i'th eigenspace of R, depending on the context.

[7] For the proof, see Bub and Clifton (1996).

[8] For a discussion of Bohr's complementarity interpretation, see Bub and Clifton (1996).

[9] See, e.g., Dirac (1958a, 46–47) and von Newmann (1955, 351, 418).

[10] Vink proposes to take *all* observables as always simultaneously determinate, with the determinate values evolving in accordance with the stochastic dynamics. There is no interpretative advantage in taking all observables as determinate, beyond the interpretative advantage (over the orthodox interpretation) in taking some appropriate single observable as always determinate – i.e., an observable that functions as a 'pointer' in all interactions that we interpret as measurements. See Bub (1995) for a discussion.

[11] For details, see Bub and Clifton (1995).

[12] See the debate on 'surrealistic Bohm trajectories' among Englert *et al.* (1992, 1993), Dürr, Fusseder, Goldstein, and Zanghì (1993) and Dewdney *et al.* (1993). See also the argument by Brown, Dewdney and Horton (1995) that in Bohmian mechanics properties like the mass and magnetic moment of a neutron cannot be regarded as localized at the hypothetical position of the particle.

ADRIAN KENT

REMARKS ON CONSISTENT HISTORIES AND BOHMIAN MECHANICS

A distressing feature of discussions of the problems of quantum theory is their tendency to transform physicists from thoughtful and sophisticated scientific critics into uncomplicated partisans or unsympathetic spectators. This holds true although the most interesting questions posed by the various formulations of quantum theory, and by rival theories, are precisely the type of scientific and technical problems which physicists are trained to address. What, for example, does any given theory or formulation allow us to predict or infer, and from what data? Which of these predictions and inferences can be tested? How precise is the mathematical formulation, and what mathematical properties does it have? To what extent is it consistent with important physical principles such as invariance under Lorentz or general coordinate transformations? How elegantly is the theory formulated? On how many arbitrary quantities does it depend?

It would be good to reach consensus. Our present lack of success seems to stem less from subtle difficulties or metaphysical differences than from the fact that scientific assessments of interpretations of quantum theory or its rivals are unfashionable. I do not want to overstate the importance of scientific appraisal. Of course, creative theoretical work is the life blood of physics; physicists need not, and perhaps generally should not, also be philosophers of science. But it is, surely, good to have a clear understanding of what current approaches to quantum theory can, or could possibly, achieve. And, in fact, I would suggest that it is now relatively easy to see that every approach to quantum theory leads to serious problems and that there remain relatively few research programs with any serious ambition of solving our difficulties. This, certainly, was the majority view at Bielefeld, though agreement on precisely which are the serious programs and problems was harder to come by. A minority view, forcefully put at the meeting, was that the problems of quantum theory have been solved by the consistent histories formalism – or at least that the form of a solution has been convincingly sketched. My impression at the end of the meeting was that, although most of the participants believed this to be false, those unpersuaded by consistent histories had reached no agreement on precisely what constitute the problems – still less the virtues – of the consistent histories program.

This article aims, no doubt optimistically, to explain both the problems and the virtues, to persuade consistent historians that their formalism is scientifically problematic, and to persuade sceptics that it is nonetheless scientifically interesting. I will try to show that the consistent histories formulation, sensibly interpreted, significantly changes our understanding

J. T. Cushing et al. (eds.), Bohmian Mechanics and Quantum Theory: An Appraisal, 343–352.

of the scientific status of quantum theory, not only because it offers a new formulation btu also because it sheds new light on earlier interpretations. In particular, I will compare and contrast the consistent histories approach, the Copenhagen interpretation *á la* Landau–Lifshitz, and Bohmian mechanics. In so doing I will argue that some important technical claims made in the consistent histories literature turn out to be simply false. When sensibly interpreted, the formalism's chief virtue turns out to be not, as advertised, that it solves the problems of quantum theory, but rather that it highlights particular scientific problems. Nonetheless, I will conclude, it can be used to give an interpretation of quantum theory which in important ways is better crafted, or at least more honest about its deficiencies, than any of the standard interpretations.

This discussion of consistent histories is drawn from recent joint work with Fay Dowker (Dowker and Kent 1995, 1996). Our conclusions are set out at length in Dowker and Kent (1996). Rather than repeating the details of the arguments here, I will try to state the main results succinctly, to add some explanatory comments, and to respond to some points raised during the Bielefeld meeting.

Our thesis is the following. We agree with other critics of consistent histories that the present interpretations of the formalism have some extremely unattractive features: indeed, we show that in the cases of Omnès and of Gell-Mann and Hartle they have more serious problems. However, we distinguish between criticism of the interpretations offered in the literature and criticisms of the consistent histories approach *per se*. To take one example, Griffiths' proposal to interpret the formalism as defining a non-classical logic is, of course, open to the usual criticisms of quantum logic. This seemed to cause confusion at Bielefeld: some took Griffiths' logic to be an essential part of the consistent histories program. yet nothing in the consistent histories formalism requires it. On the contrary, the formalism defines a sensible interpretation of quantum theory, using ordinary logic and language, which we call the Unknown Set interpretation. It is instructive to examine this interpretation, to see why it cannot be improved upon without going beyond the consistent histories formalism, and to understand its weakness as a scientific theory – for one is then forced to appreciate both that the consistent histories formalism has virtues which other approaches to quantum theory lack, and that it has defects which other formulations remedy.

Let me now try to explain the reasons for our conclusions, assuming familiarity with the basic notions of consistent histories. For definiteness, consider the non-relativistic formulation in which sets of consistent histories are defined by sequences of projective decompositions $\{\sigma_1, \ldots, \sigma_n\}$ at times $t_1, \ldots, t_n$, each σ_i comprising projections $P_i^1, \ldots, P_i^{n_i}$, together with the Gell-Mann–Hartle consistency conditions

$$(1) \quad Tr(P_n^{a_n} \ldots P_1^{a_1} \rho_i P_1^{b_1} \ldots P_n^{b_n}) = \delta_{a_1 b_1} \ldots \delta_{a_n b_n} p(a_1, \ldots, a_n),$$

this last expression defining the probability of the history. Here ρ_i is the density matrix defining initial conditions of the system, which I take to be the universe. There are other interesting formulations and consistency conditions in the literature. Moreover, the formalism admits a time-symmetric generalisation of quantum mechanics in which a second density matrix ρ_f defines final conditions. However, so far as we can tell, the basic scientific problems of the formalism are unsolved by any of these variations.

When the possible sequences of projective decompositions are suitably parametrised, the Gell-Mann–Hartle consistency conditions reduce to simple algebraic equations in the parameters. In other words, once the boundary conditions are fixed, the classification of consistent sets is a purely algebraic problem. It is hard to solve the relevant equations in any but the simplest of examples, or to prove general results about their solutions. Assume for the moment, though, that the equations have no very special properties. One would then expect that almost all the solutions can be parametrised by a number of parameters equal to the number of unknowns minus the number of consistency equations. If the number of parameters is much larger than the number of equations, one would also expect that, given an approximate solution, one can generically find an exact solution very close by.

I mention these mathematical trivialities because, if they apply to consistent historical descriptions of real world physical events or experiments, they have interesting consequences for our understanding of the theory.[1] Suppose, for example, we set up a series of N independent experiments in which (to use the standard Copenhagen language) distinct macroscopic devices measure observables of a microscopic quantum system and display the results by pointers, the experiments being complete and their results displayed at times $t_1, \ldots, t_N$ separated by macroscopic intervals. Consistent historians can, of course, reproduce the standard probabilistic predictions for the results of these experiments, and do so roughly thus. First, we identify the initial density matrix. Next, we fix orthogonal projection operators at times $t_1, \ldots, t_N$, corresponding to the possible positions of the relevant pointers and their complement. Then we argue that the decoherence effect of the environment (photons interacting with the pointers, and so on) will ensure that the set defined by these projections satisfies the consistency conditions to an extremely good approximation – the off-diagonal terms, let us say, are no larger than 10^{-40}. Finally, we take this degree of inconsistency as completely negligible, and simply use the standard decoherence functional expressions for the probabilities of the various results. The justification for this procedure, given by Gell-Mann and Hartle, is that we need not require a fundamental theory to give precisely defined probabilities, or to give probabilities which precisely obey the standard sum rules, since the purpose of theory is to calculate testable quantities and errors of 10^{-40} in our probability calculations are inconsequential in any conceivable experimental test.

If our expectations about the consistency equations are justified, we can improve on this discussion. The number of parameters needed to parametrise possible projection operators is hugely – perhaps infinitely – greater than the number of consistency equations here, and this has two important consequences.

The first is that the approximately consistent set used in these calculations could, in principle, be replaced by a very similar exactly consistent set which would produce essentially the same probabilities – indeed, we expect a parametrised family of exactly consistent sets passing close to the set we initially used. We do *not* expect the sets in this family generally to involve projections we would naturally consider: their projections will generally be onto complicated subspaces of the Hilbert space describing the apparatus and its environment. Nor do we expect any single set in the family to be picked out in any natural way. Nonetheless, the consistent histories formalism tells us that the family contains valid sets of histories with well-defined probabilities. Thus there is no need in principle ever to introduce approximately consistent sets: we can assume, without any serious fear of experimental contradiction, that exactly consistent sets are the only ones of fundamental physical relevance. Of course, this makes no practical difference, since we do not know precisely *which* exactly consistent set is relevant to any given experimental or cosmological calculation, and in practice – unless and until some rule is found which identifies the relevant consistent set for us – we would generally use the usual approximately consistent set and accept that we thereby introduce small errors. Nonetheless, on this view, the formalism defines a mathematically precise theory, and this – if elegance and lack of ad hockery are thought to be of any intrinsic merit – must surely be counted a gain.

The second and perhaps more significant consequence is that the consistent histories formalism shows that the standard Copenhagen description is chosen from a far larger class of possibilities than we previously appreciated. For we expect the parametrised family of consistent sets to be characterised by a very large number of parameters, and to be dominated by exactly consistent sets very far away from the approximately consistent set we chose initially. Now it has certainly always been understood that there are slight ambiguities in any Copenhagen description of a eries of experiments, since there is a certain freedom in the choice of the Heisenberg cuts between system and observer. What is new here is the discovery that there is a continuous family of equally valid physical descriptions of the experiments and that, while this family includes standard Copenhagen descriptions, almost all of its members involve variables quite different from, and not even approximately deterministically related to, the classical degrees of freedom used in the standard discussions. In the Copenhagen approach, any assignment of probabilities to physical events not describable by classical degrees of freedom is forbidden. If we accept the consistent histories formalism as a correct generalisation of the

Copenhagen interpretation, we have to accept that such an assignment is theoretically sensible, and we then have to understand why the Copenhagen interpretation is nonetheless all that we need for practical purposes.

This, in fact, is the key question. The formalism offers a myriad of possible variables for describing physics. Can it, suitably interpreted, explain why the world reliably continues to appear to us always to be described by the particular measure zero subset corresponding to familiar quasi-classical variables? Dowker and I argue that it cannot.

In fact, we make the following stronger claim. The scientific content of the consistent histories formalism is given by the so-called Unknown Set interpretation, which postulates that the fundamental probabilistic theory of nature is defined by a choice of initial density matrix, Hamiltonian and canonical variables – all of which we might hope to specify precisely by some elegant theory – together with some unknown and theoretically unspecifiable consistent set of histories. The histories from that set define the sample space of possible events, and the decoherence functional then defines the probability measure on that sample space in the usual way. Thus, one history from the Unknown Set is chosen randomly to be realised, and it is this history which describes all of physics.[2] No other history, from this or any other set, is realised, and no set other than the single Unknown Set is of any relevance for calculating the probabilities of physical events.

Let me emphasise at once that we are not suggesting that this interpretation is ultimately satisfactory. We put it forward to strip away what we see as inessential and sometimes confusing proposals in the literature. Our claim is that, insofar as the consistent histories literature supplies sensible interpretations of the formalism, those interpretations are almost precisely scientifically equivalent to the Unknown Set interpretation – in other words, they make almost precisely the same predictions, retrodictions and inferences. Where the literature claims to go beyond these predictions, retrodictions and inferences, it is either erroneous or else relies on significant assumptions extraneous to the consistent histories formalism.

Before making the case that the Unknown Set interpretation really does encapsulate the scientific content of the formalism, let me discuss its scientific implications. An interesting question, raised at Bielefeld by Michael Dickson, is whether so apparently weak an interpretation really deserves the title of a scientific theory. To give a fully satisfactory answer would require a general set of criteria for scientific theories. I do not have such a set of criteria. However, it seems to me that, once the initial density matrix, Hamiltonian, and canonical variables are specified, the interpretation ought comfortably to pass any reasonable test. It is well-defined, and moreover comes from a quite elegant and natural mathematical formalism; it is certainly falsifiable; one even, at present, has to admit (as a matter of logic rather than of plausibility) the possibility that it is the best purely mathematical theory of nature which can be constructed. The interpretation makes one definite prediction, which is that all the events we have observed to date, or will observe in the future, can be described by a history of non-zero probability from some consistent set. If this fails to hold, as in principle it could, then the interpretation – and, of course, the entire consistent histories formalism – must be rejected. The interpretation makes further predictions. These predications are generally probabilistic, in the same way as the Copenhagen interpretation is. What is new and peculiar to the consistent histories formalism is that they are also conditional on an unknown physical quantity: the relevant consistent set. Given a particular observed history, we can predict that if a certain event is described by one branch of a consistent extension of that history, and if that particular consistent extension turns out to be part of the Unknown Set, then the relevant event will occur with the conditional probability defined in the usual way by the decoherence functional. If – when, say, we attempt to predict the outcome of a series of experiments – we find that we do observe definite results (i.e., in this interpretation, that the corresponding projections do belong to the Unknown Set) but that the calculated conditional probabilities predict outcomes significantly different from those observed then, again, we must reject either the theory being interpreted – i.e. the specification of boundary conditions, Hamiltonian, and canonical variables – or the interpretation itself. Likewise, even before we perform experiments, we are likely to reject some aspect of the theory if the observed history to date is highly improbable. Like most probabilistic tests, these last two of course require some intuitive or theoretical method of coarse-graining events. That is, we reject the theory not because a small probability event occurs (all possible alternatives may have small probability) but because we believe we can identify a natural division of the alternatives into two classes and we find that sum of the probabilities of the events in the class to which the occurring event belongs is small.

The problem with the Unknown Set interpretation – and, of course, the reason for doubt as to whether it constitutes a theory – is that it gives no algorithm for making probabilistic predictions which depend only on the observed data. Every prediction takes the following form: "*if* the Unknown Set contains the following projective decomposition at the following future time, *then* the probability for the future event described by one of the projections is p". The interpretation does not predict that any future events will occur, or that those which do occur will be describable in terms of familiar variables. In particular, it does not predict that those variables which Gell-Mann and Hartle call quasi-classical – variables which describe macroscopic aggregates and which generally follow deterministic equations of motion to a very good approximation – will continue to be relevant. Quite the contrary: according to this interpretation, the apparent persistence of quasiclassicality is a great and inexplicable mystery. Thus, granted that the interpretation defines a scientific theory, it is a theory with a glaring weakness. For most physicists, surely, believe that they *will* continue to experience a quasiclassical world for the foreseeable future: few are startled each morning by the dawning of yet another quasiclassical day. The persistence of quasiclassical experience, in other words, is part of our theory of nature. Some well-known presentations of quantum theory assume it explicitly (Landau and Lifshitz 1958). Many do not, apparently because it has not been understood that there are well-defined interpretations of quantum theory in which quasiclassicality would not be perceived to persist, and that we need some scientific reason for rejecting such interpretations. A virtue of the consistent histories formalism, in the Unknown Set interpretation, is that it makes these points absolutely clear.

Let us now turn to the interpretations in the consistent histories literature. It is only possible to outline the arguments here, but perhaps a brief precis will be of use. I hope the reader, and those criticised, will forgive the necessarily crude summaries.

Griffiths (1993) suggests that the consistent histories formalism should be interpreted as defining a new logic adapted to propositions describing the physical world. Griffiths' logic has the property that any two propositions referring to projections belonging to different consistent sets can be true without implying that their conjunction is true. We can, for instance, predict that the detectors at CERN will function tomorrow in the ordinary way, producing quasiclassical records of the events they detect, and also predict that the detectors, and their recording devices, and much else besides, will *not* behave quasiclassically tomorrow. We cannot, however, use Griffiths' logical rules to deduce the prediction that the detectors both will and will not behave quasiclassically. We hence avoid contradiction though – as is usual with quantum logic – at the price of a theory which we simply do not understand how to interpret. Griffiths' interpretation, however, skirts the key point. We can never experience the truth or falsity of propositions from more than one consistent set. If the formalism is fundamentally correct then all our scientific endeavours will be described by one consistent set and the scientifically relevant problem is the identification of that set. We can, of course, do calculations in other sets; we can too, if we wish, manipulate propositions involving other sets according to Griffiths' logical rules – but neither of these activities are of any use in predicting the future we will actually experience. Griffiths' interpretation is scientifically equivalent to the Unknown Set interpretation.

Omnès (1991; 1992; 1994) also interprets the formalism as defining rules for the logical analysis of propositions about the physical world. Omnès' logics are conventional: propositions belonging to incompatible consistent sets simply cannot be discussed together. The significant new proposal in Omnès' interpretation is the notion of a "true proposition" – a proposition which is not given to us in the form of observed data, but is deducible from those data by a new rule appended to the consistent histories formalism. Unfortunately (Dowker and Kent 1996), as Omnès accepts, the rule he originally proposed fails to allow the intended deductions: indeed, it generally seems to allow almost no deductions.[3]

Perhaps it is a slight overstatement to say that we are left with an interpretation scientifically equivalent to the Unknown set interpretation: this depends on exactly how narrowly one defines science when its subject matter is the past. Any principle which allows even a few inferences about the past, untestable though they may be, would probably generally be regarded as scientifically useful if those inferences form part of an elegant and compelling theoretical explanation of present data. There seems to be no evidence that any inferences implied by Omnès' original criterion do so, but the possibility cannot be completely excluded. However, so far as the criteria for truth in the existing literature are concerned and insofar as they apply to predications, we must indeed conclude that they indeed do not affect the scientific status of the formalism.

Gell-Mann and Hartle's (1990; 1993; 1994) conclusions, however, certainly go beyond those implied by the Unknown Set interpretation. In fact, Dowker and I (1996) argue that those conclusions are not entirely coherent in their use of the formalism. Nonetheless, the central claim of Gell-Mann and Hartle's interpretation is tenable. This is the suggestion that quasiclassicality appears to us to persist not because quasiclassical variables play any special role in the theory, but because we ourselves have evolved organs of perception which are sensitive to those variables and a mental apparatus which represents the world in quasiclassical terms. But is this a valid argument without further assumption, or is it, like other recent ideas relating consciousness to quantum theory, a speculation?

At first sight it appears not only a valid argument but close to a truism. Almost all scientists would agree that our perceptions and our mental algorithms have evolved to become highly sophisticated at gathering and utilising quasiclassical data. This agreement, though, is predicated on the assumption that one may assume a quasiclassical description of the world. Quasiclassical variables arise naturally in higher order theories of nature such as classical mechanics, chemistry, and terrestrial biology, and evolutionary biologists take their use for granted. Likewise, our theories of brain function are classical theories and our understanding of consciousness,such as it is, is entirely based on classical models. We cannot use biological science to justify any general conclusions about evolution, perception, or consciousness form within a novel interpretation of quantum theory such as the consistent histories interpretation. For if we take seriously a theory – such as the consistent histories formalism – which describes us as being in superpositions of quasiclassical states, or in states defined in terms of entirely non-quasiclassical variables, we can make no statement about our perceptions in those states without new hypotheses. Such hypotheses would necessarily be speculative: they certainly do not follow from our conventional, quasiclassical understanding of the relation of perception to brain function; nor do they follow from any empirical data or theoretical insight presently available to us. What one would need, in fact, is a theory of consciousness written directly in the language of the formalism. It is hard to imagine, and Gell-Mann and Hartle do not try to explain, how one would presently go about trying to formulate such a theory.

We conclude, then, modulo a minor caveat about Omnès' treatment of the past, that the Unknown Set interpretation is indeed scientifically equivalent to the interpretations discussed in the literature, when they are stripped of extraneous hypotheses. Although there is currently no single canonical formulation of quantum theory, it can reasonably be argued that the consistent histories formulation is the minimal formulation of the quantum theory of a closed system which produces a well-defined scientific theory. It tells us that, even when we ignore general relativity, our theory of the macroscopic world – and in particular our expectation of its persisting quasiclassicality – involves assumptions that go beyond both quantum theory and any theory of the cosmological boundary conditions. Some thoughtful critics and advocates of orthodox quantum theory have long appreciated this. Perhaps it has remained controversial only through the wider confusion over interpretations of quantum theory. The consistent histories formalism now so clearly defines a natural interpretation of quantum theory, and spells the conclusion out so precisely, that it is hard to see how any serious controversy can persist.

As we have seen, the formalism also shows that there are equally valid, perfectly well-defined, alternatives in which quasiclassicality persists only for an interval, or never arises. This is an important new development in our understanding of quantum theory.[4]

Bohmians and collapse model theorists may happily accept that the formalism supplies new arguments against orthodox quantum theory, but will perhaps feel our analysis confirms their belief that the formalism has no positive scientific use. Is the formalism not, after all, scientifically sterile? Were the problems it illustrates not solved long ago by Bohmian mechanics? Are dynamical collapse models, with their intriguing alternative explanation of quasiclassicality, not a far more vital subject of research? I sympathise with the spirit of these questions, but let me end by explaining why I cannot dismiss the formalism so conclusively.

First, while defining the problems of quantum theory very clearly, it also suggests an interesting possible form of a solution. All one needs is a rule (perhaps probabilistic) which takes as input the dynamics and boundary conditions of a theory and produces as output a consistent set (the Unknown Set) which turns out to be quasiclassical. This, of course, begs the question of whether such a rule can be found. Here the superiority of Bohmian mechanics and of GRW-type collapse models is presently clear, since the analogous selection principle is already known in both cases and both theories explain quasiclassicality. It remains to be seen whether they give the right explanation and whether they are capable of giving any explanation in the context of relativistic quantum field theory: the same, of course, is true of the consistent histories formalism.

Second, the consistent histories formalism seems to be a strong competitor theory where cosmological applications are concerned. For example, one can easily imagine theories of structure formation involving a series of past events which can be described within a consistent set but which, even if a good Bohmian cosmological theory were to exist, could not naturally be described in terms of Bohmian trajectories.

Third, the consistent histories formalism surely ought to be explored further precisely because it is a good formulation of quantum theory. If the eventual goal is to go beyond quantum theory, it is probably as well to understand all interesting formulations and interpretations of the theory. Different formulations, after all, may inspire different post-quantum theories. We need, in particular, to understand the possible definitions of "truth" and their properties; a general treatment of quantum field theory in the consistent histories formulation; and a clearer understanding of how the formalism applies to cosmology.

University of Cambridge

NOTES

[1] Unfortunately, though it would be surprising if this discussion did not apply to real world physics, there seems to be no way to test the question directly. It would, though, be interesting to test whether the consistency equations do indeed have the expected algebraic properties in moderately-sized Hilbert spaces and whether our conclusions hold in toy models.

[2] It is often suggested that a fundamental theory which assigns probabilities to a single event in this way is problematic, or even meaningless. This is usually intended to be a criticism only of a particular type of theory, but seems in fact to imply a rejection of all probabilistic physical theories. For, practically speaking, a theory phrased in this way is no more or less testable than any other probabilistic theory, since we can only perform finitely many experiments.

[3] Could another definition of "truth" do the job? Omnès has new proposals. Dowker and I too have investigated possible alternative rules. Our tentative conclusion is that interesting rules do exist which allow at least some non-trivial inferences about the past, but we can identify no rule which allows any useful predictions of the future. We hope to give a detailed discussion elsewhere.

[4] Its novelty might be disputed. It is true that something similar occurs in interpretations of quantum mechanics in which the events at different times are entirely uncorrelated. For such interpretations arbitrary basis selection rules can be used at each point in time, and in particular one can use rules in which the system lies in an eigenstate of quasiclassical operators for a while and an eigenstate of non-quasiclassical operators thereafter. But few take such ahistorical interpretations seriously. One can probably find historical interpretations, other than the consistent histories formalism, in which quasiclassicality does not persist – for example, it ought to be possible to produce generalised Bohmian theories with this property – but I can think of no discussion of such interpretations in the literature.

GIANCARLO GHIRARDI AND RENATA GRASSI

BOHM'S THEORY VERSUS DYNAMICAL REDUCTION

1. INTRODUCTION

We share today's widespread opinion that standard quantum mechanics (SQM), in spite of its enormous successes, has failed in giving a satisfactory picture of the world, as we perceive it. The difficulties about the conceptual foundations of the theory arising, as is well known, from the so-called objectification problem, have stimulated various attempts to overcome them. Among these one should mention the search for a deterministic completion of the theory, the many worlds and many minds interpretations, the so called environment-induced superselection rules, the quantum histories approach and the dynamical reduction program. In this paper we will focus our attention on the only available and precisely formulated examples of a deterministic completion and of a stochastic and nonlinear modification of SQM (i.e., Bohm's theory and the spontaneous reduction models, respectively). It is useful to stress that while the first theory is fully equivalent, from a predictive point of view, to SQM, the second one qualifies itself as a rival of SQM, but with empirical divergence so small that it can claim all the same experimental support. Accordingly, they represent explicit answers to the conclusion reached by Bell (Bell 1987b) that *either the wave function as given by the Schrödinger equation is not everything or it is not right.*

The specific purpose of this work is to compare the two just mentioned theoretical schemes with reference to their formal structure, their conceptual implications and the position they allow one to take about physical reality. As we will see, while there are similarities (typically the privileged role assigned to positions), there are relevant differences (particularly with reference to the nonlocal characteristics) between the two theories.

Section 2 is devoted to a sketchy description of the hidden-variable program and of Bohm's pilot-wave theory. Section 3 recalls the main features of the dynamical-reduction models. In Section 4 we begin comparing the two theories with reference to their formal structure. In Section 5 we will be mainly concerned with the problem of describing physical reality. Section 6 deals with the classical and nonclassical features of the theories under examination, while Section 7 is devoted to a discussion of their nonlocal aspects and the ensuing implications for a relativistic program.

2. HIDDEN VARIABLES AND BOHM'S THEORY

The hidden variables-program consists in an attempt to make epistemic the nonepistemic probabilities of SQM, by introducing, to describe indi-

J. T. Cushing et al. (eds.), Bohmian Mechanics and Quantum Theory: An Appraisal, 353–377.

vidual systems, parameters which have to be added to or which replace the wave function in such a way that the resulting theory is fully deterministic. As is well known, any theory of this type meets some difficulties in connection with the problem of attributing objective properties to individual physical systems due to the unavoidable contextual nature of some physical quantities.

The basic principles of the pilot wave theory (Bohm 1952a,b; Bohm and Hiley, 1984) can be summarized as follows:

1. The wave function does not give a complete characterization of the state of a physical system. In Bohmian mechanics one adds to it further variables (i.e. the positions of all the particles of the system under consideration). Such parameters (which are the hidden variables of the theory) correspond to properties actually possessed by the constituents. Thus, any particle has a definite position at any time even though the wave function is not a position eigenstate and particles associated to the same wave function can *have* different positions.
2. The evolution of the wave function is governed by the linear Schrödinger equation in all instances. In particular, collapses never occur.
3. Position variables evolve according to a deterministic evolution equation, the dynamics at a given point and at a given time depending on the wave function evaluated at the point considered.

The ontology of the theory is quite simple: there are particles moving along precise trajectories $\mathbf{q}_i(t)$ and, in addition to them, there is a real[1] entity, the wave function $\Psi(\mathbf{q}_1, \ldots, \mathbf{q}_N, t)$. The evolution is governed by two fundamental equations. First of all, as already stated, there is the Schrödinger equation

$$\frac{\partial\Psi}{\partial t} = \left[-\sum_i \frac{\hbar^2}{2m_1}\Delta_i + V\right]\Psi, \tag{2.1}$$

and, secondly, there is Bohm's equation for the position variables

$$\frac{d\mathbf{q}_i}{dt} = \frac{\hbar^2}{2m_i i}\left[\frac{\Psi^*(\mathbf{q}_1, \ldots, \mathbf{q}_N, t)\nabla_i\Psi(\mathbf{q}_1, \ldots, \mathbf{q}_N, t) - \Psi(\mathbf{q}_1, \ldots, \mathbf{q}_N, t)\nabla_i\Psi^*(\mathbf{q}_1, \ldots, \mathbf{q}_N, t)}{|\Psi(\mathbf{q}_1, \ldots, \mathbf{q}_N, t)|^2}\right] \tag{2.2}$$

The right hand side of this equation is the ratio between the probability current and the probability density of SQM, evaluated at the point $q = (\mathbf{q}_1, \ldots, \mathbf{q}_N)$ of the $3N$-dimensional configuration space. The specification of the initial conditions requires one to assign the wave function $\Psi(\mathbf{q}_1, \ldots, \mathbf{q}_N, 0)$, and the positions of all particles $\mathbf{q}_i(0) = \mathbf{q}_i^0$, $(i = 1, \ldots, N)$ at time $t = 0$. The trajectories of the particles are obtained by first solving Eq. (2.1), inserting the solution in Eqs. (2.2), and solving these equations with the assigned initial values.

Obviously, when we perform a preparation procedure of the SQM-type (such procedures are the only ones which can be performed), we have no way to control the positions of the particles. It is assumed that after the preparation such positions turn out to be distributed[2] in accordance with the SQM probability density $|\Psi(\mathbf{q}_1, \ldots, \mathbf{q}_N, 0)|^2$. The very nice feature of the formalism is that, as a consequence of the evolution equation, the probability density of the positions of the particles at time t turns out to coincide with $|\Psi(\mathbf{q}_1, \ldots, \mathbf{q}_N, t)|^2$. This is the precise sense in which Bohmian mechanics can be considered as empirically equivalent to SQM. The measurement processes are described within the scheme without resorting to any additional assumption, and the pointer of the apparatus ends up in a definite position identifying the outcome obtained.

3. THE DYNAMICAL REDUCTION PROGRAM

As already mentioned, genuinely Hilbert-space models have recently been presented which, by considering nonlinear and stochastic modifications of Schrödinger's dynamics imply, without entailing any violation of established experimental facts, wave packet reduction with fixed pointer positions in measurement processes and, more generally, forbid the persistence of linear superpositions of macroscopically distinguishable states.

3a. *Quantum Mechanics with Spontaneous Localizations*

We begin by discussing a slightly modified version (in which mass has a privileged role) of the first and simplest model of this kind (Ghirardi *et al.* 1986), quantum mechanics with spontaneous localization (QMSL). The model is based ont he assumption that, besides the standard evolution, physical systems are subjected to spontaneous localizations occurring at random times and affecting their elementary constituents. Such processes, which we will call "hittings", are formally described in the following way.

We consider a system of N particles. We suppose that when the i-th particle of the system suffers a localization the wave function changes according to

$$\text{(3.1)} \quad \Psi(\mathbf{r}_1, \ldots \mathbf{r}_N) \rightarrow \Psi_{\mathbf{x}}(\mathbf{r}_1, \ldots, \mathbf{r}_N) = \frac{\Phi_{\mathbf{x}}(\mathbf{r}_1, \ldots, \mathbf{r}_N)}{||\Phi_{\mathbf{x}}||}$$

$$\Phi_{\mathbf{x}}(\mathbf{r}_1, \ldots \mathbf{r}_N) = \left(\frac{\alpha}{\pi}\right)^{\frac{3}{4}} e^{-\frac{\alpha}{2}(\mathbf{r}_i - \mathbf{x})^2} \Psi(\mathbf{r}_1, \ldots, \mathbf{r}_N).$$

The probability density of the process occurring at point $\mathbf{x}$ is given by $||\Phi_{\mathbf{x}}||^2$. For what concerns the temporal features of the processes we assume that the hittings for the various particles occur independently and at randomly distributed times with a mean frequency λ_m which depends on the mass of the considered particle. We choose $\lambda_m = (m/m_0)\lambda$, where m is the mass of the particle, m_0 is the nucleon mass and λ is of the order of 10^{-16} sec^{-1}. The localization parameter $1/\sqrt{\alpha}$ is assumed to take the value 10^{-5} cm.

Let a macroscopic system be in a state $|\Psi\rangle = |\Psi_1\rangle + |\Psi_2\rangle$ which is a superposition of two states $|\Psi_1\rangle$ and $|\Psi_2\rangle$ in which a certain number of particles are in different positions. When one of these particles suffers a hitting that localizes it in the position corresponding to the state $|\Psi_1\rangle$ ($|\Psi_2\rangle$), the other term of the superposition is exponentially suppressed. Therefore, the macroscopic system jumps either to the state $|\Psi_1\rangle$ or to $|\Psi_2\rangle$ every time one of the particles differently located in the two states suffers a hitting. This implies that the frequency leading to the suppression of the coherence between the two states increases proportionally to the masses which are displaced.

The QMSL mechanism does not respect the symmetry properties of the wave function in the case of identical constituents. Its generalization satisfying such a requirement, the continuous spontaneous localizations model (CSL), has been presented and discussed in various papers (Pearle 1989; Ghirardi, Pearle and Rimini 1990; Ghirardi and Rimini 1990).

3b. *The CSL Model*

The model is based on a linear stochastic evolution equation for the state vector. The evolution does not preserve the norm, but only the average value of the squared norm. The equation is

$$\frac{d|\Psi_w(t)\rangle}{dt} = \left[-\frac{i}{\hbar}H + \Sigma_i A_i w_i(t) - \gamma\Sigma_i A_i^2\right]|\Psi_w(t)\rangle. \tag{3.2}$$

In Eq. (3.2), the quantities $w_i(t)$ are c-number Gaussian stochastic processes satisfying

$$\langle\langle w_i(t)\rangle\rangle = 0, \qquad \langle\langle w_i(t)w_j(t')\rangle\rangle = \gamma\delta_{ij}\delta(t-t'), \tag{3.3}$$

while the quantities A_i are commuting self-adjoint operators. Let us assume, for the moment, that these operators have a purely discrete spectrum and let us denote by P_σ the projection operators on their common eigenmanifolds.

The physical meaning of the model is made precise by the following prescription: if a homogeneous ensemble (pure case) at the initial time $t = 0$ is associated to the state vector $|\Psi, 0\rangle$, then the ensemble at time t is the union of homogeneous ensembles associated with the normalized vectors $|\Psi_w(t)\rangle/\||\Psi_w(t)\rangle\|$, where $|\Psi_w(t)\rangle$ is the solution of Eq. (3.2) with the assigned initial conditions and for the specific stochastic process w which has occurred in the interval $(0, t)$. The probability density for such a subensemble is the "cooked one", given by

$$P_{\text{Cook}}[w] = P_{\text{Raw}}[w]\||\Psi_w(t)\rangle\|^2, \tag{3.4}$$

where we have denoted by $P_{\text{Raw}}[w]$ the "raw" probability density associated, according to Eq. (3.3), to the Gaussian process w in the interval $(0, t)$,

$$(3.5) \qquad P_{\text{Raw}}[w] = \frac{1}{N} e^{-\frac{1}{2\gamma}\sum_i \int_0^t d\tau w_i^2(\tau)},$$

N being a normalization factor.

One can prove that the map from the initial ensemble to the final ensemble obeys the forward time translation semigroup composition law. It is also easy to prove that the evolution, at the ensemble level, is governed by the dynamical equation for the statistical operator

$$(3.6) \qquad \frac{d\rho(t)}{dt} = -\frac{i}{\hbar}[\rho(t), H] + \gamma\Sigma_i A_i \rho(t) A_i - \frac{\gamma}{2}\{\Sigma_i A_i^2, \rho(t)\},$$

from which one immediately sees that, if one disregards the Hamiltonian evolution, the off-diagonal elements $P_\sigma\rho(t)P_\tau$ ($\sigma \neq \tau$) of the statistical operator are exponentially damped.

For our concerns, the relevant feature of the dynamical process (3.2) with the prescription (3.4) is that it drives the state vector of each individual member of the ensemble into one of the common eigenmanifolds of the operators A_i, with the appropriate probability. To make this clear, we consider (Pearle 1989) a simplified case in which only one operator A appears in Eq. (3.2). The solution of this equation corresponding to the particular initial condition (involving only two eigenmanifolds of A with eigenvalues α, β)

$$(3.7) \qquad |\Psi, 0\rangle = P_\alpha|\Psi, 0\rangle + P_\beta|\Psi, 0\rangle,$$

when the Hamiltonian is disregarded, is[3]

$$(3.8) \qquad |\Psi_B(t)\rangle = e^{\alpha B(t) - \alpha^2\gamma t}P_\alpha|\Psi, 0\rangle + e^{\beta B(t) - \beta^2\gamma t}P_\beta|\Psi, 0\rangle.$$

Here $B(t)$ is the Brownian process

$$(3.9) \qquad B(t) = \int_0^t d\tau w(\tau).$$

Taking into account Eq. (3.8) and the cooking prescription, one gets the cooked probability density for the value $B(t)$ of the Brownian process at time t:

$$(3.10) \qquad P_{\text{Cook}}[B(t)] = P_{\text{Raw}}[B(t)]\,\| |\Psi_B(t)\rangle \|^2$$

$$= \|P_\alpha|\Psi, 0\rangle\|^2 \frac{1}{\sqrt{2\pi\gamma t}} e^{-\frac{1}{2\gamma t}(B(t) - 2\alpha\gamma t)^2}$$

$$+ \|P_\beta|\Psi, 0\rangle\|^2 \frac{1}{\sqrt{2\pi\gamma t}} e^{-\frac{1}{2\gamma t}(B(t) - 2\beta\gamma t)^2}.$$

From (3.10) it is evident that for $t \to \infty$, the Brownian process $B(t)$ can assume only values belonging to an interval of width $\sqrt{\gamma t}$ around[4] either the value of $2\alpha\gamma t$ or the value $2\beta\gamma t$. The corresponding probabilities are $\|P_\alpha|\Psi, 0\rangle\|^2$ and $\|P_\beta|\Psi, 0\rangle\|^2$, respectively. The occurrence of a value near to $2\alpha\gamma t$ for the random variable $B(t)$ leads, according to Eq. (3.8), to a state vector, for $t \to \infty$, lies in the eigenmanifold corresponding to the eigenvalue a of A. In fact, one gets

$$(3.11)\quad \frac{\|P_\beta|\Psi_B(t)\rangle\|^2}{\|P_\alpha|\Psi_B(t)\rangle\|^2} \cong e^{-2\gamma t(\alpha-\beta)^2}\frac{\|P_\beta|\Psi,0\rangle\|^2}{\|P_\alpha|\Psi,0\rangle\|^2} \xrightarrow[t\to\infty]{} 0.$$

Analogously, when the random variable $B(t)$ takes a value near to $2\beta\gamma t$, for $t \to \infty$, the state vector is driven into the eigenmanifold corresponding to the eigenvalue β of A.

It is then clear that the model establishes a one-to-one correspondence between the "outcome" (the final "preferred" eigenmanifold into which an individual state vector is driven) and the specific value (among the only ones having an appreciable probability) taken by $B(t)$ for $t \to \infty$, a correspondence irrespective of what $|\Psi, 0\rangle$ is, provided only that it has non-zero amplitudes for the two considered eigenmanifolds. In the general case of several operators A_i, a similar conclusion holds for the "outcomes" α_i of A_i and the corresponding Brownian processes $B_i(t)$.

This concludes the exposition of the general structure of the CSL model. Obviously, to give a physical content to the theory one must choose the so-called preferred basis (i.e., the eigenmanifolds on which reduction takes place) or, equivalently, the set of commuting operators A_i. The specific form that has been shown to possess all the desired features has been presented in (Pearle 1989; Ghirardi, Pearle and Rimini 1990). Recently various considerations have led to a slightly modified version making the mass play a specific role (Pearle and Squires 1994; Ghirardi *et al.* 1995). In this version one identifies the discrete index i and the operators A_i of the above formulae with the continuous index $\mathbf{r}$ and the operator

$$(3.12)\quad M(\mathbf{r}) = \textstyle\sum_k m^{(k)}N^{(k)}(\mathbf{r}),$$

where $m^{(k)}$ is the mass of the particles of type k and $N^{(k)}(\mathbf{r})$ are the number density operators giving the average density of particles of type k in a volume of about 10^{-15} cm^3 around the point $\mathbf{r}$

$$(3.13)\quad N^{(k)}(\mathbf{r}) = \left[\frac{\alpha}{2\pi}\right]^{\frac{3}{2}} \sum_s \int d\mathbf{q}\, e^{-\frac{\alpha}{2}(\mathbf{q}-\mathbf{r})^2}\, a^+_{(k)}(\mathbf{q}, s)a_{(k)}(\mathbf{q}, s).$$

Here $a^+_{(k)}(\mathbf{q}, s)$ and $a_{(k)}(\mathbf{q}, s)$ are the creation and annihilation operators of a particle of type (k) at point $\mathbf{q}$ with spin component s, satisfying the canonical commutation or anticommutation relations. Correspondingly one has a continuous family of stochastic Gaussian processes satisfying

$$(3.14)\quad \langle\langle w(\mathbf{r}, t)\rangle\rangle = 0,$$

$$\langle\langle w(\mathbf{r}, t)w(\mathbf{r}', t')\rangle\rangle = \frac{\gamma}{m_0^2}\delta(\mathbf{r} - \mathbf{r}')\delta(t - t').$$

In the above formula m_0 is the nucleon mass, the parameter α is assumed to take the same value (10^{10} cm^{-2}) as in the case of QMSL and γ is related to the frequency $\lambda = 10^{-16}sec^{-1}$ of that model according to $\gamma = \lambda(4\pi/\alpha)^{3/2}$. With the above choices the theory exhibits the same features of standard CSL concerning the localization of nucleons. Obviously, also, other massive particles suffer localizations but with a rate depending on their masses. The nonhamiltonian terms in the dynamical equation lead to the objectification of position variables.

4. A FIRST COMPARISON OF THE TWO APPROACHES

In this and in the following section we will be interested in comparing the two theories, in particular by taking into account their formal structure, their achievements and the position they allow one to take with respect to the description of the world around us. As we will see, there are deep and interesting analogies as well as remarkable differences concerning all of these points.

Bohm's theory, as already mentioned, assigns an absolutely prominent role to the positions of the particles of which the physical system under consideration is made up. These positions evolve in time according to the precise equations (2.1) and (2.2) which uniquely determine, once the initial configuration is specified by the assignment of $\mathbf{q}_i^0$ and $\Psi(\mathbf{q}_i^0, 0)$, the positions at all subsequent times. The quantum-mechanical wave function, evolving according to the Schrödinger equation at all times and under all circumstances (so, no reduction ever takes place), enters the game only to the extent that it determines the particle trajectories. These trajectories are such as to reproduce at all times the quantum distribution of the outcomes of position measurements on the particles themselves. Thus, Bohm's theory can be considered in a very specific sense as "predictively equivalent" to SQM since it yields the same statistical predictions. On the other hand, it is a different theory with a precise conceptual status quite at variance with the one of SQM in all its interpretations. In particular, the theory is completely deterministic and fully reversible at the fundamental level. The positions of all of the particles of the universe and its wave function are uniquely determined at any time given their values at an earlier time.

The quantum probabilities characterizing SQM, which within such a scheme have a nonepistemic nature, acquire an epistemic status: they derive from lack of knowledge, ignorance, and not from any irreducible element of chance in the laws of nature. Obviously, such ignorance plays a fundamental role and has to be accepted as a matter of principle. In fact, should the epistemic uncertainty turn out to be controllable, one could easily prepare statistical ensembles violating the statistical predictions of SQM. But, what we cannot control, and thus we are doomed to ignorance about, is a certain intelligible and precise feature of the world, rather than an unspecified element of chance characterizing it.

On the contrary, dynamical reduction theories, which assume that the wave function represents the most exhaustive, complete specification of an individual physical system, accept that genuine chance rules natural processes. The wave function does not evolve according to the linear, deterministic and reversible laws of SQM but obeys a modified equation obtained by including stochastic and nonlinear terms yielding a fundamentally irreversible behaviour. As a consequence, the statistical predications of dynamical reduction theories differ from those of SQM. Obviously, the discrepancies are such that they do not contradict any known fact about physical processes. The irreversible nature of the evolution implies an energy increase even for isolated systems, so that the theory has an arrow of time built into its very foundations.

It has to be stressed that, in spite of the fact that in the theory, as well as in SQM, *there is nothing but the wave function* (Bell 1987b), the nonlinear and stochastic modifications of the standard evolution assign to positions an absolutely privileged role. This is a feature which is shared by both theories we are comparing. As it has been discussed many times, and as it should be obvious from a general point of view, if one tries to induce the objectification of macroscopic properties in a dynamical way, one has to make a precise choice about which macroproperties have to become objective (the so called problem of the preferred basis). At the macrolevel the choice of the positions is quite a natural one (we recall Einstein's statement (Born 1971) that *a macro-body must always have a quasi-sharply defined position in the objective description of reality*). Detailed investigations (Ghirardi *et al.* 1986) have shown that, within a conceptual framework like the one of the dynamical reduction theories, the only way of inducing precisely such a type of macro-objectification by a fundamental universal dynamics, consists in assuming that the same variables (i.e. positions) are, with extremely low probabilities objectified also at the microscopic level. Moreover, it has also been possible to prove (Benatti *et al.* 1988) that processes objectifying other microproperties do not lead to the desired amplification mechanism.

5. HOW PHYSICAL REALITY IS DESCRIBED

We take a very simple position about science. A scientific theory is a conceptual scheme dealing with formal elements which are supposed to embody some information about something *existing* and to account for the physical processes *which take place* "out there". Among these processes there are experiments performed by observers and also what have been called measurement-like events which *we are obliged to admit are going on more or less all the time, more or less everywhere* (Bell 1990).

To be something more than a simple list of facts and to allow one to go beyond a purely instrumentalist position (i.e. *to explain* rather than simply *to describe*), a theory must be supplemented with some sort of interpretation. We are interested in interpretations which allow one to make claims about properties related to specific variables, possessed by individual physical systems. With reference to this crucial point, which is the core of the debate about the foundations of quantum mechanics, we start by making precise the statement that such properties are objective.

Definition: we will claim that a property corresponding to a value (a range of values) of a certain variable in a given theory is *objectively possessed or accessible* when, according to the predictions of that theory, there is an experiment (or a physical process) yielding reliable information about the variable that would, if performed (or taking place), give an *outcome* corresponding to the claimed value.[5] Thus, the crucial feature characterizing *accessibility* (as far as statements about individual physical systems are concerned) is the matching of the claims and the *outcomes* of physical processes testing the claims.

5a. *Hidden Variables and Bohm Theory*

Historically, the search for hidden variable theories has been motivated by the desire of getting a *classical picture* of the world compatible with what had been discovered about microscopic systems. In this spirit it was required that in the case of an individual physical system to any *observable* of the theory corresponds an objectively possessed property in the above sense. The Gleason (1957) theorem and the Kochen and Specker (1967) analysis have made clear that such a program cannot be consistently pursued. In particular, it has been shown that there exist at least one (actually infinitely many) nonmaximal observable A which turns out to be a function, in the Dirac sense, of two maximal noncommuting observables ($A = f(B)$, $A = g(C)$), such that $f(b) \neq g(c)$, where b and c are the values assigned to the observables B and C, respectively, given the state of the system.

In spite of the above remarks, one can build up deterministic hidden-variable theories predictively equivalent to SQM provided one accepts the contextual nature of some observables (i.e., the fact that the complete specification of the state assigns a definite truth value to a proposition only relative to a specified context). In particular, as stressed by Bell, if one follows this line, one has to accept that the outcome of the measurement of a nonmaximal observable might depend on which other observable compatible with it is actually measured on the system (Bell 1966). From the previous discussion it should be obvious that, in general, properties referring to contextual observables cannot be objectively possessed.[6] After these general considerations, we can analyze this problem within Bohm's theory.

Suppose one wants to claim that to any observable a definite property corresponds at all times. Let us consider a particle in one dimension and its momentum variable p.

(a) If one world try to take a "classical" attitude, one could be tempted to identify the value of the momentum with $m\dot{q}$. This cannot be consistently done. In fact, let the system be described, at $t = 0$, by the non-normalizable state corresponding to the wave function

$$(5.1) \qquad \Psi(q, 0) = \frac{1}{2\sqrt{\pi\hbar}} \left(e^{i\frac{p_0 q}{\hbar}} + e^{-i\frac{p_0 q}{\hbar}}\right).$$

Since $\Psi(q, 0)$ is real, according to Eq. (2.2) $\dot{q}(0) = 0$ (independently of its actual position), for any particle of the ensemble associated with the given state. Then, at $t = 0$, the momentum of all particles would have the value $p(0) = m\dot{q}(0) = 0$. On the other hand, if one would perform a measurement of the momentum of the particles of the ensemble, one would find, according to the predictions of the theory, either the value $+p_0$ or the value $-p_0$. Therefore, one an conclude that there is no measurement procedure of the momentum leading to a distribution of outcomes agreeing with the one of $m\dot{q}$ as given by Bohm's theory.

(b) It is important to stress that, within the theory we are discussing, there are various measurement procedures of the momentum whose outcomes reproduce the quantum predictions. With reference to the previous situation, let us take into account a measurement like process in which the particle under consideration interacts with another particle acting as a measurement device for the momentum. This second particle (the pointer of the apparatus), whose position variable will be denoted as Q, before the interaction takes place is associated to a wave function which is very well localized around $Q = 0$ (the "ready state" of the pointer). Accordingly, the state of the system of the two particles at the initial time $t = 0$ is

$$(5.2) \qquad \Psi(q, Q, 0) = \frac{1}{2\sqrt{\pi\hbar}} \left[\frac{\eta}{\pi}\right]^{\frac{1}{4}} \left(e^{i\frac{p_0 q}{\hbar}} + e^{-i\frac{p_0 q}{\hbar}}\right) e^{-\frac{\eta}{2} Q^2}.$$

The evolution during the measurement time interval $(0, \bar{t})$ is governed by the interaction Hamiltonian

$$(5.3) \qquad H = cpP,$$

where c is an appropriate coupling constant which can take positive or negative values at the experimenter's whim, and P is the momentum conjugate to Q. Disregarding the kinetic energy terms, the wave function associated to the composite system at time $t \in (0, \bar{t})$ is

$$(5.4) \qquad \Psi(q, Q, t) = \frac{1}{2\sqrt{\pi\hbar}} \left[\frac{\eta}{\pi}\right]^{\frac{1}{4}} \left(e^{i\frac{p_0 q}{\hbar}} e^{-\frac{\eta}{2}(Q - cp_0 t)^2} + e^{-i\frac{p_0 q}{\hbar}} e^{-\frac{\eta}{2}(Q + cp_0 t)^2}\right).$$

The final pointer position will be centered around either $+cp_0 t$ or $-cp_0 t$. In order that the measurement outcome be reliable, the condition $\eta(cp_0 \bar{t})^2 \gg 1$ must be satisfied. Obviously, when $c > 0$, we will attribute to the momentum of the measured particle the value $+p_0$ if the pointer particle is, at the end of the process, on the positive real Q-semiaxis, while we will attribute the value $-p_0$ if it is on the negative semiaxis. If the experimenter changes the sign of the coupling constant c, the opposite association of the outcomes of the measurement to the position of the pointer must be made. No matter which choice is made, this measurement procedure gives a distribution of outcomes agreeing with the quantum mechanical one.

(c) It is important to stress and easy to show that, for any definite initial position of the measured particle (i.e., for any given value of the hidden variable), one can get different outcomes depending on the specific measurement procedure which is used. To see this with reference to our case, one has simply to take into account that symmetry considerations imply that the "pointer particle" cannot cross the origin. Thus, if it is on the positive semiaxis at $t = 0$, it will be shifted by a positive quantity of the order of $|c|p_0\bar{t}$ independently from the sign of c. This shows that the outcome of the momentum measurement on the tested particle depends on the measurement set-up.[7] Therefore, in Bohm's theory, the momentum variable is contextual and, consequently, no objectively possessed property corresponding to such variable can be attributed to the particle.

To conclude this subsection, we remark that the only consistent attitude one can take about properties within Bohmian mechanics is that of limiting his considerations to what Bell (Bell 1981a) has called the *exposed beables* of the theory (i.e., the position variables). Since such variables are noncontextual, their values are accessible according to our previous definition. Obviously, within such a scheme, one has to give the status of a beable of the theory also to the wave function since it determines the evolution of the noncontextual exposed beables.

5b. *CSL, Preliminary Considerations*

We now discuss an analogous problem within the CSL theory. To this purpose it is useful to devote some time to present a recently proposed interpretation for such a theory (Ghirardi *et al.* 1995) and to expose the reasons which have motivated such a choice. We start by pointing out the inappropriateness of the Hilbert space topology to describe the concept of similarity or difference of two macroscopic states. In fact, suppose a physical system S is an almost rigid body and let us consider the following three states: $|\phi^A\rangle$, $|\phi^B\rangle$ and $|\tilde{\phi}^A\rangle$. The state $|\phi^A\rangle$ corresponds to a definite internal state of S and to its centre of mass being well localized around A, the state $|\phi^B\rangle$ is simply the displaced of $|\phi^A\rangle$ so that it is well localized in a far region B, the state $|\tilde{\phi}^A\rangle$ differs from $|\phi^A\rangle$ simply by the fact that one or a microscopic number of its "constituents" are in states which are orthogonal to the corresponding ones in $|\phi^A\rangle$.

It is obvious that, under any reasonable assumption about similarity or difference of the states of the universe, $|\tilde{\phi}^A\rangle$ must be considered very close (identical) to $|\phi^A\rangle$, while $|\phi^B\rangle$ must be considered very different from $|\phi^A\rangle$. On the other hand, according to the Hilbert-space topology.

(5.5) $\|(|\phi^A\rangle - |\tilde{\phi}^A\rangle)\| = \||\phi^A\rangle - |\phi^B\rangle)\| = \sqrt{2}.$

This shows with striking force that the Hilbert space topology is totally inadequate for the description of the macroscopic world. Within CSL, as we will show in Subsection 5d, taking advantage of its dynamical features it becomes possible to overcome this difficulty by introducing an appropriate topology.

To pursue our analysis, let S be a physical system which constitutes our universe, and let $H(S)$ be its associated Hilbert space. Let us make an attempt to describe the physical reality of *what exist out there* by giving the status of exposed beables to the mean values of the average (over the characteristic volume of CSL) mass density operator (3.12)

(5.6) $\mathcal{M}(\mathbf{r}, t) = \langle\Psi(t)|M(\mathbf{r})|\Psi(t)\rangle.$

If one assumes, as one can consistently do within a nonrelativistic quantum framework, that the system S contains a fixed and finite number of particles, Eq. (5.6) establishes, for a given t, a mapping of the unit sphere of $H^{(S)}$ into the space of positive and bounded functions of $\mathbf{r}$.

A question naturally arises: within the conceptual framework we are interested in, does $\mathcal{M}(\mathbf{r}, t)$ represent an objectively possessed property of physical systems? Before answering this question, a digression and a comparison with the case of SQM are appropriate.

The map (5.6) is obviously many to one. To better focus on this point, as well as for future purposes, let us consider a physical system consisting of a very large number N of particles, two space regions A and B with spherical shape and radius R and two possible statevectors $|\Psi^{\oplus}\rangle$ and $|\Psi^{\otimes}\rangle$ of its Hilbert space. The state $|\Psi^{\oplus}\rangle$ is the linear superposition, with equal amplitudes, of two states $|\Psi_N^A\rangle$ and $|\Psi_N^B\rangle$ in which the N particles are well localized with respect to the characteristic length (10^{-5} cm) of the model and uniformly distributed in regions A and B, respectively, in such a way that the density turns out to be of the order of 1 gr/cm^3. On the other hand, $|\Psi^{\otimes}\rangle$ is the tensor product of two states $|\Phi_{N/2}^A\rangle$ and $|\Phi_{N/2}^B\rangle$ corresponding to $N/2$ particles being uniformly distributed in region A and $N/2$ in region B, respectively

(5.7) $|\Psi^{\oplus}\rangle = \frac{1}{\sqrt{2}}\{|\Psi_N^A\rangle + |\Psi_N^B\rangle\}, \qquad |\Psi^{\otimes}\rangle = |\Phi_{N/2}^A\rangle \oplus |\Phi_{N/2}^B\rangle.$

It is trivially seen that the two considered states give rise to the same function $\mathcal{M}(\mathbf{r})$.

Let us now investigate the status of the quantity $\mathcal{M}(\mathbf{r})$ within SQM. It is quite obvious that in the case of $|\Psi^{\oplus}\rangle$, $\mathcal{M}(\mathbf{r})$ cannot be considered as an "accessible" mass density function. To see this, let us suppose that one can use quantum mechanics to describe the gravitational interaction between massive bodies and let us consider the following *gedanken* experiment: a test mass is sent through the middle point of the line joining the centres of regions A and B with its momentum orthogonal to it. In the case of the state $|\Psi^{\otimes}\rangle$ for the system of the N particles quantum mechanics predicts that the test particle will not be deflected. On the other hand, if the same test is performed when the state is $|\Psi_N^A\rangle$ ($|\Psi_N^B\rangle$), quantum mechanics predicts an upward (downward) deviation of the test particle. Due to the linear nature of the theory, this implies that if one would be able to prepare the state $|\Psi^{\oplus}\rangle$ the final state would be

$$(5.8) \qquad |\phi\rangle = \frac{1}{\sqrt{2}} \{|\Psi_N^A\rangle \otimes |\phi^{\mathrm{UP}}\rangle + |\Psi_N^B\rangle \otimes |\phi^{\mathrm{DOWN}}\rangle.$$

with an obvious meaning for the symbols. If one includes the test particle in the "universe" and considers the mass density operator in regions corresponding to the wave packets $|o^{\mathrm{UP}}\rangle$ and $|\phi^{\mathrm{DOWN}}\rangle$, one discovers that nowhere in the universe is there a density corresponding to the density of the test particle. In a sense, if one would insist on giving a meaning to the density function, he would be led to conclude that the particle has been split by the interaction into two pieces of half its density so that the test has no definite outcome. According to our criterion, this proves clearly that the mass density distribution $\mathcal{M}(\mathbf{r})$ in the case of the state $|\Psi^{\oplus}\rangle$ is manifestly nonaccessible.

5c. *CSL and Accessible Reality*

Coming back to CSL, it should be obvious that the situation is radically different from the one in SQM since such a theory does not allow the persistence for more than a split second of states like $|\Psi^{\oplus}\rangle$ which are those giving rise to the measurement problem.

To analyze this point, it is useful to introduce a precise mathematical criterion emboding, for the mass density function, the accessibility request put forward at the beginning of this section. To this end, let us consider the mass density variance at $\mathbf{r}$ at time t defined by the following map from the unit sphere of $H^{(S)}$ into $\Re^3$

$$(5.9) \qquad \mathcal{V}(\mathbf{r}, t) = \langle\Psi(t)|[M(\mathbf{r}) - \langle\Psi(t)|M(\mathbf{r})|\Psi(t)\rangle]^2|\Psi(t)\rangle.$$

We consider the ratio

$$(5.10) \qquad \mathcal{R}^2(\mathbf{r}, t) = \mathcal{V}(\mathbf{r}, t)/\mathcal{M}^2(\mathbf{r}, t),$$

and we claim that at time t the mass density at point $\mathbf{r}$ is accessible if:

$$(5.11) \qquad \mathcal{R}(\mathbf{r}, t) \ll 1.$$

This criterion is clearly reminiscent of the probabilistic interpretation of the state vector in standard quantum mechanics. Actually, within such a theory, Eq. (5.11) corresponds to the fact that the spread of the operator $M(\mathbf{r})$ is much smaller than its mean value. Even though we take a completely different attitude with respect to the mean value of $M(\mathbf{r})$, it turns out to be useful to adopt the above criterion also within the new context. In fact, as we will discuss in what follows, when one has a space region such that for all its points relation (5.11) holds, all reliable tests aimed to ascertain the mass density value give outcomes corresponding to $\mathcal{M}(\mathbf{r})$.

With reference to the previous example we stress that in the case of $|\Psi^{\otimes}\rangle$ all points within regions A and B are such that criterion (5.11) is very well satisfied. In the case of $|\Psi^{\oplus}\rangle$ for the same points one would have

$$(5.12) \quad \mathcal{M}(\mathbf{r}) \cong \frac{\mathrm{d}}{2} m_0, \qquad \mathcal{V}(\mathbf{r}) \cong \frac{\mathrm{d}^2}{2} m_0^2,$$

where d represents the average density in a volume of 10^{-15} cm^3 around $\mathbf{r}$. There follows

$$(5.13) \quad \mathcal{R}(\mathbf{r}) \cong 1,$$

so that, as expected, the average mass density in regions A and B is non-objective.

Obviously for a microsystem, unless its wave function is localized to better than 10^{-5} cm, its average mass density turns out to be nonaccessible at almost all times since CSL does not forbid superpositions of arbitrarily far apart states. This fact does not constitute a difficulty for the proposed interpretation. On the contrary, it embodies the fact that nature has compelled us to allow *electrons* (in general microsystems) *to enjoy the cloudiness of waves*, while our experience of the world requires *tables and chairs and ourselves, and back marks on photograph to be rather definitely in one pace rather than another and to be described in classical terms* (Bell 1986). The possible nonaccessibility of the mass density distribution does not entail that it is not real. Actually, even regions in which the mass density is nonaccessible according to our mathematical criterion may play an important physical role.[8] We point out that the requirement of being able to speak, within a context like the one of CSL leading to objectification at the macroscopic level, of objectively possessed properties and of introducing an appropriate topology to describe similarity and difference of macroscopic situations, has led us to consider "two levels of reality". The attribution of a real status to $\mathcal{M}(\mathbf{r})$ even when it is not accessible does not give rise to the difficulties one meets when one tries to interpret the wave function as giving the charge or the mass density of the elementary constituents of nature. Actually, this corresponds simply to accepting the well known fact that *however far the wave function has extended, the reaction of a detector . . . remains spotty* (Bell 1990).

5d. *The Appropriate Topology for the CSL Model*

Let us denote by $\mathfrak{U}^{(S)}$ the unit sphere in $H^{(S)}$ and let us consider the non linear map $\mathcal{M}$ associating to the element $|\varphi\rangle$ of $\mathfrak{U}^{(S)}$ the element $\mathcal{M}(\mathbf{r}, |\varphi\rangle) = \langle\varphi|M(\mathbf{r})|\varphi\rangle$ of $L^{(2)}$, the Hilbert space of the real square integrable functions on $\mathfrak{R}^{(3)}$.

On $\mathfrak{U}^{(S)}$ we define a topology by introducing a mapping Δ: $\mathfrak{U}^{(S)} \otimes \mathfrak{U}^{(S)} \to \mathfrak{R}^{+}$ according to

$$\text{(5.14)} \quad \Delta(|\varphi\rangle, |\psi\rangle) = \{\int d\mathbf{r}[\mathcal{M}(\mathbf{r}, |\varphi\rangle) - \mathcal{M}(\mathbf{r}, |\psi\rangle)]^2\}^{1/2}.$$

Such a mapping is not a distance since it may happen that $\Delta(|\varphi\rangle, |\psi\rangle) = 0$ even though $|\varphi\rangle \neq |\psi\rangle$. However, Δ meets all other properties of a distance

$$\text{(5.15)} \quad \begin{aligned} &\Delta(|\varphi\rangle, |\psi\rangle) = \Delta(|\psi\rangle, |\varphi\rangle) \geq 0, \\ &\Delta(|\varphi\rangle, |\psi\rangle) \leq \Delta(|\varphi\rangle, |\chi\rangle) + \Delta(|\chi\rangle, |\psi\rangle), \end{aligned}$$

as one can easily prove.

From now on we will limit our considerations to the proper subset $\mathfrak{A}^{(S)}$ of $\mathfrak{U}^{(S)}$ of those states which are allowed by the CSL dynamics.[9] For any element $|\varphi\rangle$ of $\mathfrak{A}^{(S)}$ we then consider the set of states of $\mathfrak{A}^{(S)}$ for which $\Delta(|\varphi\rangle, |\psi\rangle) \preccurlyeq \varepsilon$. Here the quantity ε has the dimensions of a mass and is chosen of the order of $10^{18}\ m_0$, with m_0 the nucleon mass. From the properties of the map Δ it follows that

i. $\{\Delta(|\varphi\rangle, |\psi\rangle) \preccurlyeq \varepsilon$ and $\Delta(|\varphi\rangle, |\chi\rangle) \preccurlyeq \varepsilon\}$ implies $\Delta(|\chi\rangle, |\psi\rangle) \preccurlyeq \varepsilon$.
ii. $\{\Delta(|\varphi\rangle, |\psi\rangle) \succcurlyeq \varepsilon$ and $\Delta(|\varphi\rangle, |\chi\rangle) \preccurlyeq \varepsilon\}$ implies $\Delta(|\chi\rangle, |\psi\rangle) \succcurlyeq \varepsilon$.

We have introduced the parameter ε in such a way that it turns out to be sensible to consider similar to each other states whose "distance" Δ is smaller than (or of the order of) ε. More specifically, when

$$\text{(5.16)} \quad \Delta(|\varphi\rangle, |\psi\rangle) \preccurlyeq \varepsilon.$$

we will say that $|\varphi\rangle$ and $|\psi\rangle$ are "physically equivalent".

We now discuss the "distorted" (with respect to the Hilbert space one) topology associated to the "distance" Δ. First of all we stress that the two states $|\varphi^A\rangle$ and $|\tilde{\varphi}^A\rangle$ which are maximally distant in the Hilbert space topology, turn out to be equivalent (i.e. to satisfy condition (5.16) in the new topology). This represents an example showing how such a topology takes more appropriately into account the fact that under any sensible assumption, the "universes" associated with these two states are very similar.

Obviously, one problem arises. Criterion (5.16) leads one to consider equivalent states which are quite different from a physical point of view, even at the macroscopic level. To clarify this statement we take into account two states $|\varphi\rangle$ and $|\psi\rangle$ corresponding to an almost rigid body located, at $t = 0$, in the same position but with macroscopically different momenta, let us say $P = 0$ and P, respectively. Even though the two states are physically quite different, their distance at $t = 0$ is equal to zero. However, if one waits up to the time in which the state $|\psi\rangle$ has moved away from $|\varphi\rangle$, the "distance" $\Delta(|\varphi(t)\rangle, |\psi(t)\rangle)$ becomes large and the two states are no longer equivalent.

5e. *The CSL Ontology*

As we pointed out a the beginning of this section, we are interested in describing something "existing and evolving out there" in such a way that the resulting physical picture is compatible with our experience of the world. This requires, as nicely pointed out by Bell (Bell 1989), making precise what he called the kinematical and the dynamical aspects of the theory. *The kinematics requires one to identify something as being really there* (i.e., something *we are going to take seriously*), while the dynamics accounts for the way it changes with time.

For the reader who has followed the above analysis, it should be clear that, within CSL, physical systems are fully described by the state vector which evolves according to equations embodying genuine elements of chance. However, even though *the state vector is everything* (Bell 1987b), since the dynamics itself tends to make objective the mass density, it turns out to be appropriate to relate the kinematical elements in Bell's sense to the mass density distribution. This is what we have done in the previous subsections by putting forward explicit criteria allowing one to recognize its being accessible and by showing the appropriateness of resorting to the mass density function to define the idea of similarity and difference between macroscopic situations.

In a sense we could state that the exposed beables of the theory are the values of the mass density function and that, a the appropriate level, the dynamics allows one to take them seriously. All other properties which emerge as a consequence of this process derive from this basic dynamical feature concerning the mass density function. With reference to the case analyzed at the end of the previous subsection, we stress that, if one has a superposition of those two states, the macroscopic momentum of the system becomes accessible just as a consequence of the objectification of the mass density.

6. CLASSICAL AND NONCLASSICAL FEATURES

From the previous analysis it should be clear to everybody that both theories under discussion are basically nonclassical. However, they also exhibit some relevant classical features which, in particular, allow one to take a realistic position about nature. The classical aspects of CSL emerge only at the macroscopic level. On the contrary, Bohm's theory has a feature in common with classical theories even in the case of microscopic systems (i.e., the fact that particles have definite positions). In spite of this, the status of a beable given to the wave function unavoidably brings into play highly nonclassical features. We recall that, for example, in the case of one particle whose wave function is different from zero in two far-separated space regions, in spite of the fact that the particle must be claimed to be *either* in one *or* in the other region, one must attribute a real status also to the branch of the wave function corresponding to the empty region.

To see this let us consider the case (see Figure 1) of a particle whose wave function is different from zero practically only along the two paths 1 and 2, and suppose that things are arranged in such a way that one has constructive interference in one of the two final directions and destructive interference in the other. In Bohm's theory the particle follows one of the two paths. However, for particles following path 1, the insertion of an absorber along path 2, allows them to activate, in some cases, the detector corresponding to destructive interference.

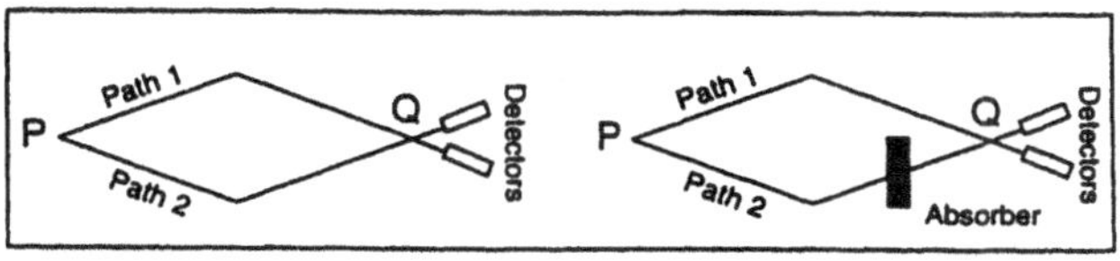

Fig. 1. An interference experiment which puts into evidence the nonclassical features of Bohmian mechanics.

Let us compare this situation with the one in which the wave function is different from zero practically only along path 1. In such a case, obviously, the insertion of the absorber does not affect, either within SQM or in Bohm's theory, the subsequent evolution of the system. The previous statement about the emergence of nonclassical features of the theory should now be clear. When standard quantum mechanics allows one to state that the particle follows path 1, the presence of the absorber along path 2 has no effect whatsoever on the particle itself. When the wave function is the superposition of two packets following the two paths, in standard quantum mechanics one cannot even think of the particle as following a definite path. On the contrary, in such a case, in Bohm's theory the particle actually follows one of the two paths. However, even for particles following path 1, the presence of the absorber has a considerably invasive effect since it may remarkably change their evolution. This shows that even when the *non-contextual* property of *being* along path 1 is possessed by an individual system, it does not exhibit the feature of a classical particle possessing such a property. It seems to be aware of "being in a superposition".

7. NONLOCAL ASPECTS AND RELATIVISTIC INVARIANCE

As is well known, the principal locality assumption needed to prove Bell's theorem for the stochastic case (Bell 1971) is equivalent to the conjunction of two other assumptions, viz., in Shimony's terminology, parameter independence and outcome independence (Suppes and Zanotti 1976; van Fraassen 1982; Jarrett 1984; Shimony 1984). In view of the experimental violation of the Bell inequality, one has to give up either or both of these assumptions. We are interested in comparing the status of the two theories under examination from this point of view.

To start with, let us fix our notation. We will denote by λ all parameters (which may include the quantum mechanical state vector or even reduce to it alone) which provide, within the theoretical scheme one is interested in, the complete specification of the state of an individual physical system. For simplicity we will refer to a standard EPR-Bohm like situation and we will denote by

(7.1) $p_\lambda^{LR}(x, y; \mathbf{n}, \mathbf{m}),$

the joint probability of getting the outcome x ($x = \pm 1$) in a measurement of the spin-component along $\mathbf{n}$ at the left (L), and $y (y = \pm 1)$ in a measurement of the spin-component along $\mathbf{m}$ at the right (R) wing of the apparatus.

We assume that the experimenter at L can make a free-will choice of the direction $\mathbf{n}$ and similarly for the experimenter at R and the direction $\mathbf{m}$. Both experimenters can also choose not to perform the measurement.

Bell's locality assumption can be expressed as

(7.2) $p_\lambda^{LR}(x, y; \mathbf{n}, \mathbf{m}) = p_\lambda^{L}(x; \mathbf{n}, *)p_\lambda^{R}(y; *, \mathbf{m}),$

where the symbol * appearing in the probability distributions a the r.h.s. denotes that the corresponding measurement is not performed.

As already remarked, condition (7.2) is equivalent to the conjunction of two logically independent conditions

(7.3.I) $p_\lambda^{L}(x; \mathbf{n}, \mathbf{m}) = p_\lambda^{L}(x; \mathbf{n}, *)$
$p_\lambda^{R}(y; \mathbf{n}, \mathbf{m}) = p_\lambda^{R}(y; *, \mathbf{m}),$

and

(7.3.II) $p_\lambda^{LR}(x, y; \mathbf{n}, \mathbf{m}) = p_\lambda^{L}(x; \mathbf{n}, \mathbf{m})p_\lambda^{R}(y; \mathbf{n}, \mathbf{m}),$

where we have denoted, for example, by the symbol $p_\lambda^{L}(x; \mathbf{n}, \mathbf{m})$ the probability of getting, for the given settings $\mathbf{n}, \mathbf{m}$, the outcome x at L. Equations (7.3.I) express *parameter independence*, the requirement that the probability of getting an outcome at L (R) is independent from the setting chosen at R (L), while Eq. (7.3.II) (*outcome independence*) expresses the requirement that the probability of an outcome at one wing does not depend on the outcome which is obtained at the other wing.

The above "splitting" of the locality requirement into two independent conditions is particularly useful to discuss their different conceptual implications with respect to relativistic requirements. In fact, as proved by Jarrett himself (Jarrett 1984), when conditions (7.3.I) are violated, if one had access to the variables λ which specify completely the state of individual physical systems, one could send faster-than light signals from R (L) to L (R). On the contrary, if only condition (7.3.II) is violated, then faster-than light signalling cannot be achieved since the stochastic outcome at a wing cannot be controlled by the experimenter.

It is well known that in a deterministic theory (i.e., one for which the range of any one of the above probability distributions is the set $\{0, 1\}$) there cannot be outcome dependence, so that violation of the locality requirement (7.2) implies parameter dependence. This fact by itself shows that deterministic theories which reproduce quantum predictions meet more serious difficulties with relativity than do stochastic ones.

Before coming to analyze these matters with reference to the two theories we are concerned with, we recall that SQM violates locality by violating only outcome independence and that it does not allow faster-than-light signalling between distant observers (Eberhard 1978; Ghirardi *et al.* 1980).

7a. *Parameter Dependence and Outcome Dependence in the Two Theories*

Concerning the pilot wave theory, we can state without further analysis that it violates parameter independence since, as we have already remarked, any deterministic theory reproducing the quantum mechanical probabilities must do so. Bohmian mechanics is a well defined theory with a precise formalism and conceptual status. If one takes seriously, as we do, the fact that the complete specification of the state of an individual physical system, at a certain time, is given by the positions of its constituents, one has to accept that the theory implies instantaneous effects at-a-distance at its fundamental level[10] Obviously, since one has no access to the "hidden" positions and, at the ensemble level, the theory gives the same predications of SQM, one cannot take advantage of this fact to actually send superluminal signals.

Coming to CSL, we first of all prove that it satisfies the condition of parameter independence and therefore, since for the experiment under discussion its predictions agree with those of SQM, it must violate the outcome independence requirement.

Since within CSL the stochastic dynamical reduction takes place just as a consequence of the fact that macroscopic objects evolve into a superposition of macroscopically different states, to perform our analysis we must take into account which apparatuses are present and how the triggering mechanism describing the micro-macro interactions works. For simplicity we will assume that such interactions taking place at L and R are governed by appropriate coupling constants g_L and g_R. To investigate whether there is parameter dependence we will then compare situations in which one of the coupling constants is made equal to zero (corresponding to no measurement being performed) with a situation in which it is non-zero.

We recall that within the CSL model the initial situation is *completely* characterized by the assignment of the initial state vector $|\Psi, 0\rangle$. As stated above, the unfolding of the individual physical processes is governed by a linear evolution equation depending on the coupling constants g_L and g_R and on a specific realization $w(\mathbf{x}, t)$ of the stochastic process. The probability of occurrence of such processes depends on the overall physical situation and, in particular, on the coupling constants being equal to zero or not. This is due to the fact that the evolution of the state vector $|\Psi, 0\rangle$ depends on the coupling constants and the state vector itself enters in determining the cooked probability density of occurrence of the stochastic processes.

We can now come to the specific discussion of the process we are interested in. Strictly speaking, one should take into account the whole physical situation and in particular one should specify the initial state of the system plus the apparatuses and study the subsequent evolution. Nevertheless, it should be clear that, for the analysis we are interested in, one can simplify the description by limiting his considerations to the spin Hilbert space only, provided one correspondingly changes the value of the parameter γ governing the reduction rate in such a way that reduction takes place within the times which are characteristic for a macroscopic measurement device.

Taking such a position, one has, in the case in which both apparatuses are switched on ($g_R \neq 0$ and $g_L \neq 0$), a linear dynamical equation analogous to (3.2)

$$\frac{\mathrm{d}|\Psi_{w_L, w_R}(t)\rangle}{\mathrm{d}t} = \{[(\sigma^L \cdot \mathbf{n})w_L(t) - \gamma] + [(\sigma^R \cdot \mathbf{m})w_R(t) - \gamma]\}|\Psi_{w_L, w_R}(t)\rangle. \tag{7.4}$$

The probability distribution of the stochastic processes is obtained through the cooking procedure prescribed by the theory

$$P_{\mathrm{Cook}}[w_L \ \&\ w_R] = P_{\mathrm{Raw}}[w_L \ \&\ w_R]\,\||\Psi_{W_L \,\&\, W_R}(t)\rangle\|^2, \tag{7.5}$$

where $P_{\mathrm{Raw}}[w_L \ \&\ w_R]$ is the probability distribution associated to a white noise with

$$\langle\langle w_L\rangle\rangle = 0; \quad \langle\langle w_R(t)\rangle\rangle = 0; \quad \langle\langle w_L(t)w_R(t')\rangle\rangle = \gamma\delta_{L,R}\delta(t - t'). \tag{7.6}$$

Obviously, the appearance of the two terms at the r.h.s. of (7.4), corresponds to the fact that both apparatuses are switched on and are triggered by the microsystems at L and R, respectively. If one wants to compare this situation with the one in which the observer at the right, freely switches off the apparatus by making $g_R = 0$, one has to consider another stochastic equation

$$(7.7)\quad \frac{\mathrm{d}|\Psi_{w_L}(t)\rangle}{\mathrm{d}t} = \{(\sigma^L \cdot \mathbf{n})w_L(t) - \gamma]|\Psi_{w_L}(t)\rangle.$$

The solutions of equations (7.4) and (7.7) at time t for the same initial conditions are[11]

$$(7.8)\quad |\Psi_{B_L, B_R}(t)\rangle = e^{F_{LB_L}(t)}e^{F_{RB_R}(t)}|\Psi, 0\rangle,$$

and

$$(7.9)\quad |\Psi_{B_L}(t)\rangle = e^{F_{LB_L}(t)}|\Psi, 0\rangle,$$

respectively. In Eqs. (7.8) and (7.9) we have put

$$(7.10)\quad F_{LB_L}(t) = \sigma^L \cdot \mathbf{n}B_L(t) - \gamma t; \qquad F_{RB_R}(t) = \sigma^R \cdot \mathbf{n}B_R(t) - \gamma t,$$

where

$$(7.11)\quad B_L(t) = \int_0^t \mathrm{d}\tau w_L(\tau), \qquad B_R(t) = \int_0^t \mathrm{d}\tau w_R(\tau).$$

We come back now to Eq. (7.4) and we evaluate the cooked probability density of occurrence of the Brownian processes $B_L(t)$ and $B_R(t)$ by multiplying the raw probability density by the square of the norm of the state vector (7.8). As usual we have

$$(7.12)\quad P_{\mathrm{Cook}}[B_L(t)\ \&\ B_R(t)] = P_{\mathrm{Raw}}[B_L(t)\ \&\ B_R(t)]\,\||\Psi_{B_L\,\&\,B_R}(t)\rangle\|^2,$$

and

$$(7.13)\quad P_{\mathrm{Raw}}[B_L(t)\ \&\ B_R(t)] = P_{\mathrm{Raw}}[B_L(t)]P_{\mathrm{Raw}}[B_R(t)].$$

Taking into account Eq. (7.8), one then gets from (7.12)

$$(7.14)\quad P_{\mathrm{Cook}}[B_L(t)\ \&\ B_R(t)] = P_{\mathrm{Raw}}[B_L(t)]P_{\mathrm{Raw}}[B_R(t)]\ \||\Psi_{B_L\,\&\,B_R}(t)\rangle\|^2$$
$$= P_{\mathrm{Raw}}[B_L(t)]\ \|e^{F_{LB_L}(t)}|\Psi, 0\rangle\|^2 P_{\mathrm{Raw}}[B_R(t)] \left\| \frac{e^{F_{RB_R}(t)}e^{F_{LB_L}(t)}|\Psi, 0\rangle}{\|e^{F_{LB_L}(t)}|\Psi, 0\rangle\|} \right\|^2.$$

Let us consider the marginal cooked probability density of $B_L(t)$

$$(7.15)\quad P^{\#}_{\mathrm{Cook}}[B_L(t)] = \int \mathrm{d}[B_R(t)]P_{\mathrm{Cook}}[B_L(t)\ \&\ B_R(t)]$$
$$= P_{\mathrm{Raw}}[B_L(t)]\ \|e^{F_{LB_L}(t)}|\Psi, 0\rangle\|^2$$
$$\cdot \int \mathrm{d}[B_R(t)]P_{\mathrm{Raw}}[B_R(t)] \left\| \frac{e^{F_{RB_R}(t)}e^{F_{LB_L}(t)}|\Psi, 0\rangle}{\|e^{F_{LB_L}(t)}|\Psi, 0\rangle\|} \right\|^2.$$

Since the equation

$$(7.16)\quad \frac{\mathrm{d}|\Psi_{W_R}(t)\rangle}{\mathrm{d}t} = \{(\sigma^R \cdot \mathbf{m})w_R(t) - \gamma\}|\Psi_{W_R}(t)\rangle,$$

preserves the stochastic average of the square of the norm of the state vector the last integral in Eq. (7.15) takes the value 1. This means that $P^{\#}_{\text{Cook}}[B_L(t)]$ turns out to equal the probability $P_{\text{Cook}}[B_L(t); *]$ – the cooked probability density of occurrence of the Brownian process $B_L(t)$ for the same initial condition if the process were described by Eq. (7.7) or, equivalently, if the apparatus at R were switched off.

As we have already remarked in Section 3, within CSL it is possible to establish a one-to-one correspondence between the outcome at left (right) and the specific value taken by the Brownian process $B_L(t)$ ($B_R(t)$) for $t \to \infty$. Then, the above proof of the equality between $P^{\#}_{\text{Cook}}[B_L(t)]$ and $P_{\text{Cook}}[B_L(t); *]$ amounts to having shown that the theory under consideration exhibits parameter independence. The analysis performed by Bell (Bell 1987b) of the QMSL model by resorting to a multi-time formalism for two distant systems shows that also this theory does not present parameter-dependent effects.

7b. *Parameter Dependence and Relativistic Invariance Requirements*

In Ghirardi *et al.* (1993) the connection between nonlocality and "genuine Lorentz invariance" has been investigated.[12] In particular, it has been proved that it is not possible to build a "genuinely Lorentz invariant" theory which exhibits parameter dependence effects and which, in the nonrelativistic limit, does not give rise to backward causation. To analyze this point, let us start by considering, with reference to an EPR-Bohm like situation, the nonrelativistic limit of the theory in a given reference frame O. In this frame we assume that the measurement at right takes place at an earlier time than the one at left (i.e., $t_R < t_L$) and that there exist a set of values of the parameters λ which have a non zero probability of occurrence and are such that

$$(7.17.\text{I}) \quad p_\lambda^L(x; \mathbf{n}, \mathbf{m}) \neq p_\lambda^L(x; \mathbf{n}, *),$$

while, the absence of backward causation implies, for all λ's

$$(7.17.\text{II}) \quad p_\lambda^R(y; \mathbf{n}, \mathbf{m}) = p_\lambda^R(y; *, \mathbf{m}).$$

In the case of a Galilean theory, there is a residue, or an analogue of Lorentz invariance, which can be taken into account in the case in which one is dealing with two widely separated systems. In particular, if one assumes that the two regions R and L are very far apart from each other, there exist reference frames (say, O') moving even with a very small velocity with respect to O, but for which the temporal order of the space-time events (R, t_R) and (L, t_L) is inverted (i.e., $t'_R > t'_L$). Since the probability of occurrence of an event is an objective fact (i.e., it cannot depend on the reference frame considered), it must hold also for O' that

$$(7.18) \quad p'^L_{\lambda'}(x'; \mathbf{n}', \mathbf{m}') \neq p'^L_{\lambda'}(x'; \mathbf{n}', *),$$

where λ' is the parameter which identifies for O' the systems which are identified by λ for O, and the primes on p, x, $\mathbf{n}$ and $\mathbf{m}$ have an obvious meaning. It is important to stress that (7.18) is simply the requirement that both observers can look at the same systems and that they agree on the objective probabilities. However, as already remarked, for O', $t'_R > t'_L$. If the theory were invariant for the transformation under consideration, and analogous situation would occur also for the observer O for some values of λ. But this contradicts the assumption (7.17.II) expressing that for O the past cannot have a parametric dependence from the future. This argument constitutes, therefore, a proof that, under the considered assumptions, there cannot be a "genuinely relativistic invariant" theory which exhibits parameter dependence. Here we understand "genuinely relativistic" rather strongly: it excludes theories (like Lorentz's classical electromagnetic theory, or the quantum field theory version of Bohm's theory (Bohm 1952b) that have a preferred frame which cannot be discovered by experiments. We favor genuinely relativistic theories. We recall that Bell, despite his deep appreciation of Bohm's theory, believed (Bell 1989) that to make the Lorentz group phenomenological in this way *is an incredible position to take – "I think it is quite logically consistent, but when one sees the power of the hypothesis of Lorentz invariance in modern physics, I think you just can't believe in it."* (Bell 1989, 13. Emphases added.)

In a recent preprint (Berndl, Dürr, Goldstein and Zanghì 1996) attention has been called to the fact that there seems to be the possibility (Dürr *et al.* 1990) of a Bohmian quantum field theory *in which a foliation of space-time into space-like hypersurfaces is an additional beable*. Because of this, such a theory has no preferred reference frame. It can circumvent our impossibility proof because, in the nonrelativistic limit, it exhibits backward causation effects. The authors assume that the hypersurfaces of the foliation do not intersect and this would be a nice feature of the model since it would guarantee that the backward causation effects implied by it would not give rise to paradoxical consequences (see below.) However, it seems to us that the model is not so precisely formulated to allow one to be sure that the above assumption is dynamically consistent.

Another attempt one could think of would be to look, for example, for a Tomonaga–Schwinger type equation for the beables f with no particular foliation $\{\sigma\}$. Obviously, one could meet serious difficulties: this equation could turn out to be non integrable, etc. However, we think that even if a theory of this type could be formulated, due to the fact that the space-like hypersurfaces cross each other and there are parameter dependent effects (which allow faster than light signalling should one have access to the hidden parameters λ), it would exhibit backward causation effects which would give rise to paradoxical situations. Typically, one could consider three space-time points A, B, C such that B is space-like with respect to both A and C, while C is in the absolute past of A. Should one have access to be hidden variables, there might be situations in which A could cause an effect at B, and B might correspondingly make C aware of what A has done. C being in the absolute past of A, he could forbid A to perform the original action.

To conclude this section we remark that, in our opinion, it is not so obvious that there could be a genuinely Lorentz invariant theory exhibiting parameter dependent effects.

8. CONCLUSIONS

In this paper we have compared Bohm's theory and the dynamical reduction theories with reference to the most relevant conceptual issues about the foundational problems of quantum theory. We have pointed out the pros and cons of both approaches. Our conclusion is that they both deserve to be taken seriously as yielding a possible realistic description of nature.

ACKNOWLEDGMENT

We are greatly indebted to Professor S. Goldstein for sharp and useful comments.

University of Trieste

NOTES

[1] It is useful to point out that at least two different attitudes can be taken about the interpretation of the theory. They have been focused in a recent paper by Bas van Fraassen (1994) and have been denoted as the Cartesian and the Newtonian construals, respectively. According to the first attitude, what is real of the theory are only the positions of the particles. Consequently, since equal positions at a given time can evolve into different positions at a subsequent time, under this construal the theory is not deterministic. In the Newtonian construal, besides the position of the particles, also the wave function is considered as real. The theory is then strictly deterministic. We will take the Newtonian attitude in what follows.

[2] Recently, some beautiful results about this point (Dürr *et al.* 1992a) have been derived. The authors shows that typical configurations of the universe as a whole imply that the positions of the particles of an ensemble associated to a given wave function Ψ are distributed according to $|\Psi|^2$ with overwhelming probability. We cannot discuss this important point here. We refer the reader to the above paper.

[3] In Eq. (3.8) and following we have changed the notation for the state vector from $|\Psi_w(t)\rangle$ used in Eq. (3.2) to $|\Psi_B(t)\rangle$ to stress the fact that, under our assumptions, the state at time t does not depend on the specific sample function $w(\tau)$ in the interval $(0, t)$ but only on its integral $B(t)$ of Eq. (3.9).

[4] Note that, even though the spread $\sqrt{\gamma t}$ tends to ∞ for $t \to \infty$, its ratio to the distance $2(\alpha - \beta)\gamma t$ between the two considered peaks of the distribution tends to zero.

[5] In a previous paper (Ghirardi *et al.* 1995) we have simply used the expression "objective" to denote what we call here "objectively possessed or accessible". We had a vague feeling that the term we were using was not completely pertinent and could give rise to misunderstandings. Professor S. Goldstein has appropriately called our attention to the fact that both usual meanings of that term (i.e., "real" or "opposite to subjective") do not fit with the sense which emerges for it from our work and has suggested the expression "accessible". In the search for an expression which would embody precisely what we had

in mind, we have considered also the possibility of resorting to the expression "empirically adequate". However, this term reminds one directly of the "empirical reality" concept introduced by B. d'Espagnat (d'Espagnat 1990). Because of this fact, the use of such term could seriously mislead the reader. In fact, according to the definition of empirical reality, due to the practical impossibility of distinguishing, at the macroscopic level, a pure state from a statistical mixture, replacing one with the other would be empirically adequate, which is completely at odds with the meaning we want to give to the expression objectively possessed or accessible.

[6] We point out that here we assume that the measurement can be chosen by free will. In any case, even if one gives up such an assumption, it turns out to be impossible to attribute to the system considered by itself objective properties (i.e., properties which do not depend on the overall experimental set up). Accordingly, property attribution becomes a relational feature.

[7] We note that if the "pointer particle" would be on the negative semiaxis at $t = 0$, it would be shifted by a negative quantity of the same order, independently of the sign of c. Thus, the outcome of the measurement of p, given a complete specification at $t = 0$ of the state of the measured particle, depends both on the sign of the coupling constant (which can be chosen at free will by the observer) and on the initial position of the "pointer particle" (a parameter which, however, cannot be controlled by the experimenter).

[8] A typical example is represented by an intergalactic cloud with a very low mass density and wave functions for its particles with a large spread. According to our strict criterion (5.8), the mass density of the cloud is nonobjective, while, if we adopt the physically meaningful criterion put forward at the beginning of this section and to test it we resort to consider its effects (e.g., on the trajectory of a far asteroid), we would be led to state that our claims about the mass density are objectively true since they agree with the "physical outcome". Here, an important distinction is appropriate. The theory contains a specific dynamical mechanism and specific parameters which allow a precise mathematical formulation of the objectivity criterion. Obviously, when one is interested in a particular class of physical processes, the objectivity requirement at the beginning of this section could be satisfied even though the strict mathematical one is not. This does not raise any problem. It finds its counterpart in the fact that to analyze a specific physical situation one can use a coarser graining than the one corresponding to the fundamental characteristic length of the theory. Accordingly, if we adapt the graining to the problem of interest, we can assert that the mass of the cloud is objective. Obviously, while for practical purposes one can change the graining according to the problem he is interested in, adopting a graining which is finer than the fundamental one of the theory makes no sense because, if one takes the theory seriously as describing the laws governing the evolution of the universe, then mass density would almost never become objective for such a graining.

[9] For a more precise definition of such a set see (Ghirardi *et al.* 1995).

[10] This attitude has the same logical status as the assertion that the probabilities concerning physical processes have an epistemic character and are not due to the presence of genuine elements of chance in nature.

[11] In Eq. (7.8) and following we make, for the same reasons discussed there, a change of notation analogous to the one made for (3.8).

[12] See below for the precise meaning we attribute to the expression in quotation marks.

BIBLIOGRAPHY

Accardi, L. (ed.) (1994), *The Interpretation of Quantum Theory: Where Do We Stand?* Roma, Istituto della Enciclopedia Italiana.

Aharonov, Y., D. Z Albert and L. Vaidman (1988), "How the Result of a Measurement of a Component of the Spin of a Spin-1/2 Particle Can Turn Out To Be 100," *Physical Review Letters* **60**, 1351–1354.

Aharonov, Y., D. Z Albert, A. Casher and L. Vaidman (1987), "Surprising Quantum Effects," *Physics Letters A* **124**, 199–203.

Aharonov, Y., J. Anandan, S. Popescu and L. Vaidman (1990), "Superpositions of Time Evolutions of a Quantum System and a Quantum Time Machine," *Physical Review Letters* **64**, 2965–2968.

Aharonov, Y., P. G. Bergmann and J. L. Lebowitz (1964), "Time Symmetry in the Quantum Process of Measurement," *Physical Review B* **134**, 1410–1416.

Aharonov, Y. and L. Vaidman (1990), "Properties of a Quantum System During the Time Interval Between Two Measurements," *Physical Review A* **41**, 11–20.

Aharonov, Y. and L. Vaidman (1993), "Measurement of the Schrödinger Wave of a Single Particle," *Physics Letters A* **178**, 38–42.

AHQP (1962–1964), *Archive for the History of Quantum Physics*, interview transcripts on deposit in the Department of History of Science, University of California, Berkeley. [A catalogue of the contents of this archive has been published by Kuhn *et al.*]

Albert, D. Z (1992), *Quantum Mechanics and Experience*. Cambridge, MA, Harvard University Press.

Albert, D. Z (1994), "Bohm's Alternative to Quantum Mechanics," *Scientific American* **270** (May), 58–67.

Albert, D. Z and B. Loewer (1988), "Interpreting the Many-Worlds Interpretation," *Synthese* **77**, 195–213.

Albert, D. Z and B. Loewer (1991), "The Measurement Problem: Some 'Solutions'," *Synthese* **86**, 87–98.

Albert, D. Z and B. Loewer (1996), "Tails of Schrödinger's Cat" in Clifton, pp. 81–92.

Albeverio, S., G. Casati, U. Cattaneo, D. Merlini and R. Moresi (eds.) (1990), *Stochastic Processes, Physics and Geometry*. Singapore, World Scientific.

Allcock, G. R. (1969), "The Time of Arrival in Quantum Mechanics I. Formal Considerations," *Annals of Physics* **53**, 253–285.

Allén, S. (ed.) (1989), *Possible Worlds in Humanities, Arts and Sciences: Proceedings of Nobel Symposium 65*. Berlin, Walter de Gruyter.

Anandan, J. (1993), "Protective Measurement and Quantum Reality," *Foundations of Physics Letters* **6**, 503–532.

Anandan, J. and H. R. Brown (1995), "On the Reality of Space-Time Geometry and the Wave Function," *Foundations of Physics* **25**, 349–360.

Arnold, V. I. (1963), "Small Denominators and Problems of Stability of Motion in Classical and Celestial Mechanics," *Russian Mathematical Surveys* **18**(6), 85–191.

Arnold, V. I. (1989), *Mathematical Methods of Classical Mechanics*. 2nd ed. New York, Springer.

Aspect, A., J. Dalibard and G. Roger (1982b), "Experimental Test of Bell's Inequalities Using Time-Varying Detectors," *Physical Review Letters* **49**, 1804–1807.

Aspect, A., P. Grangier and G. Roger (1981), "Experimental Tests of Local Realistic Theories via Bell's Theorem," *Physical Review Letters* **47**, 460–463.

J. T. Cushing et al. (eds.), Bohmian Mechanics and Quantum Theory: An Appraisal, 379–397.

Aspect, A., P. Grangier and G. Roger (1982a), "Experimental Realisation of Einstein-Podolsky-Rosen-Bohm *Gedankenexperiment*: a New Violation of Bell's Inequalities," *Physical Review Letters* **49**, 91–94.

Atmanspacher, H. and H. Scheingraber (eds.) (1991), *Information Dynamics*. New York, Plenum Press.

Bacciagaluppi, G. and M. Dickson (1995), "Modal Interpretations with Dynamics," (unpublished manuscript).

Baublitz, M. (1996), "An Appraisal of Stochastic Interpretations of Quantum Mechanics" in Jeffers *et al.* (to be published).

Baublitz, M. and A. Shimony (1996), "Tension in Bohm's Interpretation of Quantum Mechanics" in this volume.

Baz', A. I. (1967), "Lifetime of Intermediate States," *Soviet Journal of Nuclear Physics* **4**, 182–188.

Behm, R. J., N. Garcìa and H. Rohrer (eds.) (1990), *Scanning Tunneling Microscopy and Related Methods*. Dordrecht, Kluwer Academic Publishers.

Belinfante, F. J. (1973), *A Survey of Hidden-Variables Theories*. Oxford, Pergamon Press.

Bell, J. S. (1964), "On the Einstein-Podolsky-Rosen Paradox," *Physics* **1**, 195–200. Reprinted in Bell (1987a), pp. 14–21.

Bell, J. S. (1966), "On the Problem of Hidden Variables in Quantum Mechanics," *Reviews of Modern Physics* **38**, 447–452. Reprinted in Bell (1987a), pp. 1–13.

Bell, J. S. (1971), "Introduction to the Hidden-Variable Question" in d'Espagnat (1971), pp. 171–181. Reprinted in Bell (1987a), pp. 29–39.

Bell, J. S. (1975), "The Theory of Local Beables," CERN report TH-2053. Reprinted in Bell (1987a), pp. 52–62.

Bell, J. S. (1976), "How to Teach Special Relativity," *Progress in Scientific Culture* **1**(2). Reprinted in Bell (1987a), pp. 67–80.

Bell, J. S. (1977), "Free Variables and Local Causality," *Epistemological Letters* (Feb., 1977). Reprinted in Bell (1987a), pp. 100–104.

Bell, J. S. (1981a), "Quantum Mechanics for Cosmologists" in Isham *et al.*, pp. 611–637. Reprinted in Bell (1987a), pp. 117–138.

Bell, J. S. (1981b), "Bertlmann's Socks and the Nature of Reality," *Journal de Physique*, Colloque C2, suppl. au numero 3, Tome **42**, 41–61. Reprinted in Bell (1987a), pp. 139–158.

Bell, J. S. (1984), "Beables for Quantum Field Theory," CERN-TH 4035/84. Reprinted in Bell (1987a), pp. 173–180.

Bell, J. S. (1986), "Six Possible Worlds of Quantum Mechanics" in Allén (1989), pp. 359–373. Reprinted in Bell (1987a), pp. 181–195.

Bell, J. S. (1987a), *Speakable and Unspeakable in Quantum Mechanics*. Cambridge, Cambridge University Press.

Bell, J. S. (1987b), "Are There Quantum Jumps?" in Kilmister, pp. 41–52. Reprinted in Bell (1987a), pp. 201–212.

Bell, J. S. (1989), "Towards an Exact Quantum Mechanics" in Deser and Finkelstein, pp. 1–26.

Bell, J. S. (1990), "Against Measurement" in Miller, pp. 17–31. [Also appeared in *Physics World* **3**(8) (1990), 33–40.]

Beller, M. (1983a), "The Genesis of Interpretations of Quantum Physics, 1925–1927," unpublished Ph.D. dissertation, University of Maryland.

Beller, M. (1983b), "Matrix Theory Before Schrödinger: Philosophy, Problems, Consequences," *Isis* **74**, 469–491.

Beller, M. (1985), "Pascual Jordan's Influence on the Discovery of Heisenberg's Indeterminacy Principle," *Archive for History of Exact Sciences* **33**, 337–349.

Beller, M. (1988), "Experimental Accuracy, Operationalism, and the Limits of Knowledge – 1925 to 1935," *Science in Context* **2**, 147–162.

Beller, M. (1990), "Born's Probabilistic Interpretation: A Case study of 'Concepts in Flux'," *Studies in History and Philosophy of Science* **21**, 563–588.

Beller, M. (1992), "Schrödinger's Dialogue with Göttingen-Copenhagen Physicists – 'Quantum Jumps' and Realism" in Bitbol and Darrigol, pp. 277–306.

Beller, M. (1993), "Einstein and Bohr's Rhetoric of Complementarity," *Science in Context* **6**, 241–255.

Beller, M. (1996), "Rhetoric of Antirealism and the Copenhagen Spirit," *Philosophy of Science* (to be published).

Beller, M. and A. Fine (1994), "Bohr's Response to EPR" in Fay and Fosle, pp. 1–31.

Belousek, D. (1996), "Einstein's 1927 Unpublished Hidden Variable Theory: Its Background, Context and Significance," *Studies in History and Philosophy of Modern Physics* (to be published).

Beltrametti, E. G. and B. C. van Fraassen (eds.) (1981), *Current Issues in Quantum Logic*. New York, Plenum Press.

Benatti, F., G. C. Ghirardi, A. Rimini and T. Weber (1988), "Operations Involving Momentum Variables in Non-Hamiltonian Evolution Equations," *Il Nuovo Cimento B* **101**, 333–355.

Berezin, F. A. (1966), *The Method of Second Quantisation*. New York, Academic Press.

Berezin, V. A. and V. A. Rubakov (eds.) (1995), *Proceedings of the Sixth Seminar on Quantum Gravity, June 12–19, 1995, Moscow, Russia*. Singapore, World Scientific (to be published).

Berndl, K. (1995), *Zur Existenz der Dynamik in Bohmschen Systemen*, Ph. D. thesis. Aachen, Mainz Verlag.

Berndl, K. (1996), "Global Existence and Uniqueness of Bohmian Trajectories" in this volume.

Berndl, K., M. Daumer, D. Dürr, S. Goldstein and N. Zanghì (1995), "A Survey of Bohmian Mechanics," *Il Nuovo Cimento* **110B**, 737–750.

Berndl, K., D. Dürr, S. Goldstein, G. Peruzzi and N. Zanghì (1995), "On the Global Existence of Bohmian Mechanics," *Communications in Mathematical Physics* **173**, 647–673.

Berndl, K., D. Dürr, S. Goldstein and N. Zanghì (1994), "Towards a Relativistic Quantum Theory of Particles (Having Trajectories)," talk given in Cambridge at the Third UK Conference on Foundations of Quantum Theory and Relativity.

Berndl, K., D. Dürr, S. Goldstein and N. Zanghì (1996), "Nonlocality, Lorentz Invariance, and Bohmian Quantum Theory," *Physical Review A* **53**, 2062–2073.

Bitbol, M. and O. Darrigol (eds.) (1992), *Erwin Schrödinger: Philosophy and the Birth of Quantum Mechanics*. Gif-sur-Yvette Cedex, France, Editions Frontières.

Blum, W., H.-P. Dürr and H. Rechenberg (eds.) (1984), *Werner Heisenberg. Collected Works. CI*. München, Piper.

Bohm, D. (1951), *Quantum Theory*. Englewood Cliffs, NJ, Prentice-Hall.

Bohm, D. (1952a), "A Suggested Interpretation of the Quantum Theory in Terms of 'Hidden' Variables. I," *Physical Review* **85**, 166–179. Reprinted in Wheeler and Zurek, pp. 369–382.

Bohm, D. (1952b), "A Suggested Interpretation of the Quantum Theory in Terms of 'Hidden' Variables. II," *Physical Review* **85**, 180–193. Reprinted in Wheeler and Zurek, pp. 383–396.

Bohm, D. (1952c), "Reply to a Criticism of a Causal Re-Interpretation of the Quantum Theory," *Physical Review* **87**, 389–390.

Bohm, D. (1953a), "Comments on a Letter Concerning the Causal Interpretation of the Quantum Theory," *Physical Review* **89**, 319–320.

Bohm, D. (1953b), "Proof That Probability Density Approaches $|\psi|^2$ in Causal Interpretation of the Quantum Theory," *Physical Review* **89**, 458–466.

Bohm, D. (1953c), "A Discussion of Certain Remarks by Einstein on Born's Probability Interpretation of the ψ-Function" in Born (1953), pp. 13–19.

Bohm, D. (1957), *Causality and Chance in Modern Physics*. London, Routledge & Kegan Paul.

Bohm, D. (1969), "Some Remarks on the Notion of Order" in Waddington, pp. 18–40.

Bohm, D. (1961–62), "Classical and Non-Classical Concepts in the Quantum Theory" in Toulmin, pp. 197–216.

Bohm, D. (1980), *Wholeness and the Implicate Order*. New York, Routledge & Kegan Paul.

Bohm, D. (1987), "Hidden Variables and the Implicate Order" in Hiley and Peat, pp. 33–45.

Bohm, D. and B. J. Hiley (1984), "Measurement Understood Through the Quantum Potential Approach," *Foundations of Physics* **14**, 255–274.

Bohm, D. and B. J. Hiley (1989), "Non-Locality and Locality in the Stochastic Interpretation of Quantum Mechanics," *Physics Reports* **172**, 93–122.

Bohm, D. and B. J. Hiley (1993), *The Undivided Universe: An Ontological Interpretation of Quantum Mechanics*. London, Routledge.

Bohm, D., B. J. Hiley and P. N. Kaloyerou (1987), "An Ontological Basis for the Quantum Theory," *Physics Reports* **144**, 321–375.

Bohm, D. and F. D. Peat (1987), *Science, Order and Creativity*. New York, Bantam.

Bohm, D., R. Schiller and J. Tiomno (1955), "A Causal Interpretation of the Pauli Equation (A)," *Supplemento al Nuovo Cimento* **1**, 48–91.

Bohm, D. and J.-P. Vigier (1954), "Model of the Causal Interpretation of Quantum Theory in Terms of a Fluid with Irregular Fluctuations," *Physical Review* **96**, 208–216.

Bohr, N. (1928), "The Quantum Postulate and the Recent Development of Atomic Theory," *Nature* **121**, 580–590. Reprinted in Bohr (1934), pp. 52–91, and in Wheeler and Zurek, pp. 87–126.

Bohr, N. (1929), "The Atomic Theory and the Fundamental Principles Underlying the Description of Nature" in Bohr (1934), pp. 102–119. [Originally appeared in Danish in *Fysisk Tidsskrift* (1929).]

Bohr, N. (1930), "Philosophical Aspects of Atomic Theory," Address to the Royal Society of Edinburgh. Bohr Manuscript Collection, AHQP.

Bohr, N. (1931), "Space-Time Description and Conservation Principles," Bohr Manuscript Collection, AHQP.

Bohr, N. (1933), "Light in Life," *Nature* **131**, 421–423; 457–459.

Bohr, N. (1934), *Atomic Theory and the Description of Nature*. Cambridge, Cambridge University Press.

Bohr, N. (1935), "Can Quantum-Mechanical Description of Physical Reality Be Considered Complete?" *Physical Review* **48**, 696–702. Reprinted in Wheeler and Zurek, pp. 145–151.

Bohr, N. (1937a), The Hitchcock Lectures, University of California, Berkeley. Manuscript in the Raymond Birge Collection, American Institute of Physics.

Bohr, N. (1937b), "Causality and Complementarity," *Philosophy of Science* **4**, 289–298. [Originally appeared in 1936 as "Kausalität und Komplementarität" in *Erkenntnis* **6**, 293–303.]

Bohr, N. (1948), "On the Notions of Causality and Complementarity," *Dialectica* **2**, 312–319.

Bohr, N. (1949), "Discussion with Einstein on Epistemological Problems in Atomic Physics" in Schilpp, pp. 199–241.

Bohr, N. (1957), "The Philosophical Lessons of Atomic Physics," Bohr Manuscript Collection, AHQP.

Bohr, N. (1958a), "Atoms and Human Knowledge," Bohr Manuscript Collection, AHQP.

Bohr, N. (1958b), *Atomic Physics and Human Knowledge*. New York, John Wiley & Sons.

Bohr, N. (1963), *Essays 1958–1962 on Atomic Physics and Human Knowledge*. London, Interscience.

Born, M. (1926), "Quantenmechanik der Stossvorgänge," *Zeitschrift für Physik* **38**, 803–827. Translated as "Quantum Mechanics of Collision Processes" in Ludwig, pp. 206–225.

Born, M. (1949), *Natural Philosophy of Cause and Chance*. Oxford, Clarendon Press.

Born, M. (1951), *The Restless Universe*. New York, Dover Publications.

Born, M. (1953), *Scientific Papers Presented to Max Born*. New York, Hafner Publishing Co.

Born, M. (1971), *The Born-Einstein Letters*. New York, Walker and Company.

Brenig, W. and R. Haag (1959), "Allgemeine Quantentheorie der Stossprozesse," *Fortschritte der Physik* **7**, 183–242. Appears in translation as "General Quantum Theory of Collision Processes" in Ross, pp. 13–108.

de Broglie, L. (1926), "Sur la possibilité de relier phénomènes d'interférence et de diffraction à la théorie des quanta de lumière," *Comptes Rendus des Séances de l'Académie des Sciences* **183**, 447–448.

de Broglie, L. (1927), "Sur le rôle des ondes continues Ψ en Mécanique ondulatoire," *Comptes Rendus des Séances de l'Académie des Sciences* **185**, 380–382.

de Broglie, L. (1928), "La nouvelle Dynamique des Quanta" in *Electrons et Photons*, pp. 105–132.

de Broglie, L. (1953), "Vue d'Ensemble sur mes Travaux scientifiques" in George, pp. 457–486.

de Broglie, L. (1956), *Une Tentative d'Interprétation Causale et Non-Linéaire de la Mécanique Ondulatoire*. Paris, Gauthier-Villars. [Appears in English translation in de Broglie (1960).]

de Broglie, L. (1960), *Non-Linear Wave Mechanics: A Causal Interpretation*. Amsterdam, Elsevier Publishing Co.

de Broglie, L. (1970), "The Reinterpretation of Wave Mechanics," *Foundations of Physics* **1**, 5–15.

Brouard, S., R. Sala and J. G. Muga (1994), "Systematic Approach to Define and Classify Quantum Transmission and Reflection Times," *Physical Review A* **49**, 4312–4325.

Brown, H. R. (1986), "The Insolubility Proof of the Quantum Measurement Problem," *Foundations of Physics* **16**, 857–870.

Brown, H. R. (1992), "Bell's Other Theorem and its Connection with Nonlocality. Part I" in van der Merwe *et al.*, pp. 104–116.

Brown, H. R. (1996), "Bovine Metaphysics: Remarks on the Significance of the Gravitational Phase Effect in Quantum Mechanics" in Clifton, pp. 183–193.

Brown, H. R., C. Dewdney and G. Horton (1995), "Bohm Particles and Their Detection in the Light of Neutron Interferometry," *Foundations of Physics* **25**, 329–347.

Brown, H. and R. Harré (eds.) (1988), *Philosophical Foundations of Quantum Field Theory*. Oxford, Oxford University Press.

Brown, H. R. and R. Sypel (1995), "On the Meaning of the Relativity Principle and Other Symmetries," *International Studies in the Philosophy of Science* **9**, 233–251.

Brown, L. M. (1958), "Two-Component Fermion Theory," *Physical Review* **111**, 957–964.

Bub, J. (1992), "Quantum Mechanics Without the Projection Postulate," *Foundations of Physics* **22**, 737–754.

Bub, J. (1994a), "On the Structure of Quantal Proposition Systems," *Foundations of Physics* **24**, 1261–1279.

Bub, J. (1994b), "How to Interpret Quantum Mechanics," *Erkenntnis* **41**, 253–273.

Bub, J. (1995), "Interference, Noncommutativity, and Determinateness in Quantum Mechanics," *Topoi* **14**, 39–43.

Bub, J. and R. Clifton (1996), "A Uniqueness Theorem for Interpretations of Quantum Mechanics," *Studies in History and Philosophy of Modern Physics* (to be published).

Bukharin, N. I. *et al.* (1931), *Science at the Cross Roads* [new edition (1971)]. London, Cass.

Butterfield, J. (1989), "A Space-Time Approach to the Bell Inequality" in Cushing and McMullin, pp. 114–144.

Büttiker, M. (1983), "Larmor Precession and the Traversal Time for Tunneling," *Physical Review B* **27**, 6178–6188.

Büttiker, M. and R. Landauer (1982), "Traversal Time for Tunneling," *Physical Review Letters* **49**, 1739–1742.

Candelas, P. and D. W. Sciama (1984), "Is There a Quantum Equivalence Principle?" in Christensen, pp. 78–88.

Carnap, R. (1966), *Philosophical Foundations of Physics*. New York, Basic Books.

Casimir, H. B. G. (1967), "Recollections from the Years 1929–1931" in Rozental (eds.), pp. 109–113.

Cassidy, D. C. (1992), *Uncertainty: The Life and Science of Werner Heisenberg*. New York, W. H. Freeman and Co.

Choquet-Bruhat, Y. and J. W. York (1980), "The Cauchy Problem" in Held, pp. 99–172.

Christensen, S. M. (ed.) (1984), *Quantum Theory of Gravity: Essays in Honor of the 60th Birthday of Bryce S. DeWitt*. Bristol, Adam Hilger Ltd.

Clifton, R. (1993), "Getting Contextual and Nonlocal Elements-of-Reality the Easy Way," *American Journal of Physics* **61**, 443–447.

Clifton, R. (1995), "Independently Motivating the Kochen-Dieks Modal Interpretation of Quantum Mechanics," *British Journal for the Philosophy of Science* **46**, 33–57.

Clifton, R. (ed.) (1996), *Perspectives on Quantum Reality*. Dordrecht, Kluwer Academic Publishers.

Combes, J.-M., R. G. Newton and R. Shtokhamer (1975), "Scattering Into Cones and Flux Across Surfaces," *Physical Review D* **11**, 366–372.

Courant, R. and D. Hilbert (1989), *Methods of Mathematical Physics*, Vol. II. New York, John Wiley & Sons.

Cushing, J. T. (1990), *Theory Construction and Selection in Modern Physics: The S Matrix*. Cambridge, Cambridge University Press.

Cushing, J. T. (1994), *Quantum Mechanics: Historical Contingency and the Copenhagen Hegemony*. Chicago, University of Chicago Press.

Cushing, J. T. (1995a), "Quantum Tunneling Times: A Crucial Test for the Causal Program?", *Foundations of Physics* **25**, 269–280.

Cushing, J. T. (1995b), "Review of David Bohm and Basil Hiley's *The Undivided Universe: An Ontological Interpretation of Quantum Theory*," *Foundation of Physics* **25**, 507–510.

Cushing, J. T. and E. McMullin (eds.) (1989), *Philosophical Consequences of Quantum Theory: Reflections on Bell's Theorem*. Notre Dame, IN, University of Notre Dame Press.

Daneri, A., A. Loinger and G. M. Prosperi (1962), "Quantum Theory of Measurement and Ergodicity Conditions," *Nuclear Physics* **33**, 297–319.

Daneri, A., A. Loinger and G. M. Prosperi (1966), "Further Remarks on the Relations Between Statistical Mechanics and Quantum Theory of Measurement," *Il Nuovo Cimento B* **44**, 119–128.

Daumer, M. (1995), *Streutheorie aus der Sicht Bohmscher Mechanik*, Ph.D. thesis. Aachen, Mainz Verlag.

Daumer, M. (1996), "Scattering Theory from a Bohmian Perspective" in this volume.

Daumer, M., D. Dürr, S. Goldstein and N. Zanghì (1994), "Scattering and the Role of Operators in Bohmian Mechanics" in Fannes *et al.*, pp. 331–338.

Daumer, M., D. Dürr, S. Goldstein and N. Zanghì (1996), "On the Role of Operators in Quantum Theory," in preparation.

Davies, E. B. (1976), *Quantum Theory of Open Systems*. London, Academic Press.

Davies, P. C. W. (1974), *The Physics of Time Asymmetry*. Berkely, University of California Press.

Davies, P. C. W. (1986), "Measurement of the Velocity of a Dirac Particle," *Journal of Physics A: Mathematical and General* **19**, 2115–2121.

Delgado, V., S. Brouard and J. G. Muga (1995), "Does Positive Flux Provide a Valid Definition of Tunnelling Times?", *Solid State Communications* **94**, 979–982.

Derezinski, J. (1993), "Asymptotic Completeness of N-Particle Systems," *Annals of Mathematics* **138**, 427–476.

Deser, S. and R. J. Finkelstein (eds.) (1989), *Themes in Contemporary Physics II*. Singapore, World Scientific.

Destri, C. and H. J. de Vega (1987), "Light-Cone Lattice Approach to Fermionic Theories in 2D: The Massive Thirring Model," *Nuclear Physics B* **290**, 363–391.

Dewdney, C. (1985), "Particle Trajectories and Interference in a Time-Dependent Model of Neutron Single Crystal Interferometry," *Physics Letters A* **109**, 377–384.

Dewdney, C. (1992), "Constraints on Quantum Hidden-Variables and the Bohm Theory," *Journal of Physics A: Mathematical and General* **25**, 3615–3626.

Dewdney, C. (1994), "Are There Quantum Jumps?" in Laurikainen *et al.*, pp. 71–84.
Dewdney, C. (1997), *Windows on the Quantum World.* London, John Wiley & Sons (to be published).
Dewdney, C., L. Hardy and E. J. Squires (1993), "How Late Measurements of Quantum Trajectories Can Fool a Detector," *Physics Letters A* **184**, 6–11.
Dewdney, C., P. R. Holland and A. Kyprianidis (1987), "A Quantum Potential Approach to Spin Superposition in Neutron Interferometry," *Physics Letters A* **121**, 105–110.
Dewdney, C., P. R. Holland, A. Kyprianidis and J.-P. Vigier (1988), "Spin and Non-Locality in Quantum Mechanics," *Nature* **336**, 536–544.
Dewdney, C., G. Horton, M. M. Lam, Z. Malik, and M. Schmidt (1992), "Wave-Particle Dualism and the Interpretation of Quantum Mechanics," *Foundations of Physics* **22**, 1217–1265.
Dewdeny, C. and M. M. Lam (1991), "What Happens in a Quantum Transition?" in Atmanspacher and Scheingraber, pp. 329–341.
DeWitt, B. D. and N. Graham (eds.) (1973), *The Many-Worlds Interpretation of Quantum Mechanics.* Princeton, NJ, Princeton University Press.
Diacu, F. N. (1992), *Singularities of the N-Body Problem.* Montréal, Les Publications CRM.
Diacu, F. N. (1993), "Painlevé's Conjecture," *Mathematical Intelligencer* **15**(2), 6–12.
Dickson, W. M. (1995), "Faux-Boolean Algebras and Classical Models," *Foundations of Physics Letters* **8**, 401–415.
Dieks, D. (1989), "Quantum Mechanics Without the Projection Postulate and Its Realistic Interpretation," *Foundations of Physics* **19**, 1397–1423.
Dieks, D. (1994), "Modal Interpretation of Quantum Mechanics, Measurements, and Macroscopic Behavior," *Physical Review A* **49**, 2290–2300.
Dirac, P. A. M. (1935), *The Principles of Quantum Mechanics*, 2nd ed. Oxford, Oxford University Press.
Dirac, P. A. M. (1958a), *The Principles of Quantum Mechanics*, 4th ed. Oxford, Oxford University Press.
Dirac, P. A. M. (1958b), "Generalized Hamiltonian Dynamics," *Proceedings of the Royal Society of London A* **246**, 326–332.
Dirac, P. A. M. (1958c), "The Theory of Gravitation in Hamiltonian Form," *Proceedings of the Royal Society of London A* **246**, 333–343.
Dollard, J. D. (1964), "Asymptotic Convergence and the Coulomb Interaction," *Journal of Mathematical Physics* **5**, 729–738.
Dollard, J. D. (1969), "Scattering into Cones I: Potential Scattering," *Communications in Mathematical Physics* **12**, 193–203.
Dowker, F. and A. Kent (1995), "Properties of Consistent Histories," *Physical Review Letters* **75**, 3038–3041.
Dowker, F. and A. Kent (1996), "On the Consistent Histories Approach to Quantum Mechanics," *Journal of Statistical Physics* **82**, 1575–1646.
Dumont, R. S. and T. L. Marchioro (1993), "Tunneling-Time Probability Distribution," *Physical Review A* **47**, 85–97.
Dürr, D., W. Fusseder, S. Goldstein and N. Zanghì (1993), "Comment on 'Surrealistic Bohm Trajectories'," *Zeitschrift für Naturforchung* **48a**, 1261–1262.
Dürr, D., S. Goldstein and N. Zanghì (1990), "On a Realistic Theory for Quantum Physics" in Albeverio *et al.*, pp. 374–391.
Dürr, D., S. Goldstein and N. Zanghì (1992a), "Quantum Equilibrium and the Origin of Absolute Uncertainty," *Journal of Statistical Physics* **67**, 843–907.
Dürr, S., S. Goldstein and N. Zanghì (1992b), "Quantum Mechanics, Randomness, and Deterministic Reality," *Physics Letters A* **172**, 6–12.
Dürr, D., S. Goldstein and N. Zanghì (1992c), "Quantum Chaos, Classical Randomness, and Bohmian Mechanics," *Journal of Statistical Physics* **68**, 259–270.
Dürr, D., S. Goldstein and N. Zanghì (1993), "A Global Equilibrium as the Foundation of Quantum Randomness," *Foundations of Physics* **23**, 721–738.

Dürr, D., S. Goldstein and N. Zanghì (1996a), "Bohmian Mechanics as the Foundation of Quantum Mechanics" in this volume.

Dürr, D., S. Goldstein and N. Zanghì (1996b), "Bohmian Mechanics, Identical Particles, Parastatistics and Anyons," in preparation.

Earman, J. (1986), *A Primer on Determinism*. Dordrecht, Reidel Publishing Co.

Eberhard, P. H. (1978), "Bell's Theorem and the Different Concepts of Locality," *Il Nuovo Cimento B* **46**, 392–419.

Edelen, D. G. B. (1963), "On the Foundations of Relativistic Energy Mechanics," *Il Nuovo Cimento* **30**, 292–320.

Einstein, A. (1905), "Zur Electrodynamik bewegter Körper," *Annalen der Physik* **17**, 891–921. [Translated as "On the Electrodynamics of Moving Bodies" in Lorentz *et al.*, pp. 35–65.]

Einstein, A. (1926), "Letter to Max Born (December 4, 1926)" in Born (1971), pp. 90–91.

Einstein, A. (1949), "Autobiographical Notes" in Schilpp, pp. 3–95.

Einstein, A. (1953), "Elementare Überlegungen zur Interpretation der Grundlagen der Quanten-Mechanik" in M. Born (1953), pp. 33–40.

Einstein, A. (1954), *Ideas and Opinions*. New York, Crown Publishing Co.

Einstein, A., B. Podolsky and N. Rosen (1935), "Can Quantum-Mechanical Description of Physical Reality Be Considered Complete?", *Physical Review* **47**, 777–780. Reprinted in Wheeler and Zurek, pp. 138–141.

Electrons et Photons: Rapports et Discussions du Cinquième Conseil de Physique tenu á Bruxelles du 24 au 29 Octobre 1927 sous les Auspices de l'Institut International de Physique Solvay (1928). Paris, Gauthier-Villars.

Englert, B.-G., M. O. Scully, G. Süssmann and H. Walther (1992), "Surrealistic Bohm Trajectories," *Zeitschrift für Naturforschung* **47a**, 1175–1186.

Englert, B.-G., M. O. Scully, G. Süssmann and H. Walther (1993), "Reply to Comment on 'Surrealistic Bohm Trajectories'," *Zeitschrift für Naturforschung* **48a**, 1263–1264.

Enss, V. (1978), "Asymptotic Completeness for Quantum Mechanical Potential Scattering," *Communications in Mathematical Physics* **61**, 285–291.

Enss, V. and B. Simon (1980), "Finite Total Cross Sections in Nonrelativistic Quantum Mechanics," *Communications in Mathematical Physics* **76**, 177–209.

d'Espagnat, B. (ed.) (1971), *Foundations of Quantum Mechanics*. New York, Academic Press.

d'Espagnat, B. (1990), "Towards a Separable "Empirical Reality?", *Foundations of Physics* **20**, 1147–1172.

d'Espagnat, B. (1995), *Veiled Reality*. Reading, MA, Addison-Wesley.

Everett, H. (1957), "'Relative State' Formulation of Quantum Mechanics," *Reviews of Modern Physics* **29**, 454–462. Reprinted in DeWitt and Graham, pp. 141–149, and in Wheeler and Zurek, pp. 315–323.

Faddeev, L. D. (1965), *Mathematical Aspects of the Three-Body Problem in the Quantum Scattering Theory* (Academy of Sciences of the U.S.S.R., Works of the Steklov Mathematical Institute Vol. **69**). Israel, Israel Program for Scientific Translation.

Fannes, M., C. Maes and A. Verbeure (eds.) (1994), *On Three Levels: The Mathematical Physics of Micro-, Meso-, and Macro-Approaches to Physics*. New York, Plenum.

Faye, J. (1991), *Niels Bohr: His Heritage and Legacy*. Dordrecht, Kluwer Academic Publishers.

Faye, J. (1994), "Once More: Bohr-Høffding," *Danish Yearbook of Philosophy* **29**, 106–113.

Fay, J. and H. J. Folse (eds.) (1994), *Niels Bohr and Contemporary Philosophy*. Dordrecht, Kluwer Academics Publishers.

Feyerabend, P. (1978), *Against Method*. London, Verso.

Feynman, R. P. (1948), "Space-Time Approach to Non-Relativistic Quantum Mechanics," *Reviews of Modern Physics* **20**, 367–387.

Feynman, R. P. (1951), "An Operator Calculus Having Applications in Quantum Electrodynamics," *Physical Review* **84**, 108–128.

Feynman, R. P. and M. Gell-Mann (1958), "Theory of the Fermi Interaction," *Physical Review* **109**, 193–198.

Feynman, R. P., R. B. Leighton and M. Sands (1963–65), *The Feynman Lectures on Physics, Vols. I–III.* Reading, MA, Addison-Wesley.

Fine, A. (1970), "Insolubility of the Quantum Measurement Problem," *Physical Review D* **2**, 2783–2787.

Fine, A. (1973), "Probability and the Interpretation of Quantum Mechanics," *British Journal for the Philosophy of Science* **24**, 1–37.

Fine, A. (1974), "On the Completeness of Quantum Theory," *Synthese* **29**, 257–289.

Fine, A. (1982), "Joint Distributions, Quantum Correlations, and Commuting Observables," *Journal of Mathematical Physics* **23**, 1306–1310.

Fine, A. (1986), *The Shaky Game: Einstein, Realism and the Quantum Theory*. Chicago, University of Chicago Press.

Fine, A. (1989), "Correlations and Efficiency: Testing the Bell Inequalities," *Foundations of Physics* **19**, 453–478.

Fine, A. and J. Leplin (eds.) (1989), *PSA 1988: Proceedings of the 1988 Biennial Meeting of the Philosophy of Science Association*, Vol. 2 East Lansing, MI, Philosophy of Science Association.

Finkelstein, D. (1987), "All is Flux" in Hiley and Peat, pp. 289–294.

Fleming, G. N. (1989), "Lorentz Invariant State Reduction, and Localization" in Fine and Leplin, pp. 112–126.

Fleming, G. N. (1992), "The Objectivity and Invariance of Quantum Predictions" in Hull *et al.*, pp. 104–113.

Forman, P. (1971), "Weimar Culture, Causality, and Quantum Theory, 1918–1927: Adaptation by German Physicists and Mathematicians to a Hostile Intellectual Environment," *Historical Studies in the Physical Sciences* **3**, 1–115.

Forman, P. (1979), "The Reception of an Acausal Quantum Mechanics in Germany and Britain" in Mauskopf, pp. 11–50.

Forman, P. (1984), "*Kausalität, Anschaulichkeit*, and *Individualität*, or How Cultural Values Prescribed the Character and Lessons Ascribed to Quantum Mechanics" in Stehr and Meja, pp. 333–347.

Freistadt, H. (1957), "The Causal Formulation of Quantum Mechanics of Particles (the Theory of De Broglie, Bohm and Takabayasi)," *Supplemento al Nuovo Cimento* **5**, 1–70.

French, A. P., and P. J. Kennedy (eds.) (1985), *Niels Bohr: A Centenary Volume*. Cambridge, MA, Harvard University Press.

Fulling, S. A. (ed.) (1989), *Aspects of Quantum Field Theory in Curved Space-Time*. New York, Cambridge University Press.

Fulling, S. A. (ed.) (1995), *Heat Kernel Techniques and Quantum Gravity: Discourses in Mathematics and Its Applications, No. 4*. College Station, TX, Texas A & M University Department of Mathematics (to be published).

Garuccio, A. and J.-P. Vigier (1981), "Description of Spin in the Causal Stochastic Interpretation of Proca-Maxwell Waves: Theory of Einstein's 'Ghost Waves'," *Lettere al Nuovo Cimento* **30**, 57–63.

Gell-Mann, M. (1994), *The Quark and the Jaguar*. New York, W. H. Freeman and Co.

Gell-Mann, M. and M. L. Goldberger (1953), "The Formal Theory of Scattering," *Physical Review* **91**, 398–408. Reprinted in Ross, pp. 277–307.

Gell-Mann, M. and J. B. Hartle (1990), "Quantum Mechanics in the Light of Quantum Cosmology" in Kobayashi *et al.*, pp. 321–343. [This same paper appears in Zurek (1990), pp. 425–458.]

Gell-Mann, M. and J. B. Hartle (1993), "Classical Equations for Quantum Systems," *Physical Review* **47**, 3345–3382.

Gell-Mann, M. and J. B. Hartle (1994), "Equivalent Sets of Histories and Multiple Quasiclassical Domains," University of California at Santa Barbara preprint.

George, A. (ed.) (1953), *Louis de Broglie, Physicien et Penseur*. Paris, Éditions Albin Michel.

Gerver, J. L. (1991), "The Existence of Pseudocollisions in the Plane," *Journal of Differential Equations* **89**, 1–68.

Ghirardi, G. C. and R. Grassi (1996), "Bohm's Theory Versus Dynamical Reduction" in this volume.

Ghirardi, G. C., R. Grassi and F. Benatti (1995), "Describing the Macroscopic World: Closing the Circle Within the Dynamical Reduction Program," *Foundations of Physics* **25**, 5–38.

Ghirardi, G. C., R. Grassi, J. Butterfield and G. N. Fleming (1993), "Parameter Dependence and Outcome Dependence in Dynamical Models for State Vector Reduction," *Foundations of Physics* **23**, 341–364.

Ghirardi, G. C., P. Pearle and A. Rimini (1990), "Markov Processes in Hilbert Space and Continuous Spontaneous Localization of Systems of Identical Particles," *Physical Review A* **42**, 78–89.

Ghirardi, G. C. and A. Rimini (1990), "Old and New Ideas in the Theory of Quantum Measurement" in Miller, pp. 167–191.

Ghirardi, G. C., A. Rimini and T. Weber (1980), "A General Argument Against Superluminal Transmission Through the Quantum-Mechanical Measurement Process," *Lettere al Nuovo Cimento* **27**, 293–298.

Ghirardi, G. C., A. Rimini and T. Weber (1986), "Unified Dynamics for Microscopic and Macroscopic Systems," *Physical Review D* **34**, 470–491.

Ghosh, R., C. K. Hong, Z. Y. Ou and L. Mandel (1986), "Interference of Two Photons in Parametric Down Conversion," *Physical Review A* **34**, 3962–3968.

Gleason, A. M. (1957), "Measures on the Closed Subspaces of a Hilbert Space," *Journal of Mathematics and Mechanics* **6**, 885–893.

Goldstein, S. and D. N. Page (1995), "Linearly Positive Histories: Probabilities for a Robust Family of Sequences of Quantum Events," *Physical Review Letters* **74**, 3715–3719.

Goldwirth, D. S. and T. Piran (1989), "Spherical Inhomogeneous Cosmologies and Inflation: Numerical Methods," *Physical Review D* **40**, 3263–3279.

Good, R. H. and T. J. Nelson (1971), *Classical Theory of Electric and Magnetic Fields*. New York, Academic Press.

Gotsman, E. and G. Tauber (eds.) (1985), *From SU(3) to Gravity*. Cambridge, Cambridge University Press.

Grangier, P., G. Roger and A. Aspect (1986), "Experimental Evidence for a Photon Anticorrelation Effect on a Beam Splitter: A New Light on Single Photon Interferences," *Europhysics Letters* **1**, 173–179.

Green, T. A. and O. E. Lanford (1960), "Rigorous Derivation of the Phase Shift Formula for the Hilbert Space Scattering Operator of a Single Particle," *Journal of Mathematical Physics* **1**, 139–148.

Greenberger, D. M., M. A. Horne and A. Zeilinger (1989), "Going Beyond Bell's Theorem" in Kafatos, pp. 69–72.

Greenberger, D. M. and A. Zeilinger (eds.) (1995), *Fundamental Problems in Quantum Theory (Annals of the New York Academy of Sciences* **755**). New York, New York Academy of Sciences.

Griffiths, R. B. (1984), "Consistent Histories and the Interpretation of Quantum Mechanics," *Journal of Statistical Physics* **36**, 219–272.

Griffiths, R. B. (1993), "The Consistency of Consistent Histories: A Reply to d'Espagnat," *Foundations of Physics* **23**, 1601–1610.

Halbwachs, F. (1960), *Théorie relativiste des Fluides a Spin*. Paris, Gauthier-Villars.

Halpern, O. (1952), "A Proposed Re-interpretation of Quantum Mechanics," *Physical Review* **87**, 389.

Hanson, N. R. (1958), *Patterns of Discovery*. Cambridge, Cambridge University Press.

Hardy, L. (1992), "Quantum Mechanics, Local Realistic Theories, and Lorentz-Invariant Realistic Theories," *Physical Review Letters* **68**, 2981–2984.

Hardy, L. (1996), "Contextuality in Bohmian Mechanics" in this volume.

Hardy, L. and E. J. Squires (1992), "On the Violation of Lorentz-Invariance in Deterministic Hidden-Variable Interpretations of Quantum Theory," *Physics Letters A* **168**, 169–173.

Hartman, T. E. (1962), "Tunneling of a Wave Packet," *Journal of Applied Physics* **33**, 3427–3433.

Hauge, E. H., J. P. Falck and T. A. Fjeldly (1987), "Transmission and Reflection Times for Scattering of Wave Packets off Tunneling Barriers," *Physical Review B* **36**, 4203–4214.

Hauge, E. H. and J. A. Støvneng (1989), "Tunneling Times: A Critical Review," *Reviews of Modern Physics* **61**, 917–936.

Healey, R. (1989), *The Philosophy of Quantum Mechanics: An Interactive Interpretation*. Cambridge, Cambridge University Press.

Heilbron, J. L. (1988), "The Earliest Missionaries of the Copenhagen Spirit" in Ullmann-Margalit, pp. 201–233.

Heisenberg, W. (1925), "Über quantentheoretische Umdeutung kinematischer und mechanischer Beziehungen," *Zeitschrift für Physik* **33**, 879–893. [Appears as "Quantum-Theoretical Re-Interpretation of Kinematic and Mechanical Relations" in van der Waerden, pp. 261–276.]

Heisenberg, W. (1927a), "Schwankungerscheinungen und Quantenmechanik," *Zeitschrift für Physik* **40**, 501–506.

Heisenberg, W. (1927b), "Über den anschaulichen Inhalt der quantentheoretischen Kinematik und Mechanik," *Zeitschrift für Physik* **43**, 172–198. [Appears as "The Physical Content of Quantum Kinematics and Mechanics" in Wheeler and Zurek, pp. 62–84.]

Heisenberg, W. (1928), "Erkenntnistheoretische Probleme in der modernen Physik," reprinted in Blum *et al.*, pp. 22–28.

Heisenberg, W. (1930), *The Physical Principles of the Quantum Theory*. Chicago, University of Chicago Press. [Reprinted in 1949 by Dover Publications (New York).]

Heisenberg, W. (1931), "Kausalgesetz und Quantenmechanik," *Erkenntnis zugleich Annalen der Philosophie* **2**, 172–182. Reprinted in Blum *et al.*, pp. 29–39.

Heisenberg, W. (1934), "Wandlungen der Grundlagen der exacten Naturwissenschaft in jüngster Zeit," *Angewandte Chemie* **47**, 697–702. Reprinted in Blum *et al.*, pp. 96–101. [Appears in translation as "Recent Changes in the Foundations of Exact Science" in Heisenberg (1979), pp. 11–26.]

Heisenberg, W. (1936), "Prinzipielle Fragen der modernen Physik," *Neuere Fortschritte in den exakten Wissenschaften. Fünf Wiener Vorträge. Dritter Zyklus*, S. 91–102. Liepzig und Wien, Franz Deuticke. Reprinted in Blum *et al.*, pp. 108–119. [Appears in translation as "Questions of Principle in Modern Physics" in Heisenberg (1979), pp. 41–52.]

Heisenberg, W. (1946), "Dr mathematische Rahmen der Quantentheorie der Wellenfelder," *Zeitschrift für Naturforschung* **1**, 608–622.

Heisenberg, W. (1958), *Physics and Philosophy*. New York, Harper & Row.

Heisenberg, W. (1971), *Physics and Beyond*. New York, Harper & Row.

Heisenbrg, W. (1973), "Development of Concepts in the History of Quantum Theory" in Mehra, pp. 264–275.

Heisenberg, W. (1979), *Philosophical Problems of Quantum Physics*. Woodbridge, Ox Bow Press.

Heitler, W. (1954), *The Quantum Theory of Radiation, 3rd ed.* Oxford, Clarendon Press.

Held, A. (ed.) (1980), *General Relativity and Gravitation*, Vol. 1. New York, Plenum.

Hessen, B. (1931), "The Social and Economic Roots of Newton's 'Principia' " in Bukharin *et al.*, pp. 149–212.

Hiley, B. J. and F. D. Peat (eds.) (1987), *Quantum Implications: Essays in Honour of David Bohm*. London, Routledge & Kegan Paul.

Holland, P. R. (1993), *The Quantum Theory of Motion*. Cambridge, Cambridge University Press.

Home, D. and R. Chattopadhyay (1996), "DNA Molecular Cousin of Schrödinger's Cat: A Curious Example of Quantum Measurement," *Physical Review Letters* **76**, 2836–2839.

Hull, D., M. Forbes and K. Okruhlik (eds.) (1992), *PSA 1992: Proceedings of the 1992 Biennial Meeting of the Philosophy of Science Association*, Vol. 1. East Lansing, MI, Philosophy of Science Association.

Iannaccone, G. and B. Pellegrini (1994), "Characteristic Times in the Motion of a Particle," *Physical Review B* **49**, 16548–16560.

Ikebe, T. (1960), "Eigenfunction Expansions Associated with the Schrödinger Operators and Their Applications to Scattering theory," *Archive for Rational Mechanics and Analysis* **5**, 1–34.

Isham, C., R. Penrose and D. W. Sciama (eds.) (1981), *Quantum Gravity 2*. Oxford, Clarendon Press.

Itzykson, C. and J.-B. Zuber (1978), *Quantum Field Theory*. London, McGraw-Hill.

Jammer, M. (1974), *The Philosophy of Quantum Mechanics*. New York, John Wiley & Sons.

Jammer, M. (1989), *The Conceptual Development of Quantum Mechanics*, 2nd ed. New York, Tomash Publishers.

Jarrett, J. (1984), "On the Physical Significance of the Locality Conditions in the Bell Arguments," *Nous* **18**, 569–589.

Jaworski, W. and D. M. Wardlaw (1988), "Time Delay in Tunneling: Transmission and Reflection Time Delays," *Physical Review A* **37**, 2843–2854.

Jeffers, S., S. Roy and J.-P. Vigier (eds.) (1996), *The Present Status of the Quantum Theory of Light*. Dordrecht, Kluwer Academic Publishers (to be published).

Jones, R. (1991), "Realism About What?", *Philosophy of Science* **58**, 185–202.

Joos, E. and H. D. Zeh (1985), "The Emergence of Classical Properties Through Interaction with the Environment," *Zeitschrift für Physik B* **59**, 223–243.

Kafatos, M. (ed.) (1989), *Bell's Theorem, Quantum Theory, and Conceptions of the Universe*. Dordrecht, Kluwer.

Kaiser, D. (1994), "Bringing the Human Actors Back on Stage: The Personal Context of the Einstein-Bohr Debate," *British Journal for the History of Science* **27**, 129–152.

Kaloyerou, P. N. (1985), "Investigation of the Quantum Potential in the Relativistic Domain," unpublished Ph. D. dissertation, London University.

Kaloyerou, P. N. (1992), "On the Wootters-Zurek Development of Einstein's Two-Slit Experiment," *Foundations of Physics* **22**, 1345–1377.

Kaloyerou, P. N. (1994), "The Causal Interpretation of the Electromagnetic Field," *Physics Reports* **244**, 287–358.

Kaloyerou, P. N. (1995), "Causal Interpretation of the Modified Klein-Gordon Equation," *Foundations of Physics* **25**, 1413–1460.

Kamefuchi, S., H. Ezawa, Y. Murayama, M. Namiki, S. Nomura, Y. Ohnuki and T. Yojima (eds.) (1984), *Proceedings of the International Symposium on the Foundations of Quantum Mechanics in the Light of New Technology*. Tokyo, Physical Society of Japan.

Karakostas, V. (1994), "Limitations on Stochastic Localisation Models of State Vector Reduction," *International Journal of Theoretical Physics* **33**, 1645–1659.

Kato, T. (1961), "Fundamental Properties of Hamiltonian Operators of Schrödinger Type," *Transactions of the American Mathematical Society* **70**, 195–211.

Keiser, M. and R. T. Jantzen (eds.) (1995), *Proceedings of the 7th Marcel Grossmann Meeting on General Relativity*. Singapore, World Scientific (to be published).

Kellert, S. H. (1993), *In the Wake of Chaos*. Chicago, University of Chicago Press.

Kent, A. (1990), "Against Many Worlds Interpretations," *International Journal of Modern Physics A* **5**, 1745–1762.

Kernaghan, M. (1994), "Bell–Kochen–Specker Theorem with 20 Vectors," *Journal of Physics A: Mathematical and General* **27**, L829–L830.

Kilmister, C. W. (ed.) (1987), *Schrödinger: Centenary Celebration of a Polymath*. Cambridge, Cambridge University Press.

Kobayashi, S., H. Ezawa, Y. Murayama and S. Nomura (eds.) (1990), *Proceedings of the 3rd International Symposium on Foundations of Quantum Mechanics in the Light of New Technology*. Tokyo, Physical Society of Japan.

Kochen, S. (1985), "A New Interpretation of Quantum Mechanics" in Lahti and Mittelstaedt (1985), pp. 151–169.

Kochen, S. and E. P. Specker (1967), "The Problem of Hidden Variables in Quantum Mechanics," *Journal of Mathematics and Mechanics* **17**, 59–87.

Koestler, A. (1971), *The Case of the Midwife Toad*. New York, Random House.

Krips, H. (1987), *The Metaphysics of Quantum Theory*. Oxford, Clarendon Press.

Kuhn, T. S., J. Heilbron, P. Forman and L. Allen (1967), *Sources for History of Quantum Physics*. Philadelphia, American Philosophical Society.

Kyprianidis, A. (1988), "Hamilton-Jacobi Theory and Quantum Mechanics. I," *Physics Letters A* **131**, 411–417.

Lahti, P. and P. Mittelstaedt (eds.) (1985), *Symposium on the Foundations of Modern Physics*. Singapore, World Scientific Publishing Co.

Lahti, P. and P. Mittelstaedt (eds.) (1987), *Symposium on the Foundations of Modern Physics 1987*. Singapore, World Scientific Publishing Co.

Lam, M. M. and C. Dewdney (1994a), "The Bohm Approach to Cavity Quantum Scalar Field Dynamics. Part I: The Free Field," *Foundations of Physics* **24**, 3–27.

Lam, M. M. and C. Dewdney (1994b), "The Bohm Approach to Cavity Quantum Scalar Field Dynamics. Part II: The Interaction of the Field with Matter," *Foundations of Physics* **24**, 29–60.

Lanczos, C. (1986), *The Variational Principles of Mechanics*. New York, Dover Publications.

Landau, L. D. and E. M. Lifshitz (1958), *Quantum Mechanics: Non-Relativistic Theory*. Reading, MA, Addison-Wesley.

Landauer, R. and Th. Martin (1994), "Barrier Interaction Time in Tunneling," *Reviews of Modern Physics* **66**, 217–228.

Landsburg, P. T. (1984), "Two General Problems in Quantum Biology" in Löwdin *et al.*, pp. 55–61.

Laurikainen, K. V. (1988), *Beyond the Atom: The Philosophical Thought of Wolfgang Pauli*. Berlin, Springer-Verlag.

Laurikainen, K. V., C. Montonen and K. Sunnaborg (eds.) (1994), *Symposium on the Foundations of Modern Physics 1994*. Gif-sur-Yvette Cedex, France, Editions Frontières.

Leavens, C. R. (1990), "Transmission, Reflection and Dwell Times within Bohm's Causal Interpretation of Quantum Mechanics," *Solid State Communications* **74**, 923–928.

Leavens, C. R. (1993), "On the Olkhovsky–Recami Approach to the 'Tunneling Time Problem'," *Solid State Communications* **85**, 115–119.

Leavens, C. R. (1995a), "The 'Tunneling Time Problem': Fundamental Incompatibility of the Bohm Trajectory Approach with the Projector and Conventional Probability Current Approaches," *Physics Letters A* **197**, 88–94.

Leavens, C. R. (1995b), "Bohm Trajectory and Feynmann Path Approaches to the Tunneling Time Problem," *Foundations of Physics* **25**, 229–268.

Leavens, C. R. (1996), "The 'Tunneling-Time Problem' for Electrons" in this volume.

Leavens, C. R. and G. C. Aers (1990), "Tunneling Times for One-Dimensional Barriers" in Behm *et al.*, pp. 59–76.

Leavens, C. R. and G. C. Aers (1993), "Bohm Trajectories and the Tunneling Time Problem" in Wiesendanger and Güntherodt, pp. 105–140.

Leavens, C. R., G. Iannaccone and W. R. McKinnon (1995), "On the Approach to the Stationary-State-Scattering Limit within Bohmian Mechanics," *Physics Letters A* **208**, 17–24.

Leavens, C. R. and W. R. McKinnon (1994), "An Exact Determination of the Mean Dwell Time Based on the Quantum Clock of Salecker and Wigner," *Physics Letters A* **194**, 12–20.

Leibbrandt, G. (1987), "Introduction to Noncovariant Gauges," *Reviews of Modern Physics* **59**, 1067–1119.

Lewis, D. K. (1986a), "Causation" in Lewis (1986b), pp. 159–172.

Lewis, D. K. (1986b), *Philosophical Papers II*. Oxford, Oxford University Press.

Linde, A. D. (1982), "A New Inflationary Universe Scenario: A Possible Solution of the Horizon, Flatness, Homogeneity, Isotropy and Primordial Monopole Problems," *Physics Letters B* **108**, 389–393.

Lippmann, B. and J. Schwinger (1950), "Variational Principles for Scattering Processes I," *Physical Review* **79**, 469–480. Reprinted in Ross, pp. 243–276.

Littlejohn, R. G. (1992), "The van Vleck Formual, Maslov Theory, and Phase Space Geometry," *Journal of Statistical Physics* **68**, 7–50.

Lockwood, M. (1989), *Mind, Brain, and the Quantum*. Oxford, Basil Blackwell.

Lorentz, H. A., A. Einstein, H. Minkowski and H. Weyl (n.d.), *The Principle of Relativity*. New York, Dover Publications.

Löwdin, P.-O., J. R. Sabin and M. C. Zerner (eds.), *International Journal of Quantum Chemistry: Quantum Biology Symposium No. 11*. New York, John Wiley & Sons.

Ludwig, G. (1968), *Wave Mechanics*. Oxford, Pergamon Press.

Mackman, S. and E. Squires (1995), "Lorentz Invariance and the Retarded Bohm Model," *Foundations of Physics* **25**, 391–397.

Marsden, J. E. and F. J. Tipler (1980), "Maximal Hypersurfaces and Foliations of Constant Mean Curvature in General Relativity," *Physics Reports* **66**, 109–139.

Masden, M. S. (1988), "Scalar Fields in Curved Spacetimes," *Classical and Quantum Gravity* **5**, 627–639.

Mather, J. N. and R. McGehee (1975), "Solutions of the Collinear Four Body Problem Which Become Unbounded in Finite Time" in Moser (1975), pp. 573–599.

Maudlin, T. (1994), *Quantum Non-Locality and Relativity*. Oxford, Blackwell.

Maudlin, T. (1996), "Space-Time in the Quantum World" in this volume.

Mauskopf, S. H. (ed.) (1979), *The Reception of Unconventional Science, AAAS Selected Symposium 25*. Boulder, CO, Westview Press.

McKinnon, W. R. and C. R. Leavens (1995), "Distributions of Delay Times and Transmission Times in Bohm's Causal Interpretation of Quantum Mechanics," *Physical Review A* **51**, 2748–2757.

Mehra, J. (ed.) (1973), *The Physicist's Conception of Nature*. Dordrecht, D. Reidel Publishing Co.

Mermin, N. D. (1993), "Hidden Variables and the Two Theorems of John Bell," *Reviews of Modern Physics* **65**, 803–815.

Messiah, A. (1962), *Quantum Mechanics*, Vol. 1. Amsterdam, North-Holland Publishing Co.

Miller, A. I. (ed.) (1990), *Sixty-Two Years of Uncertainty*. New York, Plenum Press.

Møller, C. (1945), "General Properties of the Characteristic Matrix in the Theory of Elementary Particles I," *Kongelige Danske Videnskabernes Selskab, Matematisk-Fysiske Meddelelser* **23**, No. 1. Reprinted in Ross, pp. 109–154.

Moser, J. (1973), *Stable and Random Motions in Dynamical Systems*. Princeton, NJ, Princeton University Press.

Moser, J. (ed.) (1975), *Dynamical Systems, Theory and Applications*. Berlin, Springer-Verlag.

Mott, N. F. (1929), "Scattering of Fast Electrons by Atomic Nuclei," *Proceedings of the Royal Society A* **124**, 425–442.

Muga, J. G., S. Brouard and R. Sala (1992), "Transmission and Reflection Tunneling Times," *Physics Letters A* **167**, 24–28.

Murphy, M. P. and L. A. J. O'Neill (eds.) (1995), *What Is Life? The Next Fifty Years*. Cambridge, Cambridge University Press.

Nelson, E. (1966), "Derivation of the Schrödinger Equation from Newtonian Mechanics," *Physical Review* **150**, 1079–1085.

Nye, Mary Jo (1972), *Molecular Reality: A Perspective on the Scientific Work of Jean Perrin*. New York, American Elsevier Publishing Company.

Olkhovsky, V. S. and E. Recami (1992), "Recent Developments in the Time Analysis of Tunneling Processes," *Physics Reports* **214**, 339–356.

Olkhovsky, V. S., E. Recami and A. K. Zaichenko (1994), "Reply to the Comment by Leavens on Olkhovsky-Recami's Approach to the 'Tunneling Time Problem': More About the Time Analysis of Tunneling Processes," *Solid State Communications* **89**, 31–35.

Omnès, R. (1988), "Logical Reformulation of Quantum Mechanics I. Foundations," *Journal of Statistical Physics* **53**, 893–932.

Omnès, R. (1991), "About the Notion of Truth in Quantum Mechanics," *Journal of Statistical Physics* **62**, 841–861.

Omnès, R. (1992), "Consistent Interpretations of Quantum Mechanics," *Reviews of Modern Physics* **64**, 339–382.

Omnès, R. (1994), *The Interpretation of Quantum Mechanics*. Princeton, NJ, Princeton University Press.

Ou, Z. Y. (1988), "Quantum Theory of Fourth-Order Interference," *Physical Review A* **37**, 1607–1619.

Ou, Z. Y. and L. Mandel (1988), "Observation of Spatial Quantum Beating with Separated Photodetectors," *Physical Review Letters* **61**, 54–57.

Page, D. N. (1995a), "Probabilities Don't Matter" in Keiser and Jantzen (to be published).

Page, D. N. (1995b), "Information Loss in Black Holes and/or Conscious Beings?" in Fulling (to be published).

Page, D. N. (1995c), "Sensible Quantum Mechanics: Are Only Perceptions Probabilistic?", University of Alberta report.

Page, D. N. (1995d), "Sensible Quantum Mechanics: Are Probabilities Only in the Mind?" in Berezin and Rubakov (to be published).

Page, D. N. (1996), "Aspects of Quantum Cosmology" in Sanchez and Zichichi, pp. 431–450.

Pagonis, C. and R. Clifton (1995), "Unremarkable Contextualism: Dispositions in the Bohm Theory," *Foundations of Physics* **25**, 281–296.

Pais, A. (1982), '*Subtle is the Lord . . .*'. New York, Oxford University Press.

Pauli, W. (1945), "Niels Bohr on his 60th Birthday," *Reviews of Modern Physics* **17**, 97–101.

Pearle, P. (1989), "Combining Stochastic Dynamical State-Vector Reduction with Spontaneous Localization," *Physical Review A* **39**, 2277–2289.

Pearle, P. and E. Squires (1994), "Bound State Excitation, Nucleon Decay Experiments, and Models of Wave Function Collapse," *Physical Review Letters* **73**, 1–5.

Penrose, R. (1995), "Why New Physics Is Needed to Understand the Mind" in Murphy and O'Neill, pp. 115–129.

Percival, I. (1991), "Schrödinger's Quantum Cat," *Nature* **351**, 357.

Peres, A. (1980), "Measurement of Time by Quantum Clocks," *American Journal of Physics* **48**, 552–557.

Peres, A. (1993), *Quantum Theory: Concepts and Methods*. Dordrecht, Kluwer.

Petersen, A. (1963), "The Philosophy of Niels Bohr," *Bulletin of the Atomic Scientists* **19**(7), 8–14.

Pitowsky, I. (1994), "George Boole's 'Conditions of Possible Experience' and the Quantum Puzzle," *British Journal for the Philosophy of Science* **45**, 95–125.

Polanyi, M. (1961), *The Logic of Personal Knowledge. Essays Presented to Michael Polanyi on his Seventieth Birthday*. London, Routledge & Kegan Paul.

Qadir, A. and J. A. Wheeler (1985), "York's Cosmic Time Versus Proper Time as Relevant to Changes in the Dimensionaless 'Constants', K-Meson Decay, and the Unity of Black Hole and Big Crunch" in Gotsman and Tauber, pp. 383–394.

Reck, M., A. Zeilinger, H. J. Bernstein and P. Bertani (1994), "Experimental Realization of Any Discrete Unitary Operator," *Physical Review Letters* **73**, 58–61.

Redhead, M. L. G. (1988), "A Philosopher Looks at Quantum Field Theory" in Brown and Harré, pp. 9–23.

Redhead, M. L. G. (1989), *Incompleteness, Nonlocality, and Realism*. Oxford, Clarendon Press.

Reed, M. and B. Simon (1975), *Methods of Modern Mathematical Physics II*. London, Academic Press.

Reed, M. and B. Simon (1979), *Methods of Modern Mathematical Physics III*. London, Academic Press.

Ross, M. (1963), *Quantum Scattering Theory*. Bloomington, IN, Indiana University Press.

Rozental, S. (ed.) (1967), *Niels Bohr: His Life and Work as Seen by His Friends and Colleagues*. Amsterdam, Elsevier Science Publishers B. V.

Ruelle, D. (1969), "A Remark on Bound States in Potential-Scattering Theory," *Il Nuovo Cimento A* **61**, 655–662.

Rybachenko, V. F. (1967), "Time of Penetration of a Particle Through a Potential Barrier," *Soviet Journal of Nuclear Physics* **5**, 635–639.

Sakurai, J. J. (1967), *Advanced Quantum Mechanics*. Reading, MA, Addison-Wesley.

Salecker, H. and E. P. Wigner (1958), "Quantum Limitations of the Measurement of Space-Time Distances," *Physical Review* **109**, 571–577.

Salmon, W. C. (1984), *Scientific Explanation and the Causal Structure of the World*. Princeton, NJ, Princeton University Press.

Samols, T. M. (1995), "A Stochastic Model of a Quantum Field Theory," *Journal of Statistical Physics* **80**, 793–809.

Sanchez, N. and Zichichi (eds.) (1996), *Proceedings of the International School of Astrophysics "D. Chalonge," 4th Course: String Gravity and Physics at the Planck Energy Scale, Erice, Sicily, 8–19 September 1995*. Dordrecht, Kluwer Academic Publishers.

Sa-yakanit, V., W. Sritrakool, J.-O. Berananda, M. C. Gutzwiller, A. Inomata, S. Lundqvist, J. R. Klauder and L. Schulman (eds.) (1989), *Path Integrals from meV to MeV*. Singapore, World Scientific.

Schilpp, P. (ed.) (1949), *Albert Einstein: Philosopher-Scientist*. La Salle, IL, Open Court Publishing Co.

Schlick, M. (1931), "Causality in Contemporary Physics" in Toulmin, pp. 83–121.

Schrödinger, E. (1926), "Quantisierung als Eigenwertproblem," *Annalen der Physik* **79**, 361–376. [Appears as "Quantization as an Eigenvalue Problem" in Ludwig, pp. 94–105.]

Schrödinger, E. (1935), "Die gegenwärtinge Situation in der Quantenmechanik," *Die Naturwissenschaften* **23**, 807–812; 812–828; 844–849. [Appears as "The Present Situation in Quantum Mechanics" in Wheeler and Zurek, pp. 152–167.]

Schrödinger, E. (1944), *What is Life?* Cambridge, Cambridge University Press.

Schulman, L. S. and R. W. Ziolkowski (1989), "Path Integral Asymptotics in the Absence of Classical Paths" in Sa-yakanit *et al.*, pp. 253–278.

Schutz, B. F. (1971), "Hamiltonian Theory of a Relativistic Perfect Fluid," *Physical Review D* **4**, 3559–3566.

Schweber, S. S. (1961), *Introduction to Relativistic Quantum Field Theory*. Evanston, IL, Row, Peterson and Company.

Schweber, S. S., H. A. Bethe and F. de Hoffmann (1955), *Mesons and Fields*, Vol. 1. Evanston, IL, Row, Peterson & Co.

Sciama, D. W., P. Candelas and D. Deutsch (1981), "Quantum Field Theory, Horizons and Thermodynamics," *Advances in Physics* **30**, 327–366.

Shimony, A. (1984), "Controllable and Uncontrollable Nonlocality" in S. Kamefuchi *et al.*, pp. 225–230.

Simon, B. (1971), *Quantum Mechanics for Hamiltonians Defined as Quadratic Forms*. Princeton, NJ, Princeton University Press.

Simon, B. (1977), "An Introduction to the Self-Adjointness and Spectral Analysis of Schrödinger Operators" in Thirring and Urban, pp. 19–42.

Sklar, L. (1993), *Physics and Chance*. Cambridge, Cambridge University Press.

Sokolovski, D. and L. M. Baskin (1987), "Traversal Time in Quantum Scattering," *Physical Review A* **36**, 4604–4611.

Sokolovski, D. and J. N. L. Connor (1990), "Path-Integral Analysis of the Time Delay for Wave-Packet Scattering and the Status of Complex Tunneling Times," *Physical Review A* **42**, 6512–6525.

Sokolovski, D. and J. N. L. Connor (1991), "Negative Probability and Distributions of

Dwell, Transmission and Reflection Times for Quantum Tunneling," *Physical Review A* **44**, 1500–1504.

Sommerfeld, A. (1927) "Zum gegenwärtigen Stande der Atomphysik," *Physikalische Zeitschrift* **28**, 231–239.

Spiller, T. P., T. D. Clark, R. J. Prance and H. Prance (1990), "Barrier Traversal Time in the Quantum Potential Picture," *Europhysics Letters* **12**, 1–4.

Squires, E. J. (1990a), "Why Is Position Special?", *Foundations of Physics Letters* **3**, 87–93.

Squires, E. J. (1990b), *Conscious Mind in the Physical World*. Bristol, Adam Hilger.

Squires, E. J. (1992), "An Apparent Conflict Between the de Broglie-Bohm Model and Orthodoxy in Quantum Cosmology," *Foundations of Physics Letters* **5**, 71–75.

Squires, E. J. (1993), "A Local Hidden-Variable Theory that, FAPP, Agrees with Quantum Theory," *Physics Letters A* **178**, 22–26.

Squires, E. J. (1995), "Quantum Theory, Relativity, and the Bohm Model" in Greenberger and Zeilinger, pp. 451–463.

Squires, E. J. (1996), "Local Bohmian Mechanics" in this volume.

Squires, E. J. and S. Mackman (1994), "The Bohm Model with Fermion-Boson Correlations," *Physics Letters A* **185**, 1–4.

Stehr, N. and V. Meja (eds.) (1984), *Society and Knowledge*. New Brunswick, NJ, Transaction Books.

Steinberg, A. M. (1995a), "How Much Time Does a Tunneling Particle Spend in the Barrier Region?" *Physical Review Letters* **74**, 2405–2409.

Steinberg, A. M. (1995b), "Conditional Probabilities in Quantum Theory and the Tunneling Time Controversy," *Physical Review A* **52**, 32–42.

Suppes, P. (ed.) (1976), *Logic and Probability in Quantum Mechanics*. Dordrecht, Reidel Publishing Co.

Suppes, P. and M. Zanotti (1976), "On the Determinism of Hidden Variable Theories with Strict Correlation and Conditional Statistical Independence of Observables" in Suppes, pp. 445–455.

Teller, P. (1989), "Relativity, Relational Holism, and the Bell Inequalities" in Cushing and McMullin, pp. 208–223.

Thirring, W. (1992), *Classical Dynamical Systems*, 2nd ed. New York, Springer-Verlag.

Thirring, W. and P. Urban (eds.) (1977), *The Schrödinger Equation*. Wien, Springer-Verlag.

Toulmin, S. (ed.) (1970), *Physical Reality*. New York, Harper and Row.

Ullmann-Margalit, E. (ed.) (1988), *Science in Reflection*. Dordrecht, Kluwer Academic Press.

Vaidman, L. (1993), "About Schizophrenic Experiences of the Neutron or Why We Should Believe in the Many-Worlds Interpretation of Quantum Theory," Tel-Aviv University preprint TAUP 2058–93.

Vaihinger, H. (1925), *The Philosophy of "As if"*. New York, Harcourt, Brace and Company.

Valentini, A. (1991a), "Signal-Locality, Uncertainty, and the Subquantum H-Theorem. I," *Physics Letters A* **156**, 5–11.

Valentini, A. (1991b), "Signal-Locality, Uncertainty and the Subquantum *H*-Theorem. II," *Physics Letters A* **158**, 1–8.

Valentini, A. (1992), "On the Pilot-Wave Theory of Classical, Quantum and Subquantum Physics," unpublished Ph. D. thesis, International School for Advanced Studies, Trieste, Italy.

Valentini, a. (1996a), "Pilot-Wave Theory of Fields, Gravitation and Cosmology" in this volume.

Valentini, A. (1996b), *On the Pilot-Wave Theory of Classical, Quantum and Subquantum Physics*. Berlin, Springer-Verlag (to be published).

van der Merwe, A., F. Selleri and G. Tarozzi (eds.) (1992), *Bell's Theorem and the Foundations of Modern Physics*. Singapore, World Scientific.

van der Waerden, B. L. (ed.) (1967), *Sources of Quantum Mechanics*. Amsterdam, North-Holland.

van Fraassen, B. C. (1981), "A Modal Interpretation of Quantum Mechanics" in Beltrametti and van Fraassen, pp. 229–258.

van Fraassen, B. C. (1982), "The Charybdis of Realism," *Synthese* **52**, 25–38.

van Fraassen, B. C. (1989), *Laws and Symmetry*. Oxford, Clarendon Press.

van Fraassen, B. C. (1991), *Quantum Mechanics: An Empiricist View*. Oxford, Clarendon Press.

van Fraassen, B. C. (1994), "Interpretation of Quantum Mechanics: Parallels and Choices" in Accardi, pp. 7–14.

van Hove, L. (1952), "Les Difficultés de Divergences pour un Modèle particulier de Champ quantifié," *Physica* **18**, 145–159.

van Tiggelen, B. A., A. Tip and A. Lagendijk (1993), "Dwell Time for Light and Electrons," *Journal of Physics A: Mathematical and General* **26**, 1731–1748.

Vermaas, P. E. (1996), "Unique Transition Probabilities in the Modal Interpretation," *Studies in History and Philosophy of Modern Physics* (to be published).

Vink, J. C. (1992), "Quantum Potential Interpretation of the Wave Function of the Universe," *Nuclear Physics B* **369**, 707–728.

Vink, J. C. (1993), "Quantum Mechanics in Terms of Discrete Beables," *Physical Review A* **48**, 1808–1818.

von Neumann, J. (1938), "On Infinite Direct Products," *Composito Mathematica* **6**, 1–77. [Also reprinted in von Neumann (1961).]

von Neumann, J. (1955), *Mathematical Foundations of Quantum Mechanics*. Princeton, Princeton University Press.

von Neumann, J. (1961), *Collected Works*, Vol. III [A. H. Taub (ed.)]. London, Pergamon Press.

von Weiszäcker, C. F. (1985), "A Reminiscence from 1932" in French and Kennedy, pp. 183–190.

von Weizsäcker, C. F. (1987), "Heisenberg's Philosophy" in Lahti and Mittelstaedt (1987), pp. 277–293.

Waddington, C. H. (ed.) (1969), *Towards a Theoretical Biology*, Vol. 2. Chicago, Aldine Publishing Co.

Wald, R. M. (1994), *Quantum Theory in Curved Spacetime and Black Hole Thermodynamics*. Chicago, University of Chicago Press.

Wheeler, J. A. and W. H. Zurek (eds.) (1983), *Quantum Theory and Measurement*. Princeton, NJ, Princeton University Press.

Wiesendanger, R. and H.-J. Güntherodt (eds.) (1993), *Scanning Tunneling Microscopy III*. Berlin, Springer.

Wightman, A. S. and S. S. Schweber (1955), "Configuration Space Methods in Relativistic Quantum Field Theory. I," *Physical Review* **98**, 812–837.

Wigner, E. (1961), "The Probability of the Existence of a Self-Reproducing Unit" in Polanyi, pp. 231–238.

Wootters, W. K. and W. H. Zurek (1982), "A Single Quantum Cannot Be Cloned," *Nature* **299**, 802–803.

Xia, Z. (1992), "The Existence of Noncollision Singularities in Newtonian Systems," *Annals of Mathematics* **135**, 411–468.

Yamada, N. (1996), "Probabilities for Histories in Nonrelativistic Quantum Mechanics," *Physical Review A* **54**, 182–203.

York, J. W. (1972), "Role of Conformal Three-Geometry in the Dynamics of Gravitation," *Physical Review Letters* **28**, 1082–1085.

Zeh, H. D. (1993), "There Are No Quantum Jumps, Nor Are There Particles!", *Physics Letters A* **172**, 189–192.

Zimba, J. and R. C. Penrose (1993), "On Bell Non-Locality Without Probabilities: More

Curious Geometry," *Studies in History and Philosophy of Modern Physics* **24**, 697–720.

Zurek, W. H. (1981), "Pointer-Basis of Quantum Apparatus: Into What Mixture Does the Wave Packet Collapse?", *Physical Review D* **24**, 1516–1525.

Zurek, W. H. (1982), "Environment-Induced Superselection Rules," *Physical Review D* **26**, 1862–1880.

Zurek, W. H. (1990), *Complexity, Entropy and the Physics of Information*. Redwood City, CA, Addison-Wesley Publishing Co.

Zurek, W. H. (1993), "Preferred States, Predictability, Classicality and the Environment-Induced Decoherence," *Progress of Theoretical Physics* **89**, 281–312.

NAME INDEX

This index is to proper names mentioned in the text (except that commonly-used modifiers, as in "Schrödinger equation" or "Bohm theory", do not appear here).

105. F. Burwick (ed.): *Approaches to Organic Form.* Permutations in Science and Culture. 1987 ISBN 90-277-2541-1
106. M. Almási: *The Philosophy of Appearances.* Translated from Hungarian. 1989 ISBN 90-277-2150-5
107. S. Hook, W.L. O'Neill and R. O'Toole (eds.): *Philosophy, History and Social Action.* Essays in Honor of Lewis Feuer. With an Autobiographical Essay by L. Feuer. 1988 ISBN 90-277-2644-2
108. I. Hronszky, M. Fehér and B. Dajka: *Scientific Knowledge Socialized.* Selected Proceedings of the 5th Joint International Conference on the History and Philosophy of Science organized by the IUHPS (Veszprém, Hungary, 1984). 1988 ISBN 90-277-2284-6
109. P. Tillers and E.D. Green (eds.): *Probability and Inference in the Law of Evidence.* The Uses and Limits of Bayesianism. 1988 ISBN 90-277-2689-2
110. E. Ullmann-Margalit (ed.): *Science in Reflection.* The Israel Colloquium: Studies in History, Philosophy, and Sociology of Science. Vol. 3. 1988 ISBN 90-277-2712-0; Pb 90-277-2713-9
111. K. Gavroglu, Y. Goudaroulis and P. Nicolacopoulos (eds.): *Imre Lakatos and Theories of Scientific Change.* 1989 ISBN 90-277-2766-X
112. B. Glassner and J.D. Moreno (eds.): *The Qualitative-Quantitative Distinction in the Social Sciences.* 1989 ISBN 90-277-2829-1
113. K. Arens: *Structures of Knowing.* Psychologies of the 19th Century. 1989 ISBN 0-7923-0009-2
114. A. Janik: *Style, Politics and the Future of Philosophy.* 1989 ISBN 0-7923-0056-4
115. F. Amrine (ed.): *Literature and Science as Modes of Expression.* With an Introduction by S. Weininger. 1989 ISBN 0-7923-0133-1
116. J.R. Brown and J. Mittelstrass (eds.): *An Intimate Relation.* Studies in the History and Philosophy of Science. Presented to Robert E. Butts on His 60th Birthday. 1989 ISBN 0-7923-0169-2
117. F. D'Agostino and I.C. Jarvie (eds.): *Freedom and Rationality.* Essays in Honor of John Watkins. 1989 ISBN 0-7923-0264-8
118. D. Zolo: *Reflexive Epistemology.* The Philosophical Legacy of Otto Neurath. 1989 ISBN 0-7923-0320-2
119. M. Kearn, B.S. Philips and R.S. Cohen (eds.): *Georg Simmel and Contemporary Sociology.* 1989 ISBN 0-7923-0407-1
120. T.H. Levere and W.R. Shea (eds.): *Nature, Experiment and the Science.* Essays on Galileo and the Nature of Science. In Honour of Stillman Drake. 1989 ISBN 0-7923-0420-9
121. P. Nicolacopoulos (ed.): *Greek Studies in the Philosophy and History of Science.* 1990 ISBN 0-7923-0717-8
122. R. Cooke and D. Costantini (eds.): *Statistics in Science.* The Foundations of Statistical Methods in Biology, Physics and Economics. 1990 ISBN 0-7923-0797-6
123. P. Duhem: *The Origins of Statics.* Translated from French by G.F. Leneaux, V.N. Vagliente and G.H. Wagner. With an Introduction by S.L. Jaki. 1991 ISBN 0-7923-0898-0
124. H. Kamerlingh Onnes: *Through Measurement to Knowledge.* The Selected Papers, 1853-1926. Edited and with an Introduction by K. Gavroglu and Y. Goudaroulis. 1991 ISBN 0-7923-0825-5
125. M. Čapek: *The New Aspects of Time: Its Continuity and Novelties.* Selected Papers in the Philosophy of Science. 1991 ISBN 0-7923-0911-1

126. S. Unguru (ed.): *Physics, Cosmology and Astronomy, 1300-1700.* Tension and Accommod tion. 1991 ISBN 0-7923-1022
127. Z. Bechler: *Newton's Physics on the Conceptual Structure of the Scientific Revolution.* 199 ISBN 0-7923-1054
128. É. Meyerson: *Explanation in the Sciences.* Translated from French by M-A. Siple and D. Siple. 1991 ISBN 0-7923-1129
129. A.I. Tauber (ed.): *Organism and the Origins of Self.* 1991 ISBN 0-7923-1185-
130. F.J. Varela and J-P. Dupuy (eds.): *Understanding Origins.* Contemporary Views on t Origin of Life, Mind and Society. 1992 ISBN 0-7923-1251
131. G.L. Pandit: *Methodological Variance.* Essays in Epistemological Ontology and t Methodology of Science. 1991 ISBN 0-7923-1263
132. G. Munévar (ed.): *Beyond Reason.* Essays on the Philosophy of Paul Feyerabend. 1991 ISBN 0-7923-1272
133. T.E. Uebel (ed.): *Rediscovering the Forgotten Vienna Circle.* Austrian Studies on Ot Neurath and the Vienna Circle. Partly translated from German. 1991 ISBN 0-7923-1276
134. W.R. Woodward and R.S. Cohen (eds.): *World Views and Scientific Discipline Formatio* Science Studies in the [former] German Democratic Republic. Partly translated fro German by W.R. Woodward. 1991 ISBN 0-7923-1286
135. P. Zambelli: *The Speculum Astronomiae and Its Enigma.* Astrology, Theology and Scien in Albertus Magnus and His Contemporaries. 1992 ISBN 0-7923-1380
136. P. Petitjean, C. Jami and A.M. Moulin (eds.): *Science and Empires.* Historical Studies abo Scientific Development and European Expansion. ISBN 0-7923-1518
137. W.A. Wallace: *Galileo's Logic of Discovery and Proof.* The Background, Content, and U of His Appropriated Treatises on Aristotle's *Posterior Analytics.* 1992 ISBN 0-7923-1577
138. W.A. Wallace: *Galileo's Logical Treatises.* A Translation, with Notes and Commentary, His Appropriated Latin Questions on Aristotle's *Posterior Analytics.* 1992 ISBN 0-7923-1578
Set (137 + 138) ISBN 0-7923-1579
139. M.J. Nye, J.L. Richards and R.H. Stuewer (eds.): *The Invention of Physical Scienc* Intersections of Mathematics, Theology and Natural Philosophy since the Seventeen Century. Essays in Honor of Erwin N. Hiebert. 1992 ISBN 0-7923-1753-
140. G. Corsi, M.L. dalla Chiara and G.C. Ghirardi (eds.): *Bridging the Gap: Philosoph Mathematics and Physics.* Lectures on the Foundations of Science. 1992 ISBN 0-7923-1761
141. C.-H. Lin and D. Fu (eds.): *Philosophy and Conceptual History of Science in Taiwan.* 1992 ISBN 0-7923-1766
142. S. Sarkar (ed.): *The Founders of Evolutionary Genetics.* A Centenary Reappraisal. 1992 ISBN 0-7923-1777
143. J. Blackmore (ed.): *Ernst Mach – A Deeper Look.* Documents and New Perspectives. 1992 ISBN 0-7923-1853
144. P. Kroes and M. Bakker (eds.): *Technological Development and Science in the Industri Age.* New Perspectives on the Science–Technology Relationship. 1992 ISBN 0-7923-1898
145. S. Amsterdamski: *Between History and Method.* Disputes about the Rationality of Scienc 1992 ISBN 0-7923-1941
146. E. Ullmann-Margalit (ed.): *The Scientific Enterprise.* The Bar-Hillel Colloquium: Studies History, Philosophy, and Sociology of Science, Volume 4. 1992 ISBN 0-7923-1992

147. L. Embree (ed.): *Metaarchaeology*. Reflections by Archaeologists and Philosophers. 1992 ISBN 0-7923-2023-9
148. S. French and H. Kamminga (eds.): *Correspondence, Invariance and Heuristics*. Essays in Honour of Heinz Post. 1993 ISBN 0-7923-2085-9
149. M. Bunzl: *The Context of Explanation*. 1993 ISBN 0-7923-2153-7
150. I.B. Cohen (ed.): *The Natural Sciences and the Social Sciences*. Some Critical and Historical Perspectives. 1994 ISBN 0-7923-2223-1
151. K. Gavroglu, Y. Christianidis and E. Nicolaidis (eds.): *Trends in the Historiography of Science*. 1994 ISBN 0-7923-2255-X
152. S. Poggi and M. Bossi (eds.): *Romanticism in Science*. Science in Europe, 1790–1840. 1994 ISBN 0-7923-2336-X
153. J. Faye and H.J. Folse (eds.): *Niels Bohr and Contemporary Philosophy*. 1994 ISBN 0-7923-2378-5
154. C.C. Gould and R.S. Cohen (eds.): *Artifacts, Representations, and Social Practice*. Essays for Marx W. Wartofsky. 1994 ISBN 0-7923-2481-1
155. R.E. Butts: *Historical Pragmatics*. Philosophical Essays. 1993 ISBN 0-7923-2498-6
156. R. Rashed: *The Development of Arabic Mathematics: Between Arithmetic and Algebra*. Translated from French by A.F.W. Armstrong. 1994 ISBN 0-7923-2565-6
157. I. Szumilewicz-Lachman (ed.): *Zygmunt Zawirski: His Life and Work*. With Selected Writings on Time, Logic and the Methodology of Science. Translations by Feliks Lachman. Ed. by R.S. Cohen, with the assistance of B. Bergo. 1994 ISBN 0-7923-2566-4
158. S.N. Haq: *Names, Natures and Things*. The Alchemist Jābir ibn Ḥayyān and His *Kitāb al-Aḥjār* (Book of Stones). 1994 ISBN 0-7923-2587-7
159. P. Plaass: *Kant's Theory of Natural Science*. Translation, Analytic Introduction and Commentary by Alfred E. and Maria G. Miller. 1994 ISBN 0-7923-2750-0
160. J. Misiek (ed.): *The Problem of Rationality in Science and its Philosophy*. On Popper vs. Polanyi. The Polish Conferences 1988–89. 1995 ISBN 0-7923-2925-2
161. I.C. Jarvie and N. Laor (eds.): *Critical Rationalism, Metaphysics and Science*. Essays for Joseph Agassi, Volume I. 1995 ISBN 0-7923-2960-0
162. I.C. Jarvie and N. Laor (eds.): *Critical Rationalism, the Social Sciences and the Humanities*. Essays for Joseph Agassi, Volume II. 1995 ISBN 0-7923-2961-9
Set (161–162) ISBN 0-7923-2962-7
163. K. Gavroglu, J. Stachel and M.W. Wartofsky (eds.): *Physics, Philosophy, and the Scientific Community*. Essays in the Philosophy and History of the Natural Sciences and Mathematics. In Honor of Robert S. Cohen. 1995 ISBN 0-7923-2988-0
164. K. Gavroglu, J. Stachel and M.W. Wartofsky (eds.): *Science, Politics and Social Practice*. Essays on Marxism and Science, Philosophy of Culture and the Social Sciences. In Honor of Robert S. Cohen. 1995 ISBN 0-7923-2989-9
165. K. Gavroglu, J. Stachel and M.W. Wartofsky (eds.): *Science, Mind and Art*. Essays on Science and the Humanistic Understanding in Art, Epistemology, Religion and Ethics. Essays in Honor of Robert S. Cohen. 1995 ISBN 0-7923-2990-2
Set (163–165) ISBN 0-7923-2991-0
166. K.H. Wolff: *Transformation in the Writing*. A Case of Surrender-and-Catch. 1995 ISBN 0-7923-3178-8
167. A.J. Kox and D.M. Siegel (eds.): *No Truth Except in the Details*. Essays in Honor of Martin J. Klein. 1995 ISBN 0-7923-3195-8

Boston Studies in the Philosophy of Science

168. J. Blackmore: *Ludwig Boltzmann, His Later Life and Philosophy, 1900–1906.* Book One: A Documentary History. 1995 ISBN 0-7923-3231-
169. R.S. Cohen, R. Hilpinen and Q. Renzong (eds.): *Realism and Anti-Realism in the Philosophy of Science.* Beijing International Conference, 1992. 1995 ISBN 0-7923-3233-
170. I. Kuçuradi and R.S. Cohen (eds.): *The Concept of Knowledge.* The Ankara Seminar. 1995 ISBN 0-7923-3241-
171. M.A. Grodin (ed.): *Meta Medical Ethics*: The Philosophical Foundations of Bioethics. 1995 ISBN 0-7923-3344-
172. S. Ramirez and R.S. Cohen (eds.): *Mexican Studies in the History and Philosophy of Science.* 1995 ISBN 0-7923-3462-
173. C. Dilworth: *The Metaphysics of Science.* An Account of Modern Science in Terms of Principles, Laws and Theories. 1995 ISBN 0-7923-3693-
174. J. Blackmore: *Ludwig Boltzmann, His Later Life and Philosophy, 1900–1906* Book Two: The Philosopher. 1995 ISBN 0-7923-3464-
175. P. Damerow: *Abstraction and Representation.* Essays on the Cultural Evolution of Thinking. 1996 ISBN 0-7923-3816-
176. G. Tarozzi (ed.): *Karl Popper, Philosopher of Science.* (in prep
177. M. Marion and R.S. Cohen (eds.): *Québec Studies in the Philosophy of Science.* Part I: Logic, Mathematics, Physics and History of Science. Essays in Honor of Hugues Leblanc. 1995 ISBN 0-7923-3559-
178. M. Marion and R.S. Cohen (eds.): *Québec Studies in the Philosophy of Science.* Part II: Biology, Psychology, Cognitive Science and Economics. Essays in Honor of Hugues Leblanc. 1996 ISBN 0-7923-3560-
    Set (177–178) ISBN 0-7923-3561-
179. Fan Dainian and R.S. Cohen (eds.): *Chinese Studies in the History and Philosophy of Science and Technology.* 1995 ISBN 0-7923-3463-
180. P. Forman and J.M. Sánchez-Ron (eds.): *National Military Establishments and the Advancement of Science and Technology.* Studies in 20th Century History. 1996 ISBN 0-7923-3541-
181. E.J. Post: *Quantum Reprogramming.* Ensembles and Single Systems: A Two-Tier Approach to Quantum Mechanics. 1995 ISBN 0-7923-3565-
182. A.I. Tauber (ed.): *The Elusive Synthesis: Aesthetics and Science.* 1996 ISBN 0-7923-3904-
183. S. Sarkar (ed.): *The Philosophy and History of Molecular Biology: New Perspectives.* 1996 ISBN 0-7923-3947-
184. J.T. Cushing, A. Fine and S. Goldstein (eds.): *Bohmian Mechanics and Quantum Theory: An Appraisal.* 1996 ISBN 0-7923-4028-
185. K. Michalski: *Logic and Time.* An Essay on Husserl's Theory of Meaning. 1996 ISBN 0-7923-4082-
186. G. Munévar (ed.): *Spanish Studies in the Philosophy of Science.* 1996 ISBN 0-7923-4147-
187. G. Schubring (ed.): *Hermann Günther Graßmann (1809–1877): Visionary Mathematician, Scientist and Neohumanist Scholar.* Papers from a Sesquicentennial Conference. 1996 ISBN 0-7923-4261-

Also of interest:

R.S. Cohen and M.W. Wartofsky (eds.): *A Portrait of Twenty-Five Years Boston Colloquia for the Philosophy of Science, 1960-1985.* 1985 ISBN Pb 90-277-1971-

Previous volumes are still available.

KLUWER ACADEMIC PUBLISHERS – DORDRECHT / BOSTON / LONDON

9 780792 340287

GPSR Compliance

The European Union's (EU) General Product Safety Regulation (GPSR) is a set of rules that requires consumer products to be safe and our obligations to ensure this.

If you have any concerns about our products, you can contact us on ProductSafety@springernature.com

In case Publisher is established outside the EU, the EU authorized representative is:

Springer Nature Customer Service Center GmbH
Europaplatz 3
69115 Heidelberg, Germany

Batch number: 10809439

Printed by Printforce, the Netherlands